ENVIRONMENTAL POLLUTION and PLANT RESPONSES

edited by

Shashi Bhushan Agrawal, Ph.D.
Department of Biological Sciences
Allahabad Agricultural Institute
Allahabad, India

Madhoolika Agrawal, Ph.D.
Centre of Advanced Study in Botany
Banaras Hindu University
Varanasi, India

Library of Congress Cataloging-in-Publication Data

Agrawal, Shashi Bhushan.
Environmental pollution and plant responses / Shashi Bhushan Agrawal, Madhoolika Agrawal
p. cm.
Includes bibliographical references.
ISBN 1-56670-341-7 (alk. paper)
1. Plants, Effects of pollution on. I. Agrawal, Madhoolika. II. Title.
QK750.A34 1999
581.7—dc21 99-26504
CIP

This book contains information obtained from authentic and highly regarded sources. Reprinted material is quoted with permission, and sources are indicated. A wide variety of references are listed. Reasonable efforts have been made to publish reliable data and information, but the author and the publisher cannot assume responsibility for the validity of all materials or for the consequences of their use.

Direct all inquiries to CRC Press LLC, 2000 Corporate Blvd., N.W., Boca Raton, Florida 33431.

Lewis Publishers is an imprint of CRC Press LLC

International Standard Book Number 1-56670-341-7
Library of Congress Card Number 99-26504
Printed in the United States of America 1 2 3 4 5 6 7 8 9 0
Printed on acid-free paper

Preface

The last few decades have witnessed a rapid increase in awareness of the environmental hazards of our technological civilization leading to the pollution of the biosphere. In recent years, there have been changes in our understanding of problems related to environmental pollution with new threats of climate change effects that have triggered a wide range of apprehension among scientists and governments in different parts of the globe. Changes in environmental constituents can either have a direct impact on the inhabitants of Earth or an indirect impact via changes in climate. Countless meetings of experts, a large number of conferences and scientific congresses, and a multitude of articles in journals and books popularized the notion of pollution and helped in enforcing concrete decisions which could check the route of efficient flow of pollutants to the biosphere to a certain extent. Scientific knowledge of the different forms of environmental pollution and its effects on plants has made enormous progress over the last decade. The transformations that pollutants may undergo under different environments can facilitate the dispersal and further toxicity of pollutants. The resultant effects may be in the form of indirect and delayed action which may have far more alarming effects than direct action. In recent years, biological consequences have been viewed in a broader ecological context. Many of the aspects of environmental pollution are interlinked and are of concern to us because of their impact on human beings and on resources critical to our survival.

Environmental pollution is a rapidly expanding field and the amount of literature produced per year is immense. The basic objective of this book is to review the major research efforts that led to our current state of knowledge related to environmental pollution and its effects on individual, population, and community levels. This book comprises several articles on various aspects of environmental quality and provided the opportunity not only for evaluating the importance of effects on different levels of plant organization, but also for predicting the future consequences of the release of specified pollutants through their mechanisms of action. The problem of understanding the links between environmental quality and plants assumes enormous importance in terms of agriculture, forestry, horticulture, agronomy, etc. In-depth understanding of the links at different levels of plant organization is a prerequisite for management practices to sustain productivity under the threat of environmental pollution. The purpose of this book is to evaluate what is known, identify the information gaps, and provide directions for future research in view of the present uncertainities about the magnitude and range of potential environmental problems and consequent changes.

The book comprises 21 chapters authored by renowned scientists from various parts of the world who have long working experience in the corresponding areas and are well known for their contributions and expertise in the respective fields. The topics range from individual plant to community responses to pollutants of major concern today and in the future in different regions of the world. The chapters have been organized by treating the global climate change aspect and its impacts first, starting from current methodologies of experimentation, their mode of action, and then individual and ecological consequences. Separate chapters are devoted to the aspects of CO_2 enrichment and enhanced UV-B effects. Since developing countries are major areas of concern from an environmental pollution view point, a couple of chapters are devoted to case studies regarding air pollution and acid rain effects. In addition to plant responses, alternative methods for quantifying changes in environmental quality using plants are also discussed in few contributions. This authoritative updating of information on various aspects of environmental pollution would serve as a reference book as well as a text for students of the environmental sciences and related disciplines. This book should greatly benefit teachers and researchers, as it will not only be a source of available information to date, but will also provide possible hypotheses for future problems and directions for research. The book should be helpful to government and nongovernment organizations and to industry to acquaint them with current environmental issues and their likely effects.

We express our deep sense of gratitude to the distinguished contributors of this book for their generosity in acceding to our request to share their valuable expertise. The prompt response and encouragement received from them are gratefully acknowledged. Thanks are also due to Profs. D. P. Ormrod (Canada), J. B. Mudd (U.S.A.), C. K. Varshney (India), P. S. Dubey (India), A. C. Posthumus (the Netherlands), J. F. Bornman (Sweden), B. R. Chaudhary (India), and Dr. Vishwanath Pandey for their best wishes and help in various ways. We especially thank Dr. D. T. Krizek, of the Climate Stress Laboratory, BARC, Beltsville, MD, for his kind support and for the review of Chapter 11. We express our gratitude to Professor R. B. Lal, Director, Allahabad Agricultural Institute, Allahabad for his best wishes and keen interest in this venture. We also thank our mothers, Mrs. Prem Vati and Mrs. Prem Kumari, and daughters, Miss Richa and Miss Shivani, for their support and encouragement. The editors are also thankful to CSIR (Division of Scientific & Technical Personnel) and the Ministry of Environment & Forests, New Delhi for financial assistance. Last, but not least, the editors are grateful to Andrea H. Demby, Jane Kinney, and Robert Hauserman of CRC Press LLC for their efforts in the publication of this book.

We derive utmost pleasure in dedicating this book to commemorate the contributions of Professor Dhruva Narain Rao, M.Sc., Ph.D. (Ottawa), the former Head, Department of Botany, Banaras Hindu University, India and at present a visiting professor in Concordia University, Quebec, Canada, who has pioneered the work on air pollution and plant responses in India.

S. B. Agrawal

Madhoolika Agrawal

About the Editors

S. B. Agrawal, Ph.D., is an assistant professor in the Department of Biological Sciences of Allahabad Agricultural Institute, Allahabad. He received his M.Sc. in Botany from Agra University and Ph.D. in Botany from Banaras Hindu University. He worked as a lecturer (Botany) in the Directorate of Higher Education, Uttar Pradesh, from 1983 to 1984, and as a research associate and pool scientist of U.G.C. and C.S.I.R., New Delhi, at the Centre of Advanced Study in Botany, Banaras Hindu University, from 1984 to 1998.

Dr. Agrawal holds a National Scholarship from the government of India. He is a Fellow of the International Society for Conservation of Natural Resources. He is on the Research Board of Advisors of the American Biographical Institute. He is also an associate editor of the international journal of environmental sciences, *Vasundhara*.

Dr. Agrawal has made significant contributions on several aspects of chromosome research and environmental biology, including the effects of air pollutants and UV-B on algae and higher plants. He has published more than 72 research papers in scientific journals and edited books. He has also edited one book, *Cytology, Genetics and Molecular Biology of Algae*, in 1996, published by SPB Academic Publishing bv, Amsterdam, the Netherlands. He was a visiting scientist at USDA/ARS, Beltsville Agricultural Research Center, Maryland, during 1990.

Madhoolika Agrawal, Ph.D., is an associate professor of Botany at Banaras Hindu University. She obtained her M.Sc. and Ph.D. in Botany from Banaras Hindu University. Prior to joining B.H.U., Dr. Agrawal worked as a Scientist 'B' at the Industrial Toxicological Research Center (C.S.I.R.), Lucknow.

Dr. Agrawal has made significant contributions in the field of air pollution, with particular emphasis on the cause–effect relationship. She has conducted extensive field studies to ascertain the role of air pollutants as a stress factor for tropical deciduous vegetation in India. She has extended her research on global climate change effects with particular reference to enhanced UV-B radiation and elevated CO_2. Currently, Dr. Agrawal is investigating the atmosphere depositions and the budgeting of nutrients and trace elements around industrial sites.

Dr. Agrawal is a recipient of a Fulbright Fellowship under which she worked at Beltsville Agricultural Research Center, Maryland. She has also worked at Lancaster University, U.K., under an INSA-Royal Society exchange fellowship. In recognition of her work she received the UNESCO/ROSTCA young scientist award in 1988, and the Professor Hira Lal Chakraborty Award from the Indian Science Congress Association in 1996. She has worked on various projects with the Ministry of Environment and Forests, India, the Department of Science and Technology, New Delhi, the Council of Scientific and Industrial Research, New Delhi, the Overseas Development Authority and DIFD, U.K. She is an environmental consultant for the World Bank sponsored project on widening of highways. She is on the editorial advisory board of *Progress in Environmental Science*, Arnold Press, U.K. Dr. Agrawal has published 80 research papers in scientific journals and 17 chapters in edited books.

Contributors

Madhoolika Agrawal
Department of Botany
Banaras Hindu University
Varanasi, India

S. B. Agrawal
Department of Biological Sciences
Allahabad Agricultural Institute
Allahabad, India

Christopher J. Atkinson
Horticulture Research International
East Malling, West Malling
Kent, United Kingdom

Maurizio Badiani
Dipartimento di Agrobiologia e Agrochimica
Universita´ di Reggio Calabria
Gallina, Italy

Jeremy Barnes
Air Pollution Laboratory
Department of Agricultural
and Environmental Science
University of Newcastle
Newcastle upon Tyne
United Kingdom

J. N. B. Bell
Department of Biology
Imperial College of Science, Technology,
and Medicine
Silwood Park, Ascot, Berkshire,
United Kingdom

James P. Bennett
Biological Resources Division
University of Wisconsin
Institute for Environmental Studies
Madison, Wisconsin

Katrien Bortier
Department of Biology
Universitaire Instelling Antwerpen
Antwerp, Belgium

Patricia A. Bowen
Pacific Agri-Food Research Centre
Agriculture Canada
Agassiz, British Columbia, Canada

Reinhart Ceulemans
Department of Biology
Universitaire Instelling Antwerpen
Antwerp, Belgium

Sharad B. Chaphekar
Dhus Wadi, Thakurdwar
Mumbai, India

Richard F. E. Crang
Department of Plant Biology
University of Illinois at Urbana-Champaign
Urbana, Illinois

Ulrich Dämmgen
Institute for Agroecology
Federal Agricultural Research Center
Braunschweig, Germany

Marisa Domingos
Secao de Ecologia
Institute de Botanica
Sao Paulo, Brazil

Jixi Gao
Institute of Environmental Ecology
Chinese Research Academy of Environmental
Sciences
Beijing, China

Yingxin Gao
Institute of Environmental Ecology
Chinese Research Academy of Environmental
Sciences
Beijing, China

J. V. Groth,
Department Plant Pathology
University of Minnesota
St. Paul, Minnesota

Madeleine S. Günthardt-Goerg
Swiss Federal Institute of Forest, Snow, and Landscape Research
Birmensdorf, Switzerland

James Heath
Department of Biological Sciences
Institute of Environmental and Natural Sciences
Lancaster University
Lancaster, United Kingdom

Cao Hongfa
Institute of Environmental Ecology
Chinese Research Academy of Environmental Sciences
Beijing, China

Andreas Klumpp
Hohenheim Universität
Institut für Landschafts- und Pflanzenökologie
Stuttgart, Germany

Sagar V. Krupa
Department of Plant Pathology
University of Minnesota
St. Paul, Minnesota

Edward H. Lee
Director
Institute of Life Science
National Central University
Chung-Li, Taiwan

Tom Lyons
Air Pollution Laboratory
Department of Agricultural and Environmental Science
University of Newcastle
Newcastle upon Tyne
United Kingdom

William J. Manning
Department of Microbiology
University of Massachusetts
Amherst, Massachusetts

Terry A. Mansfield
Department of Biological Sciences
Institute of Environmental and Natural Science
Lancaster, United Kingdom

F. M. Marshall
Department of Biology
Imperial College of Science, Technology, and Medicine
Silwood Park, Ascot, Berkshire, United Kingdom

Stefan Maurer
Paul Scherrer Institut
Villigen, Switzerland

Rainer Matyssek
Lehrstuhl für Forstbotanik
Ludwig-Maximillians Universität München
Freising, Germany

Rajesh K. Mehra
Department of Entomology and Environmental Toxicology Program
University of California
Riverside, California

K. Mièieta
Department of Botany
Comenius University
Bratislava, Slovakia

Franco Miglietta
Consiglio Nazionale Della Ricerche
Istituto per l'Agrometeorologia e l'Analisi Ambientale Applicata All'Agricoltura
Firenze, Italy

G. Murín
Institute of Cell Biology
Comenius University
Bratislava, Slovakia

Anna Rita Paolacci
Dipartimento di Agrobiologia e Agrochimica
Universita´ degli Viterbo
Viterbo, Italy

Maria Luisa Pignata
Facultad de Ciencias Exactas
Fisicas y Naturales
Universidad Nacional de Córdoba
Córdoba, Argentina

Matthias Plöchl
Institute of Agricultural Engineering
Bornim (ATB)
Potsdam-Bornim, Germany

Andrea Polle
Forstbotanisches Institut
Georg-August-Universität Göttingen
Göttingen, Germany

Antonia Raschi
Consiglio Nazionale delle Ricerche
Istituto per l'Agrometeorologia e l'Analisi
Ambientale Applicata All'Agricoltura
Firenze, Italy

Victor C. Runeckles
Professor Emeritus
Department of Plant Science
University of British Columbia
Vancouver, British Columbia, Canada

Yingwa Shen
Institute of Environmental Ecology
Chinese Research Academy of Environmental Sciences
Beijing, China

Jianmin Shu
Institute of Environmental Ecology
Chinese Research Academy of Environmental Sciences
Beijing, China

Ludwig de Temmerman
Centre for Veterinary and Agrochemical Research
Tervuren, Belgium

Manfred Tevini
Botanical Institute II
University of Karlsruhe
Karlsruhe, Germany

Rudra D. Tripathi
Environmental Botany Division
National Botanical Research Institute
Rana Pratap Marg
Lucknow, India

Enikö Turcsányi
Air Pollution Laboratory
Department of Agricultural and Environmental Science
University of Newcastle
Newcastle upon Tyne
United Kingdom

Linbo Zhang
Institute of Environmental Ecology
Chinese Research Academy of Environmental Sciences
Beijing, China

Ralf Zimmerling
Agriculture Institut for Climate Research
Institut für Agrarrelevante Klimaforsch.
Müncheberg, Germany

Dedicated to Professor D. N. Rao

Contents

CHAPTER 1

Global Climate Change and Crop Responses: Uncertainties Associated with the Current Methodologies

Sagar V. Krupa and J.V. Groth

CONTENTS

1.1 INTRODUCTION

Global climate change, *global warming,* and the *greenhouse effect* are terms that are used interchangeably to describe the observed and/or predicted changes in the interactions between the chemical and physical climates of the Earth.[1] Thus, global climate change should be viewed as a system of atmospheric processes and their products (Table 1.1). Whether global air temperature will increase uniformly across all latitudes, by how much, and when, is subject to debate, but there is little doubt that tropospheric concentrations of several trace gases are increasing.[2-4]

1-56670-341-7/00/$0.00+$.50

Table 1.1 Summary of Atmospheric Processes and Products Relevant to Global Climate Change[a]

Processes	Products
Loss in stratospheric O_3	Increased UV-B penetration[b]
Increases in greenhouse gases: CO_2, O_3, CO, CH_4, CFCs, OBs, N_2O, COS, SO_2?; NO_x (NO_2 + NO); Changes in the concentrations of H_2O molecules	Changes in: Temperature, Precipitation, Radiation, Evaporation, Wind, Secondary aerosols
Increased UV-B penetration[b]	Decreased tropospheric O_3

[a] CFCs = chlorofluorocarbons; OBs = organo-bromines.
[b] Increased UV-B penetration may decrease tropospheric O_3 and affect other greenhouse gases.
Source: Runeckles, V.C. and Krupa, S.V., Environ. Pollut., 83, 191, 1994. With permission.

In addition to any consistent changes in the air temperature (and consequent alterations in the length of the growing season), increases in certain trace gases alone will have an effect on agriculture and crop productivity.[5] Traditionally, the study of the effects of physical climate parameters (air temperature, precipitation, etc.) on crops has been the domain of agricultural meteorologists and agronomists. While plant and ecophysiologists have studied the assimilation of carbon dioxide (CO_2) by vegetation, others have examined the adverse effects of trace gas air pollutants such as ozone (O_3) and sulfur dioxide (SO_2) on crop growth and productivity.[4,6-8] A number of investigators have also examined the effects of elevated ultraviolet (UV)-B radiation (280 to 315 or 320 nm) on crops.[9-11] Virtually all these efforts relate to univariate studies, although that compartmentalized view of climate–plant interactions is slowly changing, as our understanding of the dynamic interactions between the chemical and physical climate is rapidly increasing. To expand this view, in this chapter we provide a discussion of the uncertainties associated with what we know and what we need to know, in the context of global climate change and its potential impacts on crop growth and productivity.

1.2 CLIMATE CHANGE: CURRENT DATABASE*

Among the variables listed in Table 1.1, parameters most relevant to crop growth and productivity are tropospheric O_3, CO_2, UV-B, global–photosynthetically active radiation (PAR), temperature, and precipitation. As discussed elsewhere, the interactions between these variables and crop response are fundamentally stochastic in nature. Therefore, it is extremely important to understand the spatial and temporal dynamics of the growth-regulating factors.

In general, the measurements of tropospheric O_3 concentrations have been the domain of various national and local air quality regulatory agencies. It is not possible to provide a detailed compilation of the locations of these monitoring sites or the corresponding data on a global basis. It should be noted, however, that at least in the U.S., a predominant number of such measurement sites are in urban areas.[12] Therefore, real-time monitored O_3 data for crops are scarce, although regional-scale photochemical–meteorological models have been used to predict spatial-scale O_3 distribution.[13] Such models, however, provide seasonal or annual mean concentrations or geographic areas of exceedances of a given air quality standard, rather than biologically meaningful data applicable to crop responses. Similarly, although not directly applicable in crop response studies, there are TOMS (total O_3 monitoring spectrometer) measurement data available from the NIMBUS-7 satellite for

* Some useful reference search sites on the World Wide Web (WWW): for CO_2 — cdiacesd.ornl.gov/trends/co2/contents/html; for UV-B — www.wdc.rl.ac.uk/wdcmain/appendix/gdappenb2.html; for trace gases including TOMS-O_3, UV-B, air temperature, and precipitation — www.wmo.ch/web/ddbs/clmdata.html

the global distribution of tropospheric column O_3[14] (presentation of real data was not feasible in this review). However, such data, coupled with the predictions into the future,[15] can be used to identify geographic regions and time periods, where and when crops may be at risk.[8,16] Such an effort, although useful as a first attempt, is limited by our current lack of knowledge of the responses of many crop species to O_3, particularly those in developing countries.

In comparison with the discussion of O_3, there is a global-scale, integrated database for monthly mean atmospheric CO_2 concentrations.[17] Interestingly, among all the measurement sites, there are only three locations immediately relevant to agriculture. Based on annual average values, it can be concluded that atmospheric CO_2 concentrations are very comparable (360 to 365 ppm) at most measurement sites, after accounting for the balance in the strengths of the sources and sinks. In general, annual CO_2 concentrations have been increasing at a rate of ~0.4% since the early 1970s. However, as with other trace gases, there are spatial and temporal variabilities in the CO_2 levels and these features are governed by the human influence and by the behavioral patterns of high-elevation forested vs. low-elevation agricultural areas.[18]

Even though scientific concern regarding possible stratospheric O_3 depletion and the consequent increase in the transmission of UV-B radiation to the ground level was present in the mid-1970s, there has been no long-term global monitoring network for UV-B radiation,[19] although this is changing.[20] In comparison, there has been a long-term network for monitoring "global" solar radiation (also air temperature and precipitation depth — The World Radiation Network (Stanhill and Moreshet).)[21] The term *global* customarily refers to both direct-beam and diffuse solar radiation and in the context of crops, PAR constitutes about 47% of the global radiation.[22] The wavelengths of UV-B radiation (280 to 315 or 320 nm) are only a small portion (~0.5%) of the total global radiation arriving at the surface.[23] Nevertheless, in the future, significant increases in the specific wavelengths of UV-B are predicted.[24] Our current knowledge of UV-B measurements is a source of uncertainty. Frequently measurements are made with instruments that provide data on integrated, erythemal (wavelengths that result in the sunburn of Caucasian skin) solar spectrum (portions of both UV-B and UV-A, 315 or 320 to 400 nm[11, 20, 25]). Such data are not very useful in our understanding of crop responses and need to be replaced by the use of multiband instruments (see Sections 1.4 and 1.5).

1.3 EXPERIMENTAL METHODOLOGIES

As stated previously, at the present time our knowledge of the effects of elevated concentrations of trace gases and UV-B on crop productivity is almost entirely based on univariate experiments.[1] This is understandable, because multivariate studies are complex by nature and require large financial resources. Nevertheless, any study conducted in the field must have realism and relevance to the ambient conditions. Since the issue of global climate change pertains to concerns of the future, the accepted practice is to evaluate the effects of elevated levels of the climate variables (CO_2, O_3, UV-B, and temperature) on present-day crop cultivars.[9-11,16,26-29] The implications of this in future agriculture are discussed in a latter section of this chapter.

1.3.1 Ozone and Carbon Dioxide

A predominant part on our knowledge of the effects of elevated concentrations of trace gases (O_3 and CO_2) on crops is derived from the use of open-top field-exposure chambers, although some studies have used chamberless artificial exposure systems or even spatially variable ambient exposures themselves.[30,31] All these approaches have advantages and disadvantages (Table 1.2). In recent years, the chamber approach, particularly for O_3 studies, has been the subject of debate.[32,33] In addition to the differences in the microclimatological growing conditions within the chamber vs. the ambient, the unnatural, constant airflow and turbulence within the chamber are of great concern.[34] According to the aforementioned publications, the exposure–response curves derived under the chamber conditions cannot be used directly to estimate a quantitative effect under chamberless, field conditions. By contrast, a number of scientists working with the effects of elevated levels of CO_2 have been forced to use field-exposure chambers, because of the prohibitive costs for the CO_2 required to conduct chamberless, open-field studies. Similarly, our meager, current understanding of the joint effects of elevated O_3 and CO_2 is based almost solely on chamber studies, although chamberless methods are feasible.

1.3.2 Ultraviolet-B Radiation

The lack of comprehensive knowledge of the effects of UV-B, particularly under field conditions, and the limitations in our understanding of its joint effects with other environmental variables has been largely due to the difficulty in conducting realistic experiments under ambient conditions.[11] Almost all of the UV-B - crop response studies have used artificial UV radiation sources, combined with various types of filters to enhance the short-wavelength portion of an already existing source of light (Table 1.3). For example, the most frequently used filter to allow the penetration of UV-B is cellulose acetate (CA) (see Table 1.3). However, this filter does not allow wavelengths in the UV-B band below 290 nm and, in addition, permits the incidence of lower wavelengths of UV-A (315 or 320 to 400 nm).[36] Therefore, the spectral distributions of the artificial sources are not the same as that of the sun, and their relative effectiveness has to be assessed by using a biological weighting function (e.g., generalized plant damage action spectrum, Caldwell[37]) that may not be fully representative of the response of whole plants of different crop species.[38] Furthermore, in the artificial exposure systems, higher spectral irradiance is provided at the longer, rather than the shorter wavelengths of the UV-B band. In this context, based on the O_3 absorption spectrum, the opposite will be true for ambient conditions.[39] Indeed, according to Cutchis,[24] with overhead sun and typical O_3 amounts, a 10% decrease in column O_3 would result in a 20% increase in UV-B penetration at 305 nm, a 250% increase at 290 nm, and a 500% increase at 287 nm, all within the UV-B band. This type of relationship between loss in column O_3 and wavelength-dependent UV-B penetration has been reported for Toronto, Canada.[40] In this regard, different plant processes react differently to different wavelengths of radiation (action spectra). Current artificial exposure systems do not permit differential shorter-wavelength-based treatments similar to what has been predicted and/or may be measured under ambient conditions in the future. An exception to this might be the use of artificially generated ozone (has the same UV absorption spectrum as the natural O_3) flowing through a plexiglass cuvette or flow cell as a UV filter, in crop exposure experiments.[41] Overall, the methodological difficulties,[20,25,35] coupled with the existence of photo-repair (longer wavelengths of solar radiation offsetting the effects of UV-B)[25] and inductive resistance mechanisms in plants,[42] have been the sources of additional uncertainty in crop response studies.

1.4 EXPOSURE DOSIMETRY

Dose can be defined as the exact amount or extent of a given treatment to which a given receptor is subjected at one time or at stated intervals. In the present context, a derivation from the concept of dose is the separation of the terms *exposure dose*[43] and *effective dose*.[44] Exposure dose can be defined as the air concentration of a pollutant over a given duration to which a crop is subjected. To the contrary, effective dose is the actual pollutant concentration absorbed by the plant over a given duration. It is the effective dose that leads to a crop response. This concept is also true for UV-B radiation.[11,25] Certain pollutants such as gaseous sulfur compounds accumulate in the leaf tissue, and, therefore, foliar analysis for these elements performed as a time series can be used in establishing numerical relationships between cause and effect.[45] While methods are available for

Table 1.2 Comparative Advantages and Disadvantages of Some Field Exposure Methods for Gaseous Pollutants

Method	Advantage	Disadvantage
1. Open-top chambers (updraft), e.g., CO_2, O_3, and SO_2	a. Most widely used system in the U.S.; some 15 years of historical records with O_3 b. Many crops can be grown to maturity under conditions somewhat analogous to the ambient c. Effects of air pollutants can be evaluated singly or as mixtures d. Comparisons can be made between O_3 filtered (~80% pollutant removal) and unfiltered ambient air or between elevated O_3 and CO_2 concentrations e. Reasonable control on environmental variables within the chamber	a. Artificial chamber effect on plant growth and productivity b. High cost for including sufficient number of treatments and labor intensive c. Complex computer-controlled system required to mimic ambient pollutant exposure dynamics within the chamber d. Pollutant flow within the chamber artificial and not similar to the ambient e. Modifications in the microclimate within the chamber can lead to altered incidence of pathogens and pests f. Rain shadows present g. Is subject to weather hazards, including incursion of ambient air into the chamber at times
2. Open-top chambers (downdraft), e.g., O_3	a. Same as (b), (c), (d), and (e) of (1) b. Pollutant flow more realistic, from top of the plant canopy downward	a. Same as (a), (b), (c), and (e) of (1) b. Ozone exclusion from the ambient air entering the chamber varies from 25 to 70% c. Ambient rain is excluded d. As with (1), is subject to weather hazards
3. Field tracking chamber, e.g., CO_2	a. Permits study of natural vegetation and crops b. Tracks natural variation in the environment c. Whole ecosystem effects can be examined d. Integrated estimates of carbon and water balance can be made	a. Complexity of control functions in a remote setting b. Chamber effects on plant growth and productivity
4. Open-air, chamberless, artificial field exposure, e.g., CO_2, O_3, and SO_2	a. No chamber effect b. Large number of plants can be exposed to varying O_3 or CO_2 exposure regimes c. Desirable approach if (b), (d), and (e) under the disadvantages are rectified	a. Small changes in wind turbulence can cause large changes in pollutant concentrations b. High precision in a feedback control of pollutant release and intensive and extensive monitoring of the pollutant within the study plot required c. Control study plot difficult to deal with due to the omnipresence of O_3 and CO_2 d. Intensive and extensive field monitoring of other air pollutants and environmental variables required e. Powerful, multivariate, time series models required to evaluate the results fully

Table 1.2 (continued)

Method	Advantage	Disadvantage
5. Natural gradients of ambient, e.g., O_3 and SO_2	a. Evaluation of the real world b. High degree of replication possible	a. Sufficient number of treatments (varying O_3 and SO_2 exposure regimes) within a small geographic area required b. O_3 and other pollutants and environmental variables must be intensively monitored at each site c. Variability due to the influence of soil must be accounted for, unless standardized soil is used at all study sites d. Same as (e) of (4) e. Year-to-year variability in O_3 and SO_2 exposure and crop response must be accounted for
6. Chemical protectant (antioxidants), e.g., O_3	a. Close to the real world b. High degree of replication possible	a. Effect of the protectant itself on plant growth and yield possible; thus prior testing required b. The amount of protection provided by different chemical doses on different plant species not fully understood c. Same as all others listed under (5)
7. Cultivar screening, e.g., O_3	a. Closest to the real world b. No chambers, no chemical protectants	a. Differences in the chronic responses of cultivars to O_3 exposures must be known b. Same as (b), (d), and (e) listed in (5)

Source: Modified from Manning and Krupa.[30]

Table 1.3 Examples of Outdoor Supplementation Systems for Studies of Increased UV-B radiation

Crop	Lamp type	High/low frequency	Filters		Description	Ref.[b]
			Treatment	Control		
A. Square-wave exposure						
Glycine max	Westinghouse FS-40	LF	CA	PE "Mylar"	Timer controlled in two steps for 6 h daily to simulate 16 and 25% O_3 depletion at Beltsville, U.S., calculated using the model of Green et al. (1980); different dose by varying distance from lamps	Teramura et al. (1990)
	Q-Panel UVB-313	LF	CA	PE	Inside PVC open-top chambers, timer controlled, flux and duration adjusted biweekly, manual intervention if overcast; 3-year study with gaseous O_3 interaction simulating a 15, 20, and 35% O_3 depletion at Beltsville, U.S., calculated using the model of Green et al. (1980)	Booker et al. (1992); Fiscus et al. (1994)
Oryza sativa cultivars	Q-Panel UVB-313	LF	CA	PE "Mylar"	Field study at IRRI, Philippines using 6-h switched treatment to simulate 20% O_3 depletion	Dai et al. (1995); Olszyk et al. (1996)
Phaseolus vulgaris	Q-Panel UVB-313	HF	CA	PE and LNE	Manually modulated treatment for 7 h/day simulating 20% O_3 depletion at Pistoia, Italy, calculated using the model of Björn and Murphy (1985)	Zipoli and Grifoni (1995); Antonelli et al. (1996)
B. Modulated exposure						
Vicia faba	Westinghouse FS-40	HF	CA	PE "Mylar-D"; no lamp control in *Vicia faba*	Dynamic range of modulation 50:1, R-B meter measured ambient UV-B; study of *Vicia faba* simulating 6 and 32% O_3 depletion calculated using the model of Green et al. (1980)	Caldwell et al. (1983); Flint et al. (1985)

Table 1.3 (continued)

Crop	Lamp type	High/low frequency	Filters		Description	Ref.[b]
			Treatment	Control		
B. Modulated exposure (continued)						
Glycine max	Q-Panel UVB-313 and UVB-351	HF	Cellulose triacetate	PE	Four experiments using different lamps and filters around the lamps and the lamp frames to vary the ratios of UV-B:UV-A:PAR; UV-B exposure simulating 36% O3 depletion at Logan, U.S., calculated using the model of Green et al. (1980)	Caldwell et al. (1994)
Triticum aestivum and *Avena fatua*	Q-Panel UVB-313	HF	CA	PE	Mixtures and monocultures of *Triticum aestivum* and *Avena fatua*; dose equivalent to different O_3 depletions calculated for Logan, U.S., using the model of Green et al. (1980)	Barnes et al. (1985), (1994)
Glycine max	Q-Panel UVB-313	LF	CA	PE "Mylar-S"	Comparison of square-wave and modulated exposure of *Glycine max* at Beltsville, U.S.	Sullivan et al. (1994)
Pisum sativum	Philips TL40/12	LF	CA	LNE	Study of *Pisum sativum* at Wellesbourne, U.K. Simulated 15% O_3 depletion calculated using model of Björn and Murphy (1985) and following seasonal variations in the O_3 column; sensor cross-calibrated to PAS	Mepsted et al. (1996)
Oryza sativa	Philips F40UVB	LF	CA	PE "Mylar-D"	Rice cultivars; modulated increase in ambient UV-B to 2.67 times measured ambient; sensor response similar to DNA action spectrum	Nouchi and Kobayashi (1995)

[a] LF, low frequency (50/60 Hz); HF, high frequency (e.g., 30 kHz); PE, polyester; CA, cellulose diacetate; LNE, lamps not energized.
[b] See the original publication for information on specific references.

Source: Modified from McLeod.[35]

the quantification of effective ambient CO_2 dose (refer to Strain and Cure[46]), similar techniques are not available for O_3 and UV-B. Rates of uptake or absorption of these two variables must be inferred through the computation of their flux to the plant canopy or receptor site.[11,33,42,47]

A critical issue in establishing any cause–effect relationship is the numerical definition of the exposure or effective dose that captures the atmospheric dynamics or temporal variability of the cause and the corresponding features of the crop biology. Under ambient conditions, the relationships among many of the variables listed in Table 1.1 and crop responses are fundamentally stochastic in nature.[48]

1.4.1 Ozone and Carbon Dioxide

In the case of O_3, numerous investigators have used seasonal mean or summation methods of different types (e.g., sum of all hourly concentrations above 60 ppb or accumulated O_3 dose above a threshold, such as 40 ppb).[12,49] These types of season-end statistical definitions cannot capture the atmospheric dynamics of O_3. Other investigators have used weighting functions to describe the O_3 exposures (e.g., sigmoidal weighting).[50] Here, different hourly O_3 concentrations are weighted differently, with higher concentrations receiving more weight, based on the concept that progressively higher air concentrations of O_3 will have progressively higher potential to cause stress to the crop. Recently, Grünhage et al.[51] were unable to establish a consistent diurnal relationship between atmospheric conditions that favor the prevalence of high O_3 concentrations, its flux onto a grassland canopy, and plant uptake. In general, in the crop-growing regions of the low elevations, high O_3 concentrations occur during the later part of the afternoon or early evening.[13] In contrast, approximately 3 to 4 h after noon, the vapor pressure deficit reaches its maximum value and stomatal conductance reaches its low.[52,53] Furthermore, there is evidence to show that different growth stages of a given crop species respond differently to O_3 exposures.[54] In the final analysis, future efforts regarding the dosimetry of O_3 must include the dynamics of O_3 flux onto the crop canopy and uptake, coupled as a time series with the crop growth dynamics to account for the feedbacks of compensation and stress repair.

As with O_3, most of the artificial CO_2 exposure experiments have addressed the issue of elevated CO_2 concentrations beyond the present,[6,16,55] although in the case of O_3, a charcoal-filtered treatment (low O_3) is included.[27,28,30] We know of only one group of investigators that has included simultaneous high to low concentration gradients of CO_2 exposures in its study.[56] In the elevated CO_2 studies, the dosimetry is straightforward, since the gas concentration is kept constant, but at different levels in the various treatments.[31] However, as with O_3, here, too, the effective dose should be computed and applied in any effects modeling. Although spatial variability in ambient CO_2 concentrations is not addressed in many of these studies, in a 2-year comparative investigation, the maximum hourly CO_2 concentration at a background site in Alberta, Canada was 381 ppm, whereas at a semiurban site it was 487 ppm.[18] Thus, although annual average values across sites may be comparable, as with other gaseous pollutants, significant spatial and temporal variabilities should be expected for CO_2 concentrations also. These features may be relevant to crop response, because, for example, sorghum (*Sorghum vulgare L.* Moench) (a C_4 plant) showed almost no response to elevated CO_2 during its early stages of growth compared with the whole season.[16] In contrast, cotton (*Gossypium hirsutum* L.) (C_3) was very responsive in both the early stages of its growth and over the whole season. Cucumber (*Cucumis sativus* L.) (C_3) and radish (*Raphanus sativus* L.) (C_3) while responsive in their early stages of growth, showed a progressive decline in response over the whole growth season. The implications of these features with regard to *in vivo* carbon allocation among the plant organs are not fully understood at the present time.

1.4.2 Ultraviolet-B Radiation

In the absence of reproducible measured and demonstrable total column ozone reduction at all latitudes, the solar spectrum increases rapidly with increasing wavelength in the UV region. Conversely, absorption by biomolecules increases with decreasing wavelength. This is the scenario predicted by Cutchis[24] for a loss in total column ozone and change in UV-B, by decreasing wavelength. Therefore, the interest is not so much in the total UV-B irradiance, but in the effectiveness of that irradiance in initiating some photobiological effect. The biologically effective UV-B dose or UV-B(BE) is calculated through the application of weighting functions derived from action spectra (e.g., damage to DNA, inhibition of photosynthesis) to assess the relative biological effectiveness of polychromatic irradiance.[57]

Since action spectra differ greatly in their decline with increasing wavelength, it is important to decide which of the primary injury mechanisms are the most important for intact higher plants exposed to solar radiation. To make this decision one needs to know (1) the basic sensitivity of the respective targets (e.g., DNA, photosystem II, etc.) in the plant species under consideration; (2) the protection offered by the location of the target within the plant tissue and the actual effective UV-B fluence reaching it — the dose of interest is that which reaches the receptors, not the dose at the leaf surface[42]; (3) the extent of reciprocity (i.e., that response is a function of the product or integral of irradiance and time); and (4) whether radiation at different wavelengths when applied simultaneously and in proportion to their occurrence in sunlight would lead to an additive, more than additive, or less than additive effect.[57]

Coohill[58] examined several UV action spectra that typify the responses of higher plants to irradiation by wavelengths between 280 and 380 nm. He concluded that no single action spectrum or solar effectiveness spectrum can be considered to be general enough to predict the varied responses of plants at wavelengths throughout the UV-B range. Caldwell and Flint[59] have discussed the issue of monochromatic vs. polychromatic biological action spectra. They concluded that the development of bottom-up, mechanistic and top-down, polychromatic action spectra are unsatisfactory. The authors offered an intermediate approach using whole-plant morphological changes, but no supportive data.

There is a significant amount of variability in the procedures used for UV-B irradiance in artificial exposure experiments.[35,41,60,61] In addition, extrapolation of the spectra of component reactions of isolated organelles to the spectra of whole-plant response under ambient conditions is highly precarious,[57] and the application of such action spectra does not consider the role of polychromatic light and the phenomenon of photorepair. Coohill[58] correctly pointed out that, if common irradiation and response assay procedures are followed, then a general plant effect action spectrum may be forthcoming in the future. In contrast, Sutherland[62] concluded that although action spectra for UV-B radiation damage to higher organisms can give useful information, inappropriate analyses can lead to misleading and even incorrect results. According to the author, it might be useful to include both normalized and non-normalized action spectra in cases of complex biological responses.

1.4.3 Temperature

In contrast to the field experiments on the effects of elevated O_3, CO_2, and UV-B levels, our present knowledge of the effects of predicted future increases in the air temperature on crop production is mainly based on simulation modeling.[29] Alterations in the number of growing degree days and the temperature range to which a crop is adapted are the key considerations here.

1.5 MODELING THE CAUSE AND EFFECT RELATIONSHIPS

Among the many variables involved in global climate change (Table 1.1), future modifications in ambient concentrations of CO_2 and tropospheric O_3 and alterations in the patterns of UV-B radiation, air temperature, and precipitation depth are of much concern in regard to crop production and agriculture.[5] In this context, much of the research until now relates to univariate experiments (CO_2 or O_3 or UV-B, etc.).[16] Only in the recent years have scientists begun to examine the effects of two or more environmental variables on crop growth and productivity.[11,63]

There are many uncertainties associated with our current understanding of the numerical relationships between causes and effects. These uncertainties arise from inadequate (1) experimental designs, (2) artificial exposure techniques, (3) numerical definitions of the exposure dosimetry, and (4) application of cause–effect models themselves. Because of what was stated previously, most cause–effect models deal with a single independent variable, although this is changing.

There are two general approaches to cause and effect modeling: statistical or empirical and process or mechanistic, using actual measured results or through simulation. Krupa and Kickert[43] have summarized the advantages and limitations of the two types of approaches. Most investigators have used the statistical approach, because of its relative simplicity and significantly lower costs compared with the mechanistic approach. Further, in the statistical approach, many investigators have used single-point (season-end) models that cannot account for the stochasticity and feedbacks in the environment–crop interactions. A desirable and meaningful approach is the use of multipoint models that treat the dynamics of both atmospheric and biological processes as a time series.[54] Here, in addition to the application of a satisfactory definition of the dosimetry (corrected for flux and uptake as appropriate), a key, but perhaps simple crop process such as relative growth rate (RGR)[64] needs to be incorporated into the model. In this context, our current knowledge of modeling the effects of key climate variables on crops has been reviewed by several investigators: Boote et al.[65] — CO_2; Kickert and Krupa[54] — O_3; Krupa et al.[11] — UV-B; Fischer et al.[66] — temperature; and Evans[67] — precipitation.

Kickert and Krupa[68] identified 41 combination exposure treatments involving the key plant growth regulating variables in global climate change, using only high or low values for each variable. Clearly at the outset, although this alone might appear to be a daunting task for anyone to investigate, there are desktop computer software programs (e.g., Table Curve 3-D, Jandel Scientific, San Rafael, CA) that allow the exploration and representation of response surfaces for data on three factors from among 36,582 equations and then sort out the best equations according to the goodness of fit. Compounding the impacts of the abiotic, growth-regulating environmental variables are the impacts of diseases, insect pests, and crop–weed competition.[9]

The only practical solution to the serious difficulties described is the use of simulation modeling. Here, the model input parameters must be relevant to the observed behavior of the atmosphere and the outputs from such efforts must be verified (calibrated) by actual data gathered at a number of geographic locations with varying climatology, but with a common indigenous crop. Such an approach was used by Fischer et al.,[66] for example, for wheat, although their effort was directed strictly to the effects of CO_2-induced elevated air temperatures. While educational, such an effort is not very useful, because it omits consideration of the interactive effects of multiple variables on crops. Furthermore, in the context of global warming, various simulation efforts have used the outputs from different climate oriented general circulation models (GCMs) coupled to crop growth models to predict alterations in the geographic distribution and future crop production.[66,69] Here, predicted changes in the average air temperature are the key independent variables under consideration. Clearly, the results will vary depending on the GCM used, since different GCMs predict different levels of increase in the air temperature of the future.[2,3] Karl et al.[70] analyzed temperature data from 492 deurbanized historical climate network stations in the U.S. According to their results, the differences between day and night temperatures have declined since about 1950. During this period, daytime temperatures have declined slightly and nighttime temperatures have risen. Similar results were also obtained for mainland China and the former Soviet Union.[71] Most likely, increases in atmospheric sulfate aerosols and the consequent increases in cloud cover were the important factors influencing these observations by primarily trapping the outgoing heat at night and, thus, increasing the air temperature during that period. Nevertheless, current crop simulation models do not include spatial differences in such variations in the daytime-nighttime temperatures, although the issue has been recognized.[72] Similarly, while predicted increases in global-scale average air temperature may be true, because of the significant spatial variability in tropospheric air pollution, in the future specific geographic locations may become warmer, cooler, wetter, drier, or remain the same.[73] In addition, average values are not as important as their standard deviations or the frequencies of occurrences of extreme events in the context of crop biology.

There are many published computer simulation models for important crops across the world, especially for those in the temperate latitudes (among others, refer to the journals *Ecological Modeling* and *Crop Science*). A few models have been published for crop–weed competition, to simulate crop production in response to solar radiation, air temperature, moisture, and, in some cases, nutrient levels. There are only a few cases incorporating CO_2, and we know of no such cases that include both O_3 and UV-B.[11] A conceptual dynamic, compartmental, plant carbon-growth model has been described that could integrate the combined interactions of day-to-day changes in photosynthetically active solar radiation, ambient CO_2, moisture and nutrient availability, and stress effects from air pollutants on the mass balance of carbohydrate production and consumption by various plant organs.[74] Known and hypothetical interactions of UV-B with these factors, and the consequent joint effects of such interactions, would also have to be incorporated into the structure of such a model.

1.6 FUTURE AGRICULTURE

There is a major concern that the capacity to produce food will be greatly reduced in the future, because of the observed and/or predicted changes in the climatic conditions around the world that permit present-day agriculture.[5] Some believe that geographic areas not at present under cultivation, such as the taiga of the northern latitudes, may experience more favorable weather, resulting in extended growing seasons.[75] Some regions that support a particular crop at the present time (e.g., soybean in the upper Midwest of the U.S.) may become warmer and too dry, and a shift will have to be made to the production of another crop (e.g., sorghum) better suited for those conditions.[76] Accepting the uncertainties associated with the predictions of the current climate models and crop responses, what might be the general expectations to adapt agricultural practices to sustain food production? In this context, there are two general issues that deserve attention: (1) how might climate change affect the geographic distributions of the current crop species and their cropping practices and (2) what might be the limitations in the genetic base of the present-day crop species?

1.6.1 Crop Distribution and Cropping Systems: A Case of Trial and Error

By necessity, agriculture production is a conservative undertaking. Growers are always aware of the very real prospect of a crop failure. Once they have identified a set of crops that yield consistently (fail less often and, in general, provide good returns), they prefer to sustain their use. Crop rotation is practiced to achieve additional stability. Thus, cropping systems are developed and refined in a series of steps by trial and error. Here, small changes are preferred, in contrast to large shifts toward cultivation of a new crop or radical changes in the production of the existing crop, although individual growers and research organizations might seek better cropping methods. However, widespread adoption of new cropping schemes in general is neither rapid nor permanent. Frequently, an assumption is made that the crop production environment is relatively stable and, thus, that what has performed in the past will continue to do so in the future. However, individual components of the production and economic environments can change, leading to a decline in the cultivation of the crop. The occurrence of a new disease or an insect pest is sufficient to discard the production of a crop, at least for some time. Thus, any accelerated changes in the climate are likely to disrupt some of the current cropping systems. The process of developing new systems will take many years, and more so as the environment is predicted to change through time. This overall scenario is bound to be complex, because of the number of atmospheric processes and their products that are involved in climate change (Table 1.1). For example, average rainfall in a geographic region may increase or decrease. Although average precipitation depth is important, so is the distribution of moisture through the year. In temperate climate, rain-fed agriculture is only possible if rains occur frequently during the summer season, although drought conditions during fall through spring seasons could result in inadequate soil moisture during the spring planting. The relatively uniform distribution of moisture across many parts of North America and Europe[77] has shaped our agriculture and any changes toward disuniformity will affect future crop production. This is but one example in the context of overall climate change issues.[5] Nevertheless, development of new agricultural production practices will continue to take place by trial and error.

1.6.2 Limitations in Changing the Cultivation of Existing Crops

Many currently grown crops are considered to be genetically depauperate.[78] Elite lines and hybrids derived from such germ plasms are designed to yield well under relatively narrow and well-defined growing conditions. However, superior genotypes are used heavily in various crop-breeding programs, often resulting in considerable relatedness between cultivars grown across large geographic areas. Nevertheless, some crops are more narrow in their geographic distributions than others, but, in general, the narrow genetic base and specific goals of breeding of virtually all of our modern crops combine to make it unlikely that the crop breeders will be able to accommodate for climate change, particularly if such changes are large and rapid. Confounding this overall discussion are the issues of changes in the epidemiology of crop disease and insect pest incidences and crop–weed competition.

In the final analysis, if climate change occurs gradually, agriculture production will be able to adapt to such changes. This is more likely, if continued improvements are made in modifying the crop biology through the use of molecular tools. However, it is difficult to predict the future with any confidence, given the uncertainties associated with our current knowledge base.

1.7. CONCLUDING REMARKS

Our current knowledge of atmospheric processes and their products in global climate change and our ability to predict their impacts on crop growth and productivity in the future are fraught with significant uncertainties. There are several sources of these uncertainties. Atmospheric measurements of several of the crop growth-regulating parameters are inadequate to capture their spatial and temporal variabilities satisfactorily. Such information is critical in developing realistic experimental designs in establishing cause–effect relationships. Many of the experimental exposure systems and the corresponding definitions of the exposure dosimetry do not have the needed realism to capture the stochasticity of the relevant relationships. Here, a much closer working relationship is required between the atmospheric scientists and crop biologists.

Most of our knowledge of the effects of global climate change variables on crop growth and productivity are based on univariate experimental studies. In these situations, different crop growth regulating variables by themselves have been found to produce opposing effects on the biology of the same crop. The few bivariate artificial exposure studies to date have shown that the effect of one independent variable can be negated by another, the net result being essentially no change in the crop biomass.

There is a significant increase in the use of computer simulation models to predict the impacts of global climate change on future agriculture. While such efforts are very educative, they suffer from the use of input parameters that do not satisfactorily address many of the previously expressed concerns in initializing the models. Furthermore, such model outputs need validation.

In the case of domesticated plants (crops), the germ plasm is much narrower than the corresponding wild species. Over the short term, crop breeders will most likely compensate for the impacts of global climate change, if such a change is gradual. It is a given that many of the present-day crop cultivars will likely not be in use, for example, in the year 2025. The germ plasm of crops cannot be manipulated by conventional breeding, beyond some point where yield can be sustained. Here, efforts are already in progress to utilize molecular techniques. However, these overall issues are a source of significant uncertainty in predicting the future. Clearly, current studies on the impacts of global climate change on agriculture are warranted and are educative. However, future research efforts must address ways to minimize the sources and the extent of these uncertainties.

ACKNOWLEDGMENTS

We thank the University of Minnesota Agricultural Experiment Station for support-in-kind during the preparation of this chapter. We acknowledge with much appreciation the meticulous assistance of Leslie Johnson, through her word-processing and editorial support in completing this manuscript.

REFERENCES

1. Krupa, S. V., Global climate change: processes and products — an overview, *Environ. Monitor. Assess.*, 46, 73, 1997.
2. IPCC (Intergovernmental Panel on Climate Change), *Climate Change: The IPCC Scientific Assessment*, Houghton, J. T., Jenkins, G. J., and Ephraums, J. J., Eds., Cambridge University Press, Cambridge, 1990.
3. IPCC (Intergovernmental Panel on Climate Change), IPCC WGI 1995 Summary for Policymakers, 5th WGI Session, Madrid, 27–29 November 1995, IPCC WGI Technical Support Unit, 1995.
4. Krupa, S. V., The role of atmospheric chemistry in the assessment of crop growth and productivity, in *Plant Response to Air Pollution*, Yunus, M. and Iqbal, M, Eds., John Wiley & Sons, Chichester, U.K., 1996, 35.
5. Bazzaz, F. and Sombroek, W., Eds., *Global Climate Change and Agricultural Production: Direct and Indirect Effects of Changing Hydrological, Pedological and Plant Physiological Processes*, John Wiley & Sons, Chichester, U.K., 1996.
6. Kimball, B. A., Carbon dioxide and agricultural yield: an assemblage and analysis of 430 prior observations, *Agron. J.*, 75, 779, 1983.
7. Rennenberg, H., Ed., *Sulfur Nutrition and Sulfur Assimilation in Higher Plants: Fundamental Environmental and Agricultural Aspects*, SPB Academic Publishing, The Hague, the Netherlands, 1990.
8. Krupa, S. V. and Jäger, H.-J., Adverse effects of elevated levels of ultraviolet (UV)-B radiation and ozone (O_3) on crop growth and productivity, in *Global Climate Change and Agricultural Production: Direct and Indirect Effects of Changing Hydrological, Pedological and Plant Physiological Processes*, Bazzaz, F. and Sombroek, W., Eds., John Wiley & Sons, Chichester, U.K., 1996, 141.
9. Lumsden, P., Ed., *Plants and UV-B: Responses to Environmental Change*, Cambridge University Press, Cambridge, 1997.
10. Rozema, J., Gieskes, W. W. C., van der Geijn, S. C., Nolan, C., and de Boois, H., Eds., UV-B and biosphere, *Plant Ecol.*, 128, 1, 1997.
11. Krupa, S. V., Kickert, R. N. and Jäger, H.-J., *Elevated Ultraviolet (UV)-B Radiation and Agriculture*, R. G. Landes Company, Austin, TX, 1998.

12. U.S. EPA, Air Quality Criteria for Ozone and Other Photochemical Oxidants, Vol. II, EPA-600/P-93/00bF, U.S. Environmental Protection Agency, National Center for Environmental Assessment, Research Triangle Park, NC, 1996.
13. U.S. National Research Council, Atmospheric observations of VOC, NO_x and ozone, in *Rethinking the Ozone Problem in Urban and Regional Air Pollution*, National Academy Press, Washington, D.C., 1992, 211.
14. Fishman, J., Fakhruzzaman, K., Cros, B., and Nganga, D., Identification of widespread pollution in the Southern Hemisphere deduced from satellite analyses, *Science*, 252, 1693, 1991.
15. Hidy, G. M., Mahoney, J. R., and Goldsmith, B. J., International Aspects of the Long Range Transport of Air Pollutants, Final Report, U.S. Department of State, Washington, D.C., 1978.
16. Krupa, S. V. and Kickert, R. N., The Greenhouse effect: impacts of ultraviolet-B (UV-B) radiation, carbon dioxide (CO_2), and ozone (O_3) on vegetation, *Environ. Pollut.*, 61, 263, 1989.
17. Boden, T. A., Kaiser, D. P., and Stoss, F. W., Eds., Trends 93: A Compendium of Data on Global Change, ORNL/CDIAC-65, Carbon Dioxide Information Analysis Center, Oak Ridge National Laboratory, Oak Ridge, TN, 1994.
18. Legge, A. H. and Krupa, S. V., Eds., *Acidic Deposition: Sulphur and Nitrogen Oxides*, Lewis Publishers, Chelsea, MI, 1990.
19. Coldiron, B. M., Thinning of the ozone layer — facts and consequences, *J. Am. Acad. Dermatol.*, 27, 653, 1992.
20. Webb, A. R., Monitoring changes in UV-B radiation, in *Plants and UV-B: Responses to Environmental Change*, Lumsden, P. J., Ed., Cambridge University Press, Cambridge, 1997, 13.
21. Stanhill, G. and Moreshet, S., Global radiation climate changes: the world network, *Climte. Change*, 21, 57, 1992.
22. Monteith, J. L. and Unsworth, M. H., *Principles of Environmental Physics*, 2nd ed., Routledge, Chapman and Hall, New York, 1990.
23. German Bundestag, Ed., Protecting the Earth, A Status Report with Recommendations for a New Energy Policy, Deutscher Bundestag, Referat Öffentlichkeitsarbeit, Bonn, Germany, 1991.
24. Cutchis, P., Stratospheric ozone depletion and solar ultraviolet radiation on earth, *Science*, 184, 13, 1974.
25. Holmes, M. G., Action spectra for UV-B effects on plants: monochromatic and polychromatic approaches for analysing plant responses, in *Plants and UV-B: Responses to Environmental Change*, Lumsden, P. J., Ed., Cambridge University Press, Cambridge, 1997, 31.
26. Allen, L. H., Jr., Baker, J. T., and Boote, K. J., The CO_2 fertilization effect: higher carbohydrate production and retention as biomass and seed yield, in *Global Climate Change and Agricultural Production: Direct and Indirect Effects of Changing Hydrological, Pedological and Plant Physiological Processes*, Bazzaz, F. and Sombroek, W., Eds., John Wiley & Sons, Chichester, U.K., 1996, 65.
27. Heck, W. W., Taylor, O. C., and Tingey, D. T., Eds., *Assessment of Crop Loss from Air Pollutants*, Elsevier Applied Science, London, 1988.
28. Jäger, H.-J., Unsworth, M., De Temmerman, L., and Mathy, P., Eds., Effects of Air Pollutants on Agricultural Crops in Europe: Results of the European Open-Top Chambers Project, Report No. 46, Air Pollution Series of the Environmental Research Programme of the Commission of the European Communities, Directorate-General for Science, Research and Development, Brussels, Belgium, 1993.
29. Parry, M. L., *Climate Change and World Agriculture*, Earthscan in association with the International Institute for Applied Systems Analysis and United Nations Environment Programme, London, 1990.
30. Manning, W. J. and Krupa, S. V., Experimental methodology for studying the effects of ozone on crops and trees, in *Surface Level Ozone Exposures and Their Effects on Vegetation*, Lefohn, A. S., Ed., Lewis Publishers, Chelsea, MI, 1992, 93.
31. Hendrey, G. A. F., Lewin, K. F., and Nagy, J., Free air carbon dioxide enrichment: development, progress, results, in *CO_2 and Biosphere*, Rozema, J., Lambers, H., van de Geijn, S. C., and Cambridge, M. L., Eds., Kluwer Academic Publishers, Dordrecht, the Netherlands, 1993, 17.
32. Jetten, T. H., Physical Description of Transport Processes Inside an Open-Top Chamber in Relation to Field Conditions, Ph.D. dissertation, Agricultural University, Wageningen, the Netherlands, 1992.
33. Krupa, S. V. and Kickert, R. N., Considerations for establishing relationships between ambient ozone (O_3) and adverse crop response, *Environ. Rev.*, 5, 55, 1997.

34. Kimball, B. A., Pinter, P. J., Jr., Wall, G. W., Garcia, R. L., LaMorte, R. L., Jak, P. M. C., Arnoud Frumau, K. F., and Vugts, H. F., Comparisons of responses of vegetation to elevated carbon dioxide in free-air and open-top chamber facilities, in *Advances in Carbon Dioxide Effects Research*, Allen, L. H., Jr., Kirkham, M. B., Olszyk, D. M., and Whitman, C. E., Eds., ASA Special Publication No. 61, American Society of Agronomy, Crop Science Society of America, Soil Science Society of America, Madison, WI, 1997, 113.
35. McLeod, A. R., Outdoor supplementation systems for studies of the effects of increased UV-B radiation, *Plant Ecol.*, 128, 78, 1997.
36. Corlett, J. E., Stephen, J., Jones, H. G., Woodfin, R., Mepsted, R., and Paul, N. D., Assessing the impact of UV-B radiation on the growth and yield of field crops, in *Plants and UV-B: Responses to Environmental Change,* Lumsden, P. J., Ed., Cambridge University Press, Cambridge, 1997, 195.
37. Caldwell, M. M., Solar UV irradiation and the growth and development of higher plants, in *Photophysiology, Vol. VI., Current Topics in Photobiology and Photochemistry*, Giese, A. C., Ed., Academic Press, New York, 1971, 131.
38. Coohill, T. P., Stratospheric ozone depletion as it affects life on earth — the role of ultraviolet action spectroscopy, in *Impact of Global Climatic Changes on Photosynthesis and Plant Productivity*, Abrol, Y. P., Govindjee, Wattal, P. N., Ort, D. R., Gnanam, A., and Teramura, A. H., Eds., Oxford and IBH Publ. Co. Pvt. Ltd., New Delhi, India, 1991, 3.
39. Finlayson-Pitts, B. J. and Pitts, J. N., *Atmospheric Chemistry: Fundamentals and Experimental Techniques*, John Wiley & Sons, New York, 1986.
40. Kerr, J. B. and McElroy, C. T., Evidence for large upward trends of ultraviolet-B radiation linked to ozone depletion, *Science*, 262, 1032, 1993.
41. Tevini, M., Mark, U., and Saile, M., Plant experiments in growth chambers illuminated with natural sunlight, in *Environmental Research with Plants in Closed Chambers*, Payer, H. D., Pfirrmann, T., and Mathy, P., Eds., Air Pollution Research Report No. 26, Commission of the European Communities, Brussels, 1990, 240.
42. Bornmann, J. F., Reuber, S., Cen, Y.-P., and Weissenböck, G., Ultraviolet radiation as a stress factor and the role of protective pigments, in *Plants and UV-B: Responses to Environmental Change*, Lumsden, P. J., Ed., Cambridge University Press, Cambridge, 1997, 157.
43. Krupa, S. and Kickert, R. N., An analysis of numerical models of air pollutant exposure and vegetation response, *Environ. Pollut.*, 44, 127, 1987.
44. Runeckles, V. C., Dosage of air pollutants and damage to vegetation, *Environ. Conserv.*, 1, 305, 1974.
45. Guderian, R., *Air Pollution: Phytotoxicity of Acidic Gases and Its Significance in Air Pollution Control*, Ecological Studies 22, Springer-Verlag, New York, 1977.
46. Strain, B. R. and Cure, J. R., Eds., Direct Effects of Increasing Carbon Dioxide on Vegetation, DOE-ER-0238, U.S. Department of Energy, Washington, D.C., 1985.
47. Runeckles, V. C., Uptake of ozone by vegetation, in *Surface Level Ozone Exposures and Their Effects on Vegetation*, Lefohn, A. S., Ed., CRC Press, Boca Raton, FL, 1991, 153.
48. Krupa, S. V. and Teng, P. S., Uncertainties in estimating ecological effects of air pollutants, in *Proc. 75th Annu. Meetings, APCA*, New Orleans, LA, Air Pollut. Control Assoc., Pittsburgh, PA, Paper No. 82-6.1, 1982.
49. Fuhrer, J., Skärby, L., and Ashmore, M. R., Critical levels for ozone effects on vegetation in Europe, *Environ. Pollut.*, 97, 91, 1997.
50. Lefohn, A. S. and Runeckles, V. C., Establishing a standard to protect vegetation — ozone exposure/dose considerations, *Atmos. Environ.*, 21, 561, 1987.
51. Grünhage, L., Jäger, H.-J., Haenel, H.-D., Hanewald, K., and Krupa, S. V., PLATIN (**PL**ant-**AT**mosphere **IN**teraction): II. Co-occurrence of high ambient ozone concentrations and factors limiting plant absorbed dose, *Environ. Pollut.*, 98, 51, 1997.
52. Jarvis, P. G. and Mansfield, T. A., Eds., *Stomatal Physiology*, Cambridge University Press, Cambridge, 1981.
53. Meidner, H., Measurements of stomatal aperture and responses to stimuli, in *Stomatal Physiology*, Jarvis, P. G. and Mansfield, T. A., Eds., Cambridge University Press, Cambridge, 1981.
54. Kickert, R. N. and Krupa, S. V., Modeling plant response to tropospheric ozone: a critical review, *Environ. Pollut.*, 70, 271, 1991.

55. Rozema, J., Lambers, H., van de Geijn, S. C., and Cambridge, M. L., Eds., *CO_2 and Biosphere*, Kluwer Academic Publishers, Dordrecht, the Netherlands, 1993.
56. Johnson, H. B., Wayne Polley, H., and Mayeux, H. S., Increasing CO_2 and plant–plant interactions: effects on natural vegetation, *Vegetatio*, 104/105, 157, 1993.
57. Caldwell, M. M., Camp, L. B., Warner, C. W., and Flint, S. D., Action spectra and their key role in assessing biological consequences of solar UV-B radiation change, in *Stratospheric Ozone Reduction, Solar Ultraviolet Radiation and Plant Life*, Worrest, R. C. and Caldwell, M. M., Eds., Springer-Verlag, Berlin, 1986, 87.
58. Coohill, T. P., Ultraviolet action spectra (280–380 nm) and solar effectiveness spectra for higher plants, *Photochem. Photobiol.*, 50, 451, 1989.
59. Caldwell, M. M. and Flint, S. D., Uses of biological spectral weighting functions and the need of scaling for the ozone reduction problem, *Plant Ecol.*, 128, 66, 1997.
60. Caldwell, M. M. and Flint, S. D., Plant response to UV-B radiation: comparing growth chamber and field environments, in *Environmental Research with Plants in Closed Chambers*, Payer, H. D., Pfirrmann, T., and Mathy, P., Eds., Air Pollution Research Report No. 26, Commission of the European Communities, Brussels, 1990, 264.
61. Worrest, R. C. and Caldwell, M. M., Eds., *Stratospheric Ozone Reduction, Solar Ultraviolet Radiation and Plant Life*, Springer-Verlag, Berlin, 1986.
62. Sutherland, B. M., Action spectroscopy in complex organisms: potentials and pitfalls in predicting the impact of increased environmental UVB, *Photochem. Photobiol.*, 31, 29, 1995.
63. Unsworth, M. H. and Hogsett, W. E., Combined effects of changing CO_2, temperature, UV-B radiation, and O_3 on crop growth, in *Global Climate Change and Agricultural Production: Direct and Indirect Effects of Changing Hydrological, Pedological and Plant Physiological Processes*, Bazzaz, F. and Sombroek, W., Eds., John Wiley & Sons, Chichester, U.K., 1996, 171.
64. Hunt, R., *Plant Growth Curves: The Functional Approach to Plant Growth Analysis*, Arnold, London, 1982.
65. Boote, K. J., Pickering, N. B., and Allen, L. H., Jr., Plant modeling: advances and gaps in our capability, in *Advances in Carbon Dioxide Effects Research*, ASA Special Publication No. 61, American Society of Agronomy, Crop Science Society of America, Soil Science Society of America, Madison, WI, 1997, 179.
66. Fischer, G., Frohberg, K., Parry, M. L., and Rosenzweig, C., The potential effects of climate change on world food production and security, in *Global Climate Change and Agricultural Production: Direct and Indirect Effects of Changing Hydrological, Pedological and Plant Physiological Processes*, Bazzaz, F. and Sombroek, W., Eds., John Wiley & Sons, Chichester, U.K., 1996, 199.
67. Evans, T. E., The effects of changes in the world hydrological cycle on availability of water resources, in *Global Climate Change and Agricultural Production: Direct and Indirect Effects of Changing Hydrological, Pedological and Plant Physiological Processes*, Bazzaz, F. and Sombroek, W., Eds., John Wiley & Sons, Chichester, U.K., 1996, 15.
68. Kickert, R. N. and Krupa, S. V., Forest responses to tropospheric ozone and global climate change: an analysis, *Environ. Pollut.*, 68, 29, 1990.
69. Reilly, J., Climate change, global agriculture and regional vulnerability, in *Global Climate Change and Agricultural Production: Direct and Indirect Effects of Changing Hydrological, Pedological and Plant Physiological Processes*, Bazzaz, F., and Sombroek, W., Eds., John Wiley & Sons, Chichester, U.K., 1996, 237.
70. Karl, T. R., Baldwin, R. G., and Burgin, M. G., *Historical Climatology*, Series 4-5, National Climatic Data Center, Asheville, NC, 1988.
71. Karl, T. R., Kukla, G., Razuavayev, V. N., Vyacheslav, N., Changery, M., Quayle, R. G., Heim, R. R., Easterling, D. R., and Cong, B. F., Global warming: evidence for symmetric diurnal temperature change, *Geophys. Res. Lett.*, 18, 2252, 1991.
72. Rosenzweig, C. and Hillel, D., Agriculture in a greenhouse world, *Res. Explor.*, 9, 208, 1993.
73. Lodge, J. P., Jr., Climate change in the context of forest response, in *83rd Annu. Meeting, A&WMA*, Pittsburgh, PA, Air & Waste Management Association, Pittsburgh, PA, Paper No. 90-152.4, 1990.
74. Jäger, H.-J., Grünhage, L., Dämmgen, U., Richter, O., and Krupa, S., Future research directions and data requirements for developing ambient ozone guidelines or standards for agroecosystems, *Environ. Pollut.*, 70, 131, 1991.

75. Rosenzweig, C., Potential CO_2-induced climate effects on North American wheat-producing regions, *Climate Change*, 7, 367, 1985.
76. Dekker, W. L. and Achununi, V. R., Greenhouse warming and agriculture, in *83rd Annu. Meeting, A&WMA*, Pittsburgh, PA, Air & Waste Management Association, Pittsburgh, PA, Paper No. 90-151.2, 1990.
77. Leemans, R. and Cramer, W. P., The IIASA Database for Mean Monthly Values of Temperature, Precipitation, and Cloudiness on a Global Terrestrial Grid, RR-91-18, International Institute for Applied Systems Analysis, Laxenburg, Austria, 1991.
78. Hoyt, E., *Conserving the Wild Relatives of Crops*, IBPGR Headquarters, c/o FAO of the United Nations, Rome, 1988.

CHAPTER 2

Effects of Climate Change on the Behavior of Woody Perennials

Christopher J. Atkinson

CONTENTS

2.1 INTRODUCTION

2.1.1 Climate Change: Evidence and Causes

Since the late 1950s the Earth's atmospheric CO_2 concentration has been monitored at several locations. From the remote site at Hawaii's Mauna Loa Observatory, the atmospheric concentration of CO_2 can clearly be seen to have risen from a concentration of 315 vpm in 1958 to around 350 vpm in 1988.[1] Analyses of gas bubbles trapped within ice cores and $^{13}C/^{12}C$ ratios in tree rings support and extend the time sequence data describing changes in global CO_2 concentration.[2-5] From the preindustrial revolution CO_2 concentration of around 275 vpm to the present-day level we have already seen a rise of about 25%. Currently, it is believed that the concentration of CO_2 in the atmosphere is rising at a rate of around 1.5 vpm (0.48%) per year. This will yield a concentration of 420 vpm by the year 2035, or a doubling of the present concentration by the end of the next century.[6]

1-56670-341-7/00/$0.00+$.50

This increase in CO_2 concentration is highly correlated with the global increase in the consumption of fossil fuels (approximately 5.3 x 101^2 kg carbon year^{-1}) and is expected to continue to increase well into the next century.[7] There are also other important factors in the global carbon budget equation to be considered, particularly those associated with the behavior of oceans and forests as sinks and sources for the uptake and release of CO_2. The impact of forest clearance, particularly in the tropics, on global carbon budgets is a subject of considerable debate.

Changes in global "greenhouse gases," of which CO_2 is one, are particularly important, not simply because of any direct effect that they might have on life on Earth, but due to their physical properties, which are capable of inducing global changes in climate. Short-wave solar radiation enters Earth's atmosphere unabsorbed by greenhouse gases. When this energy is radiated back into space after absorption by Earth's surface, it is emitted as long-wave infrared radiation. The extent to which the Earth's atmosphere cools or warms is determined by the efficiency with which this long-wave radiation is lost to space. The amounts of the infrared absorbing "greenhouse gases" therefore have a profound influence on global climate, particularly temperature.

It has been calculated, using measured average global temperatures, that the observed increase in global CO_2 concentration is responsible for a temperature rise of between 0.5 to 0.7°C over the last 140 years.[8] The projected rise in temperature has been calculated using one of the general (atmospheric) circulation models (GCMs) of which there are several. Depending on which model is used, a wide range of possible effects on temperature and other climatological variables have been predicted.[6,9-11] Mitchell,[12] for example, suggests a warming of 1.5°C is likely over the next 40 years. The greatest extent of this warming would appear to be most likely occur toward the polar regions, whereas minimal changes are expected in equatorial regions.

Because of complex feedback interactions within the atmosphere, it is difficult to predict how increasing concentrations of greenhouse gases will modify the terrestrial environment of plants.[6] As yet, even the highly complex GCMs are unable to resolve accurately the important environmental details necessary to predict whole-plant or community responses to climate change. This inability has much to do with the uncertainties associated with assessing the effects of elevated CO_2 and temperature changes on the hydrological cycle.[13]

Woody perennials, particularly forest species, play a significant part in the global carbon cycle; they occupy about 30% of the land surface, 70% of the land biomass, and carry out 60% of the total photosynthesis. These factors, combined with their long life spans, provide them with the potential to be considerable sinks for the long-term storage of carbon. The gas exchange and carbon storage capacity of forest trees therefore play a central role in the regulation of global carbon flux and carbon sink sizes.

2.1.2 Scope of the Review

As yet, it is difficult to predict the role that woody perennials and forest trees may play as dynamic sinks for CO_2 storage. It is equally difficult to interpret how short-term exposures to an increased concentration of CO_2 will translate into longer-term changes in plant growth.[14] This issue remains unanswered, primarily because of the nature of the experiments that have been done. Frequently, experiments are done on juvenile woody plant material and for only a short period of time, although there are a few exceptions.[15-17] The physiology and growth of woody perennials are known to change dramatically with age, particularly with respect to their environmental responses.[18] Therefore, it is not a surprise perhaps that the response of trees to short-term CO_2 enrichment can be short-lived.[19,20] The important differences between short- and long-term physiological acclimation must be considered along with the longer-term selection of genotypes over generations of exposure to elevated levels of CO_2.[21] Ideally, experiments should last for a significant period of the life of a woody perennial, without the biomass below ground being restricted by a limited rooting volume. Species-specific responses to elevated CO_2, should be considered carefully because of the dependent effects that other variables, such as the availability of nutrients and water, can have when not taken into account.

It is the intention of this chapter to describe what is understood at present about the responses of woody perennials and to use this information to predict the possible responses of mature forests to climate change. Where possible, literature published on woody perennials and trees will be cited; this will be supplemented, where appropriate, with critical work on herbaceous species. This chapter will attempt to examine and describe a broad overview of woody plant responses to climate change, in terms of their physiological/biochemical function as well as structural changes in growth and morphology. The implications for changes in stomatal regulation influencing CO_2 uptake and water loss are not covered here, as they are very effectively dealt with in Chapter 3 by Heath and Mansfield. This chapter will also deal with the potential of the ecological structure of whole forest communities to respond to climate change.

2.2 CHANGES IN PLANT FUNCTIONING

2.2.1 Carbon Assimilation

The present concentration of CO_2 limits the ability of C_3 species to fix carbon, so any increase in CO_2 tends to enhance the rate of assimilation and therefore plant growth.[22,23] The reason why net photosynthesis may be enhanced is related to a number of factors connected to the characteristics of the primary carboxylating enzyme (rubisco).[23]

The photosynthetic rate of herbaceous C_3 plants, when grown at elevated CO_2, has been shown to increase on average by 52% with a doubling of the CO_2 concentration,[24] while an average 44% stimulation was apparent for trees.[14] Woody plants when exposed to elevated CO_2 for varying periods of time show not only stimulated photosynthesis, but also increased growth rate and biomass accumulation, e.g., *Liriodendron tulipifera*,[15, 25] *Betula pendula*,[17] *Nothofagus fusca*,[26] *Picea sitchensis*,[27] *Pinus contorta*,[18] *P. ponderosa*,[28] *P. radiata*,[26,29] *P. taeda*,[30] *Populus grandidentata*,[31] *Quercus alba*, [15, 32] *Q. robur*, [20,33] and six tropical species.[14,34, 35] There are, however, some exceptions, as in the case of *Castanea sativa*.[36] In a recent review, it was concluded that, for the 64 species examined, a doubling of the concentration of CO_2 induced an increase in photosynthesis that translated into a 38 and 63% average increase in the biomass of coniferous and deciduous species, respectively.[37]

Increases in assimilation rate may not be sustained, however, and may fall on average to around 29% of the original enhancement.[20,24,38,39] The return of the assimilation rate to its previous level is not always evident,[40,41] and sometimes assimilation may even remain at the increased level.[42,43] The response of a species may be inconsistent across a range of different studies. When the assimilation rate does fall, it is often associated with a decline in rubisco activity and/or an increase in the accumulation of nonstructural carbohydrate.[42,44-46] The most likely causes for this decline are either a loss of the rubisco protein or a change in the specific activity of enzyme.[19,20,38,46-48] The loss of photosynthetic activity may also be due to a CO_2-induced shift in the timing of the leaf ontogenetic processes as they impinge on assimilation and rubisco activity.[17,49]

Accumulation of nonstructural carbohydrate, particularly in leaves, occurs when plants are grown under elevated CO_2.[50-55] On returning plants to ambient CO_2, this accumulation of carbohydrate can be seen to decline.[56] It is suggested that increases in photosynthetic products, such as starch, may be due to an imbalance between the activities of source and sink.[53,57-59] In general, there would appear to be a good correlation between plant growth and the export of carbohydrate from the leaf; this suggests a feedback control. If the sink utilization rate of sucrose is lower than that of synthesis, the accumulation results in feedback on the enzymes of sucrose synthesis.[60] The enzymes associated with regulating sucrose synthesis and assimilate export are also known to change in response to assimilate demand, irrespective of the photosynthetic rate.[61,62] The export of carbohydrate from source leaves and the activity of storage and/or growing tissues to utilize it therefore appear therefore to be particularly important in determining the ability of plants to respond positively to elevated CO_2.

There is good evidence to suggest that unavailable or inactive sinks for carbohydrate storage and/or growth cause acclimation of photosynthetic rate, i.e., a source:sink interaction that causes feedback inhibition. For example, when citrus trees were exposed to elevated CO_2, their assimilation rate was shown to be maintained only in rapidly growing and fruiting trees.[63] Fruit thinning (sink removal) has also been shown to have a direct inhibitory influence on photosynthesis.[64] When the root growth (growth sink) of cotton plants grown at elevated CO_2 was restricted, the reduction in photosynthesis was not associated with a decline in stomatal conductance. In this case, only carboxylation efficiency declined.[57,65] These and other experiments support the hypothesis that a photosynthetic product causes sink-limited feedback inhibition of the soluble protein content, rubisco activity, and quantity through changes in the gene expression.[20, 56, 66, 67]

The increased growth of plants under elevated CO_2 appears to be due to their ability to provide a sufficient number of sites for carbohydrate storage. This is at least the reason the frequently cited accumulation of reserves does not occur in plants with large active storage sinks e.g., field-grown plants and natural vegetation.[68,69] Growth potential of storage organs, particularly below ground, is not found to be restricted.[69] This may also explain why CO_2 photosynthetic acclimation can occur with plants grown in pots.[56] Support for this hypothesis is provided by the response of species with indeterminate, as opposed to determinate, growth forms.[70] Some indeterminate species have been shown to have larger increases in dry weight than determinate species, but again this is correlated with the sink strength (size) and number.[71-74] It is suggested that the growth response of various poplar clones to elevated CO_2 was a reflection of their indeterminate growth habit.[75] The same argument may also explain why considerably more leaf area is produced by "free growing" species (e.g., *Eucalyptus pauciflora*) compared with many others.[37]

Very recent evidence from birch trees grown at elevated CO_2, in open-top chambers, over a 4 1/2-year period shows that the relative growth rate was unaffected by CO_2.[17] From the limited number of appropriate experimental comparisons of species which have been published, there would appear to be no evidence, however, to indicate that growth rate per se is a factor influencing sink strength and CO_2 responsiveness.[14]

2.2.2 Carbon Partitioning

There is some evidence from a few species that there may be changes in the proportional allocation of carbon to various sinks.[17,76-78] However, little is known about the direct influence of CO_2 on the enzymes involved with the partitioning of assimilate. It appears that the activity of the rate-controlling enzyme for sucrose synthesis (sucrose phosphate synthase, SPS) can be correlated negatively with starch accumulation.[79,80] Knowledge about the influence of CO_2 on assimilate transport is equally limited and also contradictory.[53]

CO_2-induced changes in the pattern of carbon allocation bring about not only alterations in whole-plant morphology, but also, perhaps, changes in cellular anatomy. For example, in some cases, the leaf anatomy of plants grown at elevated CO_2 can change, with leaves becoming thicker.[25,36,81] Leaf growth, with respect to thickening, is stimulated by a greater expansion of mesophyll cells, rather than by an increase in the numbers of cells.[82] Changes in the rate of leaf senescence may also occur. Delays in senescence have been reported for *Quercus alba*,[32,83] while enhancements, or compressed ontogeny, have also been observed with *Castanea sativa*, *Populus tremuloides*, and *Betula pendula*.[17,36,49,84]

Evidence exists to suggest that elevated CO_2 may increase the proportion of carbon allocated to root biomass as well as altering root architecture.[17,18,20,25,85-88] This will probably have a beneficial effect on the ability of the plant to acquire nutrients and water, particularly if the fine root density increases.[89] It is also likely that changes in the soil microorganisms will occur. The growth of root mycorrhizal systems appears to increase for plants grown at elevated CO_2,[90] which is due to an increase in the rate of flux of exudates from the roots of the host.[91] In deciduous trees after a few months, the CO_2-stimulated increase in the root carbon pool disappears, but the carbon efflux from the roots is enhanced.[17] As yet, we are still developing our understanding of the importance of such a response on nutrient recycling or of any cost–benefit analysis to the host as opposed to the mycorrhiza.[90]

It is as yet unclear whether elevated concentrations of CO_2 will influence tissue anatomy, particularly xylem anatomy. The processes that influence cambial differentiation determine the cellular fate of dividing cambial mother cells and their physical structure (e.g., lumen size and wall thickness). These features are important in determining the capacity of a stem to transport water to its leaves and its ability to withstand reductions in the availability of soil water and to avoid xylem cavitation and embolism.[92] In the very few cases where these effects have been studied, increases in CO_2 have produced quantitative and qualitative changes in xylem structure.[93,94] However, these changes are not always described as being obviously beneficial to the plant.[86] An increase in annual ring width may be due primarily to a more sensitive response of earlywood.[95] Tracheid wall thickening also increases, thereby enhancing wood density.[93] As suggested already, the relevance of these observations is clearly limited by the fact that they refer to short-term changes in juvenile wood. In the absence of an understanding of the mechanisms that control carbon allocation, it is difficult to predict the commercial benefits to forestry of an increase in carbon fixation. Additional photosynthate has to be allocated to xylem production in such a manner that it produces desirable changes in the qualitative aspects of the wood structure. These changes should ideally promote the suitability of the wood as a material and match it to its intended use. There may well be conflicts between such commercially desirable traits and those essential for maintaining the physiological functioning of the tree, e.g., water capture and transport, as is the case with possible increases in root production.[18,25,86] Increased root growth provides an effective store for carbon but, outside a coppicing system, would be of limited value to the production of timber.

Many plants, and in particular woody perennials, also allocate considerable quantities of carbon to the production of secondary metabolites. For example, leaves during photosynthesis release secondary metabolites in the form of hydrocarbons such as isoprene which combine with the oxides of nitrogen in the air to form atmospheric phytotoxic ozone. The capacity for isoprene production in *Quercus rubra* was doubled in response to a 26% increase in the concentration of CO_2,[96] while with *Populus tremuloides*, *Eucalyptus globulus*, *Q. alba*, and *Mucuna pruriens* emission rates were reduced at the higher CO_2 level.[96, 97]

There is frequently cited evidence of a decrease in the leaf area ratio. Plants that have been acclimated to elevated CO_2 may also show a shift in biomass allocation that could also be beneficial to the conservation or acquisition of resources. For the plant, this is a trade-off between its ability to acquire carbon vs. uptake of water and nutrients.[86] The shift in the ratio of leaf to root mass would have the potential advantage of increasing water uptake while simultaneously reducing its loss, and therefore increasing the effective response of the plant to drought.

Experiments with wheat have shown that plants grown at elevated CO_2 were able to compensate for restrictions in growth caused by soil water deficits. In these experiments, an increase in root biomass was observed at a depth of 0 to10 cm.[98] This change in biomass partitioning is also apparent for a number of other species during CO_2-enrichment.[25,26,29,39] A shift away from shoot growth may not, however, be advantageous when plants are in competition with other species, even in a water-limited habitat.[99] Perennial plants, however, develop an ability to reproduce sexually only after a number of years of vegetative growth. An unrestrained use of water may enable an individual to out compete its neighbors but, in evolutionary terms, this may only benefit the individual if it increases the potential of the individual having progeny in the next generation.

One means by which growth rates can be maintained as soil water potential falls is for the cell turgor pressure to be increased.[76,77,100] This can be achieved by an increase in the cellular solute. Plants that have been grown under elevated CO_2 have an excess accumulation of photosynthetic products that can contribute to osmotic adjustment.[76,77,101]

2.3 CHANGES IN PLANT COMMUNITY STRUCTURE

The geographic distribution of numerous species can be correlated closely with local and regional environmental factors, particularly water deficits,[102] for herbs and of trees.[103] The potential for changes in temperature to influence processes such as flowering, seed germination, whole-plant responses to water deficits and chilling is likely to be more obvious and perhaps more extensive than that for changes resulting directly from effects caused by increases in CO_2.[104] Attempts are being made to explore, through modeling, the potential impact of increases in CO_2 on vegetation diversity.[105,106] The inclusion of CO_2-related gross changes in transpiration within such models suggests that dry regions are most likely to experience the greatest increases in diversity. Measured changes in water use efficiency (WUE) have been used to calculate the potential for the extension of a special geographic range based on precipitation information.[107]

A particularly important question to answer must be to what extent the native plants of a region are likely to have the capacity to acclimate and/or adapt to future increases in CO_2.[108] From an evolutionary standpoint, species tend to be rather conservative; that is, they respond to environmental perturbations by migration rather than by evolving.[109] As Bradshaw and McNeilly[109] point out, the mechanism of the Darwinian process is constrained by the quantity of variability within a population. Despite the fact that changes in the environment can generate intense selection pressures, the pace at which natural selection can work is limited ultimately by the amount of variability. Clearly, this limitation, combined with geographical constraints on species migration, suggests that many species must inevitably become extinct.

Warning signs are present for some species as shown, for example, in a recent study of *Fagus sylvatica*.[110] As the annual growth of beech depends on the availability of soil water, this species is likely to be sensitive to a climate-induced decline in soil water. The ability of species such as beech to tolerate an increase in the frequency and/or severity of drought is limited by factors such as the shallow rooting behavior of the plant. Despite this, it was suggested that it was not soil moisture per se that was the cause of the severe responses of beech to drought in 1995. Instead, Beerling et al.[110] interpreted the response of beech to be a species-specific imbalance between evaporative demand and water supply as a result of limited stomatal closure in elevated CO_2 to compensate for the increase in leaf area.

Recent evidence has supported the suggestion that *F. sylvatica* may be an example of a vulnerable species.[111] Compared with oak, the leaf conductance of beech remained unaffected by elevated CO_2 and, during a period of drought, the plants which had been kept well supplied with nutrients also failed to show any reduction in stomatal conductance. Not surprisingly, this caused the soil to dry more quickly, particularly for plants with a greater leaf area grown under a high-nutrient regime. The resultant effect for beech plants grown at elevated CO_2 was an increase in whole-plant water use when compared with oak.[111]

Competitive interactions will be likely to be altered by changes in CO_2, as is the case for other environmental factors.[112,113] Therefore, it is likely that the early events of vegetation responses to CO_2 will be detected as perturbations in competitive hierarchies and changes in the abundance of species.[85,108,114-116] Those species that have dynamic, short life cycles are most likely to be able to adapt to environmental changes, whereas perennial, slow-maturing, long-lived species may have limited capacity to evolve. This does not preclude the ability of perennials to migrate. Tree-line vegetation, for example, can, and has, shown dramatic change, at least to climate warming, over what is a very short time period (150 years).[117] If migration is not possible, much will depend on the degree of phenotypic plasticity a species possesses.

Factors likely to affect the outcome of competitive interactions for light, particularly shoot densities,[69,118,119] are also influenced by CO_2-induced changes in canopy architecture.[120] In a comparative study with two coexisting honeysuckle species (*Lonicera*), one native and the other an introduced alien, it was shown that the alien species was more responsive to increases in CO_2.[121] This resulted from changes in carbon allocation, which was evident as an increase in leaf area and stem growth. Such events will also have a considerable influence on the reproductive capacity of species and their potential to exploit and migrate into new habitats.[76] The timing of seedling emergence is also important in determining the outcome of competitive interactions, as, of course, will be differential growth rates in early life. There is some evidence to indicate that small seeds germinate quicker when exposed to enhanced concentrations of CO_2.[122]

It would be difficult, retrospectively, to determine whether past increases in atmospheric CO_2 were responsible for species migration but, despite the obvious and complex nature of this subject, some attempts have been made to model the potential for changes in species distribution. In one such study, detailed changes in the composition and productivity of North American forests were implicated in response to increasing CO_2. It has been possible to identify species with different sensitivities to CO_2.[123] Only when this type of information is used, in conjunction with empirical measurements, are the possible influences of increases in CO_2 and climate change likely to be assessed effectively. CO_2-response studies with the American beech suggest that, at least in relative terms, this species is likely to be able to respond positively to changes in CO_2.[78] However, other work shows that this species may be particularly susceptible to rapid increases in temperature and that, because of its large seed size, it will have difficulty in migrating to cooler regions.[124]

The complexity of the issue and the possible fate of the American beech highlights an extremely important area of limited information. The observed and the proposed changes in CO_2 concentration have occurred, and will occur in a very short space of time, at least in evolutionary terms. The question that has to be asked is how will the rapidity of the predicted future increase in CO_2 constrain the ability of plants to either adapt or migrate. The data on *Fagus grandifolia* and *Tsuga canadensis* suggest that these species can migrate northward, but only at a rate of 20 to 25 km-100 years. To track the expected rate of change in temperature, they will have to migrate at a rate of 100 km per 100 years.[125] Clearly, one should not overlook the major contribution that humans have had in influencing plant migration.[106]

2.4 IMPACT OF CLIMATE CHANGE ON WOODY PERENNIALS

Attempts have also been made to draw conclusions about the effects of global increases in CO_2 on trees growing within either natural habitats or forest plantations. Because of the long-lived nature of many trees, it is possible to select those that began growing prior to the industrially induced increase in CO_2 concentration. Species that are extremely long-lived are particularly suitable for this analysis, as they provide a chronology of annual ring widths that may extend back thousands of years. This analysis has not been restricted to the use of living material; chronologies have been overlapped and extended further back into the past by the use of preserved material. From the analysis of annual tree ring growth, it is well known that temperature exerts the dominant effect on ring thickness (wood production), particularly at tree lines.[117] However, the influence of water availability will probably increase in environments that experience frequent summer water deficits. The results from tree ring analysis are complex, however, and frequently difficult to interpret.

There appear to be considerable differences between species in their potential to translate increases in global CO_2 concentration into increased wood production. Furthermore, such increases may not only be difficult to detect, but the ease of detection may vary with species. The physiological age of a tree will also determine the vigor of its growth and its carbon allocation priorities. There may also be complications with seasonal/diurnal variations in CO_2 concentration within the canopy which are different from those of the bulk air.[21,126] This type of variation in the canopy means that, as a tree develops and matures, it may experience very different CO_2 concentrations. Clearly, these factors are not independent of the climatic environment of the tree, but would have considerable influence on ring width. Even the position within the tree from which samples of wood are taken can be very important. This is evident from differences in the sensitivity of the whole-tree hierarchical carbon allocation pattern to environmental change.[127] It is therefore not surprising that the experimental evidence is conflicting.

An examination of growth trends in widely distributed sites within the Northern Hemisphere from 1950 onward, using tree-ring analysis, was unable to attribute the observed increases in ring width solely to enhanced atmospheric levels of CO_2.[128] The conclusions of this research are supported by a study using material of *Pinus balfouriana*, *P. murayana*, and *Juniperus occidentalis* collected in the Sierra Nevada.[129] Conversely, a study of ring widths from the extremely long-lived bristlecone pines (*P. longaeva* and *P. aristata*) has been used to suggest that such an increase is present.[130] The same may also be true for other high-elevation trees (*P. flexilis*) in southwestern U.S. and in the Arctic.[131-133] These data are the subject of much discussion regarding the validity of the techniques adopted and the analysis used.[134,135] Some of this discrepancy may be due to the confounding effects of changes in other biological or environmental factors of which we have limited or no knowledge. We may also expect different habitats to respond in rather different ways. Increases in CO_2 may enhance photosynthesis with a concomitant increase in WUE and such a fertilization effect may therefore be more evident in arid regions. Clearly, if our understanding and interpretation of the climate change scenario is correct, we should also find concurrent increases in ring-width in response to rising temperature.[136,137] Any tree-ring analysis would have to be able to predict accurately this response in terms of its effect on growth before the response could be removed from any direct CO_2 response analysis. The nature and complexity of this temperature response may make it difficult to expect any enlightening results from annual ring width analysis. Even geographic regional differences in temperature change may alter the rate at which any warming trends are reflected in tree-ring growth.[138]

2.5 PROSPECTS FOR REDUCING THE RATE OF INCREASE IN CO_2 CONCENTRATION AND CLIMATE CHANGE

One of the key questions asked about forest ecosystems focuses on their potential to be sinks for the removal of atmospheric CO_2. As already noted, woody perennials have the potential to store considerable quantities of carbon in their tissues, but the questions now being addressed go far beyond that.[139,140] Obviously, further reductions in the sizes of forests, and thus the potential size of the CO_2 sink, will exacerbate the rise in atmospheric CO_2 as further CO_2 is released by combustion. Despite the present increases in the area of northern and boreal forests, these increases do not balance the considerably more important clearances taking place in the tropics. It is this situation that prevails when assessments are made of the potential for reforestation to solve the problems of increasing atmospheric CO_2. Jarvis[21] suggests that young, actively growing forest, about twice the size of Europe, could potentially assimilate all the CO_2 produced at the present rate from combustion and oxidation. This, of course, would only be achieved during the early life of a forest, while it was an active sink for carbon.

It appears that, despite many difficulties, reforestation may at least in part provide a solution to the problem of increasing CO_2.[141] The use of woody perennials in the production of biomass as a replacement for fossil fuel–derived energy has been suggested as a better solution for reducing the rate of increases in atmospheric CO_2.[142] It is important that the ways in which trees and wood products are utilized to provide a carbon store must be assessed effectively and rapidly.[143] The use of clones of poplar to develop short-term rotation crops for biomass production could benefit from elevated CO_2 if the observed stimulation in leaf area can be converted into an increase in stemwood production.[75]

Attempts to provide only simple solutions, for example, by planting to reestablish forest growth on old sites, may be problematic. Dramatic changes in the soil and the above-ground microclimate are particularly apparent in the Amazon region after tree removal, and they now prevent the establishment of primary rain forest species.[144]

ACKNOWLEDGMENTS

Gratitude is due to Drs. Richard Harrison-Murray, David Dunstan, and Alwyn Thompson for their very helpful comments on an earlier version of this chapter.

REFERENCES

1. Keeling, C. D., Bacastow, R. B., Bainbridge, A. E., Ekdahl, C. A., Guenther, P. R., Waterman, L. S., and Chin, J. F. S., Atmospheric carbon dioxide variations at Mauna Loa Observatory, Hawaii, *Tellus*, 28, 538, 1976.
2. Stuiver, M., The history of the atmosphere as recorded by carbon isotopes, in *Atmospheric Chemistry*, Goldburg , E. D., Ed., Springer-Verlag, Berlin, 1982, 159.
3. Neftel, A., Oeschger, H., Schwander, J., Stauffer, B., and Zumbrunn, R., Ice sample measurements give atmospheric CO_2 content during the past 40,000 years, *Nature*, 295, 220, 1982.
4. Friedli, H., Lotscher, H., Oeschger, H., Siegenthaler, U., and Stauffer, B., Ice core record of the $13_C/12_C$ ratio of atmospheric CO_2 in the past two centuries, *Nature*, 324; 237, 1986.
5. Barnola, J. M., Raynard, D., Korotkevich, Y. S., and Lorius, C., Vostok ice core provides 160,000-year record of atmospheric carbon dioxide, *Nature*, 329, 408, 1987.
6. Crane, A. J., Possible effects of rising CO_2 on climate, *Plant Cell Environ.*, 8, 371, 1985.
7. Rotty, R. M. and Marland, G., Fossil fuel consumption: recent amounts, patterns and trends of CO_2, in *A Global Analysis*, Trabalka, J. R. and Reichie, D. E., Eds., Springer-Verlag, New York, 1986, 55.
8. NASA (National Aeronautics and Space Administration), Earth System Science, A Closer View; Report of the Earth System Sciences Committee; NASA Advisory Council, NASA, Washington, D.C., 1988.
9. Mitchell, J. F. B., The seasonal response of a general circulation model to changes in CO_2 and sea temperatures, *Q. J. R. Meteorol. Soc.*, 109, 113, 1983.
10. Roeckner, E., Past, present and future levels of greenhouse gases in the atmosphere and model projections of related climatic changes, *J. Exp. Bot.*, 43, 1097, 1992.
11. Cess, R. D., Zhang, M.-H., Potter, G. L., Barker, H. W., Colman, R. A., Bazlich, D. A., Del Genio, A. D., Esch, M., Fraser, J. R., Galin, V., Gates, W. L., Hack, J. J., Ingram, W. J., Kichl, J. T., Lacis, A. A., Le Treut, H., Li, Z.-X., Liang, X.-Z., Mahfouf, J. -F., McAvaney, B. J., Meleshko, V. P., Morcret, J.-J., Randall, D. A., Roeckner, E., Royer, J.-F., Sokolov, A. P., Sporysho, P. V., Taylor, K. E., Wang, W.-C., and Wetherald, R. T., Uncertainties in carbon dioxide relative forcing in atmospheric general circulation models, *Science*, 262, 1252, 1993.
12. Mitchell, J. F. B., The "greenhouse effect" and climate change, *Rev. Geophys.*, 27, 115, 1989.
13. Chahine, M. T., The hydrological cycle and its influence on climate, *Nature*, 359; 373, 1992.
14. Gunderson, C. A. and Wullschleger, S. D., Photosynthetic acclimation in trees rising atmospheric CO_2: a broader perspective, *Photosyn. Res.*, 39, 369, 1994.
15. Gunderson, C. A., Norby, K. J., and Wullschleger, S. D., Foliar gas exchange responses of two deciduous hardwoods during three years of growth in elevated CO_2: no loss of photosynthetic enhancement, *Plant Cell Environ.*, 16, 797, 1993.
16. Idso, S. B. , Kimball, B. A. and Allen, S. G., CO_2 enrichment of sour orange trees: 2.5 years into a long-term experiment, *Plant Cell Environ.*, 14, 351, 1991.
17. Rey, A. and Jarvis, P. G., Growth responses of young birch trees (*Betula pendula* Roth.) after four and a half years of CO_2 exposure, *Ann. Bot.*, 80, 809, 1997.
18. Higginbotham, K. O., Mayo, J. M., L'Hirondelle, S., and Krystofiak, D. K., Physiological ecology of Lodgepole pine (*Pinus contorta*) in an enriched CO_2 environment, *Can. J. For. Res.*, 15, 417, 1985.
19. Van Oosten, J.-J., Afif, D., and Dizengremel, P., Long-term effects of a CO_2 enriched atmosphere on enzymes of the primary carbon metabolism of spruce trees, *Plant Physiol. Biochem.*, 30, 541, 1992.
20. Atkinson, C. J., Taylor, J. M., Wilkins, D., and Besford, R. T., Effects of elevated CO_2 on chloroplast components, gas exchange and growth of oak and cherry seedlings, *Tree Physiol.*, 17, 319, 1997.
21. Jarvis, P. G., Atmospheric carbon dioxide and forests, *Philos. Trans. R. Soc. London* B, 324, 369, 1989.

22. Warrick, R. A., Gifford, R. M., and Parry, M. L., CO_2, climate change and agriculture. Assessing the response of food crops to the direct effects of increased CO_2 and climatic change, in *The Greenhouse Effect, Climatic Change, and Ecosystems*, Bolin, B., Doos, B. R., Jager, J., and Warrick, R. A., Eds., Scope 29, John Wiley & Sons, Chichester, U.K., 1986, 393.
23. Lawlor, D. W. and Michell, R. A. C., The effects of increasing CO_2 on crop photosynthesis and productivity: a review of field studies, *Plant Cell Environ.*, 14, 807, 1991.
24. Cure, J. D. and Acock, B., Crop response to carbon dioxide doubling: a literature survey, *Agric. For. Meteorol.*, 38, 127, 1986.
25. Norby, R. J. and O'Neill, E. G., Leaf area compensation and nutrient interactions in CO_2-enriched seedlings of yellow-poplar (*Liriodendron tulipifera* L.), *New Phytol.*, 117, 515, 1991.
26. Hollinger, D. Y., Gas exchange and dry matter allocation responses to elevation of atmospheric CO_2 concentration in seedlings of three tree species, *Tree Physiol.*, 3, 193, 1987.
27. Canham, A. E. and McCavish, W. J., Some effects of CO_2, daylength and nutrition on the growth of young forest tree plants. I. In the seedling stage, *Forestry*, 54, 169, 1981.
28. Surano, K. A., Daley, P. F., Houpis, J. L. J., Shinn, J. H., Helms, J. A., Palassou, R. J., and Castella, M. P., Growth and physiological responses of *Pinus ponderosa* Dougl. ex P. Laws to long-term elevated CO_2 concentrations, *Tree Physiol.*, 2, 243, 1986.
29. Conroy, J. P., Küppers, M., Küppers, B., Virgona, J. and Barlow, E. W. R., The influence of CO_2 enrichment, phosphorus deficiency and water stress on the growth, conductance and water use of *Pinus radiata* D. Don, *Plant Cell Environ.,* 11, 91, 1988.
30. Rogers, H. H., Bingham, G. E., Cure, J. D., Smith, J. M., and Surano, K. A., Responses of selected plant species to elevated carbon dioxide in the field, *J. Environ. Qual.*, 12, 569, 1983.
31. Jurik, T. W., Weber, J. A., and Gates, D. M., Short-term effects of CO_2 on gas exchange of leaves of bigtooth aspen (*Populus grandidentata*) in the field, *Plant Physiol.*, 75, 1022, 1984.
32. Norby, R. J., O'Neill, E. G., and Luxmore, R. J., Effects of atmospheric CO_2 enrichment on the growth and mineral nutrition of *Quercus alba* seedlings in nutrient-poor soil, *Plant Physiol.*, 82, 83, 1986.
33. Vivin, P., Gross, P., Aussenac, G., and Geuhl, J.-M., Whole-plant CO_2 exchange, carbon partitioning and growth in *Quercus robur* seedlings exposed to elevated CO_2 , *Plant Physiol. Biochem.*, 33, 201, 1995.
34. Ziska, L. H., Hogan, K. P., Smith, A. P., and Drake, B. G., Growth and photosynthesis response of nine tropical species with long-term exposure to elevated carbon dioxide, *Oecologia*, 86, 383, 1991.
35. Kramer, P. J., Carbon dioxide concentration, photosynthesis and dry matter production, *Bioscience*, 31, 29, 1981.
36. Mousseau, M. and Enoch, H. Z., Carbon dioxide enrichment reduces shoot growth in sweet chestnut seedlings (*Castanea sativa* Mill), *Plant Cell Environ.*, 12, 927, 1989.
37. Ceulemans, R. and Mousseau, M., Effects of elevated atmospheric CO_2 on woody plants, *New Phytol.*, 127, 425, 1994.
38. Sage, R. F., Sharkey, T. D., and Seeman, J. R., Acclimation of photosynthesis to elevated CO_2 in five C_3 species, *Plant Physiol.*, 89; 590, 1989.
39. Eamus, D. and Jarvis, P. G., The direct effects of increase in the global atmospheric CO_2 concentration on natural and commercial temperate trees and forests, *Adv. Ecol. Res.*, 19, 1, 1989.
40. Campbell, W. J., Allen, L. H., Jr., and Bowes, G., Effects of CO_2 concentration on rubisco activity, amount and photosynthesis in soybean leaves, *Plant Physiol.*, 88, 1310, 1988.
41. Campbell, W. J., Allen, L. H., Jr., and Bowes, G., Response of soybean canopy photosynthesis to CO_2 concentration, light, and temperature, *J. Exp. Bot.*, 41, 427, 1990.
42. Wong, S. C., Elevated atmospheric partial pressure of CO_2 and plant growth, *Oecologia*, 44, 68, 1979.
43. Hicklenton, P. R. and Jolliffe, P. A., Alterations in the physiology of CO_2 exchange in tomato plants grown in CO_2-enriched atmospheres, *Can. J. Bot.*, 58, 2181, 1980.
44. Besford, R. T. and Hand, D. W., The effects of CO_2 enrichment and nitrogen oxides on some Calvin cycle enzymes and nitrate reductase in glasshouse lettuce, *J. Exp. Bot.*, 40, 329, 1989.
45. Yelle, S., Beeson, R. C., Trudel, M. J., and Gosselin, A., Acclimation of two tomato species to high atmospheric CO_2. II Ribulose-1,5-bisphosphate carboxylase/oxygenase and phosphoenolpyruvate carboxylase, *Plant Physiol.*, 90, 1473, 1989.
46. Bowes, G., Growth at elevated CO_2: photosynthetic responses mediated through rubisco, *Plant Cell Environ.*, 14, 795, 1991.

47. Besford, R. T., The greenhouse effect: acclimation of tomato plants growing in high CO_2, relative changes in Calvin cycle enzymes, *J. Plant Physiol.*, 136, 458, 1990.
48. Besford, R. T., Ludwig, L. J., and Withers, A. C., The greenhouse effect: acclimation of tomato plants growing in high CO_2, photosynthesis and ribulose-1,5-bisphosphate carboxylase protein, *J. Exp. Bot.*, 41, 925, 1990.
49. Miller, A., Tsai, C-H., Hemphill, D., Endres, M., Rodermel, S., and Spalding, M., Elevated CO_2 effects during leaf ontogeny, *Plant Physiol.*, 115, 1195, 1997.
50. Ho, L. C., The regulation of carbon transport and the carbon balance of mature tomato leaves, *Ann. Bot.*, 42, 155, 1978.
51. Tolbert, N. E. and Zelitch, I., *Carbon metabolism, in CO_2 and Plants: The Response of Plants to Rising Levels of Atmospheric Carbon Dioxide*, Lemon, E. R., Ed., Westview Press, Boulder, CO, 1983, 21.
52. Cure, J. D., Rufty, T. W., Jr., and Israel, D. W., Assimilate utilization in the leaf canopy and whole-plant growth of soybean during acclimation to elevated CO_2, *Bot. Gaz.*, 148, 67, 1987.
53. Farrar, J. F. and Williams, M. L., The effects of increased atmospheric carbon dioxide and temperature on carbon partitioning, source-sink relations and respiration, *Plant Cell Environ.*, 14, 819, 1991.
54. Wullschleger, S. D., Norby, R. J., and Hendrix, D. L., Carbon exchange rates, chlorophyll content and carbohydrate status of two forest tree species exposed to carbon dioxide enrichment, *Tree Physiol.*, 10, 21, 1992.
55. Cure, J. D., Rufty, T. W., Jr., and Israel, D. W., Assimilate relations in source and sink leaves during acclimation to a CO_2-enriched atmosphere, *Physiol. Plant.*, 83, 687, 1991.
56. Sasek, T. W., DeLucia, E. H., and Strain, B. R., Reversibility of photosynthetic inhibition in cotton after long-term exposure to elevated CO_2 concentrations, *Plant Physiol.*, 78, 619, 1985.
57. Arp, W. J., Effects of source sink relations on photoynthetic acclimation to elevated CO_2 , *Plant Cell Environ.*, 14, 869, 1991.
58. Stitt, M., Rising CO_2 levels and their potential significance for carbon flow in photosynthetic cells, *Plant Cell Environ.*, 14, 741, 1991.
59. Baxter, R., Ashenden, T. W., Sparks, T. M., and Farrar, J. F., Effects of elevated carbon dioxide on three montane grass species. 1. Growth and dry matter partitioning, *J. Exp. Bot.*, 45, 305, 1994.
60. Woodrow, I. E., Optimal acclimation of the Cz photosynthetic system under enhanced CO_2, *Photosyn. Res.*, 39, 401, 1994.
61. Rufty, T. W., Jr., and Huber, S. C., Changes in starch formation and the activities of sucrose phosphate synthase and cytoplasmic fructose 1,6 bisphosphate in response to source sink alterations, *Plant Physiol.*, 72, 474, 1983.
62. Stitt, M. and Quick, W. P., Photosynthetic carbon partitioning: its regulation and possibilities for manipulation, *Physiol. Plant.*, 77, 633, 1989.
63. Koch, K. E., Jones, P. H., Avigne, W. T., and Allen, L. H., Growth, dry matter partitioning, and diurnal activities of RuBP carboxylase in citrus seedlings maintained at two levels of CO_2, *Physiol. Plant.*, 67, 477, 1986.
64. Flore, J. A. and Lakso, A. N., Environmental and physiological regulation of photosynthesis in fruit crops, *Hortie. Rev.*, 11, 111, 1989.
65. Thomas, R. B. and Strain, B. R., Root restriction as a factor in photosynthetic acclimation of cotton seedlings grown in elevated carbon dioxide, *Plant Physiol.*, 96, 627, 1991.
66. Sheen, J., Metabolic repression of transcription in higher plants, Plant Cell, 2, 1027, 1990.
67. Van Oosten, J.-J. and Besford, R. T., Sugar feeding mimics effects of acclimation to high CO_2-rapid down regulation of RuBisCO small subunit transcripts but not of the large subunit transcripts, *J. Plant Physiol.*, 143, 306, 1994.
68. Radin, J. W., Kimball, B. A., Hendrix, D. L., and Mauney, J. R., Photosynthesis of cotton plants exposed to elevated levels of carbon dioxide in the field, *Photosyn. Res.*, 12, 191, 1987.
69. Curtis, P. S., Drake, B. G., Leadly, P. W., Arp, W. J., and Whigham, D. F., Growth and senescence in plant communities exposed to elevated CO_2 concentrations on an estuarine marsh, *Oecologia,* 78, 20, 1989.
70. Chaves, M. M. and Pereira, J. S., Water stress, CO_2 and climate change, *J. Exp. Bot.*, 43, 1131, 1992.

71. Mauney, J. R., Guinn, G., Fry, K. E., and Hesketh, J. D., Correlation of photosynthetic carbon dioxide uptake and carbohydrate accumulation in cotton, soybean, sunflower and sorghum, *Photosynthetica*, 13, 260, 1979.
72. Paez, A., Hellmers, H., and Strain, B. R., CO_2 effects on apical dominance in *Pisum sativum*, *Physiol. Plant.*, 50, 43, 1980.
73. Paez, A., Hellmers, H. and Strain, B. R., Carbon dioxide enrichment and water stress interactions on growth of two tomato cultivars, *J. Agric. Sci.*, 102, 687, 1984.
74. Bhattacharya, S., Bhattacharya, N. C., Biswas, P. K., and Strain, B. R., Response of cowpea (*Vigna unguiculata* L.) to CO_2 enrichment environment on growth, dry-matter production and yield components at different stages of vegetative and reproductive growth, *J. Agric. Sci.*, 105, 527, 1985.
75. Gardner, S. D. L., Taylor, G., and Bosac, C., Leaf growth of hybrid poplar following exposure to elevated CO_2, *New Phytol.*, 131, 81, 1995.
76. Sasek, T. W. and Strain, B. R., Effects of carbon dioxide enrichment on the growth and morphology of kudzu (*Pueraria lobata*), *Weed Sci.*, 36, 28, 1988.
77. Sasek, T. W. and Strain, B. R., Effects of carbon dioxide enrichment on the expansion and size of kudzu (*Pueraria lobata*) leaves, *Weed Sci.*, 37; 23, 1989.
78. Bazzaz, F. A., The response of natural ecosystems to the rising global CO_2 levels, *Annu. Rev. Ecol. Syst.*, 21, 167, 1990.
79. Huber, S. C., Rogers, H., and Israel, D. W., Effects of CO_2 enrichment on photosynthesis and photosynthate partitioning in soybean (*Glycine max*) leaves, *Physiol. Plant.*, 62, 95, 1984.
80. Peet, M. M., Huber, S. C., and Patterson, D. T., Acclimation to high CO_2 in monoecious cucumbers. II Alterations in gas exchange rates, enzyme activities and starch and nutrient concentration, *Plant Physiol.*, 80, 63, 1986.
81 Radoglou, K. M. and Jarvis, P. G., Effects of CO_2 enrichment on four poplar clones. I. Growth and leaf anatomy, *Ann. Bot.*, 65, 616, 1990.
82 Radoglou, K. M. and Jarvis, P. G., Effects of CO_2 enrichment on four poplar clones. II. Leaf surface properties, *Ann. Bot.*, 65, 627, 1990.
83. Brown, K. R., Carbon dioxide enrichment accelerates the decline in nutrient status and relative growth rate of *Populus tremuloides* Michx. seedlings, *Tree Physiol.*, 8, 161, 1991.
84. Norby, R. J., Pastor, J., and Melillo, J., Carbon-nitrogen interactions in CO_2-enriched white oak: physiological and long-term perspectives, *Tree Physiol.*, 2, 233, 1986.
85. Bazzaz, F. A., Coleman, J. S., and Morse, S. R., Growth responses of seven major co-occurring tree species of the northeastern United States to elevated CO_2, *Can. J. For. Res.*, 20, 1479, 1990.
86. Norby, R. J., Gunderson, C. A., Wullschleger, S. D., O'Neill, E. G., and McCracken, M. K., Productivity and compensatory responses of yellow-poplar trees in elevated CO_2, *Nature*, 357, 322, 1992.
87. Mousseau, M. and Saugier, B., The direct effect of increased CO_2 on gas exchange and growth of forest tree species, *J. Exp. Bot.*, 43, 1121, 1992.
88. Berntson, G. M. and Bazzaz, F. A., The influence of elevated CO_2 on the allometry of root production and root loss in *Acer rubrum* and *Betula papyerifera*, *Am. J. Bot.*, 83, 608, 1996.
89. Fitter, A. H., An architectural approach to the comparative ecology of plant root systems, *New Phytol.*, 106, 61, 1987.
90. Berntson, G. M., Wayne, P. M., and Bazzaz, F. A., Below-round architectural and mycorrhizal responses to elevated CO_2 in *Betual allegheniensis* populations, *Funct. Ecol.*, 11, 684, 1997.
91. Norby, R. J., O'Neill, E. G., Hood, W. G., and Luxmore, R. J., Carbon allocation, root exudation and mycorrhizal colonisation of *Pinus echinata* seedlings grown under CO_2 enrichment, *Tree Physiol.*, 3, 203, 1987.
92. Tyree, M. T. and Sperry, J. S., Vulnerability of xylem to cavitation and embolism, *Annu. Rev. Plant Physiol. Plant Mol. Biol.*, 40, 19, 1989.
93. Conroy, J. P., Milham, P. J., Mazur, M., and Barlow, E. W. R., Growth, dry weight partitioning and wood properties of *Pinus radiata* D. Don after 2 years of CO_2 enrichment, *Plant Cell Environ.*, 13, 329, 1990.
94. Atkinson, C. J. and Taylor J. M., Effects of elevated CO_2 on the stem growth, vessel area and hydraulic conductivity of oak and cherry seedlings, *New Phytol.*, 133, 617, 1996.

95. Telewski, F. W. and Strain, B. R., Densitometric and ring width analysis of a 3-year-old *Pinus taeda* and *Liquidamber styraciflua* L. grown under three levels of CO_2 and two water regimes, in *Proc. Int. Symposium on Ecological Aspects of Tree-Ring Analysis*, Jacoby, G. C., Jr. and Hornbveck, J. W., Eds., U.S. Department of Energy, New York, 1987, 494.
96. Sharkey, T. D., Loreto, F., and Delwiche, C. F., High carbon dioxide and sun/shade effects on isoprene emissions from oak and aspen tree leaves, *Plant Cell Environ.*, 14, 333, 1991.
97. Monson, R. F., Hills, A. J., Zimmerman, P. R., and Fall, R. R., Studies of the relationship between isoprene emission rate and CO_2 or photon-flux density using a real-time isoprene analyser, *Plant Cell Environ.*, 14, 517, 1991.
98. Chaudhuri, U. N., Kirkham, M. B., and Kanesmasu, E. T., Root growth of winter wheat under elevated carbon dioxide and drought, *Crop Sci.*, 30, 853, 1990.
99. DeLucia, E. H. and Heckathorn, S. A., The effect of soil drought on water-use efficiency in contrasting Great Basin desert and Sierran montane species, *Plant Cell Environ.*, 12, 935, 1989.
100. Bhattacharya, N. C., Hileman, D. R., Ghosh, P. P., Musser, R. L., Bhattacharya, S., and Biswas, P. K., Interaction of enriched CO_2 and water stress on the physiology of and biomass production in sweet potato grown in open-top chambers, *Plant Cell Environ.*, 13, 933, 1990.
101. Idso, S.B., Three phases of plant response to atmospheric CO_2 enrichment, *Plant Physiol.*, 87, 5, 1988.
102. Grace, J., Climate tolerance and the distribution of plants, *New Phytol.*, 106, 113, 1987.
103. Hinckley, T. M., Teskey, R. O., Duhme, F., and Richter, H., Temperature hardwood forests, in *Water Deficits and Plant Growth, Vol 6, Woody Plant Communities*, Kozlowski, T. T., Ed., Academic Press, New York, 1981, 153.
104. Cannell, M. G. R., Carbon dioxide and the global carbon cycle, in *The Greenhouse Effect and Terrestrial Ecosystems of the UK*, Cannell, M. G. R. and Hooper, M. D., Eds., Institute of Terrestrial Ecology research publication no. 4, HMSO, London, 1990,6.
105. Rochefort, L. and Woodward, F. I., Effects of climate change and a doubling of CO_2 on vegetation diversity, *J. Exp. Bot.*, 43, 1169, 1992.
106. Woodward, F. I., Predicting plant responses to global environmental change, *New Phytol.*, 122, 239, 1992.
107. Polley, H. W., Johnson, H. B., Marino, B. D., and Mayeux, H. S., Increase in C_3 plant water-use efficiency and biomass over glacial to present CO_2 concentrations, *Nature*, 361, 61, 1993.
108. Strain, B. R. and Cure, J. D., Direct effects of increasing carbon dioxide on vegetation (DOE/ER-0238), U.S. Department of Energy, Springfield, VA, 1985.
109. Bradshaw, A. D. and McNeilly, T., Evolutionary response to global climate change, *Ann. Bot.*, 67, 5, 1991.
110. Beerling, D. J., Heath, J., Woodward, F. I., and Mansfield, T. A., Drought–CO_2 interactions in trees: observations and mechanisms, *New Phytol.*, 134, 235, 1996.
111. Heath, J. and Kerstiens, G., Effects of elevated CO_2 on leaf gas exchange in beech and oak at two levels of nutrient supply: consequences for sensitivity to drought in beech, *Plant Cell Environ.*, 20, 57, 1997.
112. Garbutt, K., Williams, W. E., and Bazzaz, F. A., Analysis of the differential responses of five annuals to elevated CO_2 during growth, *Ecology*, 71, 1185, 1990.
113. Ferris, R. and Taylor, G., Stomatal characteristics of four native herbs following exposure to elevated CO_2, *Ann. Bot.*, 73, 447, 1994.
114. Bazzaz, F. A. and Garbutt, K., The response of annuals in competitive neighborhoods: effects of elevated CO_2, *Ecology*, 69, 937, 1988.
115. Bazzaz, F. A., Garbutt, K., Reekie, E. G., and Williams, W. E., Using growth analysis to interpret competition between a C_3 and a C_4 annual under ambient and elevated CO_2, *Oecologia*, 79, 223, 1989.
116. Bazzaz, F. A. and Williams, W. E., Atmospheric CO_2 concentrations within a mixed forest: implications for seedling growth, *Ecology*, 72, 12, 1991.
117. MacDonald, G. M., Edwards, I. W. D., Moser, K. A., Pienitz, K., and Smol, J. P., Rapid response of tree line vegetation and lakes to past climate warming, *Nature*, 361, 243, 1993.
118. Gifford, R. M., Growth pattern, CO_2 exchange and dry weight distribution in wheat growing under differing photosynthetic environments, *Aus. J. Plant Physiol.*, 4, 99, 1977.
119. Tissue, D. T. and Oechel, W. C., Responses of *Eriophorum vaginatum* to elevated CO_2 and temperature in the Alaskan tussock tundra, *Ecology*, 68, 401, 1987.

120. Reekie, E. G. and Bazzaz, F. A., Competition and patterns of resource use among seedlings of five tropical trees grown at ambient and elevated CO_2, *Oecologia*, 79, 212, 1989.
121. Sasek, T. W. and Strain, B. R., Effects of CO_2 enrichment on the growth morphology of a native and an introduced honeysuckle vine, *Am. J. Bot.*, 78, 69, 1991.
122. Heichel, G. H. and Jaynes, R. A., Stimulating emergence and growth of Kalmia genotypes with carbon dioxide, *Hort. Sci.*, 9, 60, 1974.
123. Pastor, J. and Post, W. M., Responses of northern forests to CO_2 induced climate change, *Nature*, 334, 55, 1988.
124. Davis, M. B., Woods, K. D., Webb, S. L., and Futyma, R. P., Dispersal versus climate: expansion of *Fagus* and *Tsuga* into the upper Great Lakes region, *Vegetation*, 67, 93, 1986.
125. Davis, M. B., Lags in vegetation response to greenhouse warming, *Climate Change*, 15, 75, 1989.
126. Jarvis, P. G. and McNaughton, K. G., Stomatal control of transpiration, *Adv. Ecol. Res.*, 15, 1, 1986.
127. Larson, P. R., Stem form development in forest trees, *For. Sci. Monogr.*, 5, 1, 1963.
128. Kienast, F. and Luxmore, R. J., Tree-ring analysis and conifer growth responses to increased atmospheric CO_2 levels, *Oecologia*, 76, 487, 1988.
129. Graumlich, L. J., Subalpine tree growth, climate and increasing CO_2: An assessment of recent growth trends, *Ecology*, 72, 1, 1991.
130. LaMarche, V. C., Graybill, D. A., Fritts, H. C., and Rose, M. R., Increasing atmospheric carbon dioxide: tree-ring evidence for growth enhancement in natural vegetation, *Science*, 225, 1019, 1984.
131. Jacoby, G. C., Long-term temperature trends and a positive departure from the climate-growth response since the 1950s in high elevation lodgepole pine from California, in *Proceedings of the NASA Conference on Climate–Vegetation Interactions*, Rosenzweig, C. and Dickinson, R., Eds., 7, Boulder, CO, 1986, 81.
132. Graybill, D. A., A network of high elevation conifers in the western U.S. for detection of tree-ring growth response to increasing atmospheric carbon dioxide, in *Proceedings International Symposium on Ecological Aspects of Tree-Ring Analysis*, Jacoby, G. C., Jr, and Hornbeck, J. W., Eds., U.S. Department Energy, New York, 1987, 463.
133. Hari, P., Arovaara, H., Raunemaa, Y., and Hautojarvi, A., Forest growth and the effects of energy production: a method for detecting trends in the growth potential of trees, *Can. J. For. Res.*, 14, 437, 1984.
134. Cooper, C. F., Carbon dioxide enhancement of tree growth at high elevations, *Science*, 231, 860, 1986.
135. Gale, J., Carbon dioxide enhancement of tree growth at high elevations, *Science*, 231, 859, 1986.
136. Hansen, J., Johnson, D., Lacis, A., Lebedeff, S., Lee, P., Rind, D., and Russell, G., Climatic impact of increasing atmospheric carbon dioxide, *Science*, 213, 957, 1981.
137. Wigley, T. M. L., Angell, J. K., and Jones, P. D., Analysis of the temperature record, in *Detecting the Climatic Effects of Increasing Carbon Dioxide*, MacGracken, M. C. and Luther, F. M., Eds., U.S. Department of Energy, New York, 1985, 55.
138. Briffa, K. R., Bartholin, T. S., Eckstein, D., Jones, P. D., Karlen, W., Schweingruber, F. H., and Zetterberg, P., A 1,400-year tree-ring record of summer temperatures in Fennoscandia, *Nature*, 346, 434, 1990.
139. Woodwell, G. M., Hobbie, J. E., Houghton, R. A., Melillo, J. M., Moore, B., Peterson, B. J., and Shaver, G. R., Global deforestation: contribution to atmospheric carbon dioxide, *Science*, 222, 1081, 1983.
140. Tans, P. P., Fung, I. Y. and Takahashi, T., Observational constraints on the global atmospheric CO_2 budget, *Science*, 247, 1431, 1990.
141. Schroeder, P. and Ladd, L., Slowing the increase in atmospheric carbon dioxide: a biological approach, *Climate Change*, 19, 283, 1991.
142. Vitousek, P. M., Can planted forests counteract increasing atmospheric carbon dioxide? *J. Environ. Qual.*, 20, 348, 1991.
143. Jones, H. G., *Plants and Microclimate: A Quantitative Approach to Environmental Plant Physiology*, 2nd ed., Cambridge University Press, Cambridge, 1992.
144. Shukla, J., Nobre, C., and Sellers, P., Amazon deforestation and climate change, *Science*, 247, 1322, 1990.

CHAPTER 3

CO_2 Enrichment of the Atmosphere and the Water Economy of Plants

James Heath and Terry A. Mansfield

CONTENTS

3.1 INTRODUCTION

Evaluation of the likely impacts of future atmospheric CO_2 enrichment on the water economy of plants is of far-reaching significance for two main reasons:

1. There may be appreciable changes in the efficiency of water use, with many consequences for other aspects of the physiology of plants.
2. Any physiological response which causes a drop in evapotranspiration will affect surface temperatures and will therefore add to the physical component of global environmental change.

Despite the great importance of these consequences, there are still many uncertainties about the nature of the responses of individual species to elevated CO_2, so that scaling up to predict impacts on whole communities and regions is hazardous, and perhaps even foolhardy. We need to learn much more about the basic mechanisms behind responses of leaf conductance to CO_2 in the atmosphere, in order to understand the factors controlling variation between species and under different environmental conditions.

When we discuss the responses of plants to a greater CO_2 concentration in the atmosphere, it is almost certainly inappropriate to regard an improved supply of carbon as a desirable benefit per se.

1-56670-341-7/00/$0.00+$.50

Thomas[1] elegantly described the problems associated with superabundance of carbon that were encountered by early terrestrial plants, and he presented a long list of carbon-disposing strategies that have since evolved. He warned crop scientists not to base their hopes on attempts to improve the efficiency of carbon use because they "will always end up swimming against the evolutionary tide."

In this situation, it is not surprising to discover that many plants appear to reject the opportunity to maximize the increase in photoassimilation that presents itself when the atmospheric CO_2 concentration rises. Instead, there is often a decrease in leaf conductance per unit area, and this is brought about in one or both of two main ways:

1. There may be a reduction in stomatal density;
2. The apertures of individual stomata may be reduced.

These mechanisms are probably independent of one another, and consequently it is appropriate to discuss them separately.

3.2 CO_2-INDUCED CHANGES IN STOMATAL DENSITY

Woodward[2] first drew attention to the possibility that stomatal density (the number of stomata per unit area) in some plant species has declined over the past 200 years, during which time the CO_2 concentration in the atmosphere has steadily increased. He based his claim upon data obtained from the examination of leaves of specimens collected at known times and dried and stored in a herbarium. In the cases of seven species of temperate forest trees and one shrub, the mean reduction in density between 1787 and 1987 was 40%. The atmospheric CO_2 concentration over that time had risen by about 60 parts per million (ppm), or 21%.

Since this first publication there has been some controversy surrounding Woodward's conclusions, and it is now accepted that there is considerable variation from species to species. In some cases there is no response of stomatal density to CO_2, and in some there may even be a small increase. However, a comprehensive survey by Woodward and Kelly[3] of data collected by different authors covering 100 species, showed a reduction in density in 74% of cases, and an average reduction of 14.3% in all the species when the data were pooled. The determinations had been made both on herbarium specimens and on leaves from plants in present-day CO_2-enrichment experiments. It was clear that the bulk of the effect on stomatal density was in the range of 280 to 350 ppm CO_2, for there was an average reduction of only 9% in experiments in which the present-day ambient concentration was approximately doubled (i.e., between about 350 and 700 ppm CO_2).

Of particular interest are the changes in one species of dwarf willow, *Salix herbacea* L., observed in radiocarbon-dated specimens, in herbarium material of known age, and in today's living material.[4] During the Last Glacial Maximum 16,500 years before present (B.P.) the CO_2 concentration in the atmosphere fell to around 180 ppm. Over the next 6000 years it rose to 270 ppm where it remained fairly constant in the interglacial period until anthropogenic emissions caused the increase observed over the last 200 years. Beerling and Woodward[4] obtained fossilized samples from the British Isles and Norway covering the period 16,500 B.P. to 10,500 B.P. when the CO_2 concentration increased progressively from 180 to 270 ppm (except for a period around 11,170 B.P. when it rose temporarily to 290 ppm) and herbarium material from Sweden collected between 1845 and 1971. Stomatal conductances for the various specimens were estimated (after standardizing aperture width) taking into account atmospheric conditions at the time. There was a distinct, statistically significant decline in conductance with the increasing CO_2 concentration over the period of 16,500 years (Figure 3.1). The deduced stomatal conductance values were convincingly validated by determining carbon isotope discrimination of the different specimens. $\Delta^{13}C$ values are known to be strongly related to the gas exchange characteristics of leaves.[5] It was shown the $\Delta^{13}C$ values were very close to those that would occur if leaves did indeed have the conductance values in Figure 3.1.

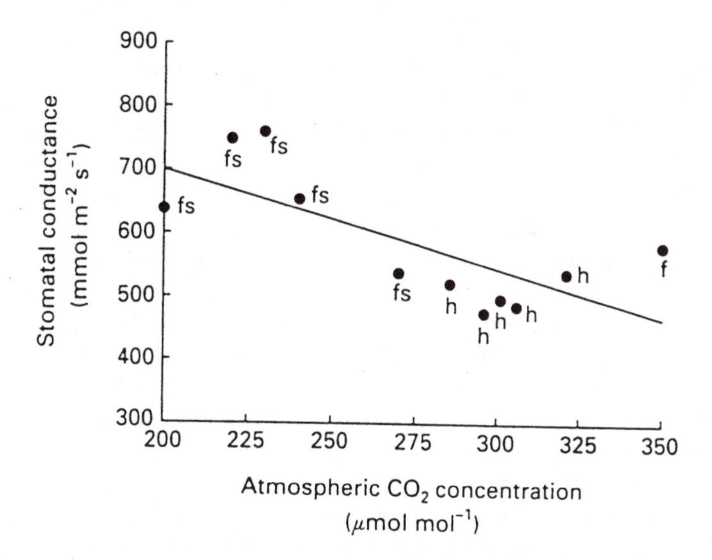

Figure 3.1 Calculated stomatal conductances of *Salix herbacea* as affected by atmospheric CO_2 concentration since the Last Glacial Maximum. Data from fossil (fs), herbarium (h), or fresh (f) leaves. (From Beerling, D. J. and Woodward, F. I., *New Phytol.*, 125, 641, 1993. With permission.)

The inevitable conclusion from these various studies, which have made such imaginative use of fossilized and dried material, is that there is a capacity in the majority of terrestrial plants (though not in all) to change stomatal density according to the CO_2 concentration of the atmosphere. When CO_2 is more abundant, stomatal density falls, and this has the effect of reducing the maximum stomatal conductance that can be attained when stomata are fully open. There are thus changes occurring during leaf development that reflect the changing physiological priorities brought about by variations in the atmospheric CO_2 concentration. It is clear that as CO_2 becomes more available, the opportunity for water conservation is of greater importance than grasping the full benefit of the more abundant supply of CO_2.

The capacity to adjust stomatal density, which involves changes at the developmental stages of leaves, is an enigmatic attribute because it does not seem relevant to plant adaptation to local environmental conditions. Plants do show considerable plasticity of development, permitting changes at whole-plant and cellular/metabolic levels which increase competitive ability in different situations. However, although there are changes in CO_2 concentration within canopies, in the atmosphere as a whole the concentration is relatively stable, apart from small seasonal variations which can be attributed to the annual cycle of photosynthesis. Are the changes in stomatal density in response to CO_2 concentration, which have been induced in many experiments with present-day plants, the result of an inherited mechanism which is sufficiently important to have been strongly conserved over evolutionary time? This may well be the case because the supply of CO_2 has varied considerably on a millennium time scale, and because the need to conserve water is of the highest priority. This issue is relevant to our understanding of the hierarchy of priorities in resource capture and utilization which has governed many aspects of plant evolution.

3.3 CO_2-INDUCED CHANGES IN STOMATAL APERTURE

The ability of stomata to respond to changes in CO_2 concentration was discovered and described by Freudenberger[6] and Heath.[7] Heath and his co-workers subsequently evaluated the control mechanisms in great detail, and many of their conclusions still provide the foundation of our present understanding. A review by Heath[8] is recommended for a comprehensive summary of the conclusions that were drawn from experiments in which attempts were made to control the internal atmosphere of the leaf (i.e., that in the intercellular spaces) by flushing with air of known CO_2 content. It was found that variations in the intercellular CO_2 concentration (C_i) played a major part in determining stomatal aperture, and that reductions in C_i as a result of mesophyll photosynthesis were partly responsible for increases in stomatal aperture in light. The CO_2 concentration in the ambient air (C_a) was shown not to be directly in control of aperture; rather it was the substomatal concentration in contact with the inner surfaces of the guard cells that was important. Changes in C_a do, however, have a major influence on stomata because of their effects on C_i.[9]

More recently, Mott[10] confirmed the importance of the inner surfaces of the guard cells as the sites of "detection" of changes in CO_2 concentration. Some limited progress has now been made in understanding the mechanism of response to CO_2, although many aspects are still uncertain. It has been suggested that malate-sensitive anion channels could provide the basis for changes in metabolism that underlie the sensing of CO_2 by the guard cells.[11] There is also evidence that CO_2 can cause rapid elevations in the amounts of cytosolic free calcium in guard cells, and it is suggested that calcium ions act as second messengers, forming part of a signal transduction pathway.[12]

Along with the cellular mechanisms of the CO_2 response itself, it is clearly important to understand any endogenous factors that may determine the stomatal sensitivity to CO_2. There is some evidence that hormones such as abscisic acid (ABA), indole-3-acetic acid (IAA) and cytokinins could be the agents that determine the magnitude of the response.[13] It is well established that ABA is a major regulator (usually an inhibitor) of stomatal opening, functioning as a hormonal signal in connection with a range of stress factors, especially drought.[14] Some observers have concluded that IAA and cytokinins do not have much influence on stomatal aperture (reviewed by Zeiger et al.[15]), and it may be true that their importance is secondary to that of ABA. Nevertheless, they do appear to be able to determine the extent of stomatal closure in response to a given concentration of CO_2, and to interact with ABA. In *Commelina communis*, IAA had no effect on stomatal aperture in CO_2-free air, but in 700 ppm CO_2 a high concentration of IAA (10^{-1} mol m^{-3}) abolished the CO_2 response and lower, more physiological, concentrations reduced it (Figure 3.2). Of particular interest was the ability of a low concentration (10^{-5} mol m^{-3}) of ABA to restore the response to CO_2 in the presence of 10^{-1} mol m^{-3} IAA.[16] In *Zea mays*, there was a similar situation except that in this species a cytokinin (*zeatin*) appeared to replace the role of IAA that had been found in *C. communis*.[17]

These control mechanisms were discovered using exogenous hormones applied to detached epidermal tissue. It has not yet been shown that endogenous hormones can function in the same way, but now that more progress has been made in unraveling the mechanism of hormone action at the cellular level, it may be profitable for these issues to be explored further. The interaction between exogenous IAA and ABA in controlling stomatal aperture has been confirmed, although not in relation to CO_2 responses.[18] Irving et al.[19] have described a role for cytosolic calcium levels in determining guard cell responses to IAA and other hormones which could provide a basis for interactions with CO_2, since CO_2 does appear to engage a signal transduction pathway involving calcium in guard cells.[12]

Before we can discuss the role of the control of stomatal aperture by CO_2 in determining water economy, it is necessary to draw attention to some complex features of the relationship between light and CO_2 in determining stomatal aperture. The effects of differences in irradiance and in the spectral quality of light in particular need to be clarified. The most significant early experiments were those of Heath and Russell,[20] in which attempts were made to control C_i by forcing air of

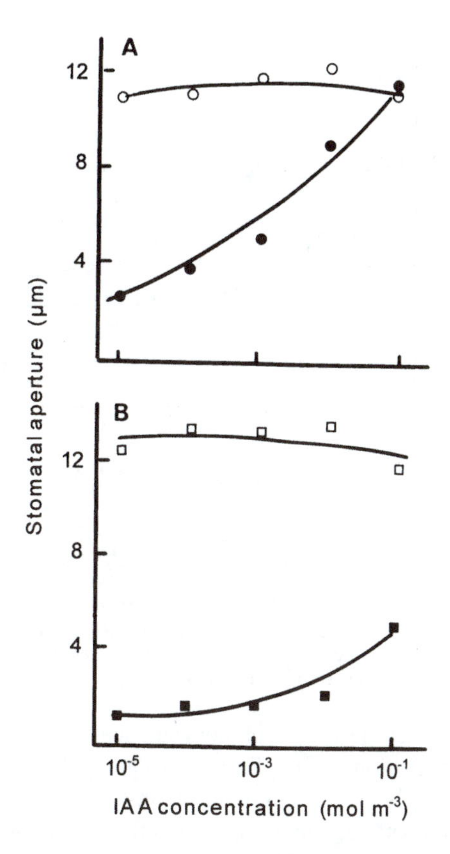

Figure 3.2 (A) Stomatal apertures (each point is a mean of 60) on abaxial epidermal strips of *C. communis* which had been incubated for 3 h in light in different concentrations of IAA. (○) zero CO_2 (▲) 700 ppm CO_2. (B) As (A), except that 10^{-5} mol m^{-3} ABA was included in the incubation medium. (□) zero CO_2 (■)700 ppm CO_2. Replotted data from Smith and Mansfield.[16]

known CO_2 concentrations continuously through the leaves. They used wheat, which has stomata on both leaf surfaces, and air was pumped through the open stomata of one surface, and it passed out through the other. They used different irradiances and also darkness, and, although the magnitude of the response to CO_2 was affected by irradiance, its nature remained very similar (Figure 3.3). Heath and Russell made some cautionary remarks about the effectiveness with which they were able to control C_i while photosynthesis was consuming CO_2, but a later study by Morison and Jarvis[21] derived a relationship between stomatal conductance and values of C_i from 0 to 500 ppm for *C. communis* which closely resembled that in Figure 3.3. There is now little doubt that in many species the stomatal response to CO_2 persists over a wide range of irradiances, and Figure 3.3 is widely representative. There are, however, some important exceptions which we will discuss later, but first it will be necessary to discuss the complex relationship between the CO_2 and light responses of stomata.

Morison and Jarvis[21] pointed out that there is not a unique relationship between C_i and stomatal conductance because light has an effect on stomata that is independent of its control of C_i. This is best illustrated by the finding that light sources of different spectral qualities, although they may establish the same value of C_i, may lead to very different stomatal apertures. In the case of *Pinus sylvestris*, for example, conductance was approximately twice as high in blue light than in red for a given value of C_i.[22] Zeiger[23] has described stomatal responses to white light in terms of at least two photoreceptor systems in guard cells: (1) a mechanism operating at the plasma membrane which is stimulated by blue light, and which leads to proton pumping and associated ionic (espe-

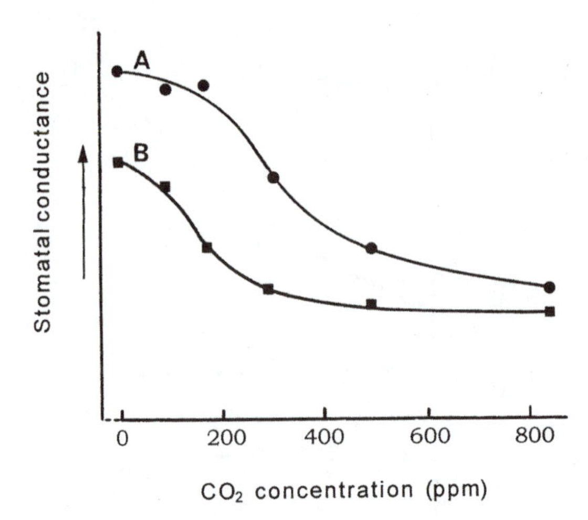

Figure 3.3 Replotted data of Heath, O. V. S. and Russell, J., *J. Exp. Bot.*, 5, 269, 1954 showing the effect of irradiance on the stomatal response to CO_2. Only data for their "high" and "medium" irradiances are shown here, which were 800 (A) and 270 foot-candles (B). These old-fashioned units correspond to about 170 and 60 μmol m^{-2} s^{-1}, respectively (but note that a precise conversion is not possible), and the treatments would be considered to represent "medium" and "low" irradiances in recent experiments. They used a viscous flow porometer to determine conductance and, because no satisfactory conversion to SI units is possible, no numerical values are attached to the ordinate.

cially potassium) fluxes; and (2) the use of photosynthetically active radiation by the guard cell chloroplasts, including some fixation of CO_2. Zeiger also pointed out that low fluences of blue light cause extensive starch hydrolysis in guard cell chloroplasts, which may lead to the production of osmotica which can assist stomatal opening independently of potassium uptake.[23]

The responses of stomata to light are thus very complex, and we understand very little about the interactions between light and CO_2 at the cellular level. The importance of the interaction is clearly illustrated by the work of Talbott et al.[24] who have also pointed to an important source of variation which will inevitably have affected many experiments without the investigators' knowledge. They found that the diurnal cycle of stomatal aperture in *Vicia faba* in growth chambers was highly variable from day to day, and this was attributable to the considerable fluctuations in ambient CO_2 concentrations in the urban environment in Los Angeles. The amplitude could exceed 100 ppm with a basal level of about 400 ppm. This finding was important because it demonstrated that many experiments in "controlled" environments have probably been conducted with much more variation in CO_2 concentration than has been recognized. From a physiological viewpoint, however, the most interesting outcome was the discovery that greenhouse-grown plants showed a much reduced stomatal sensitivity to CO_2 when compared with those in an artificially illuminated growth chamber.[24] *V. faba* plants raised in a greenhouse showed virtually no stomatal response when the CO_2 concentration was raised from 350 to 900 ppm. This contrasted with plants raised in a growth chamber, in which stomatal apertures fell from around 20 to 8 μm with the same increase in CO_2 concentration.

Talbott et al.[24] discussed these important findings in great detail, and came to the conclusion that the different photoenvironments were the most likely cause of the difference in CO_2 sensitivity. There were differences in both irradiance and spectral distribution between the growth chamber and greenhouse, and the irradiance was constant during the photoperiod in the chamber but not in the greenhouse.

These findings are of far-reaching significance because many of the conclusions about the CO_2 sensitivity of stomata have been drawn from experiments in controlled environments with artificial illumination. The photoenvironment of a greenhouse more closely resembles that out-of-doors, and if the sensitivity of stomata to CO_2 in the real world has been overestimated, some current conclusions about water conservation and climate change may be invalid.

There is some evidence that this may indeed be the case, and that photoenvironment plays an important role in determining the magnitude of the response of some species to CO_2. In a 2-year study of young plants of beech (*Fagus sylvatica*) grown in large containers in well-ventilated greenhouses, there were reductions (22% on average) in stomatal conductance in 600 ppm compared with 350 ppm CO_2 only on cloudy days when irradiance was low; on sunny days, conductance was equal to or even slightly higher in elevated CO_2 than that in ambient air.[25] However, this was not true for oak (*Quercus robur*) growing alongside the beech and exposed to precisely the same photoenvironments. Whether irradiance was low or high, the stomata of oak closed in elevated CO_2, conductance being reduced on average by 33%.[25] In a further experiment, it was shown that the degree of stomatal closure on any given day in 600 ppm relative to 350 ppm CO_2 was strongly correlated with the photosynthetic photon fluence density (PPFD) in beech and chestnut (*Castanea sativa*) (J. Heath, unpublished). When PPFD was high, the degree of the stomatal response was reduced, and sometimes conductance was significantly higher in elevated CO_2. Again, oak showed consistent reductions in stomatal conductance in elevated CO_2, and there was no significant correlation between stomatal response and PPFD. While these results are consistent with a role for photoenvironment in determining the degree or even the direction of stomatal responses to CO_2, it is impossible to separate the potential effects of other variables such as temperature and relative humidity that vary with the ambient PPFD. Furthermore, there are wide interspecific variations, shown clearly by the contrasting behavior of beech, chestnut, and oak (all members of the Fagaceae) growing in an identical environment.

As a consequence of such species- and environment-dependent variations in the stomatal response, there is at present no reliable basis for predicting the likely effects of atmospheric CO_2 enrichment on the water economy of different species, let alone whole ecosystems. However, the responses of trees appear to be generally smaller than those of crops and herbs; for example, elevated CO_2 caused an average reduction in stomatal conductance of only 23% in a survey of 23 tree species,[26] compared with an average of about 40% in herbaceous species.[27] There is some evidence to suggest that species with a higher intrinsic stomatal conductance generally show greater responses to CO_2.[28] Thus, herbs and grasses almost always show a large response to CO_2 concentration, and deciduous trees usually show some response, but conifers only show small or insignificant reductions in conductance. In a review of the most recent literature (1994 to 1997), it was found that average reductions in stomatal conductance in elevated CO_2 were just 18% for deciduous broadleaf trees and 13% for conifers.[28] In fact, increasing numbers of experiments are showing a lack of stomatal sensitivity to CO_2 in tree species; this has been attributed to the greater number of long-term studies using large trees rooted directly in the ground, rather than short-term experiments with potted seedlings.[29] Hence, the relative lack of stomatal response in many recent experiments with trees would appear to represent a realistic scenario for the future.

3.4 CO_2 ENRICHMENT AND THE WATER ECONOMY OF PLANTS

Why is it of such great importance to clarify the nature of the stomatal responses to CO_2? First, at the regional and global scale, rates of evapotranspiration from forest ecosystems are particularly important with respect to climate and hydrological cycles (see below). The assumption that rising atmospheric CO_2 concentrations will lead to a strong suppression of evapotranspiration may not hold true for forest ecosystems, due to a combination of weak stomatal responses and the increase in total leaf area that may accompany CO_2 enrichment. Consequently, it is essential to understand the reasons for the widespread variability in stomatal responses to CO_2 if future global climate is to be modeled successfully. In order to achieve this, further research is required into the underlying mechanisms of the CO_2 response itself, and the interactions with other environmental variables.

At the level of the individual plant, stomatal responses to elevated CO_2 will determine seasonal water use and soil moisture depletion. Clearly, failure to reduce stomatal conductance, combined with the increase in total leaf area that commonly occurs in elevated CO_2, will result in increased water use and therefore a greater risk of exposure to drought.[29] Furthermore, the short-term control of transpiration is of great importance in protecting trees against damage due to leaf water deficits[30] or xylem embolism during periods of high evaporative demand and drought.[31,32] Although instantaneous water use efficiency is invariably increased in elevated CO_2,[33] this may be due entirely to increased net photosynthetic rate and does not necessarily lead to improved drought tolerance.[34]

In beech (*F. sylvatica*) and oak (*Q. robur*) patterns of stomatal conductance at ambient and elevated CO_2 concentrations during periods of high evaporative demand were reflected in values of stem hydraulic conductance after three growing seasons.[35] Thus, the whole-plant hydraulic conductance of oak was reduced in elevated CO_2, whereas that of beech was unchanged. This implies that the stomatal response to CO_2 during critical periods of high transpirational demand has the potential to influence the development of hydraulic architecture and the water economy of the whole plant.[35] The study not only showed that the stomata of beech failed to close in elevated CO_2 during periods of high irradiance, but that throughout a drought lasting 4 weeks, stomatal conductance remained significantly higher in elevated CO_2, being roughly double that of the control plants by the end of the period.[25] This led to substantially greater rates of soil drying in elevated CO_2, in spite of there being no increase in total leaf area with the soil nutrient supply used here. Consequently, for any given degree of soil drought, stomatal conductance of young beech trees was maintained consistently higher in elevated CO_2 than in ambient air (Figure 3.4). Oak was shown to reduce its stomatal conductance consistently (by an average of 50%) in elevated CO_2 during drought, while, during the same period, chestnut and beech showed equal or increased conductance in elevated CO_2 (J. Heath, unpublished data). A similar contrast has been found between two *Eucalyptus* species.[36] *E. rossii* had a stronger photosynthetic and growth response to CO_2 enrichment, but showed only a small reduction in stomatal conductance, which almost disappeared by the end of a drought period. This resulted in greater rates of soil moisture depletion in elevated CO_2 due to the increase in total leaf area. In contrast, elevated CO_2 caused consistently large reductions in the stomatal conductance of *E. macrorhyncha*, and rates of soil drying were unaffected by CO_2 concentration during drought.

These results suggest that the stomatal responses of certain species may favor carbon gain at the expense of water conservation. Even closely related species — and perhaps different genotypes of the same species — may respond very differently. Given the likelihood of increased frequency and severity of drought in many regions due to global climate change, the consequences of CO_2 enrichment for plant water economy and drought tolerance is an area of research requiring much further attention. Whereas past changes in stomatal density would appear to reflect altered priorities of resource acquisition clearly, the outcome for future patterns of stomatal conductance is far less certain.

3.5 THE WIDER SIGNIFICANCE OF STOMATAL RESPONSES TO CO_2

The influence of CO_2 on stomatal aperture is of much interest to plant physiologists who are concerned with basic cellular mechanisms in plants. The topic does, however, assume much wider significance in view of anthropogenic increases in the CO_2 concentration in the Earth's atmosphere. Discussions of the impact of greater CO_2 concentrations on surface temperature have mostly concentrated on the radiative forcing that occurs because of the important role of CO_2 as a "greenhouse" gas. The warmth of the Earth's surface is due to retention of heat when some infrared radiation is prevented from passing outward through the atmosphere by CO_2, water vapor, and other gases. The familiar predictions about global warming have been almost entirely based on the physical calculations of an enhanced "greenhouse" effect when the CO_2 concentration rises. The

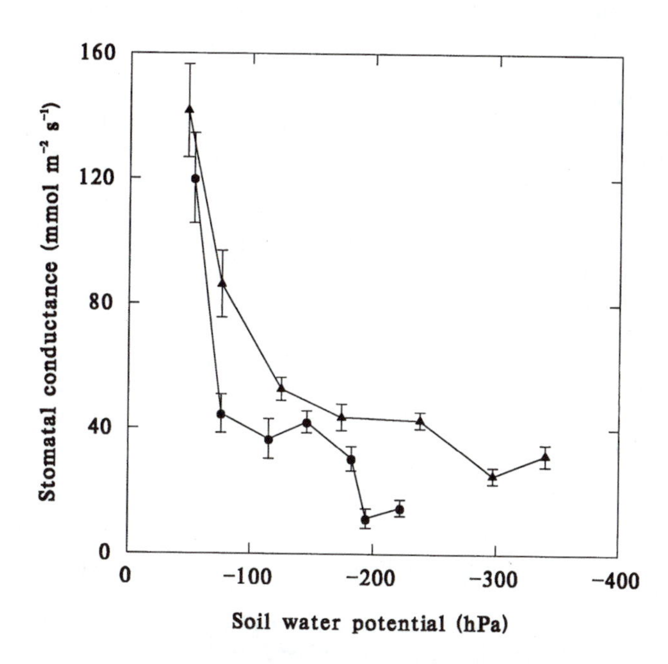

Figure 3.4 Stomatal conductance of young beech trees during 4 weeks of drought. (Replotted data from Heath, J. and Kerstiens, G., *Plant Cell Environ.*, 20, 57, 1997.) The graph shows average stomatal conductance ($n = 20$) in ambient air (●) and air enriched by 250 ppm CO_2 (▲) plotted against the soil water potential in each treatment. Error bars = 1 SE. Elevated CO_2 caused stomatal conductance to remain higher for a given degree of drought, resulting in a substantially increased rate of soil drying.

possibility that the responses of plants to CO_2 might make an independent contribution to warming has only recently come to the fore.[37, 38] Sellers et al.[37] offered a new "improved simple biosphere model" in which exchanges of energy, carbon, and water between the atmosphere and the surface are considered as tightly linked processes. The exercise involved scaling up a physiological model from the leaf to the canopy level, and it was concluded that the changes in surface air temperature due to physiological effects of CO_2 could be very important. The radiative effects of a doubling of atmospheric CO_2 concentration and the physiological effects were predicted to combine in a nonadditive manner, and in particular it was suggested that strong physiological responses occurring throughout the year in the tropics would affect the climate at high latitudes in unexpected ways, for example, a *reduction* in temperature over land in the boreal zone of the Northern Hemisphere in January.[37]

Betts et al.[38] discussed similar issues but attached much importance to CO_2-induced changes in the density of vegetation cover. They suggested that on a global scale, changes in vegetation structure could to some degree counteract physiologically driven impacts on climate.

It is clear that the debate initiated in these two seminal papers will be continued in the future, and that better understanding of the physiological impacts of CO_2 enrichment of the atmosphere will be of high priority. The responses of stomatal aperture to CO_2 will be in the foreground of the discussions, and the present degree of uncertainty will not be acceptable.

REFERENCES

1. Thomas, H., Resource rejection by higher plants, in *Resource Capture by Crops*, Monteith, J. L., Scottand, R. K., and Unsworth, M. H., Eds., Nottingham University Press, Nottingham, U.K., 1994, 375.

2. Woodward, F. I., Stomatal numbers are sensitive to increases in CO_2 from pre-industrial levels, *Nature*, 327, 617,1987.
3. Woodward, F. I. and Kelly, C. K., The influence of CO_2 concentration on stomatal density, *New Phytol.*, 131, 311, 1995.
4. Beerling, D. J. and Woodward, F. I., Ecophysiological responses of plants to global environmental change since the last glacial maximum, *New Phytol.*, 125, 641, 1993.
5. Farquhar, G. D., Ehleringer, J. R., and Hubick, K. T., Carbon isotope discrimination and photosynthesis, *Annu. Rev. Plant Physiol. Plant Mol. Biol.*, 40, 503, 1989.
6. Freudenberger, H., Die Reaktion der Schliesszellen auf Kohlensäure und Sauerstoffentzug, *Protoplasma*, 35, 15, 1940.
7. Heath, O. V. S., Control of stomatal movement by a reduction in the normal [CO_2] of the air, *Nature*,161, 179,1948.
8. Heath, O. V. S., The water relations of stomatal cells and the mechanism of stomatal movement, in *Plant Physiology, a Treatise*, Vol. 2, Steward, F. C., Ed., Academic Press, New York, 1959, 193.
9. Farquhar, G. D. and Sharkey, T. D., Stomatal conductance and photosynthesis, *Annu. Rev. Plant Physiol.*, 33, 317, 1982.
10. Mott, K. A., Do stomata respond to CO_2 concentrations other than intercellular? *Plant Physiol.*, 86, 200, 1988.
11. Hedrich, R., Marten, I., Lohse, G., Dietrich, P., Winter, H., Lohaus, G., and Heldt, H. W., Malate-sensitive anion channels enable guard-cells to sense changes in the ambient CO_2 concentration, *Plant J.*, 6, 741, 1994.
12. Webb, A. A. R., McAinsh, M. R., Mansfield, T. A., and Hetherington, A. M., Carbon dioxide induces increases in guard cell cytosolic free calcium, *Plant J.*, 9, 297, 1996.
13. Mansfield, T. A. and Davies, W. J., Mechanisms for leaf control of gas exchange, *Bioscience*, 35, 158, 1985.
14. Davies, W. J. and Zhang, J., Root signals and the regulation of growth and development of plants in drying soil, *Annu. Rev. Plant Physiol. Plant Mol. Biol.*, 42, 55, 1991.
15. Zeiger, E., Farquhar, G. D., and Cowan, I. R., *Stomatal Function*, Stanford University Press, Stanford, CA, 1987.
16. Snaith, P. J. and Mansfield, T. A., Control of the CO_2 responses of stomata by indole-3-acetic acid and abscisic acid, *J. Exp. Bot.*, 33, 360, 1982.
17. Blackman, P. G. and Davies, W. J., Modification of the CO_2 responses of maize stomata by abscisic acid and by naturally occurring and synthetic cytokinins, *J. Exp. Bot.*, 35, 174, 1984.
18. Dunleavy, P. J. and Ladley, P. D., Stomatal responses of *Vicia faba* L. to indole acetic-acid and abscisic-acid, *J. Exp. Bot.*, 46, 95, 1995.
19. Irving, H. R., Gehring, C. A., and Parish, R. W., Changes in cytosolic pH and calcium of guard cells precede stomatal movements, *Proc. Natl. Acad. Sci. U.S.A.*, 89, 1790, 1992.
20. Heath, O. V. S. and Russell, J., An investigation of the light responses of wheat stomata with the attempted elimination of control by the mesophyll, *J. Exp. Bot.*, 5, 269, 1954.
21. Morison, J. I. L. and Jarvis, P. G., Direct and indirect effects of light on stomata. I. In *Commelina communis* L., *Plant Cell Environ.*, 6, 103, 1983.
22. Morison, J. I. L. and Jarvis, P. G., Direct and indirect effects of light on stomata. I. In Scots pine and Sitka spruce, *Plant Cell Environ.*, 6, 95, 1983.
23. Zeiger, E., Light perception in stomatal guard cells. Plant, Cell Environ., 13, 739, 1990.
24. Talbott, L. D., Srivastava, A., and Zeiger, E., Stomata from growth-chamber-grown *Vicia faba* have an enhanced sensitivity to CO_2, *Plant Cell Environ.*, 19, 1188, 1996.
25. Heath, J. and Kerstiens, G., Effects of elevated CO_2 on leaf gas exchange in beech and oak at two levels of nutrient supply: consequences for sensitivity to drought in beech, *Plant Cell Environ.*, 20, 57, 1997.
26. Field, C. B., Jackson, R. B., and Mooney, H. A., Stomatal responses to increased CO_2: implications from the plant to the global scale, *Plant Cell Environ.*, 18, 1214, 1995.
27. Morison, J. I. L., Intercellular CO_2 concentration and stomatal responses to CO_2, in *Stomatal Function*, Zeiger, E., Farquhar, G. D., and Cowan, I. R., Eds., Stanford University Press, Stanford, CA, 1987, 229.
28. Saxe, H., Ellsworth, D. S., and Heath, J., Tansley review: tree and forest functioning in an enriched CO_2 atmosphere, *New Phytol.*, 139, 395, 1998.

29. Eamus, D., Responses of field grown trees to CO_2 enrichment, *Commonw. For. Rev.*, 75, 39, 1996.
30. Liang, J., Zhang, J., and Wong, M. H., Stomatal conductance in relation to xylem sap abscisic acid concentrations in two tropical trees, *Acacia confusa* and *Litsea glutinosa*, *Plant Cell Environ.*, 19, 93, 1996.
31. Cochard, H., Breda, N., and Granier, A., Whole tree hydraulic conductance and water loss regulation in *Quercus* during drought — evidence for stomatal control of embolism, *Ann. Sci. For.*, 53, 197, 1996.
32. Lu, P., Biron, P., Granier, A., and Cochard, H., Water relations of adult Norway spruce (*Picea abies* (L.) Karst) under soil drought in the Vosges mountains: whole-tree hydraulic conductance, xylem embolism and water loss regulation, *Ann. Sci. For.*, 53, 113, 1996.
33. Morison, J. I. L., Responses of plants to CO_2 under water limited conditions, *Vegetatio*, 104/105, 193, 1993.
34. Tschaplinski, T. J., Stewart, D. B., Hanson, P. J., and Norby, R. J., Interactions between drought and elevated CO_2 on growth and gas exchange of seedlings of three deciduous tree species, *New Phytol.*, 129, 63, 1995.
35. Heath, J., Kerstiens, G., and Tyree, M. T., Stem hydraulic conductance of European beech (*Fagus sylvatica* L.) and pedunculate oak (*Quercus robur* L.) grown in elevated CO_2, *J. Exp. Bot.*, 48, 1487, 1997.
36. Roden, J. S. and Ball, M. C., The effect of elevated CO_2 on growth and photosynthesis of two *Eucalyptus* species exposed to high temperatures and water deficits, *Plant Physiol.*, 111, 909, 1996.
37. Sellers, P. J., Bounoua, L., Collatz, G. J., Randall, D. A., Dazlich, D. A., Los, S. O., Berry, J. A., Fung, I., Tucker, C. J., Field, C. B., and Jensen, T. G., Comparison of radiative and physiological effects of doubled atmospheric CO_2 on climate, *Science*, 271, 1402, 1996.
38. Betts, R. A., Cox, P. M., Lee, S. E., and Woodward, F. I., Contrasting physiological and structural vegetation feedbacks in climate change simulations, *Nature*, 387, 796, 1997.

CHAPTER 4

Plants Responses to Elevated CO_2; A Perspective from Natural CO_2 Springs

Maurizio Badiani, Antonio Raschi, Anna Rita Paolacci, and Franco Miglietta

CONTENTS

1-56670-341-7/00/$0.00+$.50

ABBREVIATIONS

A_{max}	light-saturated net CO_2 assimilation rate
APX	total ascorbate peroxidase
AsA	ascorbic acid
C_a	atmospheric gas mixing ratio of carbon dioxide
CAM	Crassulacean Acid Metabolism
CAT	catalase
Chl	chlorophyll
C_i	intercellular CO_2 gas mixing ratio
DW	dry weight
g_s	stomatal conductance to water vapour
GSSG	glutathione disulfide
ϕ	maximum quantum yield for CO_2 uptake
F_V	variable Chl *a* fluorescence
F_M	maximal Chl *a* fluorescence
GSH	glutathione
J_{max}	RuBP regeneration capacity mediated by electron transport
LAI	leaf area index
LUE	light use efficiency
NADH	nicotinamide adenine dinucleotide, reduced form
NADPH	nicotinamide adenine dinucleotide phosphate, reduced form
NAR	net assimilation rate
NCDS	natural carbon dioxide springs
NSC	nonstructural carbohydrates
P_n	net CO_2 assimilation rate
PSI	photosystem I
PSII	photosystem II
qP	photochemical component of the Chl *a* fluorescence quenching
qNP	nonphotochemical component of the Chl *a* fluorescence quenching
RGR	relative growth rate
Rubisco	ribulose-1,5-bisphosphate carboxylase/oxygenase
ROI	reactive oxygen intermediates
RuBP	ribulose-1,5-bisphosphate
V_{cmax}	maximal Rubisco activitiy toward CO_2
SLA	specific leaf area
SLW	specific leaf weight
SOD	total superoxide dismutase
UV-B	ultraviolet B radiation
WUE	water use efficiency

4.1 INTRODUCTION

Atmospheric gas mixing ratio of carbon dioxide (C_a) is rising at an unprecedented rate[1] and this has prompted considerable interest in the potential impacts of elevated CO_2 on natural and managed ecosystems. Research has shown that a number of factors including increased rates of CO_2 assimilation, higher water use efficiency, decreased respiration, and the relief of nutrient stress via enhanced nutrient use efficiency and nutrient uptake will contribute to increased plant growth and productivity in the short term.[2-4] There remains, however, considerable uncertainty over the long-term responses of vegetation to increased CO_2 under field conditions.

Many controlled environment, open-top chamber, and glasshouse studies have revealed that CO_2-induced stimulation in growth and photosynthesis may not be sustained under conditions of prolonged CO_2 enrichment, because plants "acclimate" to elevated CO_2.[5,6] Yet, there is growing evidence that such "acclimation" responses are not inevitable[7] and many studies in which well-fertilized plants have been grown in the field or in containers large enough to prevent root restriction have shown little or no decrease in photosynthetic capacity in response to long-term CO_2 enrichment.[6,8] As a consequence, it is now considered that in most agricultural situations, where the soil is well tilled and nutrient supply plentiful, many crop plants will continue to respond to increased CO_2 over their entire life span.[4] Much less information is available on which to base predictions of the long-term response of natural plant communities to increased C_a, but it is believed that soil compaction combined with limited supplies of other resources (most importantly, nutrients, water, and light) may reduce the potential growth enhancement resulting from increased CO_2.[7,9,10]

Native plant material has a considerably greater ecological and physiological range than cultivated species, so the long-term responses of natural plant communities to elevated CO_2 cannot necessarily be predicted from experiments on highly selected crop plants grown under nutrient-rich conditions.[11] There are many reports showing that considerable variation exists between and within species in response to CO_2 enrichment.[12,13] Strain[14] suggests that this diversity may represent residual genetic potential for increased growth and competition under CO_2-enriched conditions, since many plant species evolved under conditions where C_a was considerably higher than it is at present. Given the potential existence of a pool of heritable variation in CO_2 responsiveness within plant species, it has been suggested that anthropogenic increases in C_a may act differentially on plant populations with the result that genotypes better able to make use of the available resources will prosper and increase in frequency at the detriment of plants unable to do so and lead to eventual genetic shifts within plant populations.[15,16] It has not, however, proved possible to test this hypothesis because of the fundamental discrepancies between the timescales of experiments and the timescales relevant in an ecological context. To date, the longest controlled exposures of natural vegetation to elevated CO_2 have lasted for no more than 4 years,[9,10] yet periods of 10 years or more will be necessary to establish how the genetic structure of even short-lived species changes in response to increased CO_2.[17] In the meantime, it has been suggested that one way of providing a more realistic picture of long-term vegetation responses to CO_2 is to identify specific characteristics or physiological traits existing in plants originating from those few special locations where atmospheric CO_2 concentrations have been naturally elevated over evolutionary timescales.

4.2 GEOTHERMAL SITES AS A RESEARCH TOOL

4.2.1 Geological Features

Geothermal sites, where CO_2 of deep origin is released at the surface by gas vents, termed natural carbon dioxide springs (NCDS), are uniquely suited for studying CO_2 effects in plants or plant communities exposed for decades or even more to elevated C_a and presumably adapted to it.[18]

Carbon dioxide in fact is abundant below the ground surface due to various geochemical and biochemical processes. Owing to the easy linking between carbon and oxygen and the terrestrial abundance of these elements, carbon dioxide occurring in the shallow environments, i.e., soil and groundwater, may have various origins. Organic origin is restricted to the shallowest part of the crust where plant respiration and bacterial activity (decomposition of organic matter) occur. Carbon dioxide in the soil air is mainly the product of microbially mediated processes. Inorganic origin of CO_2 may be both shallow and deep seated. A large amount of CO_2 derives from dissolution of carbonates by meteoric water and acids produced by sulfur oxidation. Carbon dioxide may also be produced from the oxidation of anoxic C where deep groundwaters meet shallow oxic groundwaters. Along active plate margins, high-temperature reactions (geothermal and magmatic processes) are a source of CO_2 at several levels of the crust. Finally, C_a can be released by coal deposits ignited by lightning.[19]

The CO_2 outflow at the surface may be weak, bearing concentrations in soil from the background value (i.e., biological activity) to 5 to10% v/v, or strong, with concentrations above 20 to 30% or even with direct evidence of outflow (e.g., gas vents). The mean CO_2 flux at the soil–atmosphere interface is of the order of 10^{-7} to 10^{-8} m^3 m^{-2} s^{-1} (about 600 to 6000 t Km^{-2} y^{-1}).[20] However, higher values can be detected over faults. The highest levels of CO_2 in soil and groundwater, often associated with gas vents, are typical of those areas characterized by hydrothermal circulation. Pressurized gas caps, linked to geothermal reservoirs or magmatic chambers (e.g., in volcanic areas), may give rise to strong surface emissions with visible effects on the surrounding environment.

Natural CO_2 springs used so far for biological research, or whose geological and ecological features have been characterized to this end, are present in the Czech Republic,[21] France,[22] Greece (A. Raschi, unpublished observations), Iceland,[23] Italy,[17] New Zealand,[24] Slovenia (A. Raschi, unpublished observations), Spain,[25] and the U.S.[26-28] Most of the studies cited in the present review were brought about at NCDS located in central-western Italy, where more than 100 of these have been identified and characterized so far. Of these, two in particular, Bossoleto (43° 17' N; 11° 35' E; Siena District, Tuscany) and Solfatara (42° 30' N; 12° 05' E; Viterbo District, Latium) are briefly described in the following for illustrating the many possibilities, as well as drawbacks and limitations, of NCDS as tools for elevated CO_2 research.

The enriched area of Bossoleto has the shape of a circular doline, surrounded by ancient walls that have eliminated factors of human disturbance. Gaseous emissions come out from several vents distributed on the bottom of the doline, and these emissions have been known at least as far back as the early part of the 19th century. The CO_2 from the vents is over 99% pure and has very low levels of pollutants such as hydrogen sulfide (H_2S) and sulfur dioxide (SO_2) that cannot affect the biological activity (H. Rennenberg, personal communication). At Solfatara, the emission points are dispersed over an area of about 2 Ha (hector), crossed by small streams creating seasonal swamps. As the name itself suggests, high levels of reduced sulfur compounds, especially H_2S, are co-emitted by the CO_2 vents.

4.2.2 Vegetation

The spatial arrangement, structure, and diversity of vegetation around CO_2 springs are largely determined by local ecological factors. A distributional pattern in concentric rings formed by four community types is repeated in most of the sites, despite their localization in different phytoclimatic areas. The overwhelming ecological gradient occurring in CO_2 vents involves three major factors decreasing in intensity with the distance from vents: C_a, soil moisture, and soil acidity. The combined effects of these factors exert a strong selective pressure which leads to low diversity, repetitivity, and directional variation in the qualitative and quantitative floristic composition of the vegetation.[29] At Bossoleto steep banks of limestone outcrops determine a rather different ecological situation, although the arrangement of the community is consistent with that of other springs. Small fragments of species-rich xerocalcicolous Mediterranean grassland occur behind a narrow ring of herbaceous vegetation dominated by *Agrostis stolonifera* L. closely surrounding the vents. At Solfatara, four vegetation bands occur along a soil moisture gradient: (1) one with *Scirpus lacustris* L. growing on muds submerged by bubbling water; (2) one exclusively of *Agrostis canina* L. on wet clayey soils; (3) one on raised banks with *Quercus pubescens* Willd., *Q. cerris* L., and *Q. robur* L.; and (4) one, an acidophilous shrub zone of *Cytisus scoparius* (L.) Link and *Adenocarpus complicatus* (L.) Gay.

4.2.3 Gaseous Regimes

Working with NCDS poses certain problems and questions, such as uneven spatial distribution and short-term fluctuations of CO_2 levels, coemission of phytotoxic gases, and poor soil and water qualities. In the course of experimentation, some of these were circumvented, whereas others represent inevitable constraints which can only be dealt with, at best, by accurate experimental planning and caution in the interpretation and extrapolation of the results.

Carbon dioxide emission rates vary from site to site, ranging from a few up to as much as 2000 to 5000 kg h^{-1}. Albeit, for each site, flux rate and composition at the emission points can be regarded as fairly constant over long periods of time (G. Bargagli, personal communication), direct CO_2 measurements at NCDS revealed that anomalies were occurring on rather large areas around each spring, but that the size of this enriched area was different depending on the strength of the CO_2 vents, the dimension and topography of the site, and the weather.[30] The presence of steep gradients of CO_2 concentrations with distance from the vents was also observed, indicating that plants occurring at different distances from the vents were likely exposed to different average CO_2 concentrations. Normal air turbulence causes dramatic short-term fluctuations in CO_2 concentration, due to the advection of air at normal ambient concentration. At night, moreover, the establishment of the inversion layer and the heavier-than-air nature of CO_2 causes, especially under calm conditions, the build up of extraordinarily high CO_2 concentrations, from which the common perception of NCDS as dangerous sites derives. It is still unclear if short term fluctuations (in the order of seconds or minutes) of CO_2 concentrations may have an influence on plant stomatal response, on sub-stomatal and mesophyll CO_2 concentrations and consequently on actual photosynthetic rates. On the other hand it is known that elevation of CO_2 concentration in the dark may reduce respiration.[31, 32]

Spatial and temporal variability in CO_2 concentrations occurring at NCDS prevents, in fact, a correct determination of average concentrations of CO_2 which local plant communities are exposed to. To this aim, however, valuable help was provided by the occurrence of significant anomalies in both the radiocarbon and stable carbon isotopic ratios, due to the mixing of ambient CO_2 and the CO_2 released by the vents. The radiocarbon method is based on the fact that carbon dioxide emitted from the vents, which originates from the dissolution of limestone, is completely depleted in the radioactive carbon isotope ^{14}C. When this CO_2 mixes with air, the C_a of the resulting mixture increases while its ^{14}C content decreases. This signal is detectable in plant tissues, since their radiocarbon content depends on the discrimination against ^{14}C occurring during photosynthesis and on ^{14}C abundance in source air. Discrimination against ^{14}C may be simply calculated as is currently done in radiocarbon dating methods,[33] so that ^{14}C abundance can provide an estimation of the average effective CO_2 exposure.

The stable isotope method is based on the fact that carbon released from vents is often heavier than atmospheric carbon, according to the composition of the different geologic substrates.[34] Again, the mixing of CO_2 from the vents and from the air determines an increase in C_a as well as a change in the stable carbon isotope ratio of the mixture. In this case, plants with C_4 metabolism, which are known to have a conservative discrimination vs. ^{13}C across a range of environmental conditions[35,36] may be used as biosensors to measure the average effective isotope ratio of the air near vents and from this the average CO_2 concentration.

4.2.4 Phytotoxic Gases

The composition of gas which vents at the surface in the Italian CO_2 springs is site dependent. Carbon dioxide concentrations are never lower than 90%, being the remaining part mainly composed of methane and other inert gases. H_2S and SO_2 are phytotoxic contaminants frequently encountered.[37] The percentage of H_2S is variable, ranging from less than 100 up to 1200 nmol mol^{-1}. The toxicity of H_2S to plants is well known.[38] It has been reported that concentrations as low as 30 nmol mol^{-1} may have toxic effect on sensitive species (L. De Kok, personal communication). Toxicity mechanisms are poorly understood but they are always associated to reduced growth.[39]

The presence of H_2S in the gas emissions led researchers to wonder about the occurrence of toxic effects on vegetation. To this aim, indexes of reduced sulfur accumulation and metabolic toxicity were compared in plant tissues growing at H_2S-contaminated and uncontaminated NCDS. The results obtained suggested that due to its rapid dilution and oxidation after emission from the vents, H_2S atmospheric mixing ratios at NCDS may be considered subtoxic in most cases.[40, 41]

4.2.5 Soil and Water Quality

Most Italian NCDS are characterized by poor acidic soils, with pH values ranging from 2.4 to 3.7 in different springs. This extreme acidity can explain the low content of assimilable phosphorus and may increase the mobility of metallic microelements, such as Cu and Zn, and of toxic cations, such as Al^{3+}. Many of the springs are flooded for part of the year, generally concentric to the vent area. Reduced microfloral activity is likely to cause organic matter accumulation in the soil. The presence of high CO_2 concentration in soil can *per se* prevent seed germination and seedling growth. In many of the springs a real aphytoic area can be identified, in which only soil microflora is present. In most of the springs gas emission is accompanied by water characterized by high concentrations of salts. Permanently flooded areas are also frequent, in which only halophytic plants can survive.[29]

Despite the above drawbacks, experimental flexibility afforded by NCDS remains remarkable. To mention Italian springs only, NCDS have been first used in experimental surveys in which native species were compared with their counterparts growing at selected natural sites under normal C_a, albeit finding control areas comparable to NCDS in terms of vegetation and soil type proved to be difficult. To overcome at least in part these limitations, transect experiments were run in which comparison was made among individuals of a given native species growing at increasing distances from the CO_2 vents, and thus assumed to have evolved under decreasing average C_a regimes.

4.3 PLANT RESPONSES TO ELEVATED CO_2: "SHORT-TERM" STUDIES

Short term is intended to denote those studies which, albeit based on a wealth of different experimental approaches, shared at least two common elements; i.e., an "artificial" CO_2 source and a "limited" duration. Indeed, even the longest enrichment experiments among these did not reach the minimum time span threshold thought to be required for evolutionary adaptation processes to occur, being in any case incomparably shorter than those allowed by the NCDS resources. On the other hand, all these "short-term" studies clearly benefited from the minimization of environmental disturbances and/or from the standardization of experimental factors, one of the most important being a precise control of CO_2-enrichment, not only in terms of flux rate and stability, but also in terms of gaseous impurities.

At the risk of oversimplification, only studies on terrestrial plants have been considered in the following. In pursuing the not easy task of reducing the enormous amount of existing information to a manageable size, the reader is referred to earlier fundamental literature through several excellent reviews that appeared in the first half of the present decade.[42-54] Plant function classification follows that adopted by Strain and Thomas.[42]

4.3.1 Photosynthesis and Photorespiration

4.3.1.1 Stomata

Stomata of most species close with increasing C_a and this causes a decrease in stomatal conductance (g_s). This decrease, which has been observed also in C_4 species[55,56] is commonly in the range 20 to 40%,[43,46,53,57-65] albeit higher values have been reported.[55,66] Evidence of no change[57] or even of increasing g_s under elevated CO_2[57, 65-67] has also been provided. Reported changes in stomatal density with growth at elevated C_a include increases, decreases, and no change,[43,44,57,68,69] although a decrease (14% on average) have been observed in most cases,[53,70] including those drawing on herbarium material and paleoecological evidence.[55,71] Stomatal density under elevated CO_2 has been reported to depend on leaf age and position in trees.[57,72] A few studies have been conducted on stomatal index, which does not appear to change under elevated CO_2.[69] Reductions in stomatal conductance and, possibly density explain the reduction in leaf transpiration observed, although not always,[73,74] in many different plant species growing under elevated C_a, with a 30% decrease on an average, as compared with normal CO_2.[43-45,55,57,66,75,77-80]

Since one of the best documented effects of elevated CO_2 is an increase in photosynthetic carbon fixation, the most obvious advantage of a reduced transpiration would be an enhanced water use efficiency (WUE). Indeed, instantaneous WUE has been observed to double, on an average, by doubling C_a[18,43,55,57,59,77,78,81-86] and, coherently, to decrease upon exposing plants to subambient C_a.[87] Apart from stomatal effects, CO_2-dependent stimulation of root development might also contribute to the observed increase of WUE.[88] A detrimental consequence of a reduced (evapo)transpiration might be an increased leaf temperature under elevated CO_2.[57] It has also been suggested that the CO_2-dependent reduction in leaf transpiration might be partially,[44,45,89] or even totally,[85] offset by a larger and more rapid leaf development. This would increase in turn the leaf area index (LAI)[74,85,89-91] and hence global canopy (evapo)transpiration. This issue, however, is still controversial.[57]

4.3.1.2 Photoconversion

Effects of elevated CO_2 on the foliar contents of photointercepting pigments are not straightforward, albeit reports indicating a decrease in both chlorophyll (Chl) *a* and Chl *b*[73,78,86,92,93] prevail over those reporting no change.[68,94,95] This decrease in Chl content might be compensated for by a higher number of leaves in plants exposed at elevated CO_2.[96] Pigment ratios such as Chl *a*/*b* and Chl/total carotenoids have been reported to increase,[73] to remain unchanged,[93] or even to decrease[95] under elevated CO_2. A decrease in antenna size and an increased density of reaction centers have also been reported.[97]

Uncertainty also arises from the results concerning the PSII photochemical efficiency, since elevated C_a seems to cause decrease in certain cases,[98,99] together with a deeper midday depression in F_V/F_M,[99] but not in others.[68,100] Roden and Ball[99] reported increases in qP, qNP, the xanthophyll pool and the ratios of antheraxanthin/total xanthophylls and zeaxantihn/total xanthophylls, whereas Hogan et al.[98] found no change both in the size of the xanthophyll pool and in the de-epoxidation state. No molecular symptom of photodamage was found in elevated-CO_2 grown plants when compared with the controls.[98]

CO_2-dependent changes in the composition and fluidity of the thylakoid membranes have been postulated.[101] It is possible that elevated CO_2 does not affect the structure and function of the photoconverting apparatus *per se*, rather causes acclimation of photosynthesis[57,102] or, in C_3 species, the progressive suppression of photorespiration, leading to changes similar to the ways in which excess light energy is dissipated.[98] These alterations might also have consequences on the cellular antioxidant status, since antioxidant systems are involved both in the dissipation of excess energy and in the disposal of the reactive oxygen intermediates (ROI) produced under conditions of overflow of the photosynthetic electron transport chain.

4.3.1.3 Carbon Fixation

As the C_3 photosynthetic capacity is not saturated by the current C_a, an increase in the net photosynthetic rate (P_n) is the most expected effect of increasing CO_2. Indeed, evidence that elevated CO_2, by increasing the intercellular CO_2 gas mixing ratio (C_i), stimulates P_n, is irrefutable, although not absolute.[61,63,65,78,95,103,104] This stimulation is almost[63] always observed when measuring P_n at elevated CO_2[50,99,105-109] and has also been indirectly confirmed by the depressive effect of subambient C_a.[87] It is often regarded as a short-term response,[43,58-60,65,67,73-75,85,91,110-121] although it was found to persist over several growing seasons in tree species.[122] Interestingly, stimulation of P_n may occur even in the absence of a promotive CO_2 effect on growth.[106]

Effects of elevated CO_2 might also include an enhanced maximum quantum yield for CO_2 uptake (ϕ, 20% on average[46,57,111] but also disputed [63,123]), a decreased light compensation point for P_n,[46,57] and an increased light use efficiency (LUE), which has been reported also under conditions of water shortage.[57,109] All the above CO_2 effects might be beneficial for vegetation growing in the shade (e.g. lower canopy layers and forest understory) and thus have important consequences for both managed and natural plant communities.[57] It has been reported that elevated CO_2 increases carboxylation efficiency in species of contrasting functional type (competitors, stress tolerators and ruderals).[106]

The extent of the CO_2-dependent stimulation of P_n decreases with leaf age,[117] being interpreted as the result of a CO_2-driven acceleration of the natural ontogenetic decline.[125] Stimulation is much stronger when rooting volume is large (60% on average) than when it is limiting (30% on average).[57] P_n stimulation can occur also in leaves low in nitrogen,[114,126,127] but it is negated or even changed in sign if N limitation becomes severe.[128] An increase in the P_n was observed as air temperature rises, at least within the physiological range.[57] Apart from the obvious consequences in the context of climate change and global warming, pronounced seasonal fluctuations in P_n stimulation may be expected in plants native to temperate climates.[57,122,129,130] Even soil pH has been invoked as a factor able to modulate the CO_2-dependent P_n stimulation. Accordingly, it was observed to occur in plants living on limestone but not in acidic grasslands.[130]

Most often, acclimation to elevated CO_2 is invoked as the causal agent when high-CO_2-grown plants exhibit a lower P_n than their normal-CO_2 counterparts.[57] Acclimation of P_n does not occur because of stomatal limitation, since C_i/C_a does not vary under elevated CO_2.[57] It has been found to involve decreases in the maximal Rubisco activity toward CO_2 (V_{cmax})[60,127,129,131] and in carboxylation efficiency,[116] even if not always,[73] whereas no change in the ribulose-1,5-bisphosphate (RuBP) regeneration capacity mediated by electron transport (J_{max}) has been observed.[127] Acclimatory downregulation of P_n is typically regarded as a long-term response to elevated CO_2 and has been shown to occur in a great variety of experimental circumstances.[44,45,57,59,105,112,120,132] Despite much effort in dissecting the physiological, metabolic, and even molecular features of acclimatory downregulation of P_n among species, functional types, and ecosystems, its implications and consequences are still matters of intensive debate.[18,50,73,78,100,106,113,114,133-135]

Under elevated CO_2, plants may be unable to use all the additional carbohydrates that an enhanced P_n can provide. One way to accomplish this to take advantage of the increased pCO_2/pO_2 at the sites of photoreduction, to decrease the amount/activity of Rubisco. This would allow carbohydrate supply to remain in balance with utilization capacity by sinks.[136] Earlier studies[44] identified an "acclimation syndrome," including downregulation of RuBP regeneration,[104,117] competition among P-sugars and RuBP for Rubisco sites, alteration of the normal P cycling among chloroplast and cytosol via triose phosphate synthesis inhibition, increased chloroplastic P-translocator,[137] and end-product feedback inhibition,[48,119] leading to preferential starch synthesis at the expense of sucrose synthesis (suggested also by a decline in sucrose phosphate synthase activity[138]), and ultimately resulting in chloroplast destruction by means of starch grains, possibly worsened by nitrogen and water limitations.[139]

A central event in the onset of P_n acclimation might be a downregulation of genes coding for photosynthetic enzymes, in response to an increased concentration of soluble sugars (glucose and fructose), caused in turn by an inadequate carbohydrate demand from the plant.[140] Both nuclear (*rbcS*, *rca*, *cab*, *lhcB*) and plastidial (*rbcL*, *psbA*, *psaA-B*) P_n genes might undergo downregulation under elevated CO_2.[49,57,120,121,142] This effect is reported to be stronger under nitrogen limitation,[57] whereas upregulation of genes for carbohydrate storage/utilization might occur in parallel.[102] At the signal transduction level, the sensing system might include a hexokinase and a membrane-associated glucose transporter, whose interaction might release a second messenger of unknown nature, able to interact with the regulatory sequences of P_n genes,[57,102] although this might not be the sole route required for the message to be effective.[137] Carbohydrate-induced downregulation of the above and possibly other genes would be in turn responsible for the changes occurring in chloroplast biochemistry, i.e., decline in both the amount of Rubisco protein (15% on average [51,57,105,132,143-145]) involving both its large and small subunits,[63,120,131] and in its activity (in the range 15 to 24%).[5,43,45,46,57,63,78,105,112,125,132,145,146] However, it should be mentioned that evidence negating adaptive decrease of Rubisco under elevated CO_2[117,138,147] and even an increased Rubisco activated/total ratio (25%) has been reported.[78]

Because of the high relative abundance of the Rubisco protein, its adaptive decrease under elevated CO_2 is expected to have close relationships with the whole-plant nitrogen economy. Indeed, on the one hand, Rubisco acclimation appears to be mediated by the elevated CO_2-dependent N dilution and reallocation, and happens to be more marked under low N.[57] On the other hand, it might cause per se N redistribution away from the leaf.[46,57,140,148] Adaptive decline of ancillary activities, such as carbonic anhydrase,[120] more marked under low N,[57] and in Rubisco activase,[120] has also been reported. The activity of the other primary carboxylase, namely, phosphoenolpyruvate carboxylase, has also been shown to undergo downregulation under elevated CO_2, both in C_3[137,145] and in CAM species (up to 34%).[178] This decline appears to be aggravated by low P levels.[137,145] Adaptive decrease of thylakoid proteins (PSI core protein, D1, D2, cyt *f*) have also been shown to take part in the P_n acclimatory response to elevated CO_2.[120]

If the causes of P_n acclimation to elevated CO_2 resides in nutritional and source-sink imbalances, then its occurrence/extent should be strongly dependent upon the suite of environmental factors affecting vegetative and reproductive development and should be amenable to experimental manipulation. Indeed, P_n acclimation appears to increase as N availability or rooting volume decrease, these two factors being mutually mimicking.[53,57,149] Moreover, P_n acclimation can be prevented even in plants subjected to root confinement providing that sink capacity remains high, thus preventing buildup of sugars in the leaves.[53,75,116] Coherently, P_n acclimation under elevated CO_2 might be transient during the growing season, being attenuated or even relieved as reproductive sinks (i.e., flowers/fruits) form.[119] This could be a distinct advantage for those species that form their reproductive sinks early in the growing season.[57]

4.3.1.4 Photorespiration

Photorespiration in plants growing under elevated CO_2 has seldom been measured.[147] Indeed, the established knowledge about the suppression of photorespiration in plants able to concentrate CO_2 at the Rubisco sites, such as C_4 and CAM species, leaves little doubt about the effect of increasing C_a in this respect. The activities of photorespiratory enzymes glycolate oxidase, NADH/NADPH hydroxypyruvate reductase, glutamine synthase, and catalase (CAT) have been reported either to decrease[40,145,150-152] or to be unaffected by elevated CO_2,[40,147,153] whereas P-glycolate phosphatase activity was found to increase.[147]

4.3.2. Dark (Mitochondrial) Respiration

Effects of elevated CO_2 on dark respiration are still controversial,[44] and this might also due be to inherent difficulties in discriminating it from photorespiration in C_3 species. In principle, an increased availability of respiratory substrates, enhanced growth rates, and an increased plant size, including roots,[43,115] should result in enhanced rates of both growth and maintenance respiration.[42,52,130] Although evidence in this direction has been produced, e.g., increased specific activities of the mitochondrial enzymes NAD-malic enzyme and fumarase,[145] an enhanced dark respiration under elevated CO_2, has often been negated both in leaves[120, 134, 135] and roots.[154]

On the contrary, an increasing body of evidence points to a decrease in dark respiration rates under elevated CO_2. A direct depressive effect might stem from the CO_2-dependent inhibition of respiratory enzymes such as cytochrome c oxidase and succinate dehydrogenase, observed in C_3 but not in C_4 species.[57] This direct effect, which has been observed in different plant organs as well as in soil bacteria,[18,57,85,108,111,113,115] could account for an average 20% decrease of the dark respiration rates.[57] An indirect depressive effect has also been postulated to occur, accompanying a CO_2-dependent decrease in leaf N and protein, leading in turn to a lessened requirement for metabolic energy.[57,114,154-156]

4.3.3 Photosynthates Concentration and Composition

Plant tissues exposed to elevated CO_2 accumulate nonstructural carbohydrates (NSC) in the form of starch (+160% with respect to control),[57,63,84,135,139,157] sucrose (+60% on average),[57,59,63,84,158] or fructans,[159,160] but without any change in individual fructan pattern.[159] Accumulation of NSC under elevated CO_2 has been observed also in CAM species.[46,161,162] The accumulation of NSC is more evident in fast-growing species, whereas slow-growing ones typical of certain natural environments, e.g., the Mediterranean Macchia, might invest the extra C preferentially in structural carbohydrates, i.e., fiber.[73]

Carbohydrate accumulation may be one of the concurring causes leading to downregulation of photosynthetic capacity. If this sink limitation is removed, i.e., as sink strength increases, carbohydrate accumulation may be prevented,[43,59] although this seems not always to be the case.[163]

4.3.4 Water Status

A decreased stomatal conductance in elevated CO_2-grown plants most often results in decreased evapotranspiration rates. Elevated CO_2 appears to have variable effects on hydraulic conductance,[164,165] whereas the size and number of xylem vessels in trees is either left unchanged or increased, depending on the species.[164] Less negative values for xylem pressure potential have been also reported.[55] The depression of evapotranspiration under elevated CO_2 may be partially (or even totally)[85] offset both by the resulting increase in leaf temperature, enhancing the need for evaporative cooling,[44-46] and by a larger and more rapid leaf development, in turn increasing LAI.[57, 89]

Reduced transpiration rates and higher cellular contents of soluble sugars may converge in increasing the whole-plant water potential[57,76] and the turgor potential,[42] with expected benefits under conditions of water shortage.

4.3.5 Nutrient Uptake and Distribution

The overall effect of elevated CO_2 on the plant nutrient status may critically depend on interlaced compensatory changes in the size, number, and activity of different plant parts, such as leaves and roots.[96,166] Nutrient uptake rates have been reported to vary to a considerable extent with the species, the nutrient, its chemical form, and the C fixation pathway,[80,115,166] and water availability.[80] CO_2-dependent increase in root uptake capacity for NO_3^- has been reported[166-168] but neither for NH_4^+ [169] nor for PO_4^{3-}.[166]

An apparent "dilution" of mineral nutrients in plant tissues, due to the relative increase of NSC is documented at elevated CO_2 levels.[160] This effect has been shown to occur for a number of macro- and micronutrients, including K, Mg, C_a, S, Zn, Mn, and Cu,[94,160,169,170,176,177] although evidence to the contrary,[94,170] as well as absence of any effect,[157,160,170] has also been reported.

Nitrogen levels decreased under elevated CO_2 in C_3 leaves,[84,94,111,122,160,163,166,170-173] roots, and grains,[174] but neither in C_4 species[166,175] and in N_2-fixing trees[114] nor in certain natural plant communities.[73] Depressed foliar levels of N, resulting from a decrease in the amount translocated from the roots,[168] often, but not always,[95] lead to a decrease in foliar soluble proteins,[57,105] including Rubisco.

The nutritional balance of another macronutrient P is also thought to play a role in P_n acclimation, and indeed elevated CO_2 appears to depress the content of this element in plant tissues,[94,176,177] although with some exceptions.[157,160] Decreasing levels of N and P in the presence of sustained growth rates, under elevated CO_2, imply increased nutrient utilization efficiencies, even when soil nutrient availability was limiting.[43,57,96,160,178,179] Nutrient dilution under elevated CO_2 may have important consequences for both agricultural species and natural plant communities. For the former, the benefits deriving from increasing C_a are expected to depend to a great extent on the availability/inputs of mineral nutrients, especially N.[43,178,180] For the latter, the increase of C/N and C/P ratios, a well-documented phenomenon in the tissues of C_3 species,[43,57,171,174-176,181] but controversial in C_4 ones,[181,182] may influence litter composition and hence the cycling of nutrients at ecosystem level.[174,175]

4.3.6 Vegetative and Reproductive Growth

4.3.6.1 Vegetative Growth

4.3.6.1.1 Leaf — Providing that soil fertility is not limiting,[127,183] and with certain exceptions,[60,95,184] elevated CO_2 appears to increase both the dry weight (DW),[74,78,94,185,186] and the surface area[43,45,67,73,78,91, 94,95,122,171,183,187,188] of individual leaves, leading to increase in leaf biomass.[93] In relative terms, the promotive effect has to be stronger for DW than for leaf area, as specific leaf area (SLA) is often reported to decrease[43,86,94,99,183,189] and specific leaf weight (SLW) to increase,[43,63,133] or to remain unchanged,[190] under elevated CO_2.

Effects on leaf number appear to be more controversial, as increases,[121,188] imputed to promotive effects on node[43] and shoot[111] number, tillering,[118] and branching,[43,45,171] but also no change[191] or even decreases, imputed to CO_2-driven accelerated development,[157] have been reported.

The effect on LAI is also controversial, varying from negative[57] to moderate[90,192] or even clearly stimulating.[74,91] Other putative CO_2 effects on leaf development include increases in appearance rates (decreased phyllochron intervals), in elongation rates,[85] and in area duration,[193] with an exception.[125] A few studies have been conducted on elevated CO_2 effects on leaf anatomy/histology. A reduced phloem cross-sectional area, resulting from fewer sieve cells,[194] and controversial effects on leaf epicuticular wax formation, increasing[101] or decreasing,[78] especially under conditions of low nitrogen,[195] have been reported.

4.3.6.1.2 Root — When rooting volume is not limiting, elevated CO_2 stimulates root growth and increases the root/shoot ratio,[43,60, 85,125] although not always[59,118,185,189196] and especially not in trees.[43,67,138,186] Indeed, it is not yet clear whether stimulation of root growth depends on increasing allocation of extra carbon to the roots,[85,125] whose extent appears to be in any case highly variable with the species,[166] or a secondary consequence of water and nutrient limitation, resulting from increased plant growth.[197,198] That high soil fertility depresses root responses to CO_2 and vice-versa[42,194,198] has been contradicted.[168,199]

The components of CO_2-stimulated root growth include total fresh biomass,[76,80,88,181,199-202] DW,[43,175,185,186] which may derive from an enhanced carbohydrate content,[169] relative growth rate (RGR),[200] length,[75,199,203,204] diameter (but also negated[204]), branching,[43] volume,[43,111,200,205,206] and abundance of fine roots, especially in the upper soil layers.[80,203,204] Effects on components such as root depth and half-life[63,200,207,208] formation of fine roots, mortality, and decomposition are more controversial.[199,201,204,209,210] Elevated CO_2 appears to increase the number of rhizomes[111] and of tubers.[43]

4.3.6.1.3 Whole Plant — In recent years, our ideas about the stimulatory effects of elevated CO_2 on the overall plant growth have been substantially reshaped by the evidence coming from studies on natural biodiversity and ecosystem relationships. It has become increasingly apparent that growth responses to elevated CO_2 vary to a great extent, especially in natural growth environments, because of the existence of (1) species-specific proneness in response, (2) within-population genetic variability, (3) complex interactions at the community level, and (4) intrinsic nonlinearity of the dose–response curves.[18,53,73,85,90,107,130,212-215,232] Even when attention has been focused on monospecific and genetically homogeneous plant communities, it has become increasingly evident that growth responses to CO_2 depend on the interplay of several factors such as (1) annual or perennial nature, determinate or indeterminate habit, functional type (*sensu* Grime[216]) and, more expectedly, C-fixation pathway; (2) tissue type and age; (3) growth temperature and the availability of mineral nutrients; and (4) the occurrence of acclimatory processes over time.

Increases in both fresh and dry total biomass under elevated CO_2, and especially in organs serving as long-term carbohydrate pools, such as peduncle, stem, flag leaf, have been reported by an impressive number of studies on higher plants[18,43,44,60,63,67,78,85-87,91,94,96,99,109,119,123,174,179-181,183,185,186,190,193,196,217-225] and even in bryophytes.[232] In trees, components of this stimulatory effect on growth, which is more evident in seedlings,[186] appear to include increases in shoot length and number[73] and in stem diameter.[138,225] In natural plant communities, growth effects tend to be stronger in species with high SLA.[226] Under elevated CO_2, plants not only show an increased size but also exhibit an accelerated growth,[43,45,86,94,138,157,171,190] maturation, and senescence,[43,45,75,94,105,143] although this might not hold for roots.[200] By using the weight ratio (the ratio between the biomass of high-CO_2- and control-CO_2-grown plants), an extensive literature survey conducted by Poorter et al.[218] suggested the following ranking in growth response to CO_2: C_3 crops > C_3 wild = C_3 woody > C_4 = CAM. The growth stimulation observed in C_4 species lower in extent as compared to C_3,[166,218,222] was imputed to increased WUE, rather than deriving from increased C fixation.[222] Reports are even negating any substantial effect of elevated CO_2 on both growth extent and velocity,[61,74,75,88,188,191,223] especially as far as natural or seminatural plant communities are concerned.[76,90,111,115,133,168,212,213,227] According to the functional classification proposed by Grime,[216] species classified as stress tolerators (S) and ruderals (R) are thought to be incapable, or the least capable, of responding positively in terms of growth to elevated CO_2.[106,228,229]

The increasing research interest in the effects on natural or seminatural plant ecosystems led to the common feeling that nutrient constraints (N availability) have much stronger effects on whole-plant growth than high CO_2[57,215] and that moderate or null growth responses have to be expected because nutrients in the field are always severely limited.[18,218,230] Indeed, larger growth responses to elevated CO_2 are seen under conditions of high N, P, and K levels in the soil.[157,217] Positive responses are also still observed under low levels of N and K, but not if P is limiting.[57,107,174,180,184,217] No growth stimulation is seen where and when soil nutrients are severely limiting.[128,218,230-232]

Stimulatory effects on growth have also seen to decline over time[125,161,181,233,234] and especially so in trees.[68,82,122,225,235] Under cool climate,[236] this could be due to seasonal fluctuations in temperature. Elevated CO_2 may initially increase the RGR[237,238] and the net assimilation rate (NAR)[189]; then, a decrease in leaf area ratio could first offset and then outcompete NAR, resulting in a RGR decline.[44,94,183,224,238] Even if not sustained beyond the first growing season, CO_2-dependent stimulation of RGR in perennial species can still allow increased absolute production of biomass in the following years, because of the compound interest effect of increased leaf area on the production of more new leaf area and more biomass.[122]

4.3.6.2 Reproduction

Evidence exists that elevated CO_2 might shorten the flowering time[86] and interact with photomorphogenetic plant responses, although in an unpredictable manner.[239] Increases in fruiting and reproductive output have been seen, especially in indeterminate species.[43] An increase in the number of reproductive organs, rather than in their size,[188] has been the most frequently reported response.[42,83,157,183,190,219]

4.3.6.3 Crop Yield

With certain exceptions[74] and temperature[236] and mineral nutrient (especially P[217]) limitations, increases in the marketable yield have been observed in C_3,[57,83,109,157,180,181,193,219,223, 227,235] but not in C_4 species.[223,227] Manderscheid and Weigel[95] reported greater yield effects in older than in modern wheat cultivars. No change has been reported for harvest index.[109]

4.3.6.4 Crop Quality

Lower fruit levels of N, P, and K have been reported under elevated CO_2, which may be due to a dilution effect. These high-CO_2 fruits also showed lower respiratory as well as ethylene production rates, and hence slower ripening, which may be of some commercial value.[177] An increased amylose content was found in rice grains, which could be positive for cooking quality.[240] Reductions in wheat kernel protein content were reported,[160,240-242] but also no effect[43,236,243] and no change in a series of technological indexes related to milling and baking quality.[243] To our knowledge, no CO_2-dependent change in the palatability of edible plant products has been reported to date.

4.3.7 Plant-Plant Interactions

According to current ideas, there is little doubt that increasing CO_2 will modify species composition in natural plant communities.[244] As the name itself suggest, functional competitors (C, *sensu* Grime[216]) are expected to outcompete R and S species, these last hence undergoing an indirect negative effect of increasing CO_2.[106,215,228,229,245] In general, selective advantages in the competition would be conferred by a higher CO_2 response in terms of root growth and WUE[125] and by quickness in the growth response.[246]

Under nutrient-limited growth response to elevated CO_2, another advantage of paramount competitive importance for a given species should be its trophic ability. Under limiting conditions, as those typically occurring in natural and seminatural ecosystems, nitrogen-fixing species, which can draw from a virtually unlimited reservoir in the absence of any interference from other species, are regarded as sure winners.[215,227,231,234,245,247,248] It has been suggested [43,44,247] that increasing C_a could lead C_3 species to outperform their C_4 competitors, both because inherent differences in P_n rates between the two groups would tend to level off, and, because the former do not bear additional energetic expenses for CO_2-concentrating mechanisms. However, evidence to the contrary has also been reported.[249]

Research on plant-plant interactions has focused almost exclusively on competition (interference); one study has been published concerning parasitism, reporting that the mechanisms by which the C_3 root hemiparasites *Striga* spp. affects host [*Sorghum bicolor* (L.) Moench.] growth are relatively insensitive to increased C_a, although the parasites themselves were adversely affected at elevated CO_2.[250]

4.3.8 Plant–Soil Interactions

It is generally agreed that no direct effect of elevated CO_2 on soil microbial processes can be expected, because of the already high belowground pCO_2 and due to roots/rhizosphere respiration. Hence, plant-mediated effects are supposed to occur in soils, i.e, via litter fall, root turnover, and rhizodeposition.[167] Although indirect, effects on soil biodynamics, energy flow, and nutrient cycling might be substantial because elevated CO_2 markedly affects the chemical composition of plant tissues.[173] Accordingly, soil-mediated responses are gaining increasing attention within the field of elevated CO_2 research.

4.3.8.1 Soil Inputs

Quantitative and qualitative changes in the carbon input to soil are expected under elevated CO_2.[85,124,167,168,202,206,210,244,251,252] These changes would result from the combination of (1) an increased aboveground net CO_2 uptake, leading to enhanced litter/residue production[76,181,220]; (2) a shift in carbon allocation pattern in favor of the roots[252]; (3) effects on rhizodeposition (sloughed off cells, exudates, mucillages, secretions)[43,206,253]; and (4) a decreased decomposition rate of the residues.[194] A stimulated CO_2 efflux from root/rhizosphere has also been observed in plants growing under elevated CO_2.[202]

4.3.8.2 Plant–Microbe Mutualistic Interactions (Symbiosis)

Stimulation of N_2-fixing activity and increased nodulation (imputed to an increased carbohydrate supply to the roots), but controversial effects on nitrogenase activity, have been reported.[43,114,167,231,254] Interestingly, the typical decline in N_2-fixation occuring when N_2-fixing species are grown under high N availability is not observed in the presence of elevated CO_2.[167]

Elevated CO_2-dependent changes on the size and composition of mycorrhizal microbial communities are also expected,[206,253] although their direction and extent are still controversial. Results indicating no or modest increases in mycorrhizal abundance[90,167] are balanced by the evidence of stimulatory effects on population sizes[43,115,207,251,255] and faster growth,[88,204,253] although modulated by the P availability in the soil.[167]

4.3.8.3 Microbial Decomposition/ Mineralization/ Immobilization of Plant Nutrients

Uncertainty exists for the (indirect) effects of elevated CO_2 on size, structure, and activity of the bulk soil microbial biomass.[167,168,210,256] It has been suggested that the CO_2-driven changes in the activity of microbial populations might depend critically on the previous soil nutrient status, especially in terms of nitrogen availability.[194] Stimulatory effects might therefore occur in crop soils, but not in natural ecosystems, where nutrient limitation is the rule rather than the exception.[167] There have been reports indicating increased or unchanged rates of soil C cycling,[205,210,244] stimulation of soil enzymes (phosphatases and exo- and endocellulases),[115,257] and no effect on soil N mineralization,[80,210] availability,[210,258] and turn-over rates.[244,259] Many other studies, however, showed slower nutrient cycling in soil, due to increased C/N, lignin/N, and lignin/P ratios[174,175] in litter from C_3 species.[182] The resulting inhibition of microbial decomposers[168,182,199,201] would lead to decreased N turnover rates[175,198,260,261] and increased storage/immobilization of C and N in soils.[85,124,168,175,194,201,252] Effects on soil CO_2 efflux appear to be variable with species.[181]

4.3.9 Plants–Environment Interactions

4.3.9.1 Abiotic Factors

4.3.9.1.1 Temperature — Many studies[44,45,106,129,162] have shown positive interactions between increasing CO_2 and increasing growth temperature in terms of stimulation of P_n rates and growth[57,75,86,186,221,233,248, 262,263] and of selective growth advantage for those species adapted to tropical and subtropical environments. Elevated CO_2 also appears to protect from heat stress in terms of P_n rate, PSII efficiency (A_{max}, F_V/F_M),[132] and survival.[158] However, as hyperthermic conditions increase floral sterility/abortion, the resulting in reduction of sink strength might worsen the down-regulatory effects of elevated C_a on P_n.[263]

Variable growth effects have been observed in presence of low temperatures.[43,47,78,236, 265] Boese et al.[266] found elevated C_a to protect plants from chilling-induced water stress, P_n depression, and visible damage. On the other hand, there was no effect of elevated CO_2 alone on frost hardiness; it was even detrimental when given in combination with N fertilization.[267]

4.3.9.1.2 Water Stress/Salinity — Elevated C_a not only antagonizes the deleterious effects of drought and/or excess salinity on P_n,[58,68,240] growth,[43,86,109,183,238,256,268-270] and reproductive capacity,[86] but even evokes relatively greater responses in terms of growth,[75,78,81,238,268] LUE, and yield[109] when water resources are limiting. High-CO_2 stomata appear to be more water conservative under drought,[51] and this, apart from conferring an increased drought tolerance[67] and avoidance,[219] may decrease soil water deficit.[57,66,206] This, of course, could be of particular benefit for nonmanaged plant ecosystems. In crops, the overall benefit in terms of yield may stem from the fact that elevated CO_2-grown plants are able to actively transpire and photosynthesize both further into the seasonal drought (summer) and in the event of episodic water shortage.[55,79,109] It is interesting to note that, in studies questioning the above beneficial effects of high CO_2 on drought, tree species were invariably involved.[67,68,89,271-273]

4.3.9.1.3 High/Low Irradiance — As C_3 photosynthesis is normally CO_2 limited but not light limited, positive interaction was observed between increasing CO_2 and increasing irradiance levels.[43] Indeed, elevated C_a has been suggested to antagonize high irradiance stress.[269] Beneficial effects of elevated C_a are foreseen also on the other extreme of the irradiance range, as both an enhanced survival under low light (i.e., mitigation of the shading effects)[90] and even relatively greater growth responses under reduced irradiance[78] have been reported.

4.3.9.1.4 Airborne Pollutants and UV-B — The bulk of air pollutants penetrates plant tissues *via* stomata, so that a protective effect of elevated C_a is easily anticipated due to its constrictive action on g_s.[43,44,59,61,62,131] This has been reported for tropospheric ozone (O_3), having its deleterious effects on the level of Rubisco protein and V_{cmax},[131] on P_n,[274] on biomass accumulation,[275] on root growth,[185] on senescence induction,[193] and on yield[62,276] (but contradicted by Reference 275) have been found to be antagonized, totally or in part, by elevated CO_2. Antagonistic interactions have also been observed with SO_2 on P_n[74,277] and on LAI,[174] although a case of no interaction has also been reported.[93] The relieving effects of elevated CO_2 toward air pollutant injury and damage have also been negated.[118,186,223,274,275] Since air pollutants such as ozone and SO_2 can be regarded as oxidative stressors, their interaction with elevated C_a on plant metabolism might be also mediated by combined effects on the plant antioxidant status.[59]

No clear interaction between elevated CO_2 and increasing UV-B has been observed.[115,119]

4.3.9.2 Biotic Factors

4.3.9.2.1 Plant–Microbe Phytopathogenic Interactions (Diseases) — Experimental data in this area are scanty, indicating no change or slight (nonsignificant) increase in the population size of soilborne pathogens, such as nematodes.[167] On the basis of known effects of elevated CO_2, extrapolations have been attempted suggesting stimulation of phytopathogens due to increase in plant canopy (and root) size and density, increased tissue water content, canopy humidity, leaf temperature and carbon status (starch and sugars, also in roots), increased overwintering crop residues, increased root exudation,[43,251,255] possibly decreased epicuticular wax content, and density.[195] Counteractive effects have also been envisaged, possibly due to higher plant vigor, leading to a better withstanding of infection,[255] reduced stomatal opening,[43] lower nitrogen contents, higher C/N ratio,[46] higher phenolics,[43] and accelerated ripening and senescence (i.e., shortened growth period).[251]

4.3.9.2.2 Insects — The importance of possible changes in foliar chemistry, especially allelochemicals, such as phenolics and tannins, has been related with effects of elevated CO_2 on plant–insect interactions.[163,173,278,279] The main determinant in setting the responses of insect populations might be the CO_2-driven increase in sugar content and the resulting dilution effect toward N. This would increase palatability, but decrease diet quality (lower C/N ratio), leading in turn to enhanced consumption but decreased food conversion efficiency. In turn, this would cause prolonged developmental rates and reduced adult biomass.[43,57,163,173,182,279] There are even reports indicating no change in the degree of defoliation[60] as well as in development, mortality, and adult biomass[278] for insect populations feeding on plants exposed to elevated CO_2. No effect on larval susceptibility to virus infection has been observed.[163]

4.3.9.2.3 Herbivory — In contrast to the effects on insect feeding, reduced intakes are envisaged for ruminants because of reduced N contents, increased fiber, and possibly increased contents of antinutritional factors (e.g., tannins[173]) in plant tissues, leading to a global decline in forage quality. This would negatively affect assimilation and, hence, growth and reproduction.[57,280] The resulting impact would, of course, be stronger in the absence of dietary supplementation, i.e., for wild ruminants and for domestic livestock in developing Countries.[280]

4.3.10 Plant Antioxidant Status

Being capable of impacting the plant metabolism, increasing C_a might be predicted to have a profound effect on the delicate equilibrium between prooxidants and antioxidants within the plant cell. By enhancing the pCO_2/pO_2 ratio at the sites of photoreduction, and also progressively suppressing photorespiration in C_3 species, increasing C_a could potentially reduce the basal rate of O_2 activation and ROI formation within several plant cell compartments. In the long term, this might lead to a depressed antioxidant status, as unnecessary/redundant antioxidative mechanisms might be lost in plants adapting to elevated CO_2. Such resource-saving adaptation, in principle, might turn out to be detrimental under the circumstances in which plants face environmental and developmental stimuli able to promote oxidative stress.[281] On the other hand, this might be partly or totally compensated for by the antagonistic action of elevated CO_2 toward several environmental stressors, such as drought/salinity, disthermia, high irradiance and airborne pollutants, whose damaging effects on plants appear to be mediated by ROI.

So far, the relationships between elevated C_a and the plant antioxidant status have been studied in a limited number of herbaceous and tree species, and, rather expectedly, variable and complex response patterns have been observed. Depressive effects of elevated C_a on the activity of ROI-scavenging enzymes, such as ascorbate peroxidase (APX), catalase (CAT), and superoxide dismutase (SOD), with foliar ascorbic acid (AsA) and glutathione (GSH) contents increasing or remaining unchanged, have been reported.[93,147,151-153,267,282,283] There may be a reduced need of antioxidant defense, leading to a downregulation of antioxidant status in plants exposed to increasing CO_2. In multifactorial experiments, above-ambient CO_2 increased the activities of antioxidant enzymes in the presence of elevated O_3 and/or drought, which led to the hypothesis of a better compensative ability against environmental stress in plants growing under increasing CO_2.[150,151,284] In other studies, however, the protective effect toward O_3 damage was exclusively imputed to the CO_2-dependent stomatal constriction, reducing the pollutant dose.[285]

The idea that high-CO_2 conditions attenuate the intrinsic oxidative risk for the plant cell was, however, challenged by other short-term studies, putting into evidence that a significant increase in the foliar GSSG level, causing a decrease in the GSH/GSSG ratio, and an enhanced total activity of foliar APX were almost invariably associated with the growth under elevated CO_2.[40,153,286,287] These findings suggested an enhanced consumption of nonenzymic antioxidants and the requirement for an increased H_2O_2-scavenging capacity under elevated CO_2, at least in certain species. In addition, CO_2-driven upregulation of SOD activity has also been reported.[153,277]

Since glycine is one of the building blocks of glutathione, and glycolate oxidase activity may be depressed in elevated CO_2-grown plants,[40,288] it may be speculated that the above effects might be a consequence of CO_2-driven changes on photorespiration.[286,287,289,290] Moreover, the high CO_2-dependent downregulation of the photorespiratory C cycling might well parch a source of peroxisomal H_2O_2, but at the cost of suppressing an energy-dissipating mechanism which ultimately helps in preventing chloroplastic ROI formation.[291]

4.4 PLANT RESPONSES TO ELEVATED CO_2: NATURAL CO_2 SPRINGS

Table 4.1 summarizes the evidence coming from studies on native vegetation living in the NCDS environments. In these studies, effects attributable to CO_2 enrichment have been isolated by comparing, for each species studied, populations/individuals growing at variable distances from the natural CO_2 sources, and hence supposed to have been permanently exposed to contrasting CO_2 regimes. In several of these experiments, comparison has been made *in situ*, whereas in other cases plant material from NCDS sites has been grown in controlled laboratory facilities and therein exposed to elevated CO_2, in order to get rid as much as possible of the perturbing factors occurring in the native environments.

The unique possibility of studying adaptation to elevated C_a under natural conditions and over evolutionary timescales comes at a high price in terms of site-to-site (or even within-site) variability in population genetics and edaphic factors, ever-present disturbance caused by the occurrence of environmental stress, and other unusual environmental features typical of NCDS. Moreover, the majority of the species listed in Table 4.1 are poorly characterized at the ecophysiological, biochemical, and genetic levels. Further, the NCDS case studies are relatively less, in terms of number and species coverage, than those produced by the short term approach on other plants. With the above caveats in mind, comparison between "short-term" CO_2-dependent effects and the results in Table 4.1 might aid in discriminating among responses that are presumably maintained through generations, and hence might be regarded as evolutionary modifications, from those that plants might be able to adapt to/compensate for in the long run.

The second-generation approach to CO_2 effects on plants, i.e., to focus on biodiversity and natural ecosystems, needs further scientific assessment because of variability in the responsiveness of different species to elevated CO_2, and that this inherent variance, although clearly uncomfortable for predictive purposes, might be among the most important aspects to consider. Table 4.1 confirms that one evolutionary change induced by elevated C_a might be a reduction in g_s, resulting in lessened transpiration and a better performance under water stress. Whether this would result in increased WUE would be determined by the overall long-term effect of increasing C_a on photosynthetic carbon uptake.

Table 4.1 A Synopsis of the Changes Observed in Plant–Soil Ecosystems Permanently Exposed to Natural CO_2 Enrichment, as Compared with Their Respective Controls under Present-Day CO_2.

Change Observed	Plant Species[a] and Reference
1. PHOTOSYNTHESIS AND PHOTORESPIRATION	
1.1 Stomata	
Reduction in density, index, and area	*S. herbacea* (fossil plant material),[292] *Ph. australis*,[293] *Q. pubescens*,[293] 17 NCDS species,[294,295] *E. arborea*,[41] herbaria and tombs,[26] *S. lacustris*[296], *Q. ilex*,[297] *Q. pubescens*,[298] *F. ornus*,[294] *A. unedo*,[294]
No change in number of stomatal and epidermal cell	*Q. pubescens*[295]
Decrease in stomatal conductance and transpiration rate	*Q. pubescens* (in mature trees but not in seedlings),[298-300] *A. unedo*,[301] 17 NCDS species,[294] *Q. ilex*[298,299,302,303]
Increase in WUE	*Q. ilex*,[299,304] *S. herbacea*,[304] *Q. pubescens* (in mature trees but not in seedlings)[298,299]
1.2 Photoconversion	
Decrease in PSII photochemical efficiency	*Q. pubescens*[298]
Decreased zeaxanthin levels	*Q. ilex*[305]
Increased zeaxanthin synthesis under water stress	*Q. ilex*[305]
Increased F_V/F_M under water stress	*Q. ilex*[305]
No change in leaf pigment composition	*Ph. australis* (M. Lauteri, pers. comm.)
No change in PSII photochemical efficiency	*Ph. australis* (M. Lauteri, pers. comm.)
No change in qNP and deepoxidation state	*Ph. australis* (M. Lauteri, pers. comm.)
1.3 Carbon Fixation	
Reduction in A_{max} at saturating CO_2	*Ph. australis*[293]
No change in A_{max}	*L. uraguayensis*,[27] *A. canina*,[306] *S. lacustris*[296]
No change in P_n rate	*Q. pubescens*,[298] *Q. ilex*[298]
No change in P_n if soil nitrogen not limiting	*E. arborea*[41]
Increase in carbon uptake only if nitrogen not limiting	*E. arborea*[296]
No change in carboxylation efficiency	*A. canina*[306]
No change in Rubisco content	*A. canina*,[306]
No evolutionary downregulation of Rubisco content	*A. canina*, *A. stolonifera* (S.P. Long, pers. Comm.)
Tendency to downregulation of Rubisco and protein contents	*Q. ilex*, *Q. pubescens* (S.P. Long, pers. Comm.)
1.4 Photorespiration	
No study	
2. DARK (MITOCHONDRIAL) RESPIRATION	
Tendency to increase	*Q. ilex*,[307] *Q. pubescens*[307] (probably a stage-specific effect)
3. PHOTOSYNTHATES CONCENTRATION, COMPOSITION AND ALLOCATION	
Increased starch content in herbaceous species	40 NCDS species (38 to 47% increase)[308]
Increase in NSC	*Q. pubescens* (variable throughout the year)[302,307,308]
No increase in NSC	*Q. ilex*[308]
Nonsignificant increase of NSC in phloem exudates	*Q. pubescens* (no change in xylem exudates) (A. Polle, pers. comm,)
Increase in the cell wall fraction	*A. canina*[306]
Decreased degree of lignification	*A. canina*[306]
No change in cell wall material and lignin content in leaf, bark, and wood	*Q. pubescens* (A. Polle, pers. comm.)

Table 4.1 continued

Tendency to increase in wood gravity	*Q. ilex*[307]
Increase in isoprene emission	*Q. pubescens*[298]
Increase in foliar tannins	*Q. pubescens*[298]
4. WATER STATUS	
Increased water potential	*Q. pubescens*[309] (only evident under water stress)
Decreased xylem embolism	*Q. pubescens*,[307] *P. tremula*,[307] *A. unedo*[307]
Increased osmotic potential	*Q. ilex*,[299] *Q. pubescens*[299]
5. NUTRIENTS UPTAKE AND DISTRIBUTION	
No change in leaf nitrogen	*S. lacustris*,[296] *Ph. australis*,[293] *Q. ilex*[310]
Decrease in leaf nitrogen	40 NCDS species[308]
No effect on enzymes of nitrogen and sulfur assimilation	*Q. pubescens* (Ch. Brunold, pers. Comm.)
Decrease in total and soluble proteins	*Ph. australis* (M. Lauteri, pers. comm.)
6. VEGETATIVE AND REPRODUCTIVE GROWTH	
6.1 Vegetative Growth	
6.1.1 Leaf	
Decreased leaf area	*Q. ilex*[304]
Decreased SLA	*A. canina*,[306] *P. major*[811]
Increased SLW	*Q. pubescens*[298]
6.1.2 Root	
Increased partitioning of dry matter to the roots	*A. canina*,[306] *P. major*[811]
6.1.3 Whole Plant	
Greater CO_2-sensitivity of growth	
in terms of growth speed (RGR)	*A. canina*,[306] *P. major*[811]
in terms of total biomass	*B. cylindrica*,[26] *A. canina*,[306] *P. major*,[311] *Q. ilex*[310]
No evidence of faster growth	40 NCDS specis[308]
No change in plant size	40 NCDS species,[308] *Q. ilex*[310]
No change in dry mass of leaf, bark, and wood	*Q. pubescens* (A. Polle, pers. comm)
Decreased branching	*Q. ilex*[304]
6.2 Reproduction	
Seed weight positively correlated to C_a	*A. canina*[306]
No acceleration of flowering	40 NCDS species[308]
7. PLANT–PLANT INTERACTIONS	
No study	

Table 4.1 continued

8. PLANT–SOIL INTERACTIONS	
8.1 Soil Inputs	
Lower nitrogen content in litter, causing higher C/N and lignin/N ratios	*Q. pubescens*[312]
8.2 Plant–Microbe Mutualistic Interactions (Symbiosis)	
No study	
8.3 Microbial Decomposition/ Mineralisation /Immobilization	
No direct change in soil organisms	*Q. pubescens*[312]
Slower litter decay due to lower nitrogen in leaves and roots causing higher C/N and lignin/N ratios	*Q. pubescens*[312]
9. PLANT-ENVIRONMENT INTERACTIONS	
9.1 Abiotic Factors	
9.1.1. Temperature	
Protection against high temperature at the photochemical level	*Q. ilex*[302]
9.1.2. Water Stress/Salinity	
Stronger CO_2 effect under water stress	*Q. ilex, A. unedo*[301,302,305,310]
9.1.3. High/Low Irradiance	
Protection against high irradiance at the photochemical level	*Q. ilex*[302,305]
9.1.4. Airborne Pollutants	
No study	
9.2 Biotic Factors	
9.2.1 Plant–Microbe Phytopathogenic Interactions (Diseases)	
Increased disease severity	Various pathosystems[313]
9.2.2 Insects	
No study	
9.2.3 Herbivory	
No study	
10. PLANT ANTIOXIDANT STATUS	
Increased foliar levels of oxidized glutathione	*A. stolonifera*,[287] *A. canina*[288]
Increased foliar ascorbate peroxidase activity	*A. canina*[288]

Note: For abbreviations, see text.

[a] Plant species cited: *Agrostis canina* ssp. *monteluccii* Selvi; *Agrostis stolonifera* L.; *Arabidopsis thaliana* (L.) Heynh.; *Arbutus unedo* L.; *Boehmeria cylindrica*; Erica *arborea* L.; *Fraxinus ornus* L.; *Ludwigia uraguayensis*; *Phragmites australis* (Cav.) Trin. ex Steud.; *Plantago major* L.; *Populus tremula* L.; *Quercus ilex* L.; *Quercus pubescens* Willd.; *Salix herbacea* L.; *Scirpus lacustris* L.

Rather unexpectedly, evidence from NCDS studies suggests no evolutionary tendency to Pn downregulation, in terms of photosynthetic capacity, carboxylation efficiency, and Rubisco content (Table 4.1). This is even more remarkable if one considers that acclimation-promotive conditions, first of all poor availability of mineral nutrients, are inherent features of NCDS sites. This most important aspect clearly needs to be studied in a much more systematic manner; results from NCDS experiments suggest that current ideas about acclimatory mechanisms and processes need to be reexamined. If a reduction in g_s is not associated with a worsening of the photosynthetic performance, also deduced by the lack of substantial changes in the leaf photochemistry (Table 4.1), then an increase in WUE may be expected, and indeed this appeared to be the case for certain NCDS species (Table 4.1).

The observed lack of P_n acclimatory responses immediately leads to wonder about routes of allocation of extra C in native species permanently exposed to elevated CO_2. *In situ* comparison between NCDS plants and their controls produced another puzzling piece of evidence, i.e., the lack of typical short-term responses to elevated C_a such as accelerated vegetative and reproductive development and increased plant size (Table 4.1). Also another classical effect, i.e., DW accumulation, appears to be restricted to leaves, whose thickness tends to be increased by a reduction in SLA and an increase in SLW (Table 4.1). In a tree species at least, leaf area tended to be reduced, probably incorporating depressive effects on branching (Table 4.1).

Growth chamber experiments under elevated C_a showed that NCDS plants grow faster and produce more biomass than controls (Table 4.1). On one side, this implies that a genetic potential for a greater CO_2 sensitivity has been acquired by NCDS plants in the course of evolution. On the other side, it becomes apparent that powerful environmental constraints must exist, which prevent or restrict the full expression of this potential in plants growing *in situ*. Among the possible counteracting factors occurring in the NCDS environments, nutritional limitations/imbalances appear to be the most obvious. The extra C gained by plants living under NCDS regimes might be invested in NSC, cell wall polymers, or secondary C-rich molecules, such as polyphenols and isoprene, but not in lignin (Table 4.1). Accumulation of NSC, occurring only in herbaceous species, in which it might cause N dilution (Table 4.1), is another factor which in theory, but not in practice, as previously seen, would favor P_n downregulation.

Even more than under short-term conditions, a key element for interpreting long-term responses in NCDS species would be an adequate comprehension of belowground processes. Unfortunately, information in this direction is scarce. As shoot/root relationships are concerned, the only study available revealed that, at least in captivity, herbaceous NCDS species allocate more dry matter to the roots when exposed to elevated C_a (Table 4.1), thus reproducing a response typically observed in many short-term experiments. Even if, at present, no information from *in situ* studies is available, preferential allocation of C to belowground sinks could be of evolutionary value as one of the mechanisms preventing feedback inhibition of P_n capacity in NCDS species. Of even more ecological and evolutionary value, and also contributing in explaining the lack of P_n acclimation, would be if future research might assess, that root/rhizosphere of NCDS species is particularly efficient, not only in preventing the deleterious effect of abnormal levels of phytotoxic elements, but also in the extraction, mobilization, and uptake of mineral nutrients from their uniquely featured soil environments, allowing a quasi-normal nutritional status within plant tissues. It is striking that plants growing over NCDS soils never show symptoms of mineral deficiency/imbalances. Indirect evidence about these unique capacities of certain NCDS species already exists. Selvi[29] pointed out the striking fitness of *Agrostis canina* L., an endemism of Italian NCDS, to a range of extreme soil environments, under varying climatic conditions. Rather coherently, it is a local popular notion that tree essences living at the border of the aphytoic zone in the Caldara di Manziana, one of the largest NCDS in central Italy, do not survive when transplanted in a foreign soil, even when appropriate transplanting practices, best-fertilized substrates, and optimal climatic conditions are adopted. How most NCDS species succeed in providing themselves with adequate levels of mineral nutrients remains a mystery. Indeed, very slow or almost nil mineralization of soil organic matter has been reported which is often caused by prohibitive soil pH, and could be further exacerbated by the deposition of plant residues whose C/N and lignin/N ratios would be increased by elevated CO_2 (Table 4.1), an effect largely predicted also by most short-term studies.

Finally, the results obtained from NCDS studies confirmed that, under enhanced CO_2, foliar antioxidant defenses are preserved from downregulation. Since antioxidant changes shown in Table 4.1 did not revert when vegetative progeny of NCDS plants was grown under normal C_a in the glasshouse,[286] thus appearing to be inheritable, adaptive roles of evolutionary value are suggested for them.

4.5 CONCLUSIONS AND PERSPECTIVES

The blanks appearing under several headings in Table 4.1 are intended to indicate those fundamental research areas which, although not yet covered, would benefit to the greatest extent from being pursued by means of the NCDS approach. However, even in the areas for which some information is already available, it will be a long time before a body of evidence is accumulated that could be comparable in size and completeness with the knowledge obtained from short-term studies. Because of this lack of comparability, results from NCDS studies should not be regarded as predictions, but rather as a sort of useful extrapolations when the factor "time of exposure" tends to infinity.

In summary, results from NCDS studies lend further support to the hypothesis that in an increasing CO_2 world plants may become in general more parsimonious and more efficient in water use. No doubt, a benefit would be expected for both those natural plant communities facing a future exacerbation of drought and hyperthermia, and also for crops and seminatural plants ecosystems (i.e., grasslands), under the widespread circumstances in which ever-present or recurring low water availability acts as the main constraint for agricultural productivity. Results from NCDS studies also corroborated the emerging view that elevated CO_2 effects on plant growth might vary between species, being their extent primarily set, on the long run, by the co-occurring growth conditions. Of these, the most powerful in modulating long-term plant responses might be the availability of mineral nutrients.

Whether, how, and to what extent nutritional conditions might be associated with the observed lack of photosynthetic acclimation in NCDS species remain to be established. In this respect, and on the basis of the results obtained, it could be tempting to speculate that the downregulation of photosynthetic capacity might not be a successful adaptive strategy for coping with elevated CO_2 level over an evolutionary timescale. Alternatively, the lack of P_n downregulation could be regarded as an NCDS-specific adaptation, reflecting the need of adequate flexibility to cope, for example, with huge and frequent diurnal fluctuations of local C_a conditions. Whatever the case, answering the question, "how do NCDS species avoid the occurrence of photosynthetic acclimation?" would contribute to shed further light on this fundamental and controversial issue.

Although NCDS are imperfect tools, their experimental use should continue. Artificial and natural CO_2 enrichment approaches need each other, because lessons from short-term and evolutionary responses may mutually integrate, with a synergistic effect in terms of predictive ability. At NCDS, nature did a timeless CO_2 enrichment experiment. Nowadays, far-sighted and resourceful humans are called to take advantage of this.

ACKNOWLEDGMENTS

This work contributes to the Research Project "Microevolutionary Adaptation of Plants to Elevated CO_2" (MAPLE), for which the financial support of the European Commission (Contract No. EV5VCT940432) is gratefully acknowledged.

REFERENCES

1. Conway, T. J., Trans, P., Waterman, L. S., Thoning, K. W., Masarie, K. A., and Gammon, R. M., Atmospheric carbon dioxide measurements in the remote global troposphere, 1981–1984, *Tellus*, 40, 81, 1988.
2. Eamus, D. and Jarvis, P. G., The direct effects of increase in the global atmospheric CO_2 concentration on natural and commercial temperate trees and forests, *Adv. Ecol. Res.*, 19, 1, 1989.
3. Allen, L. H., Jr., Plant responses to rising carbon dioxide and potential interactions with air pollutants, *J. Environ. Qual.*, 19, 15, 1990.
4. Kimball, B. A., Rosenberg, N. J., Heichel, G. H., Stuber, C. W., Kissel, D. E., and Ernst, S., *Impact of Carbon Dioxide, Trace Gases, and Climate Change on Global Agriculture*, American Society of Agronomy, Madison, WI, 1990.
5. Stitt, M., Rising CO_2 levels and their potential significance for carbon flow in photosynthetic cells, *Plant Cell Environ.*, 14, 741, 1991.
6. Arp, W. J., Effects of source-sink relations on photosynthetic acclimation to elevated CO_2, *Plant Cell Environ.*, 14, 869, 1991.
7. Petterson, R. and McDonald, J. S., Effects of nitrogen supply on the acclimation of photosynthesis to elevated CO_2, *Photosynth. Res.*, 39, 389, 1994.
8. Barnes, J. D., Ollerenshaw, J. H., and Whitfield, C. P., Effects of elevated CO_2 and/or O_3 on growth, development and physiology of wheat (*Triticum aestivum* L.), *Global Change Biol.*, 1, 129, 1995.
9. Grulke, N. E., Riechers, G. H., Oechel, W. C., Hjelm, U., and Jaeger, C., Carbon balance in tussock tundra under ambient and elevated atmospheric CO_2, *Oecologia*, 83, 485, 1990.
10. Drake, B. G. and Leadley, P. W., Canopy photosynthesis of crops and native plant communities exposed to long-term elevated CO_2, *Plant Cell Environ.*, 14, 853, 1991.
11. Maxon-Smith, J. W., Selection for response to CO_2-enrichment in glasshouse lettuce, *Hortic. Res.*, 17, 15, 1977.
12. Wulff, A. and Alexander, H. M., Intraspecific variation in the response to CO_2 enrichment in seeds and seedlings of *Plantago lanceolata*, *Oecologia*, 66, 458, 1985.
13. Strain, B. R. and Cure, J. D., *Direct Effects of Increasing Carbon Dioxide on Vegetation*, DOE/ER-0238. U.S. Department of Energy, Carbon Dioxide Research Division, Washington, D.C., 1985.
14. Strain, B. R., Possible genetic effects of continually increasing atmospheric CO_2, in *Ecological Genetics and Air Pollution*, Taylor, G. E., Pitelka, L. F., and Clegg, M. T., Eds., Springer-Verlag, Berlin, 1991, 237 .
15. Strain, B. R., Direct effects of increasing atmospheric CO_2 on plants and ecosystems, *Trends Ecol. Evol.*, 2, 18, 1987.
16. Friend, A. D. and Woodward, F. I., Evolutionary and ecophysiological responses of mountain plants to the growing season environment, *Adv. Ecol. Res.*, 20, 59, 1990.
17. Miglietta, F., Raschi, A., Bettarini, I., Resti, R., and Selvi, F., Natural CO_2 springs in Italy: a resource for examining long-term response of vegetation to rising atmospheric CO_2 concentrations, *Plant Cell Environ.*, 16, 873, 1993.
18. Amthor J. S., Terrestrial higher-plant response to increasing atmospheric CO_2 in relation to the global carbon cycle, *Global Change Biol.*, 1, 243, 1995.
19. Etiope, G., Migration in the ground of CO_2 and other volatile contaminants. Theory and survey, in *Plant Responses to Elevated Carbon Dioxide: Evidence from Natural Springs*, Raschi, A., Miglietta, F., Tognetti, R., and van Gardingen, P. R., Eds., Cambridge University Press, Cambridge, 1997, 7.
20. Kanemasu, E. T., Powers, W. L., and Sij, J. W., Field chamber measurements of CO_2 flux from soil surface, *Soil Sci.*, 118, 233, 1974.
21. Paces, T., Carbon dioxide springs in Czech Republic and associated ecosystems, presented at Carbon Dioxide Springs and their Use in Biological Research, International Workshop, San Miniato, Pisa, 1993.
22. Garrec, J. P., Le Thiec, D., Dixon, M., Prigel, B. and Rose, C., Trees experiments in open-top chamber near natural spring of CO_2 (Massif Central, France), presented at Carbon Dioxide Springs and their Use in Biological Research, International Workshop, San Miniato, Pisa, 1993.

23. Cook, A. C., Oechel, W. C., and Sveinbjornsson, B., Using Icelandic CO_2 springs to understand the long-term effects of elevated atmospheric CO_2, in *Plant Responses to Elevated Carbon Dioxide: Evidence from Natural Springs*, Raschi, A., Miglietta, F., Tognettia, R., and van Gardinger, P. K., Eds., Cambridge University Press, Cambridge, 1997, 87.
24. Newton, P. C. D., Bell, C. C., and Clark, H., Carbon dioxide emissions from mineral springs in Northland and the potential of these sites for studying the effects of elevated carbon dioxide on pastures, *N. Z. J. Agric. Res.*, 39, 33, 1996.
25. Piñol, J. and Terradas, J., Preliminary results on dissolved inorganic ^{13}C and ^{14}C content of a CO_2-rich mineral spring of Catalonia (NE Spain) and of plants growing in its surroundings, in *Plant Responses to Elevated Carbon Dioxide: Evidence from Natural Springs*, Raschi, A., Miglietta, F., Tognettia, R., and van Gardinger, P. K., Eds., Cambridge University Press, Cambridge, 1997, 165.
26. Woodward, F. I. and Beerling, D. J., Plant CO_2 responses in the long term: plants from CO_2 springs in Florida and tombs in Egypt, in *Plant Responses to Elevated Carbon Dioxide: Evidence from Natural Springs*, Raschi, A., Miglietta, F., Tognetti, R., and van Gardinger, P. K., Eds., Cambridge University Press, Cambridge, 1997, 103.
27. Koch, G. W., The use of natural situation of CO_2 enrichment in studies of vegetation response to increasing atmospheric CO_2, in Design and Execution of Experiments on CO_2 Enrichment, Schulze, E. D. and Mooney, H. A., Eds., Commission of the European Communities, Ecosystems Research Report No. 6, Bruxelles, 1993, 381.
28. Ehleringer, J. R., Sandquist, D. R., and Philips, S. L., Burning coal seams in southern Utah: a natural system for studies of plant response to elevated CO_2, in *Plant Responses to Elevated Carbon Dioxide: Evidence from Natural Springs*, Raschi, A., Miglietta, F., Tognettia, R., and van Gardinger, P. K., Eds., Cambridge University Press, Cambridge, 1997, 56.
29. Selvi, F., Acidophilic grass communities of CO_2 in central Italy: composition, structure and ecology, in *Plant Responses to Elevated Carbon Dioxide: Evidence from Natural Springs*, Raschi, A., Miglietta, F., Tognettia, R., and van Gardinger, P. K., Eds., Cambridge University Press, Cambridge, 1997, 114.
30. van Gardingen, P. R., Grace, J., Harkness, D. D., Miglietta, F., and Raschi, A., Carbon dioxide emissions at an Italian mineral spring: measurements of average CO_2 concentration and air temperature, *Agric. For. Meteor.*, 73, 17, 1995.
31. Reuveni, J. and Gale, J., The effect of high levels of carbon dioxide on dark respiration and growth of plants, *Plant Cell Environ.*, 8, 623, 1985.
32. Amthor, J. S., Koch, G. W., and Bloom, A. J., CO_2 inhibits respiration in leaves of *Rumex crispus*, *Plant Physiol.*, 98, 757, 1992
33. Geyh, M. A. and Scheicher, H., *Absolute Age Determination: Physical and Chemical Dating Methods and their Application*, Springer-Verlag, Heidelberg, 1990.
34. Panichi, C. and Tongiorgi, E., Carbon isotopic composition of CO_2 from springs, fumaroles mofettes and travertines of central and southern Italy: a preliminary prospections method of geothermal area, in *2nd U.N. Symposium on the Development and Use of Geothermal Resources*, Vol. 1, San Francisco, 815.
35. Farquhar, G. D., On the nature of carbon isotope discrimination in C_4 species, *Aust. J. Plant Physiol.*, 10, 205, 1983.
36. Polley, H. W., Johnson, H. W., Marino, B. D., and Mayeux, H. S., Increase in C_3 plant water-use efficiency and biomass over glacial to present CO_2 concentrations, *Nature*, 361, 61, 1993.
37. Duchi, V., Minissale, A., and Romani, L., Studio geochimico su acque e gas dell'area geotermica Lago di Vico-M. Cimini (Viterbo), *Att. Soc. Toscana. Sci. Nat. Mem. Ser.* A, 92, 237, 1985.
38. De Kok, L. J., Sulfur metabolism in plants exposed to atmospheric sulfur, in *Sulfur Nutrition and Sulfur Assimilation in Higher Plants*, Rennenberg, H., Ed., SPB Academic Publishing, The Hague, 1990, 111.
39. Maas, F. M., De Kok, L. J., Peters, J. L., and Kuiper, P. J. C., A comparative study on the effects of H_2S and SO_2 fumigation on the growth and accumulation of sulphate and sulphydryl compounds in *Trifolium pratense* L., *Glycine max* Merr. and *Phaseolus vulgaris* L., *J. Exp. Bot.*, 38, 1459, 1987.
40. Badiani, M., Paolacci, A. R., D'Annibale, A., Miglietta, F., and Raschi, A., Can rising CO_2 alleviate oxidative risk for the plant cell? Testing the hypothesis under natural CO_2 enrichment, in *Plant Responses to Elevated Carbon Dioxide: Evidence from Natural Springs*, Raschi, A., Miglietta, F., Tognetti, R., and van Gardinger, P. K., Eds., Cambridge University Press, Cambridge, 1997, 221

41. Bettarini, I., Calderoni, G., Miglietta, F., Raschi, A., and Ehleringer, J., Isotopic carbon discrimination and leaf nitrogen content of *Erica arborea* L. along a CO_2concentration gradient in a CO_2 spring in Italy, *Tree Physiol.*, 15, 327, 1997.
42. Strain, B. R. and Thomas, R. B., Anticipated effects of elevated CO_2 and climate change on plants from Mediterranean-type ecosystems utilising results of studies in other ecosystems, in *Global Changes and Mediterranean-Type Ecosystems*, Moreno, J. M. and Oechel, W. C., Eds., Springer, New York, 1995, 121.
43. Rogers, H. H., Brett Runion, G., and Krupa, S. V., Plant responses to atmospheric CO_2 enrichment with emphasis on roots and the rhizosphere, *Environ. Pollut.*, 83, 155, 1994.
44. Allen, L. H., Jr., Carbon dioxide increase: direct impacts on crops and indirect effects mediated through anticipated climatic changes, in *Physiology and Determination of Crop Yield*, Boote, K. J., Bennett, J. M., Sinclair, T. R., and Paulsen, G. M., Eds., ASA\CSSA\SSSA, Madison, WI, 1994, 425.
45. Baker, J. T. and Allen, L. H., Jr., Assessment of the impact of rising carbon dioxide and other potential climate changes in vegetation. *Environ. Pollut.*, 83, 223, 1994.
46. Long, S. P., The potential effects of concurrent increases in temperature, CO_2 and O_3 on net photosynthesis, as mediated by Rubisco, in *Plant Responses to the Gaseous Environment*, Alscher, R. G. and Wellburn, A. R., Eds., Chapman & Hall, London, 1994, 21.
47. Potvin, C., Interactive effects of temperature and atmospheric CO_2 on physiology and growth, in *Plant Responses to the Gaseous Environment*, Alscher, R. G. and Wellburn, A. R., Eds., Chapman & Hall, London, 1994, 39.
48. Sharkey, T. D., Socias, X., and Loreto, F., CO_2 effects on photosynthetic end product synthesis and feedback, in *Plant Responses to the Gaseous Environment,* Alscher, R. G. and Wellburn, A. R., Eds., Chapman & Hall, London, 1994, 55.
49. Spalding, M. H., Effects of altered carbon dioxide concentrations on gene expression in *Plant Responses to the Gaseous Environment*, Alscher, R. G. and Wellburn, A. R., Eds., Chapman & Hall, London, 1994, 79.
50. Gunderson, C. A. and Wullschleger, S. D., Photosynthetic acclimation in trees to rising atmospheric CO_2: a broader perspective, *Photosynth. Res.*, 39, 369, 1994.
51. Sage, R. F., Acclimation of photosynthesis to increasing atmospheric CO_2: the gas exchange perspective, *Photosynth. Res.*, 39, 351, 1994.
52. Wullschleger, S. D., Ziska, L. H., and Bunce, J. A., Respiratory responses of higher plants to atmospheric CO_2 enrichment, *Physiol. Plant.*, 90, 221, 1994.
53. Sage, R. F. and Reid, C. D., Photosynthetic response mechanisms to environmental change in C_3 plants, in *Plant-Environment Interactions,* Wilkinson, R. E., Ed., Marcel Dekker, New York, 1994, 413.
54. Amthor, J. S., Plant respiratory responses to the environment and their effects on the carbon balance, in *Plant-Environment Interactions*, Wilkinson, R. E., Ed., Marcel Dekker, New York, 1994, 501.
55. Owensby, C. E., Ham, J. M., Knapp, A. K., Bremer, D., and Auen, L. M., Water vapour fluxes and their impact under elevated CO_2 in a C_4-tallgrass prairie, *Global Change Biol.*, 3, 189, 1997.
56. Knapp, A. K., Fahnestock, J. T., and Owensby, C. E., Elevated atmospheric CO_2 alters stomatal responses to variable sunlight in a C_4 grass, *Plant Cell Environ.*, 17, 189, 1994.
57. Drake, B. G., Gonzàlez-Meler, M. A., and Long, S. P., More efficient plants: a consequence of rising atmospheric CO_2, *Annu. Rev. Plant Physiol. Plant Mol. Biol.*, 48, 609, 1997.
58. Liang, N. and Maruyama, K., Interactive effects of CO_2 enrichment and drought stress on gas exchange and water use efficiency in *Alnus firma*, *Environ. Exp. Bot.*, 35, 353, 1995.
59. Barnes, J. D., Ollerenshaw, J. H., and Whitfield, C. P., Effects of elevated CO_2 and/or O_3 on growth, development and physiology of wheat (*Triticum aestivum* L.), *Global Change Biol.*, 1, 129, 1995.
60. Pearson, M. and Brooks, G. L., The effect of elevated CO_2 and grazing by *Gastrophysa viridula* on the physiology and regrowth of *Rumex obtusifolius*, *New Phytol.*, 133, 605, 1996.
61. Volin, J. C. and Reich, P. B., Interaction of elevated CO_2 and O_3 on growth, photosynthesis and respiration of three perennial species grown in low and high nitrogen, *Physiol. Plant.*, 97, 674, 1996.
62. Fiscus, E. L., Reid, C. D., Miller, J. E., and Heagle, A. S., Elevated CO_2 reduces O_3 flux and O_3-induced yield losses in soybeans: possible implications for elevated CO_2 studies, *J. Exp. Bot.*, 48, 307, 1997.
63. Sicher, R. C., Kremer, D. F., and Bunce, J. A., Photosynthetic acclimation and photosynthate partitioning in soybean leaves in response to carbon dioxide enrichment, *Photosynth. Res.*, 46, 409, 1995.

64. Bunce, J. A., Wilson, K. B., and Carlson, T. N., The effect of doubled CO_2 on water use by alfalfa and orchard grass: simulating evapotranspiration using canopy conductance measurements, *Global Change Biol.*, 3, 81, 1997.
65. Picon, C., Guehl, J. M., and Ferhi, A., Leaf gas exchange and carbon isotope composition responses to drought in a drought-avoiding (*Pinus pinaster*) and a drought-tolerant (*Quercus petraea*) species under present and elevated atmospheric CO_2 concentrations, *Plant Cell Environ.*, 19, 182, 1996.
66. Bremer, D. J., Ham, J. M., and Owensby, C. E., Effect of elevated atmospheric carbon dioxide and open-top chambers on transpiration in a tallgrass prairie, *J. Environ. Qual.*, 25, 691, 1996.
67. Hibbs, D. E., Chan, S. S., Castellano, M., and Niu, C.-H., Response of red alder seedlings to CO_2 enrichment and water stress, *New Phytol.*, 129, 569, 1995.
68. Dixon, M., Le Thiec, D., and Garrec, J. P., The growth and gas exchange response of soil-planted Norway spruce [*Picea abies* (L.) Karst] and red oak (*Quercus rubra* L.) exposed to elevated CO_2 and to naturally occurring drought, *New Phytol.*, 129, 265, 1995.
69. Estiarte, M., Peñuelas, J., Kimball, B. A., Idso, S. B., LaMorte, R. L., Pinter, P. J. Jr., Wall, G. W., and Garcia, R. L., Elevated CO_2 effects on stomatal density of wheat and sour orange trees, *J. Exp. Bot.*, 45, 1665, 1994.
70. Woodward, F. I. and Kelly, C. K., The influence of CO_2 concentration on stomatal density, *New Phytol.*, 131, 311, 1995.
71. Wagner, F., Below, R., De Klerk, P., Dilcher, D. L., Joosten, H., Kürschner, W. M., and Visscher, H., A natural experiment on plant acclimation: Lifetime stomatal frequency response of an individual tree to annual atmospheric CO_2 increase, *Proc. Natl. Acad. Sci.U.S.A.*, 93, 11705, 1996.
72. Ceulemans, R., Van Praet, L., and Jiang, X. N., Effects of CO_2 enrichment, leaf position and clone on stomatal index and epidermal cell density in poplar (*Populus*), *New Phytol.*, 131, 99, 1995.
73. Scarascia-Mugnozza, G., De Angelis, P., Matteucci, G. and Kuzminsky, E., Carbon metabolism and plant growth under elevated CO_2 in a natural *Quercus ilex* L. "Macchia" stand, in *Carbon Dioxide, Populations, and Communities*, Korner, C. and Bazzaz, F. A., Eds., Academic Press, San Diego, 1996, 209.
74. Lee, E. H., Pausch, R. C., Rowland, R. A., Mulchi, C. L., and Rudorff, B. F. T., Responses of field-grown soybean (cv.Essex) to elevated SO_2 under two atmospheric CO_2 concentrations, *Environ. Exp. Bot.*, 37, 85, 1997.
75. Kimball, B. A., Pinter, P. J., Jr., Garcia, R. L., LaMorte, R. L., Wall, G. W., Hunsaker, D. J., Wechsung, G., Wechsung, F., and Kartschall, T., Productivity and water use of wheat under free-air CO_2 enrichment, *Global Change Biol.*, 1, 429, 1995.
76. Roy, J., Guillerm, J. L., Navas, M. L., and Dhillion, S., Responses to elevated CO_2 in mediterranean old-field microcosms: species, community, and ecosystem components, in *Carbon Dioxide, Populations, and Communities*, Korner, C. and Bazzaz, F. A., Eds., Academic Press, San Diego, 1996, 123.
77. Baker, J. T., Allen, L. H. Jr., Boote, K. J., and Pickering, N. B., Rice responses to drought under carbon dioxide enrichment. 2. Photosynthesis and evapotranspiration, *Global Change Biol.*, 3, 129, 1997.
78. Graham, E. A. and Nobel, P. S., Long-term effects of a doubled atmospheric CO_2 concentration on the CAM species *Agave deserti*, *J. Exp. Bot.*, 47, 61, 1996.
79. Field, C. B., Lund, C. P., Chiariello, N. R., and Mortimer, B. E., CO_2 effects on the water budget of grassland microcosm communities, *Global Change Biol.*, 3, 197, 1997.
80. Van Vuuren, M. M. I., Robinson, D., Fitter, A. H., Chasalow, S. D., Williamson, L., and Raven, J. A., Effects of elevated atmospheric CO_2 and soil water availability on root biomass, root length, and N, P and K uptake by wheat, *New Phytol.*, 135, 455, 1997.
81. Samarakoon, A. B. and Gifford, R. M., Soil water content under plants at high CO_2 concentration and interactions with the direct CO_2 effects: a species comparison, *J. Biogeogr.*, 22, 193, 1995.
82. Gorissen, A., Kuikman, P. J., and Van de Beek, H., Carbon allocation and water use in juvenile Douglas fir under elevated CO_2, *New Phytol.*, 129, 275, 1995.
83. Mayeux, H. S., Johnson, H. B., Polley, H. W., and Malone, S. R., Yield of wheat across a subambient carbon dioxide gradient, *Global Change Biol.*, 3, 269, 1997.
84. Tuba, Z., Szente, K., Nagy, Z., Csintalan, Z. and Koch, J., Responses of CO_2 assimilation, transpiration and water use efficiency to long-term elevated CO_2 in perennial C_3 xeric loess steppe species, *J. Plant Physiol.*, 148, 356, 1996.

85. Schapendonk, A. H. C. M., Dijkstra, P., Groenwold, J., Pot, C. S., and Vandegeijn, S. C., Carbon balance and water use efficiency of frequently cut *Lolium perenne* L. swards at elevated carbon dioxide, *Global Change Biol.*, 3, 207, 1997.
86. Carter, E. B., Theodorou, M. K. and Morris, P., Responses of *Lotus corniculatus* to environmental change. 1. Effects of elevated CO_2, temperature and drought on growth and plant development, *New Phytol.*, 136, 245, 1997.
87. Polley, H. W., Johnson, H. B., Mayeux, H. S., and Tischler, C. R., Are some of the recent changes in grassland communities a response to rising CO_2 concentrations?, in *Carbon Dioxide, Populations, and Communities*, Korner, C. and Bazzaz, F. A., Eds., Academic Press, San Diego, 1996, 177.
88. Ineichen, K., Wiemken, V., and Wiemken, A., Shoots, roots and ectomycorrhiza formation of pine seedlings at elevated atmospheric carbon dioxide, *Plant Cell Environ.*, 18, 703, 1995.
89. Beerling, D. J., Heath, J., Woodward, F. I., and Mansfield, T. A., Drought-CO_2 interactions in trees: observations and mechanisms, *New Phytol.*, 134, 235, 1996.
90. Korner, C., Bazzaz, F. A. and Field, C. B., The significance of biological variation, organism interactions, and life histories in CO_2 research, in *Carbon Dioxide, Populations, and Communities*, Korner, C. and Bazzaz, F. A., Eds., Academic Press, San Diego, 1996, 443.
91. Chen, S. G., Impens, I., and Ceulemans, R., Modelling the effects of elevated atmospheric CO_2 on crown development, light interception and photosynthesis of poplar in open top chambers, *Global Change Biol.*, 3, 97, 1997.
92. Wullschleger, S. D., Norby, R. J., and Hendrix, D. L., Carbon exchange rates, chlorophyll content and carbohydrate status of two forest tree species exposed to carbon dioxide enrichment, *Tree Physiol.*, 10, 21, 1992.
93. Tausz, M., De Kok, L. J., Stulen, I., and Grill, D., Physiological responses of Norway spruce trees to elevated CO_2 and SO_2, *J. Plant Physiol.*, 148, 362, 1996.
94. Mjwara, J. M., Botha, C. E. J., and Radloff, S. E., Photosynthesis, growth and nutrient changes in non-nodulated *Phaseolus vulgaris* grown under atmospheric and elevated carbon dioxide conditions, *Physiol. Plant.*, 97, 754, 1996.
95. Manderscheid, R. and Weigel, H. J., Photosynthetic and growth responses of old and modern spring wheat cultivars to atmospheric CO_2 enrichment, *Agric. Ecosyst. Environ.*, 64, 65, 1997.
96. Idso, S. B., Kimball, B. A., and Hendrix, D. L., Effects of atmospheric CO_2 enrichment on chlorophyll and nitrogen concentrations of sour orange tree leaves, *Environ. Exp. Bot.*, 36, 323, 1996.
97. Meinander, O., Somersalo, S., Holopainen, T., and Strasser, R. J., Scots pines after exposure to elevated ozone and carbon dioxide probed by reflectance spectra and chlorophyll a fluorescence transients, *J. Plant Physiol.*, 148, 229, 1996.
98. Hogan, K. P., Fleck, I., Bungard, R., Cheeseman, J. M. and Whitehead, D., Effect of elevated CO_2 on the utilization of light energy in *Nothofagus fusca* and *Pinus radiata*, *J. Exp. Bot.*, 48, 1289, 1997.
99. Roden, J. S. and Ball, M. C., Growth and photosynthesis of two eucalypt species during high temperature stress under ambient and elevated CO_2, *Global Change Biol.*, 2, 115, 1996.
100. Wang, K. Y. and Kellomaki, S., Effects of elevated CO_2 and soil-nitrogen supply on chlorophyll fluorescence and gas exchange in Scots pine, based on a branch-in-bag experiment, *New Phytol.*, 136, 277, 1997.
101. He, P., Radunz, A., Bader, D. P., and Schmid, G. H., Quantitative changes of the lipid and fatty acid composition of leaves of *Aleurites montana* as a consequence of growth under 700 ppm CO_2 in the atmosphere, *Z. Naturforsch. Teil.*, 51, 833, 1996.
102. Van Oosten, J. J. and Besford, R. T., Acclimation of photosynthesis to elevated CO_2 through feedback regulation of gene expression: climate of opinion, *Photosynth. Res.*, 48, 353, 1996.
103. Tissue, D. T. and Oechel, W. C., Response of *Eriophorum vaginatum* to elevated CO_2 and temperature in the Alaskan tussock tundra, *Ecology*, 68, 401, 1987.
104. Miglietta, F., Giuntoli, A., and Bindi, M., The effect of free air carbon dioxide enrichment (FACE) and soil nitrogen availability on the photosynthetic capacity of wheat, *Photosynth. Res.*, 47, 281, 1996.
105. Sicher, R. C. and Bunce, J. A., Relationship of photosynthetic acclimation to changes of Rubisco activity in field-grown winter wheat and barley during growth in elevated carbon dioxide, *Photosynth. Res.*, 52, 27, 1997.

106. Stirling, C. M., Davey, P. A., Williams, T. G., and Long, S. P., Acclimation of photosynthesis to elevated CO_2 and temperature in five British native species of contrasting functional type, *Global Change Biol.*, 3, 237, 1997.
107. Whitehead, S. J., Caporn, S. J. M., and Press, M. C., Effects of elevated CO_2, nitrogen and phosphorus on the growth and photosynthesis of two upland perennials: *Calluna vulgaris* and *Pteridium aquilinum*, *New Phytol.,* 135, 201, 1997.
108. Idso, S. B. and Kimball, B. A., Effects of atmospheric CO_2 enrichment on net photosynthesis and dark respiration rates of three Australian tree species, *J. Plant Physiol.*, 141, 166, 1993.
109. Clifford, S. C., Stronach, I. M., Mohamed, A. D., Azam-Ali, S. N., and Crout, N. M. J., The effects of elevated atmospheric carbon dioxide and water stress on light interception, dry matter production and yield in stands of groundnut (*Arachis hypogaea* L.), *J. Exp.Bot.*, 44, 1763, 1993.
110. Idso, S. B., Wall, G. W., and Kimball, B. A., Interactive effects of atmospheric CO_2 enrichment and light intensity reductions on net photosynthesis of sour orange tree leaves, *Environ. Exp. Bot.*, 33, 367, 1993.
111. Drake, B. G., A field study of the effects of elevated CO_2 on ecosystem processes in Chesapeake Bay, *Aust. J. Bot.*, 40, 579, 1992
112. Freeden, A. L., Koch, G. W., and Field, C. B., Effects of atmospheric CO_2 enrichment on ecosystem CO_2 exchange in a nutrient and water limited grassland, *J. Biogeogr.*, 22, 215, 1995.
113. Downton, W. J. S. and Grant, W. J. R., Photosynthetic and growth responses of variegated ornamental species to elevated CO_2, *Aust. J. Plant Physiol.,* 21, 273, 1994.
114. Vogel, C. S. and Curtis, P. S., Leaf gas exchange and nitrogen dynamics of N_2-fixing, field-grown *Alnus glutinosa* under elevated atmospheric CO_2, *Global Change Biol.*, 1, 55, 1995.
115. Sonesson, M., Callaghan, T. V., and Carlsson, B. A., Effects of enhanced ultraviolet radiation and carbon dioxide concentration on the moss *Hylocomium splendens*, *Global Change Biol.*, 2, 67, 1996.
116. Schaffer, B., Searle, C., Whiley, A. W., and Nissen, R. J., Effects of atmospheric CO_2 enrichment and root restriction on leaf gas exchange and growth of banana (*Musa*), *Physiol. Plant.*, 97, 685, 1996.
117. Hibberd, J. M., Richardson, P., Whitebread, R., and Farrar, J. P., Effects of leaf age, basal meristem and infection with powdery mildew on photosynthesis in barley grown in 700 μmol mol^{-1} CO_2, *New Phytol.*, 134, 317, 1996.
118. Balaguer, L., Barnes, J. D., Panicucci, A., and Borland, A. M., Production and utilization of assimilates in wheat (*Triticum aestivum* L.) leaves exposed to elevated O_3 and/or CO_2, *New Phytol.*, 129, 557, 1995.
119. Visser, A. J., Tosserams, M., Groen, M. W., Magendans, G. W. H., and Rozema, J., The combined effects of CO_2 concentration and solar UV-B radiation on fava bean grown in open-top chambers, *Plant Cell Environ.*, 20, 189, 1997.
120. Van Oosten, J. J. and Besford, R. T., Some relationships between the gas exchange, biochemistry and molecular biology of photosynthesis during leaf development of tomato plants after transfer to different carbon dioxide concentrations, *Plant Cell Environ.*, 18, 1253, 1995.
121. Garcia, R. L., Idso, S. B., Wall, G. W., and Kimball, B. A., Changes in net photosynthesis and growth of *Pinus eldarica* seedlings in response to atmospheric CO_2 enrichment, *Plant Cell Environ.*, 17, 971, 1994.
122. Tissue, D. T., Thomas, R. B., and Strain, B. R., Atmospheric CO_2 enrichment increases growth and photosynthesis of *Pinus taeda*: a 4 year experiment in the field, *Plant Cell Environ.*, 20, 1123, 1997.
123. Sullivan, J. H. and Teramura, A. H., The effects of UV-B radiation on loblolly pine. 3. Interaction with CO_2 enhancement, *Plant Cell Environ.*, 17, 311, 1994.
124. Pinter, P. J., Jr., Kimball, B. A., Garcia, R. L., Wall, G. W., Hunsaker, D. J., and LaMorte, R. L., Free-air CO_2 enrichment: responses of cotton and wheat crops, in *Carbon Dioxide and Terrestrial Ecosystems*, Koch, G. W. and Mooney, H. A., Eds., Academic Press, San Diego, 1996, 215.
125. Pearson, M. and Brooks, G. L., The influence of elevated CO_2 on growth and age-related changes in leaf gas exchange, *J. Exp. Bot.*, 46, 1651, 1995.
126. Nijs, I., Behaeghe, T., and Impens, I., Leaf nitrogen content as a predictor of photosynthetic capacity in ambient and global change conditions, *J. Biogeogr.,* 22, 177, 1995.
127. Curtis, P. S., Vogel, C. S., Pregitzer, K. S., Zak, D. R. and Teeri, J. A., Interacting effects of soil fertility and atmospheric CO_2 on leaf area growth and carbon gain physiology in *Populus* x *euramericana* (Dode) Guinier, *New Phytol.*, 129, 253, 1995.

128. Bowler, J. M. and Press, M. C., Effects of elevated CO_2, nitrogen form and concentration on growth and photosynthesis of a fast- and slow-growing grass, *New Phytol.*, 132, 391, 1996
129. Lewis, J. D., Tissue, D. T., and Strain, B. R., Seasonal response of photosynthesis to elevated CO_2 in loblolly pine (*Pinus taeda* L.) over two growing seasons, *Global Change Biol.*, 2, 103, 1996.
130. Wolfenden, J. and Diggle, P. J., Canopy gas exchanges and growth of upland pasture swards in elevated CO_2, *New Phytol.*, 130, 369, 1995.
131. McKee, I. F., Farage, P. K., and Long, S. P., The interactive effects of elevated CO_2 and O_3 concentration on photosynthesis in spring wheat, *Photosynth. Res.*, 45, 111, 1995.
132. Faria, T., Wilkins, D., Besford, R. T., Vaz, M., Pereira, J. S., and Chaves, M. M., Growth at elevated CO_2 leads to down-regulation of photosynthesis and altered response to high temperature in *Quercus suber* L seedlings, *J. Exp. Bot.*, 47, 1755, 1996.
133. Jackson, R. B., Luo, Y., Cardon, Z. G., Sala, O. E., Field, C. B,. and Mooney, H. A., Photosynthesis, growth and density for the dominant species in a CO_2-enriched grassland, *J. Biogeogr.,* 22, 221, 1995.
134. Ceulemans, R., Taylor, G., Bosac, C., Wilkins, D., and Besford, R. T., Photosynthetic acclimation to elevated CO_2 in poplar grown in glasshouse cabinets or in open top chambers depends on duration of exposure, *J. Exp.Bot.*, 48, 1681, 1997.
135. Will, R. E. and Ceulemans, R., Effects of elevated CO_2 concentration on photosynthesis, respiration and carbohydrate status of coppice *Populus* hybrids, *Physiol. Plant.*, 100, 933, 1997.
136. Barrett, D. J. and Gifford, R. M., Photosynthetic acclimation to elevated CO_2 in relation to biomass allocation in cotton, *J. Biogeogr.*, 22, 331, 1995.
137. Riviere-Rolland, H., Contard, P., and Betsche, T., Adaptation of pea to elevated atmospheric CO_2: Rubisco, phosphoenolpyruvate carboxylase and chloroplast phosphate translocator at different levels of nitrogen and phosphorus nutrition, *Plant Cell Environ.*, 19, 109, 1996.
138. Pushnik, J. C., Demaree, R. S., Houpis, J. L. J., Flory, W. B., Bauer, S. M., and Anderson, P. D., The effect of elevated carbon dioxide on a Sierra-Nevadan dominant species: *Pinus ponderosa*, *J. Biogeogr.*, 22, 249, 1995.
139. Pritchard, S. G., Peterson, C. M., Prior, S. A., and Rogers, H. H., Elevated atmospheric CO_2 differentially affects needle chloroplast ultrastructure and phloem anatomy in *Pinus palustris*: interactions with soil resource availability, *Plant Cell Environ.*, 20, 461, 1997.
140. Medlyn, B. E., The optimal allocation of nitrogen within the C_3 photosynthesis system at elevated CO_2, *Aust. J. Plant Physiol.,* 23, 593, 1996.
141. Van Oosten, J.-J. and Besford, R. T., Sugar feeding mimics effect of acclimation to high CO_2-rapid down regulation of RuBisCO small subunit transcripts but not of the large subunit transcripts, *J. Plant Physiol.*, 143, 306, 1994.
142. Webber, A. N., Nie, G.-Y., and Long, S. P., Acclimation of photosynthetic proteins to rising atmospheric CO_2, *Photosynth. Res.,* 39, 413, 1994.
143. Nie, G. Y., Long, S. P., Garcia, R. L., Kimball, B. A., LaMorte, R. L., Pinter, P. J. Jr., Wall, G. W., and Webber, A. N., Effects of free-air CO_2 enrichment on the development of the photosynthetic apparatus in wheat, as indicated by changes in leaf proteins, *Plant Cell Environ.*, 18, 855, 1995.
144. Balaguer, L., Valladares, F., Ascaso, C., Barnes, J. D., De Los Rios, A., Manrique, E., and Smith, E. C., Potential effects of rising tropospheric concentrations of CO_2 and O_3 on green-algal lichens, *New Phytol.*, 132, 642, 1996.
145. Van Oosten, J.-J., Dizengremel, P., Laitat, E., and Impens, R., Too much of a good thing? Long-term exposure to elevated CO_2 decreases carboxylating and photorespiratory enzymes and increases respiratory enzymes in spruce, in *Interacting Stresses on Plants in a Changing Climate*, Jackson, M. B. and Black, C. R., Eds., Springer-Verlag, Berlin, 1993, 185.
146. Sicher, R. C. and Kremer, D. F., Rubisco activity is altered in a starchless mutant of *Nicotiana sylvestris* grown in elevated carbon dioxide, *Environ. Exp. Bot.*,36, 385, 1996.
147. Thibaud, M. C., Cortez, N., Rivière, H., and Betsche, T., Photorespiration and related enzymes in pea (*Pisum sativum*) grown in high CO_2, *J. Plant Physiol.*, 146, 596, 1995.
148. Woodrow, I. E., Optimal acclimation of the C_3 photosynthetic system under enhanced CO_2, *Photosynth. Res.*, 39, 401, 1994.
149. Jacob, J., Greitner, C., and Drake, B. G., Acclimation of photosynthesis in relation to rubisco and nonstructural carbohydrate contents and *in situ* carboxylase activity in *Scirpus olneyi* grown at elevated CO_2 in the field, *Plant Cell Environ.*, 18, 875, 1995.

150. Schwanz, P., Häberle, K. H., and Polle, A., Interactive effects of elevated CO_2, ozone and drought stress on the activities of antioxidative enzymes in needles of Norway spruce trees (*Picea abies*, [L] Karst.) grown with luxurious N-supply, J. *Plant Physiol.*, 148, 351, 1996.
151. Schwanz, P., Picon, C., Vivin, P., Dreyer, E., Guehl, J. M., and Polle, A., Responses of antioxidative systems to drought stress in pendunculate oak and maritime pine as modulated by elevated CO_2, *Plant Physiol.*, 110, 393, 1996.
152. Polle, A., Eiblmeier, M., Sheppard, L., and Murray, M., Responses of antioxidative enzymes to elevated CO_2 in leaves of beech (*Fagus sylvatica* L.) seedlings grown under a range of nutrient regimes, *Plant Cell Environ.*, 20, 1317, 1997.
153. Badiani, M., D'Annibale, A., Paolacci, A. R., Miglietta, F., and Raschi, A., The antioxidants status of soybean (*Glycine max* Merrill) leaves grown under natural CO_2 enrichment in the field, *Aust. J. Plant Physiol.*, 20, 275, 1993.
154. Poorter, H., Gifford, R. M., Kriedemann, P. E., and Chin Wong, S., A quantitative analysis of dark respiration and carbon content as factors in the growth response of plants to elevated CO_2, *Aust. J. Bot.*, 40, 501, 1992.
155. Hirose, T., Ackerly, D. D., and Bazzaz, F. A., CO_2 elevation and canopy development in stands of herbaceous plants, in *Carbon Dioxide, Populations, and Communities*, Korner, C. and Bazzaz, F. A., Eds., Academic Press, San Diego, 1996, 413.
156. Amthor, J. S., Plant respiratory responses to elevated carbon dioxide partial pressure, in *Advances in Carbon Dioxide Effects Research*, Allen, L. H., Jr., Kirkham, M. B., Olszyk, D. M., and Whitman, C. E., Eds., ASA\CSSA\SSSA, Madison, WI, 1997, 35.
157. Seneweera, S., Milham, P., and Conroy, J. P., Influence of elevated CO_2 and phosphorus nutrition on the growth of a short-duration rice (*Oryza sativa* L. cv. Jarrah), *Aus. J. Plant Physiol.*, 21, 281, 1994.
158. Rowland-Bamford, A. J., Baker, J. T., Allen, L. H., Jr., and Bowes, G., Interactions of CO_2 enrichment and temperature on carbohydrate accumulation and partitioning in rice, *Environ. Exp. Bot.*, 36, 111, 1996.
159. Hibberd, J. M., Whitbread, R., and Farrar, J. F., Carbohydrate metabolism in source leaves of barley grown in 700 μmol mol^{-1} CO_2 and infected with powdery mildew, *New Phytol.*, 133, 659, 1996.
160. Fangmeier, A., Gruters, U., Hogy, P., Vermehren, B., and Jager, H. J., Effects of elevated CO_2 nitrogen supply and tropospheric ozone on spring wheat. 2. Nutrients (N, P, K, S, C_a, Mg, Fe, Mn, Zn), *Environ. Pollut.*, 96, 43, 1997.
161. Cui, M. and Nobel, P. S., Gas exchange and growth responses to elevated CO_2 and light levels in the CAM species *Opuntia ficus-indica*, *Plant Cell Environ.*, 17, 935, 1994.
162. Idso, S. B., Kimball, B. A., and Hendrix, D. L., Air temperature modifies the size-enhancing effects of atmospheric CO_2 enrichment on sour orange tree leaves, *Environ. Exp. Bot.*, 33, 293, 1993.
163. Lindroth, R. L., Roth, S., Kruger, E. L., Volin, J. C., and Koss, P. A., CO_2-mediated changes in aspen chemistry: effects on gypsy moth performance and susceptibility to virus, *Global Change Biol.*, 3, 279, 1997.
164. Atkinson, C. J. and Taylor, J. M., Effects of elevated CO_2 on stem growth, vessel area and hydraulic conductivity of oak and cherry seedlings, *New Phytol.*, 133, 617, 1996.
165. Bunce, J. A., Growth at elevated carbon dioxide concentration reduces hydraulic conductance in alfalfa and soybean, *Global Change Biol.*, 2, 155, 1996.
166. Bassirirad, H., Reynolds, J. F., Virginia, R. A., and Brunelle, M. H., Growth and root NO^-_3 and PO^{3-}_4 uptake capacity of three desert species in response to atmospheric CO_2 enrichment, *Aust. J. Plant Physiol.*, 24, 353, 1997.
167. Paterson, E., Hall, J. M., Rattray, E. A. S., Griffiths, B. S., Ritz, K., and Killham, K., Effect of elevated CO_2 on rhizosphere carbon flow and soil microbial processes, *Global Change Biol.*, 3, 363, 1997.
168. Van Ginkel, J. H., Gorissen, A., and Van Veen, J. A., Carbon and nitrogen allocation in *Lolium perenne* in response to elevated atmospheric $CO_2$with emphasis on soil carbon dynamics, *Plant Soil*, 188, 299, 1997.
169. Seegmüller, S., Schulte, M., Herschbach, C. and Rennenberg, H., Interactive effects of mycorrhization and elevated atmospheric CO_2 on sulphur nutrition of young pedunculate oak (*Quercus robur* L) trees, *Plant Cell Environ.*, 19, 418, 1996.
170. Penuelas, J., Idso, S. B., Ribas, A., and Kimball, B. A., Effects of long-term atmospheric CO_2 enrichment on the mineral concentration of *Citrus aurantium* leaves, *New Phytol.*, 135, 439, 1997.

171. Ceulemans, R., Jiang, X. N., and Shao, B. Y., Effects of elevated atmospheric CO_2 on growth, biomass production and nitrogen allocation of two *Populus* clones, *J. Biogeogr.*, 22, 261, 1995.
172. Conroy, J. and Hocking, P., Nitrogen nutrition of C_3 plants at elevated atmospheric CO_2 concentrations, *Physiol. Plant.*, 89, 570, 1993.
173. Lindroth, R. L. Consequences of elevated atmospheric $CO_2$for forest insects, in *Carbon Dioxide, Populations, and Communities*, Korner, C. and Bazzaz, F. A., Eds., Academic Press, San Diego, 1996, 347.
174. Conroy, C. P., Influence of elevated atmospheric CO_2 concentrations on plant nutrition, *Aust. J. Bot.*, 40, 445, 1992.
175. Bernston, G. M. and Bazzaz, F. A., Nitrogen cycling in microcosms of yellow birch exposed to elevated CO_2: simultaneous positive and negative below-ground feedbacks, *Global Change Biol.*, 3, 247, 1997.
176. Pfirrmann, T., Barnes, J. D., Steiner, K., Schramel, P., Busch, U., Kuchenhoff, H., and Payer, H.-D., Effects of elevated CO_2, O_3 and K deficiency on Norway spruce (*Picea abies*): nutrient supply, content and leaching, *New Phytol.*, 134, 267, 1996.
177. Behboudian, M. H. and Tod, C., Postharvest attributes of 'Virosa' tomato fruit produced in an enriched carbon dioxide environment, *Hortic. Sci.*, 30, 490, 1995.
178. Chin Wong, S., Kriedemann, P. E., and Farquhar, G. D., CO_2 x nitrogen interaction on seedling growth of four species of eucalypt, *Aust J. Bot.,* 40, 457, 1992.
179. Lewis, J. D. and Strain, B. R., The role of mycorrhizas in the response of *Pinus taeda* seedlings to elevated CO_2, *Global Change Biol.*, 133, 431, 1996.
180. Fangmeier, A., Gruters, U., Hertstein, U., Sandhage-Hofmann, A., Vermehren, B. and Jäger, H. -J., Effects of elevated CO_2, nitrogen supply and tropospheric ozone on spring wheat. I. Growth and yield, *Environ. Pollut.*, 91, 381, 1996.
181. Prior, S. A., Rogers, H. H., Runion, G. B., Torbert, H. A., and Reicosky, D. C., Carbon dioxide-enriched agroecosystems: influence of tillage on short-term soil carbon dioxide effluex, *J. Environ. Qual.*, 26, 244, 1997.
182. Ball, A. S., Microbial decomposition at elevated CO_2 levels: effect of litter quality, *Global Change Biol.,* 3, 379, 1997.
183. Prior, S. A. and Rogers, H. H., Soybean growth response to water supply and atmospheric carbon dioxide enrichment, *J. Plant Nutr.*, 18, 617, 1995.
184. Newbery, R. M. and Wolfenden, J., Effects of elevated CO_2 and nutrient supply on the seasonal growth and morphology of *Agrostis capillaris*, *New Phytol.,* 132, 403, 1996.
185. Reinert, R. A. and Ho, M. C., Vegetative growth of soybean as affected by elevated carbon dioxide and ozone, *Environ. Pollut.*, 89, 89, 1995.
186. Mortensen, L. M., Effect of carbon dioxide concentration on biomass production and partitioning in *Betula pubescens* Ehrh. seedlings at different ozone and temperature regimes, *Environ. Pollut.*, 87, 337, 1995.
187. Gardner, S. D., Taylor, G., and Bosac, C., Leaf growth of hybrid poplar following exposure to elevatd CO_2, *New Phytol.*, 131, 81, 1995.
188. Reddy, K. R., Hodges, H. F., and McKinion, J. M., Carbon dioxide effect on pima cotton growth, *Agric. Ecosyst. Environ.*, 54, 17, 1995.
189. Roumet, C., Bel, M. P., Sonie, L., Jardon, F. and Roy, J., Growth response of grasses to elevated CO_2: a physiological plurispecific analysis, *New Phytol.*, 133, 595, 1996.
190. Weigel, H. J., Manderscheid, R., Jäger, H.-J., and Mejer, G. J., Effects of season-long CO_2 enrichment on cereals. I. Growth performance and yield, *Agric. Ecosyst. Environ.*, 48, 231, 1994.
191. Rawson, H. M., Plant responses to temperature under conditions of elevated CO_2, *Aust. J. Bot.*, 40, 473, 1992.
192. Long, S. P., Modification of the response of photosynthetic productivity to rising temperature by atmospheric CO_2 concentrations: has its importance been underestimated? *Plant Cell Environ.*, 14, 729, 1991.
193. Mulholland, B. J., Craigon, J., Black, C. R., Colls, J. J., Atherton, J., and Landon, G., Effects of elevated carbon dioxide and ozone on the growth and yield of spring wheat (*Triticum aestivum* L), *J. Exp. Bot.*, 48, 113, 1997.
194. Gorissen, A., Elevated CO_2 evokes quantitative and qualitative changes in carbon dynamics in a plant/soil system: mechanisms and implications, *Plant Soi*l, 187, 289, 1996.

195. Prior, S. A., Pritchard, S. G., Runion, G. B., Rogers, H. H., and Mitchell, R. J., Influence of atmospheric CO_2 enrichment, soil N, and water stress on needle surface wax formation in *Pinus palustris* (Pinaceae), *Am. J. Bot.*, 84, 1070, 1997.
196. Hibberd, J. M., Whitebread, R., and Farrar, J. F., Effect of 700 μmol mol^{-1} CO_2 and infection with powdery mildew on the growth and carbon partitioning of barley, *New Phytol.*, 134, 309, 1996.
197. Stulen, I. and den Hertog, J., Root growth and functioning under atmospheric CO_2 enrichment, *Vegetatio*, 104/105, 99, 1993.
198. Lambers, H., Stulen, I., and Van der Werf, A., Carbon use in root respiration as affected by elevated atmospheric CO_2, *Plant Soil*, 187, 251, 1996.
199. Jongen, M., Jones, M. B., Hebeisen, T., Blum, H., and Hendrey, G., The effects of elevated CO_2 concentrations on the root growth of *Lolium perenne* and *Trifolium repens* in a FACE system, *Global Change Biol.*, 1, 361, 1995.
200. Wechsung, G., Wechsung, F., Wall, G. W., Adamsen, F. J., Kimball, B. A., Garcia, R. L., Pinter, P. J., Jr., and Kartschall, T., Biomass and growth rate of a spring wheat root system grown in free-air CO_2 enrichment (FACE) and ample soil moisture, *J. Biogeogr.*, 22, 623, 1995.
201. Van Ginkel, J. H., Gorissen, A., and Van Veen, J. A., Long-term decomposition of grass roots as affected by elevated atmospheric carbon dioxide, *J. Environ. Qual.*, 25, 1122, 1996.
202. Griffin, K. L., Bashkin, M. A., Thomas, R. B., and Strain, B. R., Interactive effects of soil nitrogen and atmospheric carbon dioxide on root/rhizosphere carbon dioxide efflux from loblolly and ponderosa pine seedlings, *Plant Soil*, 190, 11, 1997.
203. Day, F. P., Weber, E. P., Hinkle, C. R., and Drake, B. G., Effects of elevated atmospheric CO_2 on fine root length and distribution in an oak-palmetto scrub ecosystem in central Florida, *Global Change Biol.*, 2, 143, 1996.
204. Tingey, D. T., Phillips, D. L., Johnson, M. G., Storm, M. J., and Ball, J. T., Effects of elevated CO_2 and N fertilization on fine root dynamics and fungal growth in seedling *Pinus ponderosa*, *Environ. Exp. Bot.*, 37, 73, 1997.
205. Tate, K. R. and Ross, D. J., Elevated CO_2 and moisture effects on soil carbon storage and cycling in temperate grasslands, *Global Change Biol.*, 3, 225, 1997.
206. Sadowsky, M. J. and Schortemeyer, M., Soil microbial responses to increased concentrations of atmospheric CO_2, *Global Change Biol.*, 3, 217, 1997.
207. Tingey, D. T., Johnson, M. G., Phillips, D., and Storm, M. J., Effects of elevated CO_2 nitrogen on ponderosa pine fine roots and associated fungal components, *J. Biogeogr.*, 22, 281, 1995.
208. Fitter, A. H., Self, G. K., Wolfenden, J., Van Vuuren, M. M. I., Brown, T. K., Williamson, L., Graves, J. D., and Robinson, D., Root production and mortality under elevated atmospheric carbon dioxide, *Plant Soil*, 187, 299, 1996.
209. Pregitzer, K. S., Zak, D. R., Curtis, P. S., Kubiske, M. E., Teeri, J. A., and Vogel, C. S., Atmospheric CO_2, soil nitrogen and turnover of fine roots, *New Phytol.*, 129, 579, 1995.
210. Ross, D. J., Saggar, S., Tate, K. R., Feltham, C. W., and Newton, P. C. D., Elevated CO_2 effects on carbon and nitrogen cycling in grass/clover turves of a Psammaquent soil, *Plant Soil*, 182, 185, 1996.
211. Gwynn-Jones, D., Björn, L. O., Callaghan, T. V., Gehrke, C., Johanson, U., Lee, J. A., and Sonesson, M., Effects of enhanced UV-B radiation and elevated concentrations of CO_2 on a subarctic heathland, in *Carbon Dioxide, Populations, and Communities*, Korner, C. and Bazzaz, F. A., Eds., Academic Press, San Diego, 1996, 197.
212. Leadley, P. W. and Stocklin, J., Effects of elevated CO_2 on model calcareous grasslands: community, species, and genotype level responses, *Global Change Biol.*, 2, 389, 1996.
213. Ackerly, D. D. and Bazzaz, F. A., Plant growth and reproduction along CO_2 gradients: nonlinear responses and implications for community change, *Global Change Biol.*, 1, 199, 1995.
214. Körner, C. and Bazzaz, F. A., Preface, in *Carbon Dioxide, Populations, and Communities*, Korner, C. and Bazzaz, F. A., Eds., Academic Press, San Diego, 1996, xix.
215. Schenk, U., Jäger, H.-J., and Weigel, H. J., The response of perennial ryegrass/white clover swards to elevated CO_2 concentrations. 1. Effects on competition and species composition and interaction with N supply, *New Phytol.*, 135, 67, 1997.
216. Grime, J. P., Vegetation classification by reference to strategies, *Nature*, 250, 26, 1974.
217. Wolf, J., Effects of nutrient supply (NPK) on spring wheat response to elevated atmospheric CO_2, *Plant Soil*, 185, 113, 1996.

218. Poorter, H., Roumet, C., and Campbell, B. D., Interspecific variation in the growth response of plants to elevated CO_2: a search for functional types, in *Carbon Dioxide, Populations, and Communities*, Korner, C. and Bazzaz, F. A., Eds., Academic Press, San Diego, 1996, 375.
219. Baker, J. T., Allen, L. H., Jr., Boote, K. J., and Pickering, N. B., Rice responses to drought under carbon dioxide enrichment. 1. Growth and yield, *Global Change Biol.*, 3, 119, 1997.
220. Idso, S. B. and Kimball, B. A., Effects of atmospheric CO_2 enrichment on biomass accumulation and distribution in Eldarica pine trees, *J. Exp. Bot.*, 45, 1669, 1994.
221. Idso, S. B. and Kimball, B. A., Effects of atmospheric CO_2 enrichment on regrowth of sour orange trees (*Citrus aurantium*; Rutaceae) after coppicing, *Am .J. Bot.*, 81, 843, 1994.
222. Amthor, J. S., Mitchell, R. J., Runion, G. B., Rogers, H. H., Prior, S. A., and Wood, C. W., Energy content, construction cost and phytomass accumulation of *Glycine max* (L.) Merr. and *Sorghum bicolor* (L.) Moench grown in elevated CO_2 in the field, *New Phytol.*, 128, 443, 1995.
223. Rudorff, B. F. T., Mulchi, C. L., Lee, E. H., Rowland, R., and Pausch, R., Effects of enhanced O_3 and CO_2 enrichment on plant characteristics in wheat and corn, *Environ. Pollut.*, 94, 53, 1996.
224. Badger, M., Manipulating agricultural plants for a future high CO_2 environment, *Aust. J. Bot.*, 40, 421, 1992.
225. Norby, R. J., Wullschleger, S. D., Gunderson, C. A., and Nietch, C. T., Increased growth effiiency of *Quercus alba* trees in a CO_2-enriched atmosphere, *New Phytol.*, 131, 91, 1995.
226. Roumet, C. and Roy, J., Prediction of the growth response to elevated CO_2: a search for physiological criteria in closely related grass species, *New Phytol.*, 134, 615, 1996.
227. Clark, H., Newton, P. C. D., Bell, C. C., and Glasgow, E. M., Dry matter yield, leaf growth and population dynamics in *Lolium perenne/Trifolium repens*-dominated pasture turves exposed to two levels of elevated CO_2, *J. App. Ecol.*, 34, 304, 1997.
228. Campbell, B. D. and Hart, A. L., Competition between grasses and *Trifolium repens* with elevated atmospheric CO_2, in *Carbon Dioxide, Populations, and Communities*, Korner, C. and Bazzaz, F. A., Eds., Academic Press, San Diego, 1996, 301.
229. Grime, J. P., Hodgson, J. G., and Hunt, R., *Comparative Plant Ecology: A Functional Approach to Common British Species*, Unwin Hyman, London, 1988.
230. Lee, H. S. J. and Jarvis, P. G., Trees differ from crops and from each other in their responses to increases in CO_2 concentration, *J. Biogeogr.*, 22, 323, 1995.
231. Hartwig, U. A., Zanetti, S., and Hebeisen, T., Symbiotic nitrogen fixation: one key to understand the response of temperate grassland ecosystems to elevated CO_2?, in *Carbon Dioxide, Populations, and Communities*, Korner, C. and Bazzaz, F. A., Eds., Academic Press, San Diego, 1996, 253.
232. Stulen, I., Den Hertog, J., and Jansen, K., The influence of atmospheric CO_2 enrichment on allocation patterns of carbon and nitrogen in plants from natural vegetations, in *Photosynthesis: Photoreactions to Plant Productivity*, Abrol, Y. P., Mohanty, P., and Govindjee, P., Eds., Kluwer Academic, Dordrecht, the Netherlands, 1993, 509.
233. Nijs, I. and Impens, I., An analysis of the balance between root and shoot activity in *Lolium perenne* v. Melvina. Effects of CO_2 concentration and air temperature, *New Phytol.*, 137, 81, 1997.
234. Hebeisen, T., Luscher, A., Zanetti, S., Fischer, B. U., Hartwig, U. A., Frehner, M., Hendrey, G. R., Blum, H., and Nosberger, J., Growth response of *Trifolium repens* L. and *Lolium perenne* L. as monocultures and bi-species mixture to free air CO_2enrichment and management, *Global Change Biol.*, 3, 149, 1997.
235. Idso, S. B. and Kimball, B. A., Effects of long-term atmospheric CO_2 enrichment on the growth and fruit production of sour orange trees, *Global Change Biol.*, 3, 89, 1997.
236. Saebo, A. and Mortensen, L. M., Growth, morphology and yield of wheat, barley and oats grown at elevated atmospheric CO_2 concentration in a cool, maritime climate, *Agric. Ecosyst. Environ.*, 57, 9, 1996.
237. Fonseca, F., Den Hertog, J., and Stulen, I., The response of *Plantago major* ssp. *pleiosperma* to elevated CO_2 is modulated by the formation of secondary roots, *New Phytol.*, 133, 627, 1996.
238. Townend, J., Effects of elevated CO_2 water and nutrients on *Picea sitchensis* (Bong.) Carr. seedlings, *New Phytol.*, 130, 193, 1995.
239. Reekie, E. G., The effect of elevated CO_2 on developmental processes and its implications for plant-plant interactions, in *Carbon Dioxide, Populations, and Communities*, Korner, C. and Bazzaz, F. A., Eds., Academic Press, San Diego, 1996, 333.

240. Conroy, J. P., Seneweera, S., Basra, A. S., Rogers, G., and Nissen-Wooller, B., Influence of rising atmospheric CO_2 concentrations and temperature on growth, yield and grain quality of cereal crops, *Aust. J. Plant Physiol.*, 21, 741, 1994.
241. Hocking, P. J. and Meyer, C. P., Carbon dioxide enrichment decreases critical nitrate and nitrogen concentrations in wheat, *J. Plant Nutr.*, 14, 571, 1991.
242. Hocking, P. J. and Meyer, C. P., Effects of CO_2 enrichment and nitrogen stress on growth, partitioning of dry matter and nitrogen in wheat and maize, *Aust. J. Plant Physiol.*, 18, 339, 1991.
243. Rudorff, B. F. T., Mulchi, C. L., Fenny, P., Lee, E. H. and Rowland, R., Wheat grain quality under enhanced tropospheric CO_2 and O_3 concentrations, *J. Environ. Qual.*, 25, 1384, 1996.
244. Newton, P. C. D., Clark, H., Bell, C. C., Glasgow, E. M., Tate, K. R., Ross, D. J., Yeates, G. W., and Saggar, S., Plant growth and soil processes in temperate grassland communities at elevated CO_2, *J. Biogeogr.*, 22, 235, 1995.
245. Lüscher, A., Hebeisen, T., Zanetti, S., Hartwig, U. A., Blum, H., Hendrey, G. R., and Nörsberger, J., Differences between legumes and nonlegumes of permanent grassland in their responses to free-air carbon dioxide enrichment: its effect on competition in a multispecies mixture, in *Carbon Dioxide, Populations, and Communities*, Korner, C. and Bazzaz, F. A., Eds., Academic Press, San Diego, 1996, 287.
246. Teughels, H., Nijs, I., Van Hecke, P., and Impens, I., Competition in a global change environment: the importance of different plant traits for competitive success, *J. Biogeogr.*, 22, 297, 1995.
247. Reynolds, H., Effects of elevated CO_2 on plants grown in competition, in *Carbon Dioxide, Populations, and Communities*, Korner, C. and Bazzaz, F. A., Eds., Academic Press, San Diego, 1996, 273.
248. Clark, H., Newton, P. C. D., Bell, C. C., and Glasgow, E. M., The influence of elevated CO_2 and simulated seasonal changes in temperature on tissue turnover in pasture turves dominated by perennial rye grass (*Lolium perenne*) and white clover (*Trifolium repens*), *J. Appl. Ecol.*, 32, 128, 1995.
249. Bazzaz, F. A. and McConnaughay, K. D. M., Plant-plant interactions in elevated CO_2 environments, *Aust. J. Bot.*, 40, 547, 1992.
250. Watling, J. R. and Press, M. C., How is the relationship between the C_4 cereal *Sorghum bicolor* and the C_3 root hemi-parasites *Striga hermonthica* and *Striga asiatica* affected by elevated CO_2? *Plant Cell Environ.*, 20, 1292, 1997.
251. Manning, W. J. and Tiedemann, V. A., Climate change: potential effects of increased atmospheric carbon dioxide, ozone, and ultraviolet-B radiation on plant disease, *Environ. Pollut.*, 88, 219, 1995.
252. Canadell, J. G., Pitelka, L. F., and Ingram, J. S. I., The effects of elevated CO_2 on plant-soil carbon below-ground: a summary and synthesis, *Plant Soil*, 187, 391, 1996.
253. Ringelberg, D. B., Stair, J. O., Almeida, J., Norby, R. J., O'Neill, E. G., and White, D. C., Consequences of rising atmospheric carbon dioxide levels for the belowground microbiota associated with white oak, *J. Environ. Qual.*, 26, 495, 1997.
254. Zanetti, S., Hartwig, U. A., Lüscher, A., Hebeisen, T., Frehner, M., Fischer, B. U., Hendrey, G. R., Blum, H., and Nösberger, J., Stimulation of symbiotic N_2 fixation in *Trifolium repens* L under elevated atmospheric pCO_2 in a grassland ecosystem, *Plant Physiol.*, 112, 575, 1996.
255. Sanders, I. R., Plant-fungal interactions in a CO_2-rich world, in *Carbon Dioxide, Populations, and Communities*, Korner, C. and Bazzaz, F. A., Eds., Academic Press, San Diego, 1996, 265.
256. Newton, P. C. D., Clark, H., Bell, C. C., and Glasgow, E. M., Interaction of soil moisture and elevated CO_2 on the above-ground growth rate, root length density and gas exchange of turves from temperate pasture, *J. Exp. Bot.*, 47, 771, 1996.
257. Moorhead, D. L. and Linkins, A. E., Elevated CO_2 alters belowground exoenzyme activities in tussock tundra, *Plant Soil*, 189, 321, 1997.
258. Arnone, J. A. III, Indices of plant N availability in an alpine grassland under elevated atmospheric CO_2, *Plant Soil*, 190, 61, 1997.
259. Rouhier, H., Billes, G., El Kohen, A., Mousseau, M., and Bottner, P., Effect of elevated CO_2 on carbon and nitrogen distribution within a tree (*Castanea sativa* Mill.)-soil system, *Plant Soil*, 162, 281, 1994.
260. Boerner, R. E. J. and Rebbeck, J., Decomposition and nitrogen release from leaves of three hardwood species grown under elevated O_3 and/or CO_2, *Plant Soil*, 170, 149, 1995.
261. Zak, D. R., Pregitzer, K. S., Curtis, P. S., Teeri, J. A., Fogel, R., and Randlett., D. L., Elevated atmospheric CO_2 and feedback between carbon and nitrogen cycles, *Plant Soil*, 151, 105, 1993.

262. Ziska, L. H. and Bunce, J. A., The role of temperature in determining the stimulation of CO_2 assimilation at elevated carbon dioxide concentration in soybean seedlings, *Physiol. Plant.*, 100, 126, 1997.
263. Lin, W. H., Ziska, L. H., Namuco, O. S., and Bai, K., The interaction of high temperature and elevated CO_2 on photosynthetic acclimation of single leaves of rice *in situ*, *Physiol. Plant.*, 99, 178, 1997.
264. Campbell, B. D., Laing, W. A., Greer, D. H., Crush, J. R., Clark, H., Williamson, D. Y., and Given, M. D. J., Variation in grassland populations and species and the implications for community responses to elevated CO_2, *J. Biogeogr.*, 22, 315, 1995.
265. Nijs, I. and Impens, I., Effects of elevated CO_2 concentration and climate-warming on photosynthesis during winter in *Lolium perenne*, *J. Exp. Bot.*, 47, 915, 1996.
266. Boese, S. R., Wolfe, D. W., and Melkonian, J. J., Elevated CO_2 mitigates chilling-induced water stress and photosynthetic reduction during chilling, *Plant Cell Environ.*, 20, 625, 1997.
267. Taulavuori, E., Taulavuori, K., Laine, K., Pakonen, T., and Saari, E., Winter hardening and glutathione status in the bilberry (*Vaccinium myrtillus*) in response to trace gases (CO_2, O_3) and nitrogen fertilization, *Physiol. Plant.*, 101, 192, 1997.
268. Ferris, R. and Taylor, G., Contrasting effects of elevated CO_2 and water deficit on two native herbs, *New Phytol.*, 131, 491, 1995.
269. Chaves, M. M. and Pereira, J. S., Water stress, CO_2 and climate change, *J. Exp.Bot.*, 43(253), 1131, 1992.
270. Ball, M. C. and Munns, R., Plant responses to salinity under elevated atmospheric concentrations of CO_2, *Aust. J. Bot.*, 40, 515, 1992
271. Tschaplinski, T. J., Stewart, D. B., and Norby, R. J., Interactions between drought and elevated CO_2 on osmotic adjustement and solute concentrations of tree seedlings, *New Phytol.*, 131, 169, 1995.
272. Tschaplinski, T. J., Stewart, D. B., Hanson, P. J., and Norby, R. J., Interactions between drought and elevated CO_2 on growth and gas exchange of seedlings of three deciduous tree species, *New Phytol.*, 129, 63, 1995.
273. Ball, M. C., Cochrane, M. J., and Rawson, H. M., Growth and water use of the mangroves *Rhizophora apiculata* and *R. stylosa* in response to salinity and humidity under ambient and elevated concentrations of atmospheric CO_2, *Plant Cell Environ.*, 20, 1158, 1997.
274. Rudorff, B. F. T., Mulchi, C. L., Rowland, R., and Pausch, R., Photosynthetic characteristics in wheat exposed to elevated O_3 and CO_2, *Crop Sci.*, 36, 1247, 1996.
275. McKee, I. F., Bullimore, J. F., and Long, S. P., Will elevated CO_2 concentrations protect the yield of wheat from O_3 damage? *Plant Cell Environ.*, 20, 77, 1997.
276. Hertstein, U., Grunhage, L., and Jager, H.-J., Assessment of past, present and future impacts of ozone and carbon dioxide on crop yields, *Atmos. Environ.*, 29, 2031, 1995.
277. Niewiadomska, E. and Miszalski, Z., Does CO_2 modify the effect of SO_2 on variegated leaves of *Chlorophytum comosum* (Thunb) Bak? *New Phytol.*, 130, 461, 1995.
278. Docherty, M., Hurst, D. K., Holopainen, J. K., Whittaker, J. B., Lea, P. J., and Watt, A. D., Carbon dioxide-induced changes in beech foliage cause female beech weevil larvae to feed in a compensatory manner, *Global Change Biol.*, 2, 335, 1996.
279. Salt, D. T., Brooks, G. L., and Whittaker, J. B., Elevated carbon dioxide affects leaf-miner performance and plant growth in docks (*Rumex* spp.), *Global Change Biol.*, 1, 153, 1995.
280. Owensby, C. E., Cochran, R. C., and Auen, L. M., Effects of elevated carbon dioxide on forage quality for ruminants, in *Carbon Dioxide, Populations, and Communities*, Korner, C. and Bazzaz, F. A., Eds., Academic Press, San Diego, 1996, 363.
281. Foyer, C. H., Descourvières, P., and Kunert, K.-J., Protection against oxygen radicals: an important defense mechanism studied in transgenic plants, *Plant Cell Environ.*, 17, 507, 1994.
282. Schwanz, P., Kimball, B. A., Idso, S. B., Hendrix, D. L., and Polle, A., Antioxidants in sun and shade leaves of sour orange trees (*Citrus aurantium*) after long-term acclimation to elevated CO_2, *J. Exp. Bot.*, 305, 1941, 1996.
283. Polle, A., Pfirrmann, T., Chakrabarti, S., and Rennenberg, H., The effects of enhanced ozone and enhanced carbon dioxide concentrations on biomass, pigments and antioxidative enzymes in spruce needles (*Picea abies* L.), *Plant Cell Environ.*, 16, 311, 1993.
284. Rao, M. V., Hale, B. A., and Ormrod, D. P., Amelioration of ozone-induced oxidative damage in wheat plants grown under high carbon dioxide — role of antioxidant enzymes, *Plant Physiol.*, 109, 421, 1995.

285. McKee, I. F., Eiblemeier, M., and Polle, A., Enhanced ozone-tolerance in wheat grown at an elevated CO_2 concentration: ozone exclusion and detoxification, *New Phytol.*, 137, 275, 1997.
286. Badiani, M., Paolacci, A. R., Fusari, A., Bettarini, I., Brugnoli, E., Lauteri, M., Miglietta, F., and Raschi, A., The foliar antioxidant status of plants from high-CO_2 natural sites, *Physiol. Plant.*, 104, 765, 1998.
287. Badiani, M., Paolacci, A. R., Fusari, A., Bettarini, I., Brugnoli, E., Lauteri, M., Miglietta, F., and Raschi, A., Natural CO_2 enrichment and foliar antioxidants, in *Microevolutionary Adaptation of Plants to Elevated CO_2*, European Commission, Brussels, in press.
288. Booker, F. L., Reid, C. D., Brunschön-Harti, S., Fiscus, E. L., and Miller, J. E., Photosynthesis and photorespiration in soybean [*Glycine max* (L.) Merr.] chronically exposed to elevated carbon dioxide and ozone, *J. Exp. Bot.*, 48, 1843, 1997.
289. Noctor, G., Arisi, A. C. M., Jouanin, L., Valadier, M. H., Roux, Y. and Foyer, C. H., The role of glycine in determining the rate of glutathione synthesis in poplar. Possible implications for glutathione production during stress, *Physiol. Plant.*, 100, 255, 1997.
290. Noctor, G., Arisi, A. C. M., Jouanin, L., Valadier, M. H., Roux, Y., and Foyer C. H., Light-dependent modulation of foliar glutathione synthesis and associated amino acid metabolism in poplar overexpressing gamma-glutamylcysteine synthetase, *Planta*, 202, 357, 1997.
291. Tolbert, N. E., The C_2 oxidative photosynthetic carbon cycle, *Annu. Rev. Plant Physiol. Plant Mol. Biol.*, 48, 1, 1997.
292. Beerling, D. J. and Woodward, F. I., Ecophysiological responses of plants to global environmental change since the Last Glacial Maximum, *New Phytol.*, 125, 641, 1993.
293. Miglietta, F., Badiani, M., Bettarini, I., van Gardingen, P. R., Selvi, F., and Raschi, A., Preliminary studies of the long-term CO_2 response of Mediterranean vegetation around natural CO_2 vents, in *Global Changes and Mediterranean-Type Ecosystems*, Moreno, J. M. and Dechela, W.C., Eds., Springer-Verlag, New York, 1995, 102 .
294. Bettarini, I., Vaccari, F. P., and Miglietta, F., Elevated CO_2 concentration and stomatal density: observation from 17 plant species growing in a CO_2 spring in central Italy, *Global Change Biol.*, 4, 17, 1998.
295. van Gardingen, P. R., Grace, J., Jeffree, C. E., Byari, S. H., Miglietta, F., Raschi, A. and Bettarini, I., Long-term effects of enhanced CO_2 concentrations on leaf gas exchanges: research opportunities using CO_2 springs, in *Plant Responses to Elevated Carbon Dioxide: Evidence from Natural Springs*, Raschi, A., Miglietta, F., Tognetti, R., and van Gardingen, P. R., Eds., Cambridge University Press, Cambridge, 1997, 69.
296. Bettarini, I., Miglietta, F., and Raschi, A., Studying morpho-physiological response of *Scirpus lacustris* from naturally CO_2-enriched environments, in *Plant Responses to Elevated Carbon Dioxide: Evidence from Natural Springs*, Raschi, A., Miglietta, F., Tognetti, R., and van Gardingen, P. R., Eds., Cambridge University Press, Cambridge, 1997, 134.
297. Paoletti, E., Miglietta, F., Raschi, A., Manes, F., and Grossoni, P., Stomatal numbers in holm oak (*Quercus ilex* L.) leaves grown in naturally and artificially CO_2-enriched environments, in *Plant Responses to Elevated Carbon Dioxide: Evidence from Natural Springs*, Raschi, A., Miglietta, F., Tognetti, R., and van Gardingen, P. R., Eds., Cambridge University Press, Cambridge, 1997, 197.
298. Johnson, J. D., Michelozzi, M. and Tognetti, R., Carbon physiology of *Quercus pubescens* Wild. growing at the Bossoleto CO_2 spring in central Italy, in *Plant Responses to Elevated Carbon Dioxide: Evidence from Natural Springs*, Raschi, A., Miglietta, F., Tognetti, R., and van Gardingen, P. R., Eds., Cambridge University Press, Cambridge, 1997, 148.
299. Tognetti, R., Giovannelli, A., Longobucco, A., Miglietta, F., and Raschi, A., Water relations of oak species growing in the natural CO_2 spring of Rapolano (central Italy), *Ann. Sci. For.*, 53, 475, 1996.
300. Tognetti, R., Johnson, J. D., Michelozzi, M., and Raschi, A., Foliage metabolism in *Quercus pubescens* and *Quercus ilex* mature trees with lifetime exposure to local greenhouse effect, *Environ. Exp. Bot.*, 39, 233, 1998.
301. Jones, M. B., Clifton Brown, J., Raschi, A., and Miglietta, F., The effects on *Arbutus unedo* L. of long-term exposure to elevated CO_2, *Global Change Biol.*, 1, 295, 1995.
302. Chaves, M. M., Pereira, J. S., Cerasoli, S., Clifton-Brown, J., Miglietta, F., and Raschi, A., Leaf metabolism during summer drought in *Quercus ilex* trees with lifetime exposure to elevated CO_2, *J. Biogeogr.*, 22, 255, 1995.

303. Tognetti, R., Longobucco, A., Miglietta, F., and Raschi, A., Stomatal behaviour and transpiration of *Quercus ilex* plants during summer and in a Mediterranean carbon dioxide spring, *Plant Cell Environ.*, 1998, 21, 613, 1998.
304. Hättenschwiler, S., Miglietta, F., Raschi, A., and Körner, C., Morphological adjustments of mature *Quercus ilex* trees to elevated CO_2, *Acta Oecol.*, 18, 361, 1997.
305. Faria, T., Cerasoli, S., Garcia-Plazaola, J. I., Guimaraes, M. P., Abadia, A., Raschi, A., Miglietta, F., Pereira, J. S., and Chaves, M. M., Photochemical response to summer drought in *Q. ilex* trees growing in a naturally CO_2 enriched site, in *Impacts of Global Change on Tree Physiology and Forest Ecosystems*, Moheren, G. M. J., Kramer, K., and Sabate, S., Eds., Kluwer Academic Publishers, Dordrecht, the Netherlands, 1998, 119.
306. Fordham, M., Barnes, J. D., Bettarini, I., Polle, A., Slee, N., Raines, C., Miglietta, F., and Raschi, A., The impact of elevated CO_2 on growth and photosynthesis in *Agrostis canina* subsp. *monteluccii* adapted to contrasting atmospheric CO_2 concentrations, *Oecologia*, 110, 169, 1997.
307. Tognetti, R., Raschi, A., and Longobucco, A., Seasonal embolism and xylem vulnerability in deciduous and evergreen tree species of Mediterranean carbon dioxide springs, *Tree Physiol.*, 1998, submitted.
308. Körner, C. and Miglietta, F., Long-term effects of naturally elevated CO_2 on mediterranean grassland and forest trees, *Oecologia*, 99, 343, 1994.
309. Tognetti, R., Longobucco, A., Miglietta, F., and Raschi, A., Stomatal behaviour and transpiration of *Quercus pubescens* plants during summer and in a Mediterranean carbon dioxide spring, *Tree Physiol.*, 1998, submitted.
310. Hättenschwiler, S., Miglietta, F., Raschi, A., and Körner, C., Thirty years of *in situ* tree growth under elevated CO_2: a model for future forest responses? *Global Change Biol.*, 3, 463, 1997.
311. Fordham, M. C., Barnes, J. D., Bettarini, I., Griffiths, H. G., Miglietta, F. and Raschi, A., The impact of elevated CO_2 on the growth of *Agrostis canina* and *Plantago major* adapted to contrasting CO_2 concentrations, in *Plant Responses to Elevated Carbon Dioxide: Evidence from Natural Springs*, Raschi, A., Miglietta, F., Tognetti, R., and van Gardingen, P. R., Eds., Cambridge University Press, Cambridge, 1997, 174.
312. Ineson, P. and Cotrufo, F., Increasing concentration of atmospheric CO_2 and decomposition processes in forest ecosystems, in *Plant Responses to Elevated Carbon Dioxide: Evidence from Natural Springs*, Raschi, A., Miglietta, F., Tognetti, R., and van Gardingen, P. R., Eds., Cambridge University Press, Cambridge, 1997, 242.
313. Lorenzini, G., Panattoni, A., and Badiani, M., Effects of elevated CO_2 levels on the parasitism of some phytopathogen fungi under natural conditions, in *Microevolutionary Adaptation of Plants to Elevated CO_2*, European Commission, Brussells, 1999, in press.

Chapter 5

UV-B Effects on Plants

Manfred Tevini

CONTENTS

5.1 INTRODUCTION

Increases in UV-B radiation have been observed in Antarctica during ozone hole development,[1-3] as well as in the Northern Hemisphere.[4-7] Reviews of UV-B effects in terrestrial plants have been published during the last decade.[8-28] The UV-B effects have been studied during the last 20 years in growth chambers, greenhouses, or in the field mainly by irradiating plants with supplemental artificial UV-B radiation simulating different ozone (O_3)-depletion scenarios.

In growth chamber and greenhouse studies, reductions of plant size, leaf area, fresh and dry weight, lipid content, and photosynthetic activity have often been found in UV-B-sensitive plant species. Alterations of leaf surface epicuticular waxes, UV-B absorbing pigments, and the diffusive conductance of water vapor through the stomata have also been reported. However, only a few realistic field studies have been conducted that can confirm or contradict results obtained in growth chambers or greenhouses with artificial UV-B radiation. Thus, the implications of current and future changes of the O_3 layer for crops and other terrestrial plants need to be assessed in detail and existing studies need to be critically evaluated in order to predict possible consequences of enhanced UV-B radiation in the future.

1-56670-341-7/00/$0.00+$.50

5.2 UV-B RADIATION, ACTION SPECTRA, SPECTRAL BALANCE, AND OZONE-DEPLETION SCENARIOS

UV-B radiation can be applied by artificial UV-B sources to plants in growth chambers, in greenhouses, or in the field as supplemental solar UV-B radiation. In all cases, the biological effectiveness of the UV-B radiation needs to be determined in order to compare the various simulated O_3 depletion scenarios and to assess morphological, physiological, and biochemical consequences of solar UV-B radiation in plants due to O_3 column changes. Various action spectra derived with either monochromatic or polychromatic radiation indicate UV at shorter wavelengths to be more effective than at longer wavelengths[29-34] as shown in Figure 5.1. Ozone depletion results in more UV-B flux, and this effect is more pronounced at shorter wavelengths. Depending upon the shape of the action spectrum that is used as a weighting function in calculating "effective" radiation increase, the ozone simulation scenarios vary. With action spectra that drop off more steeply at longer wavelengths, such as the DNA-damage spectrum,[29] (see Figure 5.1) the relative increase in "effective" solar UV-B is greater than when calculated with spectra that do not decrease as abruptly with increasing wavelength. A new action spectrum for DNA pyrimidine dimer induction in intact alfalfa leaves has a more gradual slope than the DNA-damage spectrum, perhaps due to the shielding of the shorter wavelengths by the plant tissues that cover the DNA.[34] This consequently lowers the predicted increase of effective UV-B with O_3 depletion.

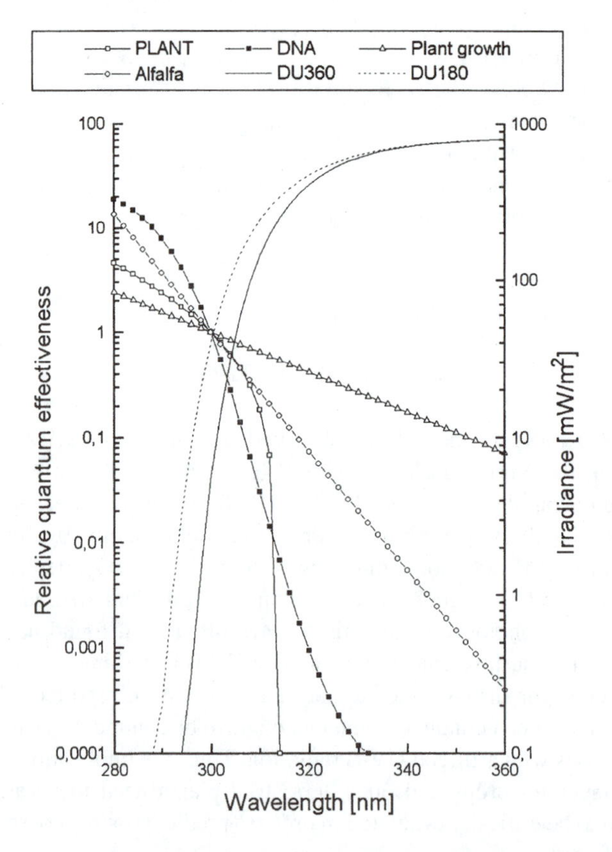

Figure 5.1 Action spectra for DNA damage,[29] pyrimidine dimer induction in alfalfa seedlings,[34] generalized plant damage,[30] and plant growth.[35] Spectral irradiance at 360 and 180 DU (Dobson units) calculated for 49°N, 21 June at local noon according to Green.[123]

Most action spectra have been developed with monochromatic radiation, but organisms in their natural environment are exposed to polychromatic radiation. A few action spectra have been developed using polychromatic radiation, and these indicate a greater "effectiveness" at longer wavelength[31,35] than the generalized PLANT action spectrum[30] or DNA-damage spectrum and thus decreasing the relative UV-B effect.

A recent field experiment demonstrates the efficacy of both visible radiation and UV-A in ameliorating growth-inhibition effects of UV-B radiation.[15] A UV-B-sensitive variety of soybean was grown under field conditions with different combinations of UV-A and visible radiation, but with the same UV-B and control (no UV-B) treatments. This was achieved by using filtered solar radiation and UV-B and UV-A from lamp systems electronically modulated to be constantly proportional to prevailing solar radiation. Plant growth was only inhibited by UV-B when both UV-A and visible radiation were low. In the three other situations, where either visible radiation was high or UV-A was moderately elevated, the plants exhibited no apparent UV-B sensitivity. Thus, in laboratory or greenhouse experiments, even if realistic levels of UV-B are used in simulating O_3 reduction, the plant response may be exaggerated relative to field conditions. Even under field conditions, if applied UV-B is not adjusted downward during cloudy periods, the UV-B sensitivity may be unduly pronounced. In alfalfa seedlings, it was clearly demonstrated that outdoor-grown plants are more resistant to DNA damage than plants grown in UV-B-free environmental chambers.[36]

Laboratory and greenhouse experiments are useful in elucidating mechanisms of UV action; however, they are probably less reliable in providing estimates of UV-B response sensitivity. Unfortunately, the most expensive and difficult experiments, i.e., those conducted in the field with modulated UV-B supplements to adjust for cloudiness and other atmospheric conditions, are seldom undertaken. As indicated in subsequent sections, another approach that achieves appropriate spectral balance, is to filter existing solar radiation to modify the UV-B component. This has been done either with plastic material or using O_3 as a filter. A new O_3 filter technique was developed.[37] In identical growth chambers different UV-B levels were attenuated by passing O_3 through two layers of UV-transparent Plexiglas covering one of the growth chambers. Plants in that chamber therefore received less than ambient UV-B compared to that chamber with no O_3 in the Plexiglass cuvette. A difference in solar UV-B radiation up to 25% can be achieved with this O_3 filter technique. In such experiments, it is possible to test the effect of attenuating solar UV-B, but not enhancing it. Nevertheless, surprising sensitivity is sometimes indicated in such experiments.

5.3 UV-B EFFECTS ON GROWTH

Many biological effects in terrestrial plants exposed to UV-B radiation have been reported, but much is yet to be learned. Most of the work has been conducted using crop plants, primarily from temperate regions, whereas tropical plants and nonagricultural plants have been more or less neglected. Studies of more than 400 plant species and cultivars (out of about 300, 000 seed plant species) have been carried out and about 50% of these plants have been considered as UV-sensitive, where sensitivity is defined as any negative morphological, physiological, or biochemical change induced by UV-B. Tree species, especially conifers in the seedling stage, are also susceptible to enhanced UV-B radiation.[38] Out of 10 conifer species examined so far, four were susceptible to UV-B in terms of reductions in height or biomass.[39]

Only a few multiseason experiments including tree species have been carried out. In a 3-year field study on loblolly pine (*Pinus taeda* L.), plants from different seed sources were grown under conditions of 16 and 25% stratospheric O_3 reduction. Initially, variations of UV-B effects among seed sources were apparent; however, at the final stage of the experiment all plants exhibited biomass reductions of 12 to 20% under the highest UV-B supplement. This suggests cumulative UV-B effects in loblolly pine which might lead to growth reductions over its lifetime.[38]

In many sensitive plant species, reduced leaf area and/or stem growth have been found, for example, in wheat, rice, maize, rye, soybean, sunflower, and cucumber.[9,10,20,40-44] For leaf area and hypocotyl length of cucumber and sunflower seedlings grown in growth chambers under artificial UV-B radiation, a fluence response relationship was demonstrated.[37,45]

Growth, physiological functions and biochemical composition of sunflower and corn seedlings were intensively studied over several summer seasons using the O_3 filter technique.[37,46,47] For both sunflower and corn seedlings, growth parameters like plant height and leaf area were significantly reduced under the higher UV-B radiation. These observations confirm results obtained in previous studies on other plant species with artificial UV-B applications in greenhouses and in the field[48] or in growth chambers.[45]

The O_3-filter cuvettes are small, and, therefore, experiments are limited to a small group of seedlings and/or juvenile plants. To provide differently filtered solar UV-B on a larger spatial scale, different thicknesses of UV-transparent Plexiglas have been used in greenhouses. In a recent study, two greenhouses located in Portugal were covered with either 3 or 5 mm Plexiglas, providing a 10% difference in weighted solar UV-B. Even with these small UV-B differences, reductions in growth of four bean cultivars were observed under higher solar UV-B. Plant height was significantly reduced under higher UV-B throughout plant development in all cultivars. Leaf area and dry mass were mainly affected at an early developmental stage. Later on, the differences disappeared. However, the reproductive phase was delayed as indicated by greater bud numbers and lower pod numbers as seen in 7-week-old plants grown in the greenhouse with higher UV-B transmittance.[49]

Recently, two comparable studies on rice cultivars of different geographic regions have been performed in greenhouses using 19 KJ m^{-2} d^{-1} (generalized plant spectrum weighted, normalized at 300 nm) as an enhanced daily dose[50] and 15.7 KJ m^{-2} d^{-1},[42] respectively, compared to about 11 KJ m^{-2} d^{-1} typically calculated in rice growing areas near the equator.[51] From 16 rice cultivars native to the Philippines, India, Thailand, China, Vietnam, Nepal, and Sri Lanka, about one third showed statistically significant decreases in total biomass and leaf area. Tiller number, highly correlated with yield, was reduced in 6 of these 16 cultivars. The Sri Lanka cultivar Kurkaruppan, however, showed increases in both total biomass and tiller number indicating that selective breeding might be a successful tool for obtaining UV-B-tolerant cultivars.[42] In the second study with four cultivars from the Philippines, total biomass changes were different among cultivars, with IR74 being the most sensitive and IR64 the least sensitive.[50] Another study with four Japanese rice cultivars revealed similar results, but unfortunately it did not state the UV radiation used in the "Phytotron" (= growth chamber) in biologically effective units.[52] Apparently, the UV treatments also included some UV-C radiation would not be present in solar radiation, even if the ozone layer were reduced. An extended study with 188 rice cultivars in phytotrons resulted in reduced plant height (143 cultivars), biomass (61 cultivars) and lower tiller number (41 cultivars) at elevated UV-B.[43] Therefore, at least in rice, it seems questionable whether increased UV-B is a threat to crop productivity.[53] However, field experiments with IR74 also in the Philippines using modulated UV-B lamp systems showed no significant effects of enhanced UV-B either in growth parameters or in biomass. In contrast, absorbance of protecting pigments increased up to 25% in that cultivar.[44]

Photomorphogenetic plant growth responses were also found in wild-type and stable-phytochrome-deficient mutants of cucumber, which indicated that growth is regulated by a as-yet uncharacterized UV-B receptor, rather than by phytochrome.[54] In etiolated tomato seedlings showing inhibition of hypocotyl elongation, the responsible chromophore of the photoreceptor system seems to be a flavin.[55]

In a growth chamber experiment, cultured shoots of pear (cv. Doyenne d'Hiver) exposed to UV-B radiation responded with apical necrosis, leaf abscission, and increased ethylene evolution compared with controls.[56] Furthermore, the polyamine putrescine content increased, which may represent a protective response, since polyamines have been reported to stabilize membrane structure, inhibit lipid peroxidation, and scavenge free radicals. Uptake of CO_2 decreased initially with

UV-B, but after 3 days photosynthetic recovery was observed. The molecular basis for growth reductions may be in part attributed to changes in DNA and/or phytohormones.

Phytohormones can affect growth by altering their concentration in the growth-sensitive tissue and by changing phytohormone-dependent processes. Elongation is related to indole acetic acid (IAA), which absorbs in the UV-B range and is readily photodegraded by UV-B *in vitro* and *in vivo* (if fluxes are sufficient) in sunflower seedlings under low white light conditions. Furthermore, plastic epidermal cell wall extensibility, which is enhanced in auxin-induced elongation growth, was also reduced.[57] Oxidative enzymes such as plant peroxidase can reduce cell elongation by working as IAA-oxidase, by cross-linking wall material and the epidermal cell wall, and by ethylene formation. Peroxidase activity was enhanced in UV-B-irradiated sunflower[57] and sugar beet plants.[58] Another phytohormone, ethylene, which causes greater radial growth and less elongation, is produced to a greater extent in UV-B irradiated sunflower seedlings than in controls.[57]

5.4 UV-B EFFECTS ON COMPETITION

Competition plays a large role in determining the predominance of different plant species in nonagricultural ecosystems and the balance between weeds and crops in agricultural systems. Photomorphological responses such as increased branching and number of leaves along with reduced leaf length can lead to changes in competition for light in mixtures of wheat and wild oat.[59,60] In a field study repeated over 5 years, the crop species gained the advantage over the weed, wild oat when the plant mixtures were given supplemental UV-B using modulated lamp systems. By using a canopy microclimate, radiation interception, and photosynthesis, these competitive differences could be quantitatively explained by small shifts in growth and allocation of these species.[61] As mentioned earlier, when either the wheat or the wild oat were grown alone, the photomorphological changes caused by UV-B resulted in no change in productivity. Furthermore, even with exaggerated UV-B exposures neither species exhibited any sensitivity in photosynthesis. Thus, indirect consequences of plant growth changes influencing the balance of competition may be more important than direct effects on productivity

5.5 UV-B EFFECTS ON FLOWERING

It has been demonstrated that exclusion of UV-B radiation by plastic films or glass stimulated flowering in *Melilotus*, *Trifolium dasyphyllum*, and *Tagetes*.[62-64] The inhibition of photoperiodic flower induction was found to be dependent on UV-B fluence rate. In a long-day plant *Hyoscyamus niger*,[65] a 20% decrease of flowering resulted from irradiation with 100 mW m^{-2} UV-B and a 50% reduction from 300 mW m^{-2} UV-B (all unweighted) compared with plants merely irradiated with white light. This suppression of flowering is correlated with changes in gibberellic acid content. Even if the total amount of flowering and fruit production is ultimately little influenced by increased solar UV-B, differences in timing of flowering[49] may have significant effects on competition for pollinators, etc. Like changes in the balance of competition, these more complex ecosystem effects may be of much greater importance than changes in agricultural productivity. However, such indirect changes are also much more difficult to predict.

Both pollen and ovules may in many cases be rather well shielded from solar UV. But, anther walls can absorb more than 98% of incident UV-B.[66] In addition, the pollen wall contains UV-B-absorbing compounds affording protection during pollination. Only after transfer to the stigma, the pollen might be susceptible to solar UV-B. At this time, the assessment of the general impact of UV-B radiation on plant pollen is difficult as only few *in vitro* experiments have been carried out. It is thought that ovaries offer sufficient protection for ovules against solar UV-B radiation.[67]

5.6 UV-B EFFECTS ON PHOTOSYNTHESIS, RESPIRATION, AND TRANSPIRATION

Photosynthesis is one of the most-studied physiological processes in the context of UV-B radiation and is often accompanied with growth experiments. When high UV-B irradiances were used in combination with low levels of white light (the usual conditions in growth chambers), effects on photosynthesis measured as CO_2 assimilation have been generally deleterious in UV-B-sensitive plants.[68] However, even in the presence of higher levels of white light found in greenhouses and the field, reductions in photosynthesis of up to 17% were reported in the UV-B-sensitive soybean cultivar "Essex" when supplied with UV-B equivalent to a 16% O_3 depletion.[69] In contrast, no change in net carbon exchange rate was found for this cultivar when grown in open-top chambers with UV-B lamp systems at lower dose levels.[70] In both experiments, modulated UV-B supplements were not employed.

Sunflower and maize seedlings, grown under different UV-B levels using the ozone-filter technique with solar radiation in Portugal,[71] also exhibited a significant difference of about 17% in photosynthetic activity (on a chlorophyll or a per plant basis) when a 12% difference in the O_3 column was simulated.

The photosynthetic properties of four age groups of needles from 3-year-old field-grown loblolly pine under enhanced UV-B (unmodulated) were examined.[72] Determination of the ratio of variable to maximum fluorescence (F_v/F_m) following dark adaptation showed reductions of 6% only in the youngest age class. Maximum photosynthesis, apparent quantum yield, and dark respiration remained unaffected by enhanced UV-B radiation. However, reductions of total biomass up to 20% and reductions in needle length in three of the four age groups were observed. The overall growth reductions were probably due to decreases in leaf surface rather than impaired photosynthetic capacity.

Multiple sites of UV-B inhibition have been demonstrated throughout the years with the most sensitive site around Photosystem II.[73-76] Photosystem I seems much more resistant to UV-B radiation.[73,75-77] In contrast, ribulose 1,5-bisphosphate carboxylase (Rubisco), the key enzyme of the Calvin cycle has recently been shown to be very UV sensitive.[78,79] For example, 1 day of UV-B exposure to relatively high artificial UV-B radiation declined the enzyme activity by 40%, the large and small subunits of Rubisco by 10 to 15%.[80]

In the field, when longer UV-B irradiation periods together with higher temperatures around 35°C occur, photoinhibition and UV-B damage may be additive and, thus, increase the impact on net photosynthesis in some species.[81] This was also deduced from the inhibition of violaxanthin deepoxidation by high UV-B flux in laboratory studies.[82] Deepoxidation of violaxanthin is thought to have a protective role against photoinhibition by dissipating excessive light energy as heat.

5.7 UV-B EFFECTS ON YIELD

Since photosynthesis is essential for plant productivity, UV-induced damage to photosynthetic function can manifest itself in less biomass or lower yield. However, growth and yield may also be influenced by UV-B without being mediated by changes in photosynthesis as mentioned earlier. It is well documented that the effectiveness of UV-B radiation is magnified when given levels of radiation are below the saturation level of photosynthesis. The reasons for this phenomenon are not completely known; however, it is thought that natural UV-protective mechanisms do not become fully developed in low levels of visible light. Studies on six cultivars of soybean were conducted under conditions of a 16% O_3 depletion simulation using unmodulated UV-B lamps in the field.[83] In all cultivars, no change in response to enhanced UV-B was observed. Also, in a recent study,[70] suppression of yield by UV-B radiation was not found for any of the three soybean cultivars "Essex," "Coker 6955," and "S53-34" with UV-B biological effective enhancements up to 5.37 KJ m^{-2} d^{-1} simulating a 37% O_3 depletion.[84] Using modulated UV-B lamp systems in field experiments, no depressions of yield were found with a simulated O_3 reductions of 6 to 32% in bean[85] or in monocultures of wheat or wild oat with simulated O_3 reductions ranging from 16 to 40%.[59]

This brief summary indicates a high degree of interspecific variability complicated further by differences in experimental conditions including artificial light sources, use of modulated vs. unmodulated lamp systems, climate, soil quality, day length, etc. In addition to this interspecific variability, there is a high degree of intraspecific variability among cultivars even when the same experimental procedures were used.[48,86] In the soybean cultivar "Essex," a 25% reduction of photosynthesis was coupled with the same reduction in soybean yield in some years when the plants were irradiated with unmodulated supplemental UV-B in the field.[87] A 20% reduction in yield was further found in some years of a 6-year field study for "Essex" when simulating a 25% O_3 reduction (5.1 KJ m^{-2} d^{-1} supplemental UV-B), whereas in the cultivar "Williams" the yield either was not influenced or even increased under the same conditions.[40]

Bean cultivars grown in two greenhouses covered with 3-mm and 5-mm Plexiglas (lower solar UV-B) showed reductions in yield up to 50 % under the higher UV-B when harvested after 7 weeks. The reason for this is a delay in development of pods in favor of flowers and buds.[49] A later harvest would minimize these differences, since the total number of buds, flowers, and pods was not changed under higher UV-B at a later stage.

Changes in allocation patterns induced by UV-B may also affect yield. For example, changes in carbon assimilate allocation in the tropical agronomic crop cassava were seen when plants received 13.9 KJ m^{-2} d^{-1} UV-B_{BE} ($\cong$15% O_3 depletion at 0°N), compared with ambient UV-B levels of 8.4 KJ m^{-2} UV-B_{BE} at the equator. This allocation change increased shoot growth in favor of root growth. A significant decrease in root weight of 32% as a response to elevated UV-B was observed after 95 days. If this growth response may persist over a prolonged period of time, a decrease in yield might be possible, since the roots are the harvestable part of this species.[81] Another yield determining factor for crop plants is the competition with weed species.[59] In mixed stands, wheat increased its competitiveness over wild oat, which is apparently not the result of differential damage to the two species, since photosynthesis was unaffected and yields of monocultures of these two species were not affected by the elevated UV-B. Rather, the shifts appear to be due to small changes in growth and allocation. Although these changes in growth form are not of consequence for production in monocultures, in mixtures there appears to be a sufficient shift in irradiance interception by the two species to result in an alteration in competition for light and photosynthetic carbon gain.[61]

5.8 UV PROTECTION AND ADAPTATION

During evolution, plants have apparently adapted to different levels of UV-B radiation, by developing at least two protection mechanisms: photoreactivation and accumulation of UV-absorbing pigments such as phenylpropanoids and flavonoids. Photoreactivation is directed toward the repair of thymine dimers of the cyclobutane type which may be responsible for much of the UV-induced damage of the DNA in plants. Recently, it has been shown that the UV-B fluence response for the dimer formation is linear to an unweighted fluence of up to 690 J m^{-2} as applied to alfalfa leaves.[88] The photolyase responsible for this repair is light driven (UV-A and visible) and is UV-B inducible as shown in *Arabidopsis thaliana* [89] but phytochrome regulated in bean hypocotyls.[90] Several studies have confirmed the importance of high white light levels in the reduction of UV-B-induced damage normally found under low white light levels.[52,58,91-93] Although this is unlikely to be totally attributable to photoreactivation, this process probably does play a role in ameliorating UV-B effects when UV-A and/or visible light are high. However, high levels of UV-B may have damaging effects on the enzyme activity itself (by destruction of UV-absorbing amino acids) and thus reduce the repair capacity, especially when the photolyase is located in the epidermal layer of the leaves.

UV-resistant species may be well protected by shielding pigments synthesized via the phenylpropanoid pathway in the epidermal layer or in leaf hairs.[94-97] Measurements of UV-B penetration through the epidermal layer into the mesophyll using a fiber-optic microprobe have confirmed the importance of absorbing material in the epidermal layers.[98] Two subalpine conifer species which exhibit no UV-B inhibition of growth under supplemental UV-B radiation[39] and 11 species from the Rocky Mountains had no measurable 300 nm radiation penetration into the mature needles.[99,100] In contrast, in newly developing leaves, there was small, measurable UV-B flux in the mesophyll, indicating that there are lesser amounts of screening properties in early needle stages.[101]

The protective function of flavonoids in ameliorating UV-B damage to photosynthesis has been clearly demonstrated in rye seedlings, which accumulate isovitexins and other phenylpropanoids exclusively in the epidermal layer.[46] The biosynthesis of these compounds is regulated by UV-B-induced isomerization processes of precursor compounds.[102] *Arabidopsis* mutants that lack flavonoids are extremely sensitive to solar UV radiation.[103,104] Protection from UV-induced DNA damage by flavonoids has been shown in maize seedlings.[105]

5.9 INTERACTIONS OF UV-B AND OTHER ENVIRONMENTAL FACTORS

Plants growing in the field are usually subjected to several potential stress factors. Thus, it is important to consider the interacting effects of UV-B with other environmental factors. Water-stress is one of the most common stress factors. In a growth chamber study, water loss in cucumber seedlings was adversely affected when a 12% O_3 depletion was simulated, as water-stressed plants irradiated with higher levels of UV-B lost their capacity to close stomata with increasing UV-B dose.[106] In contrast, radish seedlings were shown to be much less sensitive to UV-B radiation under water stress than were the cucumber. The reason for this is probably greater concentrations of UV-absorbing flavonoids in the leaf epidermis of radish leaves.[107] In the 6-year field study of Teramura and colleagues,[106] in years of low water supply, the response to UV-B appeared to be masked. Also, in a specific test of water-stress UV-B interaction, at a 25% O_3 depletion simulation[108] reductions in growth, dry weight, and net photosynthesis were found in the well-watered soybean cultivar "Essex," but not when the plants were water stressed.

Increases of CO_2 and temperature are anticipated in future climate change scenarios. Model calculations predict approximate average temperature increase of 1 to 5° C and a doubling of the ambient CO_2 by the middle of the next century.[109] It is therefore of interest to investigate the interactions of these factors on plant performance. Over the last 3 years some progress has been made in this area.

By using the O_3 filter technique to simulate a 12% O_3-depletion scenario, doubling of CO_2 did not increase the biomass of maize seedlings which was normally reduced under higher UV-B radiation. In contrast, UV-B-induced biomass reductions in sunflower could be ameliorated by CO_2 doubling (Table 5.1). This could be expected since C_3 plants in contrast to C_4 plant species do respond to enhanced CO_2.[110,111] However, an increase in temperature of 4°C over the normal daily temperature course compensated for biomass reductions normally found at lower temperatures and enhanced UV-B both in sunflower and maize seedlings.[47]

Several studies describe the combined effects of enhanced UV-B and CO_2. Wheat, rice, and soybean were grown in greenhouses at 350 or 650 ppm CO_2 at 8.8 and 15.7 KJ m^{-2} d^{-1} of weighted UV-B. As expected for C_3 plants, an increase in CO_2 increased light-saturated photosynthesis, apparent quantum efficiency, water use efficiency, biomass, and seed yield in all three species. Enhanced UV-B had no significant damaging effect on seed yield, total biomass, and photosynthesis of wheat and rice. However, increased UV-B radiation reduced the CO_2-induced increase in biomass and seed yield of wheat and rice, but not for soybean.[41]

As discussed earlier, UV-B may affect photosynthesis both at Photosystem II and at CO_2-fixation enzymes. At least in rice, this damage of the photosynthetic apparatus cannot be ameliorated

Table 5.1 Leaf Area, Biomass, and Light-Saturated Photosynthetic Rates of 18-day-old Sunflower and Maize Seedlings Grown in Ambient and Elevated CO_2 at Different Temperatures (28° and 32°C) with Ambient (+UV-B) and "Reduced" (Control) UV-B Radiation

	Leaf Area, cm^2 $plant^{-1}$	Change, %	Biomass, mg $plant^{-1}$	Change, %	Photosynthesis, μmol m^{-2} s^{-1}	Change, %
Sunflower						
Control	29,78a		136.70a		15.46a	
+UV-B	24,67b	-17	116.70b	-14	17.32b	+12
+UV-B + CO_2	31,90ac	+7	200.73d	+47	24.63c	+59
+UV-B + Temp.	34,82c	+16	160.60c	+17	14.59a	-5
+UV-B + CO_2 + Temp.	33,06c	+11	203.70d	+49	15.39a	-1
Maize						
Control	83,42a		289.10a		9.44ab	
+UV-B	65,01b	-22	221.00b	-25	8.98ab	-4
+UV-B + CO_2	70,31b	-16	241.08b	-17	10.01a	+6
+UV-B + Temp.	88,70a	+6	379.10c	+31	7.59b	-19
+UV-B +.CO_2 + Temp.	87,76a	+5	382.30c	+32	6.16c	-32

Note: Means with the same letter are not significantly different at the 95% level according to the Fisher LSD test.
Source: Mark, U. and Tevini, M., *Plant Ecol.*, 128, 224, 1997.[47]

by increased CO_2. It has been reported[112] that in *Elymus athericus*, a C_3 grass, the UV-induced dry weight reductions could not be significantly ameliorated by doubling of CO_2. Pea and tomato also showed the same lack of amelioration.[113] In all the studies, this lack of response may be due to extremely high UV-B levels used in these experiments.

Ozone in combination with UV-B reduces pollen tube growth more than either stress alone. A UV exposure of 3 W m^{-2} for 30 min followed by O_3 at 120 ppb for 3 h reduced pollen growth of *Nicotiana tabacum* by 79% and *Petunia hybrida* by 75%. The effect appeared to be additive, indicating that different targets may be affected by the two environmental factors.[114] In field experiments with open-top fumigation chambers equipped with UV-B lamps, the combined effects of enhanced UV-B radiation and higher ground-level O_3 concentrations revealed no persistent effects of UV-B (simulating a 37% O_3 reduction) on growth of the soybean cultivar "Essex." Ozone treatment resulted in visible injury, but interactions between ground-level O_3 and UV-B radiation were not found.[70]

Mineral nutrient supply, especially nitrogen, can modify the influence of UV-B on plant morphology and physiology. In greenhouses sunflower and maize plants responded to UV-B unfavorably and the degree of which depended on nitrogen supply. UV-B-reduced growth was found at 30 and 120 mg N kg^{-1} soil but not at 270 mg N kg^{-1} soil.[115] Phosphorus deficiency in combination with UV-B (simulating a 16% O_3 depletion) showed no additional damaging effects to growth and photosynthesis, but net photosynthesis was reduced by about 20% in plants sufficiently supplied with phosphorus.[116] Heavy metals are usually toxic to plants. Dube and Bornman[117] used 5 m*M* cadmium and UV-B radiation of only 6.17 KJ m^{-2} d^{-1} with *Picea abies* L. and found that the combined treatment was greater than the depressive effects on growth and photosynthesis of either stress applied alone.

Cercospora leaf spot disease induced by the fungus *Cercospora beticola* Sacc. can result in substantial losses of sugar beet harvest. With UV-B, the severity of this disease increased in one study.[118] In one cucumber cultivar, susceptible to *Colletotrichum lagenarium* or *Cladosporium cucumerinum*, preinfection treatment with UV-B (11.6 KJ m^{-2} d^{-1}) led to a greater disease development, especially on the cotyledons but not on the true leaves, whereas postinfection UV-B irradiation did not have this effect.[119]

5.10 UV-B EFFECTS ON NATURAL ECOSYSTEMS

Only a few studies on natural ecosystems exist. The first field studies on subarctic heath and moss vegetation show UV-B effects similar to those in agricultural plants such as reduced growth, changed leaf thickness, and decreased reproductivity.[120,121] In addition, quantitative differences are obvious between ecologically different species. For example, the shoot growth of two evergreen species (*Vaccinium vitis-idaea* and *Empetrum hermaphroditum*) was smaller after UV irradiation (simulating a 15% O_3 destruction) than the growth of two leaf-losing species (*Vaccinium myrtillus* and *Vaccinium uliginosum*). Over a longer timescale this could lead to immense changes in biodiversity of plant ecosystems. On the other hand, recent studies of four coastal grassland species naturally growing in the Netherlands exhibited no UV effects on biomass, photosynthesis, and morphology when higher solar UV-B was compared with lower UV-B irradiation.[122] Therefore, ameliorating effects of solar UV-B can also be expected in comparison to artificial UV radiation used in many studies so far. However, artificially increased UV-B radiation according to expected O_3-depletion scenarios may over top adaptation limits especially in ecosystems at northern latitudes.

5.11 CONCLUDING ASSESSMENT

Considering the research thus far conducted, primarily with temperate-latitude crop species, there is sufficient cause for concern about the consequences of stratospheric O_3 depletion. Perhaps the yield of many temperate-latitude crop species may not be greatly reduced, especially if sensitive cultivars are avoided. Yet, little is known about tropical crops and ecosystems that exist in areas of currently intense solar UV-B. Furthermore, the many potential indirect effects of elevated UV-B and the complications of interactions with other environmental factors at all latitudes remain ominous. These have been scarcely explored with crops and even less in nonagricultural systems. The database of realistic and ecologically relevant UV-B studies is very limited when compared with other fields of environmental research, such as with air pollutants. Furthermore, interpretation of the various existing UV-B experiments is far from straightforward.

REFERENCES

1. Lubin, D., Frederick, J. E., Booth, C. R., Lucas, T., and Neuschuler, D., Measurements of enhanced springtime ultraviolet radiation at Palmer Station, Antarctica, *Geophys. Res. Lett.*, 16,783, 1989.
2. Stamnes, K., Jin, Z., Slusser, J., Booth, C., and Lucas T., Several-fold enhancement of biologically effective ultraviolet radiation levels at MCMurdo Station Antarctica during the 1990 ozone "hole," *Geophys. Res. Lett.*, 19, 1013, 1992.
3. Frederick, J. E., Ultraviolet sunlight reaching the earth's surface: a review of recent research, *Photochem. Photobiol.*, 57, 175, 1993.
4. Blumthaler, M. and Ambach, W., Indication of increasing solar ultraviolet-B radiation flux in alpine regions, *Science*, 248, 206, 1990.
5. Blumthaler, M., Solar UV measurements, in *UV-B Radiation and Ozone Depletion, Effects On Humans, Animals, Plants, Microorganisms and Materials*, Tevini, M., Ed., Lewis Publishers, Boca Raton, FL, 1993, chap. 6.
6. Kerr, J. B. and McElroy, C. T., Evidence for large upward trends of ultraviolet radiation linked to ozone depletion, *Science*, 262, 1032, 1993.
7. Mayer, B., Seckmeyer, G., and Kyling, A., Systematic long-term comparison of spectral UV measurements and UVSPEC modelling results, *J. Geophys. Res.*, 102, 8755, 1997.
8. Caldwell, M. M., Teramura, A. H., and Tevini, M., The changing solar ultraviolet climate and the ecological consequences for higher plants, *Trends Ecol. Evol.*, 4, 363, 1989.
9. Tevini, M. and Teramura, A. H., UV-B effects on terrestrial plants, *Photochem. Photobiol.*, 50, 479, 1989.

10. Krupa, S. V. and Kickert, R. N., The greenhouse-effect-impacts of ultraviolet-B (UV-B) radiation, carbon-dioxide (CO_2), and ozone (O_3) on vegetation, *Environ. Pollut.*, 61, 263, 1989.
11. Stapleton, A. E., Ultraviolet radiation and plants: burning questions, *Plant Cell,* 4, 1353, 1992.
12. Tevini, M., Effects of enhanced UV-B radiation on terrestrial plants, in *UV-B Radiation and Ozone Depletion, Effects on Humans, Animals, Plants, Microorganisms and Materials*, Tevini, M., Ed., Lewis Publishers, Boca Raton, FL, 1993, chap. 5.
13. Bornman, J. F. and Teramura, A. H., Effects of ultraviolet-B radiation on terrestrial plants, in *Environmental UV-Photobiology*, Young, A. R., Björn, L. O., Moan, J., and Nultsch, W., Eds., Plenum Press, New York, 1993, chap. 14.
14. Caldwell, M. M. and Flint, S. D., Solar ultraviolet radiation and ozone layer changes: implications for crop plants, in *Physiology and Determination of Crop Yield*, Boote, K., Sinclair, T. R., Bennett J. M., and Paulsen G. M., Eds., ASA-CSSA-SSSA, Madison, WI, 1994, 487.
15. Caldwell, M. M., Flint, S. D., and Searles, P. S., Spectral balance and UV-B sensitivity of soybean: a field experiment, *Plant Cell Environ.*, 17, 267, 1994.
16. Tevini, M., UV-B Effects on terrestrial plants and aquatic organisms, *Prog Bot.*, 55, 174, 1994.
17. Strid, A., Chow, W. S., and Anderson, J. M., UV-B damage and protection at the molecular level in plants, *Photosynth. Res.*, 39, 475, 1994.
18. Britt, A. B., Repair of DNA damage induced by ultraviolet radiation, *Plant Physiol.*, 108, 891, 1995.
19. Bornman, J. F. and Sundby-Emanuelsson, C., Response of plants to UV-B radiation: some biochemical and physiological effects, in *Environment and Plant Metabolism, Flexibility and Acclimation*, Smirnoff, N., Ed., BIOS, 245, 1995.
20. Caldwell, M. M., Teramura, A. H., Tevini, M., Bornman, J. F., Björn, L. O., and Kulandaivelu, G., Effects of increased solar ultraviolet radiation on terrestrial plants, *Ambio*, 24, 166,1995.
21. Björn, L. O., Effects of ozone depletion and increased UV-B on terrestrial ecosystems, *Int. J Environ. Stud.*, 51, 217, 1996.
22. Tevini, M., Erhöhte UV-B-Strahlung: Ein Risiko für Nutzpflanzen? *Biol. unserer Zeit.*, 4, 246, 1996.
23. Jordan, B. R., The effects of ultraviolet-B radiation on plants: a molecular perspective, *Adv. Bot. Res.*, 22, 98, 1996.
24. Teramura, A. H. and Ziska, L. H., Ultraviolet-B radiation and photosynthesis, in *Photosynthesis and the Environment, Advances in Photosynthesis*, Baker, R., Ed., Vol. 5, Kluwer, Dordrecht, the Netherlands, 1996, 435.
25. Rozema, J., van de Staaij, J. W. M., Björn, L. O., and Caldwell, M. M., UV-B as an environmental factor in plant life: stress and regulation, *Tree*, 12, 22, 1997.
26. Rozema, J., van de Staaij, J. W. M., and Tosserams, M., Effects of UV-B radiation on plants from agro- and natural ecosystems, in *Plant and U-VB: Responses to Environmental Change*, Lumsden, P. J., Ed., Society for Experimental Biology seminar series 64, CambridgeUniversity Press, Cambridge, 1997, 213.
27. McLeod, A. R. and Newsham, K. K., Impacts of elevated UV-B on forest ecosystems, in *Plant and UVB: Responses to Environmental Change*, Lumsden, P. J., Ed., Society for Experimental Biology seminar series 64, Cambridge University Press, Cambridge, 1997, 247.
28. Corlett, J. E. Stephen, J., Jones, H. G., Woodfin, R., Mepsted, R., and Paul, N. D., Assessing the impact of UV-B radiation on the growth and yield of field crops, in *Plants and UV-B:Response to Environmental Change*, Lumsden, P. J., Ed., Society for Experimental Biology seminar series 64, Cambridge University Press, Cambridge, 1997, 195.
29. Setlow, R. B., The wavelength in sunlight effective in producing skin cancer: a theoretical analysis, *Proc. Natl. Acad. Sci. U.S.A.*, 71, 3363, 1974.
30. Caldwell, M. M., Solar UV irradiation and the growth and development of higher plants, in *Photophysiology VI*, Giese, A. C., Ed., Academic Press, New York, 1971,131.
31. Caldwell, M. M., Camp, L. B., Warner , C. W., and Flint, S. D., Action spectra and their key role in assessing biological consequences of solar UV-B radiation change, in *Stratospheric Ozone Reduction, Solar Ultraviolet Radiation and Plant Life*, Worrest, R. C. and Caldwell, M. M., Eds., NATO ASI Series G: Ecological Sciences, Vol. 8, Springer-Verlag, Berlin, 1986, 87.
32. Coohill, T. P., Ultraviolet action spectra and solar effectiveness spectra for higher plants, *Photochem. Photobiol.*, 50, 451, 1989.
33. Coohill, T. P., Photobiology School, action spectra again? *Photobiology*, 54, 859, 1991.

34. Quaite, F. E., Sutherland, B. M., and Sutherland , J. C., Action spectrum for DNA damage in alfalfa lowers predicted impact of ozone depletion, *Nature*, 358, 576, 1992.
35. Steinmüller, D., On the effect of ultraviolet radiation (UV-B) on leaf surface structure and on the mode of action of cuticular lipid biosynthesis in some crop plants, in *Karls. Beitr. Entw. Ökophysiol.*, Tevini, M., Ed., Botanical InstituteII, Karlsruher, 6, 1986, 1.
36. Takayanagi, S., Trunk, J. G., Sutherland, C., and Sutherland, B. M., Alfalfa seedlings grown outdoors are more restistant to UV-induced DNA damage than plants grown in a UV-free environmental chamber, *Photochem. Photobiol.*, 60, 363, 1994.
37. Tevini, M., Mark, U., and Saile, M., Plant experiments in growth chambers illuminated with natural sunlight, in *Environmental Research with Plants in Closed Chambers*, Payer, H. D., Pfirrmann, T., and Mathy, P., Eds., *Air Pollut. Res. Rep.*, 26, 1990, 240.
38. Sullivan, J. H. and Teramura, A. H., The effects of UV-B radiation on loblolly pine. 2. Growth of field-grown seedlings, *Trees*, 6, 115, 1992.
39. Sullivan, J. H. and Teramura, A. H., Effects of ultraviolet-B irradiation on seedling growth in the Pinaceae, *Am. J. Bot.*, 75, 225,1988.
40. Teramura, A. H., Sullivan, J. H., and Lydon, J., Effects of UV-B radiation on soybean yield and seed quality: a 6-year field study, *Physiol. Plant.*, 80, 5, 1990.
41. Teramura, A. H., Sullivan, J. H., and Ziska, L. H., Interaction of elevated ultraviolet-B radiation and CO_2 on productivity and photosynthetic characteristics in wheat, rice and soybean, *Plant Physiol.*, 94, 470, 1990.
42. Teramura, A. H., Ziska, L. H., and Szetin, A. E., Changes in growth and photosynthetic capacity of rice with increased UV-B radiation, *Physiol. Plant.*, 83, 373, 1991.
43. Dai, Q., Peng, S., Chavez, A. Q., and Vergara, B. S., Intraspecific responses of 188 rice cultivars to enhanced UV-B radiation, *Environ. Exp. Bot.*, 34, 433, 1995.
44. Dai, Q., Effects of enhanced ultraviolet-B radiation on growth and production of rice under greenhouse and field conditions, in *Climate Change and Rice*, Peng, S., Ingram, K. T., Neue, H.-U.. and Ziska L. H., Eds., Springer, Berlin, 1995189.
45. Tevini, M. and Iwanzik, W., Effects of UV-B radiation on growth and development of cucumber seedlings, in *Stratospheric Ozone Reduction, Solar Ultraviolet Radiation and Plant Life*, Worrest, R. C. and Caldwell, M. M., Eds., NATO ASI Series G: Ecological Sciences, - Verlag, Berlin, 8,1986, 271.
46. Tevini, M., Mark, U. and Saile-Mark, M., Effects of enhanced solar UV-B radiation on growth and function of crop plant seedlings, in *Current Topics in Plant Biochemistry and Phys.*, Randall, D. D. and Blevius, D. G., Eds., University of Missouri, Columbia, 9, 1991, 13.
47. Mark, U. and Tevini, M., Effects of elevated ultraviolet-B-radiation, temperature and CO_2 on growth and function of sunflower and maize seedlings, *Plant Ecol.*, 128, 224, 1997.
48. Teramura, A. H. and Murali, N. S., Intraspecific differences in growth and yield of soybean *Glycine max* exposed to UV-B radiation under greenhouse and field conditions, *Environ. Exp. Bot.*, 26, 89, 1986.
49. Saile-Mark, M. and Tevini, M., Effects of solar UV-B radiation on growth, flowering and yield of central and southern European bush bean cultivars (*Phaseolus vulgaris* L.), *Plant Ecol.*, 128, 114, 1997.
50. Dai, Q., Coronel, P. V. P., Vergara, B. S., Barnes, P. W., and Quintos, A. T., Ultraviolet-B radiation effects on growth and physiology of four rice cultivars, *Crop Sci.*, 32, 1269, 1992.
51. Bachelet, D., Barnes, P. W., Brown, D., and Brown, M., Latitudinal and seasonal variations in calculated ultraviolet-B irradiance for rice-growing regions of Asia, *Photochem. Photobiol.*, 54, 411, 1991.
52. Kumagai, T. and Sato, T., Inhibitory effects of increase in near-UV radiation on the growth of Japanese rice cultivars (*Oryza sativa* L.) in a phytron and recovery by exposure to visible radiation, *Jpn. J. Breed.*, 42, 545, 1992.
53. Fiscus, E. L. and Booker, F. L., Increased UV-B a threat to crop photosynthesis and productivity? *Photosynth. Res.*, 43, 81, 1995.
54. Ballaré, C. L., Barnes, P. W., and Kendrick, R. E., Photomorphogenic effects of UV-B radiation on hypocotyl elongation in wild type and stable-phytochrome-deficient mutant seedlings of cucumber, *Physiol. Plant.*, 83, 652, 1991.
55. Ballaré, C. L., Barnes, P. W., and Flint, S. D., Inhibition of hypocotyl elongation by ultraviolet-B radiation in de-etiolating tomato seedlings. I. The photoreceptor, *Physiol. Plant.*, 93, 584, 1995.

56. Predieri, S., Krizek, D. T, Wang, C. Y., Mirecki, R. M., and Zimmermann, R. H., Influence of UV-radiation on developmental changes, ethylene, CO_2 flux and polyamines in cv. *Doyenne d'Hiver* pear shoots grown *in vitro*, *Physiol. Plant.*, 87, 109, 1993.
57. Ros, J., Zur Wirkung von UV-Strahlung auf das Streckungswachstum von Sonnenblumenkeimlingen (*Helianthus annuus* L.) [On the effect of UV-radiation on elongation growth of sunflower seedlings (*Helianthus annuus* L.)], Thesis in *Karls. Beitr. Entw. Ökophysiol*, Tevini, M., Ed., Botanical Institure II, Karlsruhe, 8, 1990, 1.
58. Panagopoulos, I., Bornman, J. F., and Björn, L. O., Effects of ultraviolet radiation and visible light on growth, fluorescence induction, ultraweak luminescence and peroxidase activity in sugar beet plants, *J. Photochem. Photobiol. B Biol.*, 8, 73, 1990.
59. Barnes, P. W., Jordan, P. W., Gold, W. G., Flint, S. D., and Caldwell, M. M., Competition morphology and canopy structure in wheat (*Triticum aestivum* L.) and wild oat (*Avena fatua* L.) exposed to enhanced ultraviolet-B radiation, *Funct. Ecol.*, 2, 319, 1988.
60. Ballaré, C. L., Barnes, P. W, Flint, S. D. and Caldwell, M. M., Morphological responses of crop and weed species of different growth forms to ultraviolet-B radiation, *Am. J. Bot.*, 77,1354,1990 .
61. Ryel, R., Barnes, P. W., Beyschlag ,W., Caldwell, M. M., and Flint, S. D., Plant competition for light analysed with a multispecies canopy model. I. Model development and influence of enhanced UV-B conditions on photosynthesis in mixed wheat and wild oat canopies, *Oecologia*, 82, 304, 1990.
62. Kasperbauer, L. and Loomis, W., Inhibition of flowering by natural daylight on an inbred strain of *Melilotus*, *Crop Sci.*, 5, 193, 1965.
63. Klein, R., Edsall, P., and Gentile, A., Effects of near-ultraviolet and green radiations on plant growth, National Science Foundation and Contract AT(30-1-2587) from the Atomic Energy Commission, 1965.
64. Caldwell, M. M., Solar ultraviolet radiation as an ecological factor for alpine plants, *Ecol. Monogr.*, 38, 243, 1968.
65. Rau, W., Hoffmann, H., Huber-Willer, A., Mitzke-Schnabel, U., and Schrott, E., Die Wirkung von UV-B auf photoregulierte Entwicklungsvorgänge bei Pflanzen, Gesellschaft für Strahlen- und Umweltforschung mbH., München Abschlußbericht, 1988.
66. Flint, S. D. and Caldwell, M. M., Partial inhibition of *in vitro* pollen germination by simulated ultraviolet-B radiation, *Ecology*, 65, 792, 1984.
67. Flint, S. D. and Caldwell, M. M., Influence of floral optical properties on the ultraviolet radiation environment of pollen, *Am. J. Bot.*, 70, 1416, 1983.
68. Teramura, A. H., Interaction between UV-B radiation and other stresses in plants, in *Stratospheric Ozone Reduction, Solar Ultraviolet Radiation and Plant Life*, Worrest R. C. and Caldwell, M. M., Eds., 375, NATO, Advanced Science Institutes Series G, Ecological Sciences, 8, 1986, 327.
69. Murali, N. S., Teramura, A. H., and Randall, S. K., Response differences between two soybean cultivars with contrasting UV-B radiation sensitivities, *Photochem. Photobiol.*, 48, 653, 1988.
70. Miller, J. E., Booker, F. L., Fiscus, E. L., Heagle, A. S., Pursley, W. A., Vozzo, S. F., and Heck, W. W., Effects of ultraviolet-B radiation and ozone on growth, yield and photosynthesis of soybean, *J. Environ. Qual.*, 23, 83, 1994.
71. Tevini, M., Mark, U., Fieser, G., and Saile, M., Effects of enhanced solar UV-B radiation on growth and function of selected crop plant seedlings, in *Photobiology*, Riklis, E., Ed., Plenum, New York, 1991, 635.
72. Naidu, S. L., Sullivan, J. H., Teramura, A. H., and DeLucia, E. H., The effects of ultraviolet-B radiation on photosynthesis of different aged needles in field grown loblolly pine, *Tree Physiol.*, 12, 151, 1993.
73. Iwanzik, W., Tevini, M., Dohnt, G., Voss, M., Weiss, W., Gräber, P., and Renger, G., Action of UV-B radiation on photosynthetic primary reactions in spinach chloroplasts, *Physiol. Plant.*, 58, 401, 1983.
74. Bornman, J. F., Target sites of UV-B radiation in photosynthesis of higher plants, *J. Photochem. Photobiol. B Biol.*, 4, 145, 1989.
75. Renger, G., Eckert, H. J., Fromme, R., Gräber, P., Volker, M., and Hohmveit, S., On the mechanism of photosystem II deterioration by UV-B irradiation, *Photochem. Photobiol.*, 49, 97, 1989.
76. Strid, A., Chow, W. S., and Anderson, J. M., Effects of supplementary ultraviolet-B radiation on photosynthesis in *Pisum sativum*, *Biochim. Biophys, Acta*, 1020, 260, 1990.
77. Prasil, O., Adie, N., and Ohad, I., Dynamics of photosystem II: mechanisms of photoinhibition and recovery processes, in *The Photosystems. Structure, Function and Molecular Biology*, Barber, J., Ed., Elsevier, Amsterdam, 1992, 295.

78. Jordan, B. R., He, J., Chow, W. S., and Anderson, J. M., Changes in mRNA levels and polypeptide subunits of ribulose 1,5-bisphosphate carboxylase in response to supplementary ultraviolet-B radiation, *Plant Cell Environ.*, 15, 91,,1992.
79. Baker, N., Nogués, S., and Allen, D. J., Photosynthesis and Photoinhibition, in *Plant and UV-B: Responses to Environmental Change*, Lumsden, P. J., Ed., Society for Experimental Biology Seminar Series 64, Cambridge University Press, Cambridge, 1997, 233.
80. Jordan, B. R., The molecular biology of plants exposed to ultraviolet-B radiation and the interaction with other stresses, in *Interacting Stresses on Plants in a Changing Climate*, Jackson, M. B. and Black, C. R., Eds., NATO Advanced Workshop, measurements, Sol. Energy, 49, 1993, 535.
81. Ziska, L. H., Teramura, A. H., Sullivan, J. H., and McCoy, A., Influence of ultraviolet-B (UV-B) radiation on photosynthetic and growth characteristics in field grown cassava (*Manihot esculentum* Crantz), *Plant Cell Environ.*, 16, 73, 1993.
82. Pfündel, E. E., Pan, R. S., and Dilley, R. A., Inhibition of violaxanthin deepoxidation by ultraviolet-B radiation in isolated chloroplasts and intact leaves, *Plant. Physiol.*, 98, 1372, 1992.
83. Sinclair, T. R., N'Diaye, O., and Biggs, R. H., Growth and yield of field-grown soybean in response to enhanced exposure to ultraviolet-B radiation, *J. Environ. Qual.*, 19, 478, 1990.
84. Björn, L. O. and Murphy, T. M., Computer calculation of solar ultraviolet radiation at ground level, *Physiol. Veg.*, 23, 555, 1985.
85. Flint, S. D., Jordan, P. W., and Caldwell, M. M., Research note: plant protective response to enhanced UV-B radiation under field conditions: leaf optical properties and photosynthesis, *Photochem. Photobiol.*, 41, 95, 1985.
86. Biggs, R. H., Kossuth, S. V., and Teramura, A. H., Response of 19 cultivars of soybeans to ultraviolet-B irradiance, *Physiol. Plant.*, 53, 19, 1981.
87. Teramura, A. H. and Sullivan, J. H., Effects of ultraviolet-B radiation on soybean yield and seed quality: a six-year field study, *Environ. Pollut.*, 53, 466, 1988.
88. Quaite, F. E., Sutherland, B. M., and Sutherland, J. C., Quantitation of pyrimidine dimers in DNA from UVB-irradiated alfalfa (*Medicago sativa* L.) seedlings, *Appl. Theor. Electrophor.*, 2, 171, 1992.
89. Pang, Q. and Hays, J. B., UV-B-inducible and temperature-sensitive photoreactivation of cyclobutane pyrimidine dimers in *Arabidopsis thaliana*, *Plant Physiol.*, 95, 536, 1991.
90. Langer, B. and Wellmann, E., Phytochrome induction of photoreactivating enzyme in *Phaseolus vulgaris* L. seedlings, *Photochem. Photobiol.*, 52, 861, 1990.
91. Warner, C. W. and Caldwell, M. M., Influence of photon flux density in the 400–700 nm waveband on inhibition of photosynthesis by UV-B (280–320 nm) irradiation in soybean leaves: separation of indirect and immediate effects, *Photochem. Photobiol.*, 38, 341, 1983.
92. Mirecki, R.M. and Teramura, A. H., Effects of ultraviolet-B irradiance on soybean. V. The dependence of plant sensitivity on the photosynthetic photon flux density during and after leaf expansion, *Plant. Physiol.*, 74, 475, 1984.
93. Cen, Y.-P. and Bornman, J. F., The response of bean plants to UV-B radiation under different irradiances of background visible light, *J. Exp. Bot.*, 41, 1489, 1990.
94. Tevini, M., Braun, J., and Fieser, G., The protective function of the epidermal layer of rye seedlings against ultraviolet-B radiation, *Photochem. Photobiol.*, 53, 329,1991.
95. Karabourniotis, G., Papadopoulos, K., Papamarkou, M., and Manetas, Y., Ultraviolet-B radiation absorbing capacity of leaf hairs, *Physiol. Plant.*, 86, 414, 1992.
96. Reuber, S., Bornman, J. F., and Weissenböck, G., A flavonoid mutant of barley (*Hordeum vulgare* L.) exhibits increased sensitivity to UV-B radiation in the primary leaf, *Plant Cell Environ.*, 19, 593, 1996.
97. Bornman J. F., Reuber, S., Cen, Y.-P. and Weissenböck, G., Ultraviolet radiation as a stress factor and the role of protective pigments, in *Plant and UV-B: Responses to Environmental Change*, Lumsden, P. J., Ed., Society for Experimental Biology seminar series 64, Cambridge University Press, Cambridge, 1997, 157.
98. Bornman, J. F. and Vogelmann, T. C., Effects of UV-B radiation on leaf optical properties measured with fibre optics, *J. Exp. Bot.*, 42, 547, 1991.
99. DeLucia, E. H., Day, T. A, and Vogelmann, T. C., Ultraviolet-B radiation and Rocky Mountain environment: measurement of incident light and penetration into foliage, in *Current Topics in Plant Biochemistry and Physiology*, Randell, D. D., Blevins, D. G., and Miles, C. D., Eds., University of Missouri, Columbia, 10, 1991, 32.

100. Vogelmann, T. C., Plant tissue optics, *Annu. Rev. Plant Physiol. Mol. Biol.*, 44, 231,1993.
101. DeLucia, E. H., Day, T. A., and Vogelmann, T. C., Ultraviolet-B and visible light penetration into needles of two species of subalpine conifers during foliar development, *Plant Cell Environ.*, 15, 921, 1992.
102. Braun, J. and Tevini, M., Regulation of UV-protective pigment synthesis in the epidermal layer of rye seedlings (*Secale cereale* L. cv Kustro), *Photochem. Photobiol.*, 57, 318, 1993.
103. Li, J., Ou-Lee, T. M., Raba, R., Amundson, R. G., and Last, R. L., *Arabidopsis* flavonoid mutants are hypersensitive to UV-B irradiation, *Plant Cell*, 5, 171, 1993.
104. Britt, A. B., Chen, J. J., Wykoff, D., and Mitchell, D., A UV-sensitive mutant of *Arabidopsis* defective in the repair of pyrimidine-pyrimidione (6-4) dimers, *Science*, 261,1571,1993.
105. Stapleton, A. E. and Walbot, V., Flavonoid can protect maize DNA from the induction of ultraviolet radiation damage, *Plant. Physiol.*, 105, 881, 1994.
106. Teramura, A. H., Tevini, M., and Iwanzik, W., Effects of ultraviolet-B irradiance on plants during mild water stress, Vol. I. Effects on diurnal stomatal resistance, *Physiol. Plant.*, 57, 175, 1983.
107. Tevini, M., Iwanzik, W., and Teramura, A. H., Effects of UV-B radiation on plants during mild water stress, II. Effects on growth, protein and flavonoid content, *J. Pflanzenphysiol.*, 110, 459, 1983.
108. Murali, N. S. and Teramura, A. H., Effectiveness of UV-B radiation on the growth and physiology of field-grown soybean modified by water stress, *Phytochem. Phytobiol.*, 44, 215, 1986.
109. Frior, J., *The Changing Atmosphere. A Global Challenge*, Yale University Press, New Haven, CT, 1990.
110. Edwards, G. E. and Walker, D. A., *C_3, C_4, Mechanisms and Cellular and Environmental Control of Photosynthesis*, Blackwell Science Publications, London, 1983.
111. Cure, J. D. and Acock, B., Crop responses to carbon dioxide doubling: a literature survey, *Agric. For. Meteor.*, 38, 127, 1986.
112. van de Staaij, J. W. M., Lenssen, G. M., Stroetenga, M. and Rozema, J., The combined effects of elevated CO_2 levels and UV-B radiation on growth characteristics of *Elymus athericus* (=*E. pycnanathus*), *Vegetatio*, 104/105, 433, 1993.
113. Rozema, J., Lenssen, G. M. and van de Staaij, J. W. M., The combined effect of increase atmospheric CO_2 and UV-B radiation on some agricultural and salt marsh species, in *The Greenhouse Effect and Primary Productivity in European Agroecosystems*, Goudriaan, J., van Keulen, H. and van Laar, H. H., Eds., Pudoc, Wageningen, 1990, 68.
114. Feder, W. A. and Shrier, R., Combination of UV-B and ozone reduces pollen tube growth more than either stress alone, *Environ. Bot.*, 30, 451, 1990.
115. Ros, J., Synergistic and/or antagonistic effects of enhanced (reduced) UV-B radiation (artificial, solar) at various nitrogen levels in crop plants, in *European Symposium on Effects of Environmental UV Radiation*, Nolan, C. and Bauer, H., Eds., European Commission (Brussels) and the GSF (Munich), 1995, 245.
116. Murali, N. S. and Teramura, A. H., Intensity of soybean photosynthesis to ultraviolet-B radiation under phosphorus deficiency, *J. Plant. Nutr.*, 10, 501, 1987.
117. Dube, L. S. and Bornman, J. F., The response of young spruce seedlings to simultaneous exposure of ultraviolet-B radiation and cadmium, *Plant Physiol. Biochem.*, 30, 761, 1992.
118. Panagopoulos, I., Bornman, J. F., and Björn, L. O., Response of sugar beet plants to ultraviolet-B (280–320 nm) radiation and *Cercospora* leaf spot disease, *Physiol. Plant.*, 84, 140, 1992.
119. Orth, A. B., Teramura, A. H., and Sisler, H. D., Effects of ultraviolet-B radiation on fungal disease development in *Cucumis sativus*, *Am. J. Bot.*, 77, 1188, 1990.
120. Gehrke, C., Johanson, U., Callaghan, T., Chadwick, D., and Robinson, C. H., The impact of enhanced ultraviolet-B radiation on litter quality and decomposition processes in *Vaccinium* leaves from the Subarctic, *Oikos*, 72, 213, 1995.
121. Johanson, U., Gehrke, C., Björn, L. O., and Callaghan, T. V., The effects of enhanced UV-B radiation on the growth of dwarf shrubs in a sub-arctic heathland, *Funct. Ecol.*, 9, 713, 1995.
122. Tosserams, M., Pais de Sà, A., and Rozema, J., The effect of solar UV radiation on four plant species occurring in a caostal grassland vegetation in the Netherlands, *Physiol. Plant.*, 97, 731, 1996.
123. Green, A. E. S., Cross, K. R., and Smith, L. A., Improved analytic characterization of ultraviolet skylight, *Photochem. Photobiol.*, 31, 59, 1980.

CHAPTER 6

Field Studies on Impacts of Air Pollution on Agricultural Crops

J. N. B. Bell and F. M. Marshall

CONTENTS

6.1 INTRODUCTION

Studies on the effects of air pollution on crops date back almost a century, but the great bulk of research has been conducted within the last 40 years, mainly in North America and Western Europe. Over the years the experimental techniques for conducting such investigations have become increasingly sophisticated. Early studies were largely conducted in the field with plants exposed directly to air pollution. Subsequently, more complex systems involving various types of chambers became employed, with comparisons of crop growth in ambient compared with filtered air and/or addition of controlled levels of pollutants. Initially, the chambers were closed, with a single pass of air through them, sometimes at unrealistically low flow rates, leading to a high aerodynamic resistance to uptake of pollutants by the crop, thereby resulting in misleading assumptions concerning thresholds for response.[1] Subsequently, technology improved, with better control of exposure conditions. Closed chambers have been used extensively outdoors, in glasshouses, and in controlled environment conditions. A major breakthrough in chamber technology occurred 25 years ago, with the development of open-top systems,[2] which reduced the "greenhouse effect" of closed chambers and resulted in internal micrometeorological conditions approximating more closely to the ambient, thereby lowering the possibility of interactions between these and pollution impacts. Despite continued development of open-top chambers and their utilization in extensive national and international research programs, such as the U.S. National Crop Loss Assessment Network[3]

1-56670-341-7/00/$0.00+$.50

and the European Open Top Chambers Programme[4] there have remained concerns about their level of modification of microclimate, in particular altering fluxes of pollutants to foliage.[5] This has led to the development of even more advanced technology, in the form of open-air fumigation or exclusion systems, where computer-controlled application of pollutants takes place on field crops,[6] or else comparisons made between ambient and clean air by blowing filtered air over low-growing species.

All these experimental techniques have their own merits and disadvantages. Field studies without chambers represent the most realistic conditions, but each experiment inevitably reflects the net results of the pollutant regime and prevailing environmental conditions during the period concerned, thereby raising problems of interpretation and repeatability. In contrast, the more controlled an experiment, the farther is it removed from the real world, but the more readily it is interpreted and repeated. Ideally, combinations of both field- and laboratory-based experiments should be used to determine the magnitude of effects of ambient air pollution on crops, and to elucidate possible environmental interactions. However, such an approach is limited to a relatively small number of research programs, in view of the high costs entailed.

The progression in recent years toward more advanced exposure technology has in many ways detracted from the simple field-based approach, involving measurements of performance of crops at locations with different levels of air pollution. Often, but not always, such studies are conducted along transects representing gradients of air pollution, with plants grown in containers under as standardized culture regimes as possible. Such transects may be a relatively short distance away from point or line sources or of some considerable length from the center of large cities out into the surrounding rural areas. In this paper, a brief review will be made of research programs of this type in which the effects of ambient air pollution on crops have been determined. The case will be presented that such simple and cheap approaches have much to offer, particularly if the results are analyzed and interpreted in an appropriate manner, and their particular value for research in developing countries will be highlighted.

6.2 STUDIES ALONG GRADIENTS OF AIR POLLUTION IN AND AROUND CITIES

The classic early studies along urban gradients of air pollution are described in a multidisciplinary book: *Smoke: A Study of Town Air*,[7] which is concerned with an extensive research program conducted over some years prior to World War I, in the northern English industrial city of Leeds. Standard cultures of radish (*Raphanus sativus*), lettuce (*Lactuca sativa*), and spring cabbage (*Brassica oleracea*) were grown at six sites with a marked gradient of particulate and acid deposition, covering at least a tenfold and eightfold range, respectively. The effects of pollution were dramatic, with yield reductions being closely correlated with levels of deposition (Table 6.1) and maximum reductions in dry weight occurring at the most polluted site, compared with the clean site, of 55, 76, and 81% for radish, lettuce, and spring cabbage, respectively. An intermediate planting employed winter cabbage planted out in late September. In this case a large number of plants failed to survive the winter, with all specimens dying at the most polluted site by mid-November, whereas at the cleanest site all were still alive in March. This latter observation foreshadowed remarkably the clear demonstration some 80 years later by a number of workers that the impacts of low-temperature stress on plants can be increased markedly as a result of exposure to a range of air pollutants.[8] A similar observation was made by Bleasdale[9] who exposed plants at sites with different levels of coal-smoke pollution in and outside Manchester in the early 1950s, as a supporting study to his pioneering filtration experiments in that city.[10] In this exercise, spring cabbages grown at a clean air site in the countryside 27 km from the city center showed a 96% survival rate, whereas at a heavily polluted site 1.6 km from the center only 37% survived the winter. This reflected effects on *Myosotis vulgaris* (forget-me-not) and *Pinus sylvestris* (Scots pine) in the same study, where intermediate values were found at a moderately polluted site 4 km from the city center, where, unfortunately, the cabbages were eaten by a rabbit. Such effects were ascribed to a frost/pollution interaction.

Table 6.1 Effects of Urban Coal-Smoke Pollution in Leeds on Vegetable Crops 1911 – 1917

Location	Solid Deposition, kg ha^{-1}	Acidity, kg ha^{-1}	*Raphanus sativus*, dry wt. g	*Lactuca sativa*, dry wt. g	*Brassica oleracea*, dry wt. g
Garforth	—	—	5.5	18.2	349
Weetwood Lane	178	12	5.0	15.1	497
Headingley	306	21	4.4	11.4	390
University	447	29	3.3	10.2	365
Park Square (city center)	949	50	2.7	5.3	154
Hunslet (central industrial)	1754	101	2.5	4.4	66

Source: Adapted from Cohen and Ruston, 1925.[7]

In the studies described so far there was overwhelming evidence of polluted air reducing the growth of crops in the urban environment. However, the coal-smoke-polluted atmosphere of both Leeds and Manchester would have contained a mix of pollutants, thereby hindering precise attribution of causality, although Bleasdale's filtration work suggested strongly that SO_2 was the prime culprit.[10] Furthermore, transects from city centers outward do not only lie along gradients of air pollution, but also cover gradients of climatic parameters, particularly temperature which increases in urban areas. In order for the confounding effects of mixtures of pollutants and climatic factors to be elucidated, it is necessary to subject the experimental data to multivariate analysis. An early study in this respect was carried out with peas (*Pisum sativum*) around the northern English city of Newcastle upon Tyne.[11] Here, plants were exposed at 12 sites with different pollution levels, where SO_2 and fluoride deposition rates and minimum and maximum temperatures, as well as visible foliar injury and plant dry weights, were measured. The data were subjected to canonical correlation analysis, which showed a clear correlation between SO_2 levels and the appearance of visible necrotic injury on the plants, but not with fluoride, across a range of temperatures, which in themselves accounted for about 25% of the variation in crop performance.

A more ambitious transect study, involving a multivariate approach which attempted to separate out effects of SO_2, NO_2, O_3 and climatic variables was reported by Ashmore et al.[13,14] In this case 18 sites were employed from central London out to Ascot, in a rural location 37.5 km away, where peas,[14] barley (*Hordeum vulgare*), and red clover (*Trifolium pratense*)[13] were grown to maturity along an air pollution gradient. Pollutant levels varied along the transect, with NO_2 showing a strong negative correlation with distance from central London and some indications (nonsignificant) of a similar pattern for SO_2, while O_3 (measured on the basis of injury on the tobacco bioindicator cv. Bel-W3) showed a general but not significant increase in the opposite direction. As well as for the pollutants, measurements were carried out at each site of maximum and minimum temperatures and rainfall, and estimates made of the degree of exposure and of shading. In the case of peas, there was a significant positive correlation between seed dry weight of 4 cultivars and distance from London,[14] data for one of which are illustrated in Figure 6.1. The study was conducted during periods of high O_3 concentrations, and it was suggested that the decrease in growth toward central London was the result of impacts of this pollutant being enhanced by SO_2 and NO_2.[15] Support for this hypothesis was provided in the case of SO_2 by a complementary closed chamber fumigation experiment, where the addition of only 8 ppb of this pollutant to ambient air, containing elevated levels of O_3, produced substantial yield reduction compared with filtered air, but no effect when added to the latter.[14] The results of the transect study on red clover and barley reported by Ashmore et al.,[13] were subjected to multiple regression, which included both pollutant and climatic variables. The resulting analysis produced some very important results in that they showed a very clear influence of both SO_2 and NO_2 in reducing performance of the crops, although a complex pattern emerged with different effects (including some stimulations) according to pollutant, species, cultivar, and growth yield component concerned. During this experiment, ambient O_3 levels were much lower than in the case of the pea studies, and relatively few effects of this pollutant were detected from the analysis. An important aspect of this study is that it indicated impacts of SO_2 and NO_2 at less than 20 and 25 ppb, respectively. While the analysis permitted the separation of pollutant impacts from the confounding influence of climatic variables along the transect, it did not allow for possible synergistic interactions between the pollutants, as postulated in the study with peas. However, it is an excellent example of the way in which a simple transect study can be utilized to determine impacts of low levels of pollutants under field conditions, in the absence of any artificial constraints imposed by chambers.

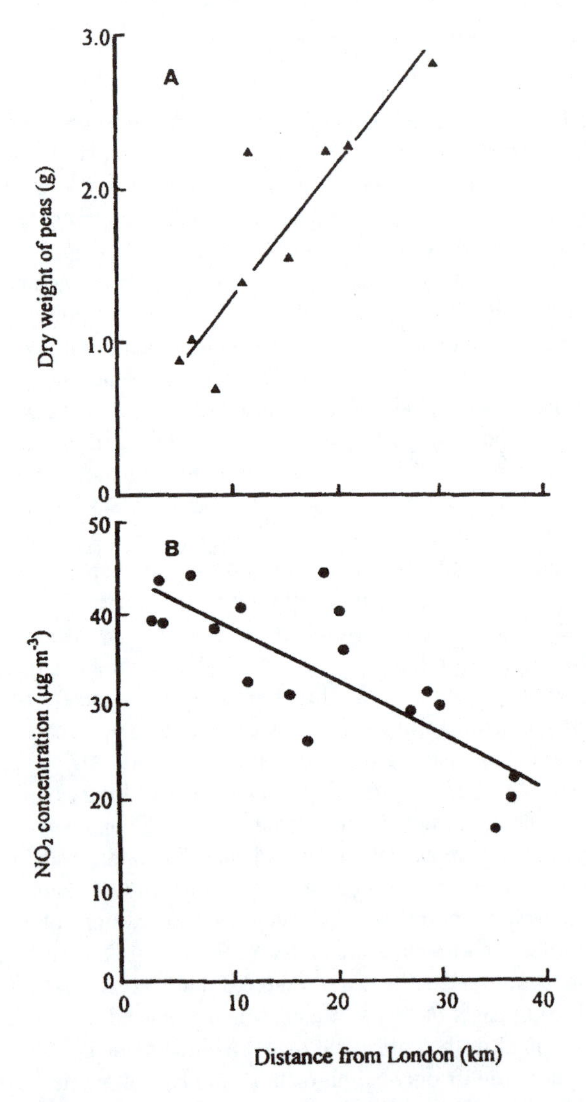

Figure 6.1 Dry weight (A) of *Pisum sativum* cv. Progreta grown along a 37.5-km transect out of central London and accompanying mean NO_2 concentrations (B). (From Mansfield and Lucas,[12] using data of Ashmore et al.,[13] and Ashmore and Dalpra.[15] With permission).

6.3 STUDIES ON THE IMPACTS OF O_3 IN THE FIELD

Field studies on the impacts of O_3 are hindered by the invariable lack of steep gradients due to the secondary nature of this pollutant. The most appropriate approach in this case is to utilize a chemical protectant against O_3 in field plots (see below). However, the work of Oshima et al.,[16,17] has shown that meaningful results can be obtained in which crop yield is related to O_3 dose over a large enough area to result in a substantial range of the latter. In this case alfalfa[16] and tomato,[17] plants were exposed in standard plots across the South Coast Air Basin of California. A clear significant negative relationship was obtained for the yield of both crops and the O_3 dose above 100 ppb, this being illustrated for alfalfa in Figure 6.2.

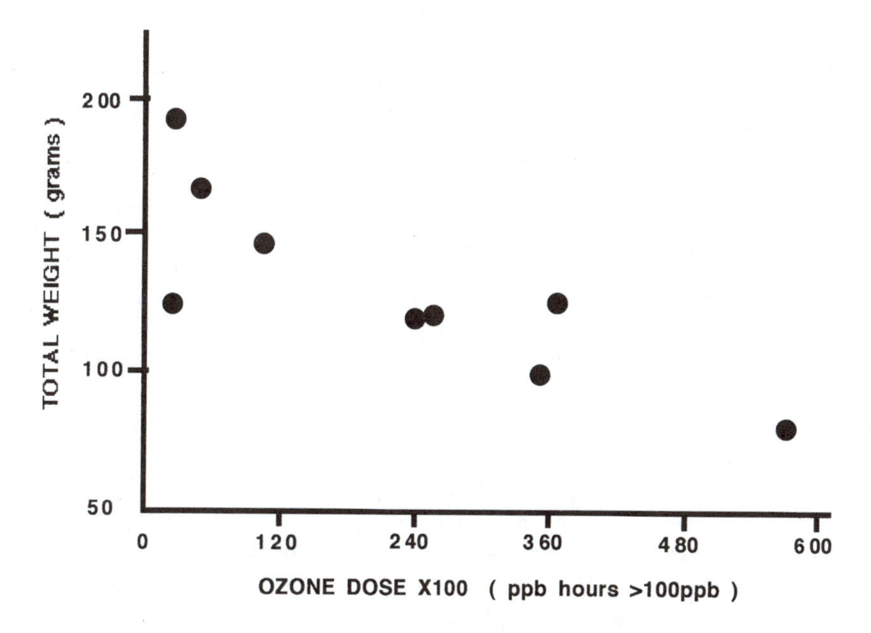

Figure 6.2 Total dry weight of *Medicago sativa* grown at field sites in Southern California in relation to ambient O_3 dose. (Adapted from Oshima, R. J. et al., *J. Air. Pollut. Control Assoc.*, 26, 861, 1976.)

For some 30 years or so it has been recognized that certain broad-spectrum fungicides, such as benomyl, can provide some degree of protection for crops against O_3. However, the use of such chemicals as protectants against ambient O_3 in the field in order to provide control plants in experimental studies was inevitably fraught with problems of interpretation due to possible impacts on growth via reduction in the adverse effects of fungal pathogens. In the late 1970s a major breakthrough occurred in this approach, with the development by E.I. Dupont de Nemours & Co. (Inc.) of the O_3-protectant chemical *N*-[2-(2-oxo-1-imadazolidinyl)ethyl]-*N* phenylurea, which is usually referred to as ethylenediurea (abbreviated as EDU). This proved to be capable of providing partial or total protection against O_3, when applied to soil or foliage, in the absence of fungicidal action, and has subsequently been utilized extensively as an experimental tool for field studies. The development of EDU proved to be the stimulus for a renaissance of field studies on crops in the developed world. Initially, such research was conducted almost entirely in North America, with crops including soybean (*Glycine max*), *Phaseolus* bean, potato (*Solanum tuberosum*), and radish.

The value of this approach did not pass unnoticed in Europe, where in the 1980s there was growing interest in the possibility of widespread reductions in crop growth due to O_3. The United Nations Economic Commission for Europe (UNECE) established a major field research program with EDU, under the auspices of its Long Range Transboundary Air Pollution Convention aimed at assessing impacts of O_3 on model crops across Europe. This involved growth of beans, clover, tobacco (*Nicotiana tabacum*) watermelon (*Citrullus lanatus*), and radishes in up to 21 countries, under standard conditions, with or without application of EDU. The results of this study are of considerable significance as effects of EDU in reducing both visible injury and biomass or yield were demonstrated at many locations, particularly in the case of the former parameter, with clear exceedance of the critical levels for O_3 developed to protect crops against this pollutant in Europe.[18]

6.4 STUDIES ON AIR POLLUTION IMPACTS NEAR POINT SOURCES

There is a substantial literature describing air pollution impacts on crops at different distances from point sources of pollutants, which examine changes in various parameters over short transects. Many of these have been conducted in India, which is the only developing country to have made any major attempt to address the problem of air pollution in relation to crops. However, in the developed world, a number of studies have utilized steep gradients of pollution from point sources in an endeavor to elucidate secondary impacts of pollutants. Several studies of this type have been carried out in the U.K., concerned with evolution of tolerance to air pollution in grasses and with the influence of pollution on herbivorous insect pests.

In the 1970s research developed into the evolution of tolerance to air pollution in plants growing in polluted areas, under the selective pressure of this stress. Subsequently, strong evidence was produced of the development of tolerance in a number of species, mainly grasses, to SO_2 and NO_2,[19] and hydrogen fluoride,[20] while more recently a major research program has developed in the U.K. into O_3 and *Plantago major*.[21] The early demonstration of a correlation between ambient SO_2 levels and tolerance to this pollutant of local pasture grass populations[22] was subsequently supported in terms of causality by a point source transect study.[23] A 2-km transect was constructed downwind of a point source of SO_2 in the form of a smokeless fuel works in a rural area of northern England and *Festuca rubra* (red fescue) populations sampled at 8 sites along this, from locations with different levels of pollution. When the populations were fumigated with a high level of SO_2, it was found that there was a marked fall in the amount of visible injury continuously along the transect toward the factory. Thus, this simple transect approach showed the selective pressure of different concentrations of SO_2 on a pasture grass species.

Isolated point sources represent an excellent opportunity for studying the impacts of pollutants on pathogens and pests of crops under natural conditions. Many observations have been made over the years on changes in the level of attack by insect pests and fungal pathogens on vegetation in polluted areas, which have led to controlled studies that demonstrate causality. However, surprisingly, relatively few field studies have been carried out into this phenomenon along transects representing gradients of air pollution. One such study was carried out over a 12-km transect downwind of a point source of SO_2 in the form of a cokeworks in northern England. Here potted *Vicia faba* (broad bean) and *Triticum aestivum* (wheat) plants were exposed at four locations along the transect for 42 days and then transferred to a controlled temperature room, where they were infested with the major aphid pests *Aphis fabae* and *Rhopalosiphon padi*, respectively. Both aphids showed an increase in mean relative growth rate (MRGR) on plants which had been grown closer to the point source, with a significant positive correlation with the SO_2 concentrations at the sites concerned (Houlden, McNeill, and Bell, unpublished). A more ambitious study over a longer transect was conducted by Bell et al.,[8] who exposed barley and wheat plants at seven locations along the transect utilized by Ashmore et al.[13,14] The plants were left in the field for 6 weeks and then transferred to a constant temperature room where the MRGR was determined over 4 days for *R. padi* and *Metapolophium dirhodum* feeding on wheat and barley, respectively. Both aphids showed a substantial increase in MRGR toward central London, with significant positive correlation with both SO_2 and NO_2 (data for *M. dirhodum* are shown in Figure 6.3). This remarkable demonstration of an impact of relatively low levels of pollutants on the performance of two important pests via changes in the host crops, provided strong evidence that similar results obtained in complementary studies with the same species combinations, using fumigations[24] and filtrations,[25] held good under field conditions. As part of the same study, a very broad field-based approach was adopted, where natural aphid populations sampled over 6 years at 18 locations around Britain, were related to prevailing SO_2 concentrations at the sites concerned.[8] This demonstrated that out of 83 aphid species examined 73 showed a positive correlation with SO_2 levels, 39 of which were significant. This again illustrates the value of a field-based approach, in this case at a national scale, in combination with more controlled experimental studies.

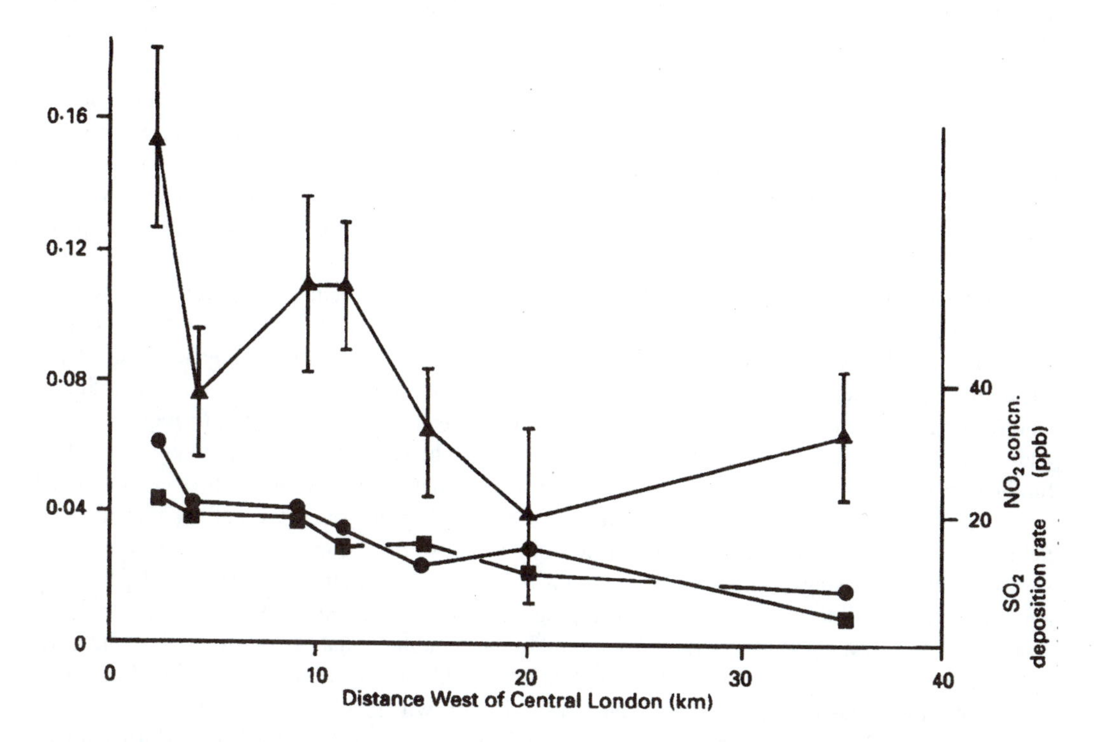

Figure 6.3 MRGR over 4 days (95% confidence limits) of *Metapolophium dirhodum* on barley previously grown for 6 weeks along an air pollution gradient from central London. ▲ MRGR; ■ SO_2 deposition rate; ● NO_2 concentration. (From Bell, J. N. B. et al., *Parasitology,* 106, 2410, 1983. With permission.)

6.5 FIELD STUDIES ON IMPACTS OF AIR POLLUTION ON AGRICULTURAL CROPS IN DEVELOPING COUNTRIES

Crop yield reductions due to air pollution in developing countries may have major implications for food security, nutrition, and economic output. Clearly, there is a need to assess the impacts of air pollution in field conditions along with other environmental threats to sustaining and/or increasing crop production, and to compare losses with other more widely recognized contraints on crop production.

The majority of research into air pollution impacts on crops that has been done outside North America, Europe, and Japan has taken place in India, where there are a number of very active research teams. A recent review[26] cited over 50 studies on 20 major crops in India over the past 20 years. Among these, there are a number of notable examples of field studies to assess the impacts of gaseous pollutants around point sources in India, such as chemical factories, thermal power plants, and petroleum refineries. The review identified nine such studies in India. However, the majority of the studies on air pollution impacts have been carried out in controlled fumigation systems using sulfur dioxide, where pollutant concentrations are often considerably higher than would be experienced in ambient conditions.

One of the most comprehensive field-based studies concerned with gaseous pollutants around a point source was carried out as a component of a major project to assess the environmental degradation in the Obra-Renukoot-Singrauli area of Uttar Pradesh, India.[27] This project included an assessment of the impacts of air pollutants on wheat in the vicinity of the Obra thermal power plant. Sulfur dioxide, NO_2, and total suspended particulate were measured at distances of 1.5 to 5.0 km from the source, mainly in the direction of the prevailing wind, and control plants were placed 22 km from the source in the opposite direction. Multiple regression analysis revealed that SO_2 was the major pollutant responsible for the observed yield reductions of between 29% (3 km NE of the plant) and 47% (1.5 km SE of the plant).

The linear relationship between wheat yield reduction and ambient sulfur dioxide concentration shown in this study was used in a calculation to estimate yield losses from sulfur dioxide around other thermal power plants in India, as part of a risk assessment exercise.[28] This was an illustrative exercise to show the potential yield losses within a 10-km range of the plants. The results ranged from an estimated loss of 13% for a power plant of 440 MW output capacity to 59% for a plant with a capacity of 2050 MW.

Studies of this type help to highlight the crops, areas, and pollutants of concern, but need to be followed up by detailed and extensive field-based investigations. In India, the ultimate aim is to establish guidelines for ambient air quality standards that are appropriate to agricultural crops; these will complement the existing standards which are based on protection of human health and cultural heritage. In order to achieve this, reliable dose-response relationships are required for a wide range of crops. This will require a program of field research to complement controlled fumigation studies. Both should be carried out using standard methodologies, in areas impacted by different levels of air pollutants and in different agroecological zones. One of the priorities for field research is to establish a clearer definition of the areas affected around cities and industries.

In the face of rising urban populations, and the increasing reliance on urban and periurban agriculture for food security in many developing countries, these areas are a priority for future research. A forthcoming study by the present authors and Indian collaborators funded by the U.K. Department for International Development will involve field investigations of crop yield losses due to air pollution in Delhi and Varanasi. The Varanasi work will involve a transect study from the central urban area to the peri-urban fringe, while in Delhi the emphasis will be on periurban areas. This is part of a wider study to assess the impacts of losses in agricultural production due to air pollution on livelihoods in developing country cities, to identify the causes of these impacts and to recommend policy measures to alleviate the effects.

Recent research efforts in China have also focused on urban areas. An important transect study,[29] indicating major impacts of ambient air pollution on both crops and their associated pests, has been carried out in the heavily polluted Chinese city of Chongqing where both SO_2 and particulate arising from coal production are at particularly high levels. Four vegetable species showed a marked decline in yield as pollutant levels increased and this was accompanied by a rise in infestation by insect pests, thus reflecting experience with transect studies reported in the U.K.; however, the design of the study did not permit separation of direct effects of pollutants on the crops and that resulting from pollution-induced stimulation of performance.

Where ozone is the major phytotoxic pollutant of concern, EDU is in many ways the ideal approach for assessing impacts on crops in developing countries. The simplicity of this technique, avoiding the expense and logistical complexities of chambers, pumps, filters, and associated equipment, has much to recommend it for such circumstances. However, before any reasonably quantitative estimate can be made of effects of ambient O_3 on crop yield using EDU, it is necessary to test the effects of this protectant under controlled conditions, in order to determine the correct dose for full (if possible) protection and also to ascertain that no effects occur on the species/cultivar concerned when growing in clean air. This in itself necessitates the use of O_3-fumigation facilities, which are often not available in developing countries, indicating the desirability of partnerships with institutions in developed nations that can provide these. EDU is also not available commercially and has to be synthesized in suitable laboratories, requiring expertise and finance. In this respect, recent experience at Imperial College has indicated that the production of 800 g of EDU requires about 30 person-hours of technician time and about £400 worth of chemicals (R. Lincoln, personal communication).

A number of field studies have been carried out with EDU in developing countries, which provide some of the best evidence for O_3 affecting crop yield in such places. A pioneering study was carried out by Bambawale,[30] who demonstrated that application of EDU to potatoes growing in the Indian Punjab reduced visible O_3-like symptoms on the foliage and improved growth, thus demonstrating the potential for ozone problems for crop production in the subcontinent. Also in the 1980s a study[31] was carried out in the vicinity of Mexico City, in an area known to be severely impacted by oxidant pollution: EDU was applied to *Phaseolus* beans and showed an increase in yield. Recently, such studies have been extended to the Middle East, with EDU experiments[32] in Egypt — a country in which oxidant problems could be expected in view of the high levels of emissions of precursors in the proximity of irrigated intensive crop production, with climatic conditions particularly conducive to photochemical O_3 production. A local cultivar of radish and of turnip (*Brassica rapa*) were grown at a rural site in the Delta some 35 km away from Alexandria. A fumigation study in Britain prior to the start of the study indicated no effect of EDU on either species in clean air and total or partial suppression of pollutant effects on yield when grown in the presence of O_3, depending on the parameters measured. When EDU was applied as a soil drench at the field sites, it almost eliminated the visible injury symptoms on both radish and turnip which were higher in the former species (Table 6.2). This differential sensitivity was also seen in the effects of EDU on yield where root and shoot weight of radish were reduced by 30 and 17% in its absence, with corresponding figures for turnip of 17 and 11%, respectively (Table 6.2).

Table 6.2 Effect of EDU on Visible Symptoms and Growth Parameters on Turnips and Radishes at a Rural Site in Egypt

	Radish		Turnip	
	-EDU	+EDU	-EDU	+EDU
Degree of visible injury (0–5 scale)	1.33	0.15	0.48	0.17
Root dry wt. (g)	0.58	0.79	0.67	0.81
Shoot dry wt. (g)	2.09	2.50	1.90	2.10

Note: All effects of EDU significant at $P < 0.05$.

Source: Adapted from Hassan, I. A. et al., *Environ. Pollut.*, 89, 107, 1995.

Very recently, a study has been carried out in Pakistan that provides further strong evidence for O_3 impacts in a developing country on an important crop. In this case a local soybean cultivar was grown to maturity, with or without a soil drench of EDU (which had already been tested on this cultivar under controlled conditions in the U.K.), at three sites: on the urban fringe of Lahore and at two rural locations in agricultural areas some 35 km away from the city center, one of which was close to a major road and industry. This study was primarily aimed at determining whether, and to what extent, O_3 impacted crops in rural areas compared with the urban fringe site, where filtration experiments in open-top chambers had already demonstrated large effects on the yield of rice[33] and wheat.[34] The results of this program not only confirmed that effects of O_3 occurred well out into the major crop-growing areas of the Punjab, but also that yield reductions increased compared with the urban fringe site, rising to around 60% in seed weight (Wahid, Shamsi, Milne, Marshall, and Ashmore, unpublished). It is noteworthy that these impacts occurred in the Pakistan

Punjab, not far from the location where Bambawale[30] showed effects of EDU on the Indian side of the border, suggesting that O_3 impacts may be widespread in this part of the subcontinent. Currently, further studies are in progress in and around Delhi, where EDU responses are being detected in tomato and soybean (C. K. Varshney and C. Rout, personal communication).

6.6 CONCLUSIONS

The use of field studies involving growth of plants at locations with different levels of air pollution represents a simple, effective, and yet underutilized tool for determining the effects of ambient pollution on crop yield. Inevitably, studies over large areas, transect or otherwise, will be subject to confounding impacts of changes in other environmental variables. Some of these can readily be eliminated by use of standardized growing conditions in terms of water supply, soil type, and containers. However, climatic variables present a much more intractable problem, with likely differences in temperature, humidity, exposure, and insolation between experimental sites, particularly in urban areas where the "heat island" and other effects will influence crop growth. In this case the establishment of a large number of sites and the concomitant measurement of all important variables likely to influence crop growth are vital. Sadly, such an approach has been ignored in some studies conducted in developing countries where comparisons of plant performance between a single "polluted" and "unpolluted" site often indicate possible important effects of pollutants, but the lack of intermediate locations and of measurements of environmental parameters other than air pollution results in a high level of uncertainty in interpreting the results. The presence of a number of variables that may influence crop growth calls for the employment of multivariate analysis of data from field studies, using techniques such as canonical correlation analysis and multiple regression. It is surprising that more use has not been made of such approaches, which can give a much clearer indication of causality, including the role of different pollutants, although interactions between these will not be revealed.

A further value of field studies across areas with different levels of pollutants is that they can be used to generate dose–response relationships, including thresholds for crop response. Such data are obtained under realistic field conditions, devoid of the constraints imposed by even the most sophisticated chamber systems. At the same time, the data must be viewed as representing the net effect of the particular environmental conditions prevailing during the experimental period, with dose–response relationships likely to vary between studies carried out on different occasions. Such problems also occur with outdoor chamber experiments, and thus understanding of the magnitude of interactions between pollutant and climatic variables requires complementary experiments to be carried out under controlled environments. The use of protectant chemicals as a control against pollution impacts in field studies has much to recommend it, in terms of demonstrating effects of individual pollutants. However, such an approach requires thorough testing of the chemical concerned for possible impacts under unpolluted conditions, and for the degree of protection provided under different pollutant and protectant concentrations. Thus, there must always be a degree of uncertainty in the use of this method for quantifying accurately the level of crop yield reduction taking place in the field. Furthermore, O_3 is the only pollutant that has so far been addressed successfully in this manner. The development of protectant chemicals for use with other major phytotoxic pollutants would be very useful and could enable the development of field studies to separate out the effects of individual gases.

Short gradients of pollutants away from point or line sources represent an excellent opportunity to study impacts on crops where changes in other environmental parameters are likely to be minimal. Such studies are particularly valuable where only a single phytotoxic pollutant is involved, thereby greatly facilitating interpretation of the results. The absence of any sort of enclosure around the experimental plants permits their exposure to all the normal stresses that a crop experiences in the field. Thus, transect studies over relatively short distances have great potential for elucidating interactions between air pollution and frost damage, infestation by invertebrate pests and infection by fungal, bacterial, and viral pathogens. Again, this is a seriously underexploited approach; yet the few studies to date indicate that it can yield valuable information, although controlled studies are still necessary for confident interpretation of the results. Transect studies also have potential for use in research on changes in species composition of agricultural pasture swards, and have been used to demonstrate impacts on intraspecific tolerance in grasses.

The greatest value of field studies lies in assessment of crop losses in the developing world. The generally simple and cheap characteristics of the technique are particularly appropriate in such circumstances, where a lack of funds, facilities, and trained personnel hinder the use of sophisticated chamber or outdoor exposure systems. The growing evidence of major impacts of air pollution on crops in developing countries,[35] where the necessity to increase food production in the face of rapid population increase is of paramount importance, underlines the value of using field studies to improve understanding of this problem. In the developed world, the value of field studies has been demonstrated over a range of scales from very local to continental, yet remains relatively little exploited. The possible reasons for this are manifold, but there must be some suspicion that such a simple technique does not recommend itself to either the scientist or the funding agencies, in an era when good and exciting science is increasingly seen as necessitating complex and sophisticated experimentation.

REFERENCES

1. Roberts, T. M., Darrall, N. M., and Lane, P., Effects of gaseous air pollutants on agriculture and forestry in the UK, *Adv. App. Biol.*, 9, 1, 1983.
2. Heagle, A. S., Philbeck, R. B., and Heck, W. W., An open-top chamber to assess the impact of air pollution on plants, *J. Environ. Qual.*, 2, 365, 1973 .
3. Adams, R. M., Glyer, J. D., and McCarl, B. A., The NCLAN economic assessment: approach, findings and conclusions, in *Assessment of Crop Loss from Air Pollutants*, Heck, W. W., Taylor, O. C., and Tingey, D. T., Eds., Elsevier Applied Science, New York, 1988, 473.
4. Jäger, H.-J., Unsworth, M., De Temmerman, L., and Mathy, P., Effects of Air Pollution on Agricultural Crops in Europe, Air Pollution Report 46, CEC, Brussels, 1994.
5. Fuhrer, J., Skärby, L., and Ashmore, M. R., Critical levels for ozone effects on vegetation in Europe, *Environ. Pollut.*, 97, 91, 1997.
6. McLeod, A. R. and Baker, C. K., The use of open field systems to assess yield response to gaseous air pollutants, in *Assessment of Crop Loss from Air Pollutants*, Heck, W. W., Taylor, O. C., and Tingey, D. T., Eds., Elsevier Applied Science, New York, 1988, 181.
7. Cohen, J. B. and Ruston, A. G., *Smoke: A Study of Town Air*, Edward Arnold, London, 1925.
8. Bell, J. N. B., McNeill, S., Houlden, G., Brown, V. C., and Mansfield, P. J., Atmospheric change: effect on plant pests and diseases, *Parasitology*, 106, S11, 1993.
9. Bleasdale, J. K. A., Atmospheric Pollution and Plant Growth, Ph.D. thesis, University of Manchester, U.K., 1952.
10. Bleasdale, J. K. A., Effects of coal-smoke pollution gases on the growth of rye grass (*Lolium perenne* L.), *Environ. Pollut.*, 5, 275, 1973.
11. Young, J. E. and Matthews, P., Pollution injury in south east Northumberland: the analysis of field data using canonical correlation analysis, *Environ. Pollut, (Ser. B).*, 2, 353, 1981.
12. Mansfield, T. A. and Lucas, P. W., Effects of gaseous pollutants on crops and trees, in *Pollution, Causes, Effects and Control*, Harrison, R. M., Ed., The Royal Society of Chemistry, Cambridge, 1996, 266.
13. Ashmore, M. R., Bell, J. N. B., and Mimmack, A., Crop growth along a gradient of ambient air pollution, *Environ. Pollut.*, 53, 99, 1988.
14. Ashmore, M. R., Mimmack, A., and Bell, J. N. B., Effects of ambient air pollution on crop species in and around London., in *Air Pollution and Ecoystems*, Mathy, P., Ed., D. Reidel, Dordrecht, the Netherlands, 1988, 641.

15. Ashmore, M. R. and Dalpra, C., Effects of London's air on plant growth, *London Environ. Bull.*, 3, 4, 1985.
16. Oshima, R. J., Poe, M., Braegelmann, P. K., and Baldwin, D. W., Ozone dosage – crop loss function for alfalfa: a standardised method for assessing crop losses from air pollutants, *J. Air Pollut. Control Assoc.*, 26, 861, 1976.
17. Oshima, R. J., Braegelmann, P. K., Baldwin, D. W., van Way, V., and Taylor, O. C., Reduction of tomato fruit size and yield by ozone, *J. Am. Soc. Hortic. Sci.*, 102, 289, 1977.
18. Benton, J., Fuhrer, J., Gimeno, B. S., Skärby, L., and Sanders, G., Results from the UN/ECE ICP — crops indicate the extent of exceedance of the critical levels of ozone in Europe, *Water Air Soil Pollut.*, 85, 1473, 1995.
19. Bell, J. N. B., Ashmore, M. R., and Wilson, G. B., Ecological genetics and chemical modifications of the atmosphere, in *Ecological Genetics and Air Pollution*, Taylor, G. E., Pitelka, L. F., and Clegg, M. T., Eds., Springer-Verlag, New York, 1991, 33.
20. Horner, J. M. and Bell, J. N. B., Evolution of fluoride tolerance in *Plantago lanceolata, Sci. Total Environ.*, 159, 163, 1995.
21. Reiling, K. and Davison, A. W., Spatial variations in ozone resistance of *Plantago major* after summers with high ozone concentrations, *New Phytol.*, 131, 337, 1992 .
22. Ayazloo, M. and Bell, J. N. B., Studies on the tolerance to sulphur dioxide of grass populations in polluted areas. I. Identification of tolerant populations, *New Phytol.*, 88, 203, 1981.
23. Wilson, G. B. and Bell, J. N. B., Studies on the tolerance to sulphur dioxide of grass populations in polluted areas. IV. The spatial relationship between tolerance and a point source of pollution, *New Phytol.*, 102, 563, 1986.
24. Houlden, G., McNeill, S., Aminu-Kano, M., and Bell, J. N. B., Air pollution and agricultural aphid pests. I. Fumigation experiments with SO_2 and NO_2, *Environ. Pollut.*, 67, 305, 1990.
25. Houlden, G., McNeill, S., Craske, A., and Bell, J. N. B., Air pollution and agricultural aphid pests. II. Chamber filtration experiments, *Environ. Pollut.*, 72, 45, 1991.
26. Varshney, C. K., Agrawal, M., Ahmad, K. J., Dubey, P. S., and Raza, S. H., Effect of Air Pollution on Indian Crop Plants. Project Report submitted to Imperial College Center for Environmental Technology, London. Jawaharlal Nehru University, New Delhi, 1997.
27. Singh, J. S., Singh, K. P., and Agrawal, M., Environmental Degradation of Obra-Renukoot-Singrauli Area and its Impact on Natural and Derived Ecosystems, Project report submitted to Ministry of Environment and Forests, Government of India, 14/167/84MAB/EN-2 RE, 1990.
28. Ashmore, M. R. and Marshall, F. M., The Impacts and Costs of Air Pollution on Agriculture in Developing Countries, Final technical report submitted to the Department for International Development, Environment Research Program, Project number ERP 6289, Imperial College of Science Technology and Medicine, London, 1997.
29. Zheng, Y., Last, F. T., Xu, Y., and Meng, M., The effects of air pollution climate in Chongqing on four species of vegetable, *Chongqing Environ. Sci.*, 18, 29, 1996.
30. Bambawale, O. M., Evidence of ozone injury to a crop plant in India, *Atmos. Environ.*, 20, 1501, 1986.
31. Laguette Rey, H. D., de Bauer, L. I., Shibata, J. K., and Mendoza, N. M., Impacto de los oxidante ambtales en el cultivo de frijol, en Montecillos, estado de Mexico, *Centro do Fitopatologia*, 66, 83, 1986.
32. Hassan, I. A., Ashmore, M. R., and Bell, J. N. B., Effect of ozone on radish and turnip under Egyptian field conditions, *Environ. Pollut.*, 89, 107, 1995.
33. Wahid, A., Maggs, R., Shamsi, S. R. A., Bell, J. N. B., and Ashmore, M. R., Effects of air pollution on rice yield in the Pakistan Punjab, *Environ. Pollut.*, 90, 323, 1995.
34. Wahid, A., Maggs, R., Shamsi, S. R. A., Bell, J. N. B., and Ashmore, M. R., Air pollution and its impacts on wheat yield in the Pakistan Punjab, *Environ. Pollut.*, 88, 247, 1995.
35. Marshall, F. M., Ashmore, M. R., and Hinchcliffe, F., A Hidden Threat to Food Production: Air Pollution and Agriculture in the Developing World; Gatekeeper Series No. SA73, International Institute for Environment and Development, London, 1997.

CHAPTER 7

Air Pollution and Vegetation Damage in South America — State of Knowledge and Perspectives

Andreas Klumpp, Marisa Domingos, and María Luisa Pignata

CONTENTS

7.1 INTRODUCTION

Since the beginning of European colonization of the South American subcontinent at the end of the 15th century, the economy of the colonies and later the independent states has essentially been based on the exploitation of their natural resources, mainly the export of agricultural products and minerals. It was only after the end of World War II that an intense industrialization of South American countries started. The process of accelerated industrialization has continued to the present day, with South American countries such as Brazil and Argentina now ranking among the world's leading producers of industrial goods.[1] Parallel to the economic transformation, the region experienced a strong population growth from approximately 190 million in 1970 to 327 million in 1995 accompanied by a concentration of the increasing population in urban areas. The percentage of urban population grew from 60% in 1970 to 79% in 1995, and today São Paulo, Buenos Aires, and Rio de Janeiro, three South American cities, are among the so-called megacities with more than 10 million inhabitants.[2]

1-56670-341-7/00/$0.00+$.50

Due to the severe internal socioeconomic problems, a higher priority has been given to industrial development than to environmental protection in these countries. Without paying attention to the environmental consequences, polluting industries have been transferred from the industrialized countries to the developing countries where the lack of environmental regulations and control is used as a locational advantage. As a consequence, in the developing countries environmental problems have appeared similar to those in the Northern Hemisphere years or decades before, history seems to "repeat itself."[3] However, for a long time the deterioration of environmental quality due to industrialization and urbanization was considered not very serious as the population density of the whole subcontinent is still low in comparison to Europe or Asia and natural resources seemed to be inexhaustible. Environmental awareness of the public is generally not well developed. Recently, environmental issues have been gaining increasing significance in the public and in the media. National and international preoccupation is growing due to the decaying living conditions, particularly in the big urban agglomerations.[2,3]

Significant scientific activities aimed at solutions to the environmental problems in South America started in the 1970s. A review on this topic,[4] counted 437 publications in the period of 1970 to 1990, 95% of them originating from only four countries (Brazil, Argentina, Chile, and Venezuela). Publications on environmental pollution problems, however, focused essentially on the emission of air pollutants from industrial and traffic sources, the ambient concentrations of air pollutants in urban and industrial areas, and measurements of bulk and dry deposition. Information on effects of pollutants on humans, fauna, flora, and materials in South America is still very limited, as has also been shown in an overview of the use of bioindicator plants in tropical and subtropical regions.[5] This is in contrast to the situation in countries of comparable developmental status such as India and Mexico where considerable knowledge of pollution effects has been acquired.[6-15]

The present chapter intends to give an overview of the publications that deal with air pollution effects on vegetation in South America. Aspects of water and soil pollution that are also of increasing importance were not considered or only briefly mentioned in order to restrict the size of the chapter. Obviously, this review cannot be complete, as additional data may be found in internal reports and regional and national journals of limited distribution and access. The authors understand this contribution as an information basis to stimulate discussion of environmental problems in South America and a more intensive cooperation between the countries of the region and with the nations of the Northern Hemisphere in the future.

7.2 EMISSIONS AND AMBIENT CONCENTRATIONS

Due to the low population density, the total air pollutant emissions of the South American subcontinent are still low in comparison with North America, Europe, and highly industrialized parts of Asia. However, they are rapidly increasing.[16,17] Emission densities may regionally even exceed those of developed countries. Detailed emission inventories of urban and industrial centers are being compiled by state environmental agencies,[18-20] but they are scarce and inefficient.

Although a significant deterioration of ambient air quality, i.e., an increase of atmospheric concentrations, caused by the locally and regionally high air pollutant emissions from different kinds of sources has been observed for a long time, the establishment of environmental agencies and the installation of air-monitoring networks in several South American countries did not take place until the 1970s.[2,4] Major air pollution problems are occurring at urban and industrial centers, increasing pollution levels, however, can also be observed at remote sites as a consequence of agricultural practices and mineral mining and processing. Motor vehicle traffic is the main contributor to the deterioration of air quality in the urban centers. The high average age of the fleet, the use of obsolete technologies, poor fuel quality, insufficient car maintenance, and high concentrations of vehicles in areas with inadequate infrastructure all contribute to the high pollutant load in the developing countries.[21,22]

Other important pollutant sources are industrial activities and in some cases domestic heating or waste incineration. The most severe problems are occurring at the metropolitan areas of São Paulo and Rio de Janeiro (Brazil), Santiago de Chile, Lima (Peru) and Caracas (Venezuela), but other cities such as Belo Horizonte (Brazil), Bogotá (Colombia), and Quito (Ecuador) are also characterized by poor air quality.[21] Of the total urban population of Latin America (including Mexico), 26% lives in cities where WHO guidelines for air quality are frequently exceeded.[1]

At São Paulo, the largest South American city, air pollution monitoring started in the late 1960s. Due to strong efforts to reduce emissions, ambient concentrations of SO_2 and suspended particulate matter (SPM) dropped significantly during the last two decades.[2,23] However, there are still 128,000 tons of SOx and 95,000 tons of SPM (65% and 41% from traffic sources, respectively) being emitted per year,[20] and WHO guidelines as well as national air quality standards for SPM and inhalable particles are clearly exceeded.[1,20,23] The same holds true for NO_x and O_3. Already, São Paulo is considered heavily O_3-polluted and, due to the ongoing growth of the vehicle fleet, even an increase of ambient concentrations of O_3 and its precursors is expected.[20,23,24]

The use of ethanol and alcohol-containing gasoline as automobile fuels has modified pollution characteristics in Brazilian cities. Reduced emissions of NO_x and CO were achieved by this measure, whereas the release of acetaldehyde and unburned alcohol from alcohol-fueled cars is now resulting in increased formation of peroxyacetyl nitrate (PAN) and O_3.[25,26]

Although the population and the vehicle fleet of Santiago are much smaller than those of Mexico City or São Paulo, the Chilean capital is currently considered the second most polluted city of Latin America. Specific topographical and meteorological factors, such as its location in a basin between mountain ranges, the frequent occurrence of inversion conditions, and the low annual precipitation worsen the air quality situation with respect to inhalable particulate and O_3.[27-29]

In Buenos Aires, the Argentinean capital, monitoring of air pollution concentrations has been rare and sporadic, and the quality of the measurements performed is not high. Therefore, there is no reliable information on the main sources of air pollutants and their concentrations. It is assumed that the city of Buenos Aires presents the same characteristics as other big cities, with traffic being the main pollution source. Due to its geographic and climatic conditions, the situation is not as critical as in other megacities. The average and daily maximum concentrations of SPM, however, exceeded WHO guidelines during sporadic measuring campaigns in 1985 to 86.[30] On the other hand, NO_x and SO_2 concentrations measured in 1994 were comparatively low.[31] Córdoba, the second city of Argentina, by contrast, faces problems with air pollution — mainly during wintertime. High emissions from traffic sources together with the topography hindering dispersion of pollutants in the atmosphere and the frequent occurrence of thermic inversions are responsible for increasing urban air pollution problems that have been the focus of recent investigation.[32]

Several studies on the chemical composition of the SPM have been performed paying special attention to the heavy metal content, particularly Pb.[33-38] In Brazil, the partial substitution of leaded gasoline by alcohol-blended gasoline and ethanol as fuels from the late 1970s on has led to a significant reduction of the ambient lead concentrations.[2,21] High concentrations of Pb and Zn in SPM have frequently been detected in studies performed in Caracas, Venezuela. Whereas the use of leaded gasoline was identified as Pb source, Zn seems to be emitted to the environment from domestic waste incinerators.[34,39] Elevated levels of Pb and hydrocarbons in SPM and bulk deposition in the same city were attributed to automobile exhaust.[40,41] Transport of polluted air masses from Caracas to cloud forests in the surroundings of the city resulted in high concentrations of Pb and Zn in fog water samples.[42]

Air quality data from such other cities as Rio de Janeiro, Lima, and Bogotá are rather scarce and incomplete; however, it is clear that these urban agglomerations are also facing severe air pollution problems. It is a common feature that WHO and national guidelines for SPM are exceeded,[1,2] and a trend to even increasing pollutant concentrations is expected.[23,28] The same holds true for nitrogen oxides and photo-oxidants. The poor air quality frequently recognized in urban areas affects not only human health,[43-45] but may also cause vegetation damage in the cities and, due to transport of polluted air masses, at remote sites. Major air pollution problems also occur in the vicinity of single industrial plants and particularly of big industrial complexes. The great majority of emission sources, particularly those of small or medium size, has never been investigated or at least data are difficult to obtain. As a consequence, and due to insufficient environmental legislation and control of emissions, many of these facilities work without any filtering techniques, severely threatening the population and the environment in the surroundings.

Studies performed in the late 1950s in two strongly industrialized regions of Colombia, Medellin Valley and Cauca Valley, revealed that the two cities Medellin and Cali were heavily polluted by heavy metals originating from industrial emissions, partly also from traffic emissions and application of fungicides in agricultural areas (see Section 7.3.3).[46]

An emission inventory of the industrial complex of Ciudad Guayana in southeastern Venezuela has shown that 3.4 tons of particulate matter are hourly released to the environment, giving rise to very high ambient concentrations up to a distance of about 20 km downwind.[47] The Venezuelan oil production in the Maracaibo Lake Basin is another focus of environmental problems in that country. Emission inventories of NO_x–N and SO_x–S showed that most of the air pollutant emissions occur in this part of the country.[48] A strong acidification of rainwater by air pollution from oil fields and petrochemical plants was described.[49] Similar to Venezuela, the big petrochemical industries installed in Argentina and Brazil also represent a source of environmental problems.[1]

Although Argentina is one of the most industrialized countries of the subcontinent, little information on air pollution problems is available. Elevated concentrations of polycyclic aromatic hydrocarbons (PAH) in airborne particulate matter and a strong mutagenic potential of air samples were detected in one of the industrial centers including an oil refinery, a steel mill, and six petrochemical plants at the Río de La Plata estuary.[50,51] Other studies dealt with fluoride emissions from an aluminum smelter in an arid zone of southern Argentina.[52,53]

The industrial complex of Cubatão, in the state of São Paulo, SE Brazil, is the most intensively studied emission source in Brazil or even in South America. From the early 1950s on, this industrial center had been erected at the coastal plain near to the seaport of Santos and the metropolitan area of São Paulo city, squeezed between the mangroves and the steep slopes of the coastal mountain range. At the end of the industrial expansion in 1975, the complex consisted of 23 industries including a refinery, a steel mill, and several petrochemical and fertilizer production plants, manufacturing a high percentage of the total national production of a big variety of petrochemical and chemical substances and fertilizers.[54,55] In 1984, when the state program for air pollution control started, very high quantities of air pollutants were emitted from the approximately 230 sources. Due to the implementation of a series of mitigation measures, emission load has drastically decreased in the meantime. The reliability of those official data, however, has repeatedly been called in question.[54,55,57] The present SO_2 emissions of Cubatão are estimated to represent still 2% of the total sulfur emissions of Brazil or 1% of the sulfur emissions of South America.[58]

Studies on ambient air quality in the Cubatão region were performed by the State Environmental Agency (CETESB) as early as 1972, and regular monitoring stations have been maintained since the beginning of the 1980s. During the decade from 1981 to 1990, as a consequence of emission abatement programs, a strong reduction of the ambient SO_2 concentrations from 87 to 18 $\mu g\ m^{-3}$ (annual mean values) was achieved. Highly elevated SPM concentrations with a high percentage of ammonium sulfate were the reason for the frequent violation of national air quality standards and the proclamation of emergency situations, although a trend to reduced SPM concentrations can also be observed. Episodic measurements performed by the same institution revealed high ambient concentrations of gaseous and solid fluorides which clearly fell in that decade but still remained at disquieting levels. The national ozone standard (160 $\mu g\ m^{-3}$ for 1h) is frequently exceeded, particularly at higher elevations and in greater distance to the sources of the precursor substances.[59] The results of intense measurements by means of continuously operating air-monitoring stations and field campaigns performed between 1991 and 1996 permit one to draw the conclusion that "the entire atmosphere around the industrial complex is strongly polluted with distinct differences in pollution load between ground level and higher altitudes."[58] Laser remote sensing supported these findings showing that highest pollutant concentrations occur between elevations of 200 to 600 m above sea level (asl). The phytotoxic relevance of fluorides and photo-oxidants such as O_3 and PAN is emphasized in the same report.[60] Concentrations of SPM are still alarming, and inhalable particles even show a tendency to increasing levels.[20]

Monitoring of the air quality has rarely been made in some petrochemical centers, although the installation of the air quality monitoring network at the largest industrial complex of Latin America in the city of Camaçari, Bahia has been a recent attempt.[61] Other areas of interest are the numerous aluminum smelters scattered around the country[62-64] and the coal-fired power plants in the South.[65] Another center of environmental preoccupation is the region of Guanabara Bay/Sepetiba Bay in the state of Rio de Janeiro,[66-68] as well as the Vale do Aço in Minas Gerais State with its metal smelters.[69]

In addition to the emissions of pollutants from industrial processing of ores, mining activities themselves affect the environment through degradation of the landscape as well as through contamination of soils, water, and air. The Andes Mountain chain is one of the regions rich in ore deposits, particularly the Central Andean Plateau in southern Peru, Bolivia, northern Argentina, and parts of the Chilean Andes. Information on environmental problems due to mining activities there, however, has rarely been published.[70-72] The same holds true for the mining areas in the Brazilian state of Minas Gerais or the coal-mining areas in the Jacuí region of southern Brazil.

The mining of gold and silver also has a long tradition in South America and is giving rise to special concern in relation to its environmental consequences. Estimates suggest that about 195,000 tons of mercury have been released into the environment in South and Central America since the 16th century, about 60% of which is thought to have been emitted into the atmosphere contributing to the high mercury fluxes in the Americas and on a global scale.[73] The recent "gold rush" in the Amazon and the Pantanal wetlands in Brazil has already caused extremely increased levels of atmospheric Hg concentrations of up to 290 $\mu g\ m^{-3}$ in the vicinity of the extraction areas[68] and represents a serious threat to the ecosystems and to the food chain.[74-76]

The intense use of mineral fertilizers and pesticides in modern agriculture, which is a source for the contamination of soils and water and, at least in restricted areas, also of the air, is also increasing. The environmental dispersion, e.g., of Cu originating from the use of cuprous fungicides in cocoa production, has been studied in the state of Bahia, NE Brazil.[77,78] Environmental hazards related to the intensive and widespread application of pesticides in the food production of tropical countries have recently been reviewed.[79,80]

Detailed studies on atmospheric chemistry of remote sites in the Amazon have given strong evidence that common agricultural practices are contributing to increased tropospheric O_3 concentrations in the Southern Hemisphere. Normal background levels of O_3 are assumed to be lower at tropical latitudes than at midlatitudes, which could be proved also by historical data obtained from passive sampling methods.[81] Recently, however, increasing O_3 concentrations are being reported not only from megacities like São Paulo and Mexico City, but also from cities of medium size such as Natal, Campinas, and Cuiabá in Brazil and from remote sites.[82-84] Natural fires in grasslands and, particularly, man-made biomass burning in order to clear land for shifting cultivation, to transform forest into agricultural lands, and to remove dry vegetation and crop residues are considered important sources of emissions of NO_x, hydrocarbons, and CO, thus contributing to the photochemical ozone formation.[85-87] Besides these sources, transport of O_3-polluted air masses from urban centers to sites covered, for example, by savannah[84] or rain forests[58] certainly contribute to the rise of O_3 levels in those regions.

From the examples presented in this chapter it becomes clear that many urban agglomerations and industrial centers in South America are heavily polluted and that in some cases the situation is worsened due to particular topographic and climatic conditions unfavorable to the dispersion of the pollutants in the atmosphere. Studies are usually restricted to areas relatively close to the emission sources, and therefore little is known about the transport of polluted air masses from the sources to remote areas and possible consequences on the air quality in natural ecosystems and agricultural lands.

7.3 EFFECTS ON VEGETATION

7.3.1 Argentina

In Argentina, the first effect-related investigations were performed in the vicinity of an aluminum smelter in the south of the country that emitted about 20 tons of gaseous fluorides every month. Soon after of the smelter began operation,[52] field studies were started to trace the emitted fluorides and they found the fluoride concentrations in plants, litter, soil, and sediments to decline with increasing distance from the emission source. Soils, seawater, and sediments proved to be the major fluoride sinks. Further, it was demonstrated that the uptake of gaseous and particulate fluorides by the xerophytic vegetation of that region can be explained by adsorption onto the external leaf surfaces and subsequent diffusion through the cuticle rather than by stomatal uptake as assumed for mesophytic species.[88] The same authors studying coastal ecosystems concluded that the fluoride input from the emissions of the aluminum plant did not affect planktonic communities.[53]

Pine species were used as bioindicators in order to compare the air quality of urban and rural sites of the regions of Mendoza (Argentina) and Leipzig-Halle (Germany). The main differences in the heavy metal burden were found with respect to Pb, which ranged between 3.5 and 37.1 μg g^{-1} dw in the center of Great Mendoza and 0.9 and 3.4 μg g^{-1} dw in Leipzig-Halle. This striking result was attributed to the use of leaded fuel in Argentina.[89] The same survey revealed unusually high levels of chlorinated insecticides and PAH in urban parks of Mendoza when compared with German sites. By contrast, hexachlorbenzene (HCB) levels of German samples exceeded Argentinean values by a factor of 3 to 5.[90]

In the city of Córdoba, studies were made in order to assess the effects of air pollutants originating from different emission sources on two varieties of *Ligustrum lucidum* and seven species of transplanted lichens. Comparing sites with high and low levels of industrial and traffic pollution, respectively, it was observed that leaves of *L. lucidum* Ait. and *L. lucidum* f. *tricolor* (Rehd.), tree species commonly used in urban areas, accumulated significantly more sulfur at high traffic sites than at sites with low traffic intensity. In *L. lucidum*, sulfur accumulation was accompanied by a decrease in pigment concentrations and an increase in specific leaf area. *L. lucidum* f. *tricolor* was more tolerant to traffic-related pollution showing S accumulation without reduction of pigment contents. Although the relationship between S accumulation and traffic intensity has already been mentioned for lichens,[91] it has not been previously reported in higher plants.[92]

Correlation analyses between the chemical parameters quantified in leaves of *L. lucidum* varieties, emission sources, and environmental conditions showed that the response to pollutants was strongly associated with the conditions of the microarea where the sampling was carried out: traffic, position of the sampling point, type of industry, distance from other trees, and topographic level were the environmental variables related to the response. Determination of oxidation products appeared to be an excellent tool for monitoring the early phytotoxic action of pollutants, malondialdehyde (MDA) being the most sensitive indicator of traffic pollution, followed by the less sensitive hydroperoxyconjugated dienes (HPCD), which only increased significantly when other symptoms of foliar damage, such as pigment degradation, had already occurred.[93 94]

In the city of Córdoba, active biomonitoring studies have been carried out for the past few years using transplants of seven lichen species. For *Punctelia subrudecta* (Nyl.) Krog., transplanted to an urban-industrial region characterized by small and medium-sized mainly metallurgical plants, an increase in the MDA, sulfur and soluble protein contents and a decrease in Chl *a* concentrations were observed at all sites in comparison to freshly picked material. There was an inverse relationship between Chl *a* and aluminum and Chl *a* and S and, on the other hand, the MDA concentration was positively correlated to the S content, phaeophytin *a*/Chl *a*/ratio, Al, and Pb. At most sites with a high level of vehicular traffic, metal contents were elevated. The main source of Pb released into the air is leaded gasoline, which is still in use in Argentina; the main sources of aluminum are small metallurgical plants.[95]

In order to estimate the air pollution-induced damage to transplanted lichens, a pollution index (PI) was proposed that allows indication of different air qualities in active biomonitoring programs established from the chemical response of the biomonitor.[95] In *P. subrudecta* samplings transplanted to urban sites, the PI value based on the chemical response of the lichen was higher in the presence of high humidity, which made it possible to infer that high humidity enhances the pollution effects in this species. The response of this species was principally associated with emissions from industries as well as from a power station, whereas different categories of vehicular traffic did not affect the lichens. Significant differences were observed in S content, MDA concentration, and PI values in lichen material transplanted to sites with different industrial density, the higher values for these parameters being found at sites with high industrial level.[96]

For *Ramalina ecklonii* (Spreng.) Mey. & Flot., the chemical response was ascribed mainly to air pollutants released by traffic. In an active biomonitoring study carried out in Córdoba, the chlorophyll concentrations (*a*, *b*, and total) in the control samples were higher than those in lichens transplanted to a transect characterized by intense traffic, starting in the urban center and covering 10 km. Phaeophytin *a*/Chl *a* ratio, sulfur accumulation, HPCD, and dw/fw ratio showed higher values in samples transplanted to urban sites than those exposed in the control area.[91]

In another biomonitoring study with transplanted *R. ecklonii* in urban areas, sulfur contents, and PI parameters showed high values in samples transplanted to locations with high traffic levels, indicating high SO_2 emissions by traffic.[97] In this work, due to the fact that the different biomonitoring sites represented various combinations of traffic and industry, the effects of the different levels of industrial density were analyzed for sites characterized by low traffic. For sites with high industrial activity, the lichen samples showed higher values of the phaephytin *a*/Chl *a* ratio. At these sites, the increase in phaeophytinization indicated that this parameter is sensitive to lichen damage caused by industrial emissions.

Lichens of the species *Parmotrema austrosinense* (Zahlbr.) Hale and *P. conferendum* Hale taken from a nonpolluted area were transplanted within the nonpolluted area and to a downtown site in Córdoba for 6 months and the same chemical parameters were analyzed. At the urban site chlorophyll concentration was lower and in *P. conferendum*, MDA concentration was significantly higher than at the control site. In general, comparison of samples transplanted to the nonpolluted site and to the urban site showed that pigments were sensitive to atmospheric pollutants.[98] *P. uruguense* (Kremplh.) Hale transplanted to nonpolluted environments and to two urban sites — one of high, the other of low pollution — showed a temporal variation of its chemical parameters throughout the study period, depending on the transplantation site of the lichens. The lower contents of carotenoids and Chl *a* as well as the higher values of the Chl *b*/Chl *a* ratio and HPCD found in the downtown of Córdoba (high pollution site) were attributed to the impact of atmospheric pollutants on the transplanted lichens. The studies confirmed that the low Chl *a* concentrations were not due to lower water contents of the thalli exposed at the urban sites but reflected an alteration of the normal physiological processes of *P. uruguense* under air pollutant stress.[99] Contrary to the results obtained with *P. conferendum*,[96] in *P. uruguense* Chl *b* was not altered by the presence of air pollutants.

After 30 days of exposure of *Usnea* sp. transplants at sites provided with air monitoring equipment, significant correlations were detected between some air pollutants and lichen damage. Thus, Chl *b*/Chl *a*, phaeophytin *a*/Chl *a*, and HPCD showed a positive correlation with particulate matter (PM_{10}); Chl *b*/Chl *a* and dw/fw with hydrocarbons; phaeophytin *a*/Chl *a* and HPCD with O_3.[100]

In a study performed at Nahuel Huapi National Park, northwestern Patagonia, the element composition of four foliose and two fruticose lichen species from pristine areas of the National

Park and from urban and periurban areas of the small, nonindustrial city of Bariloche was analyzed. Generally, the element concentrations of all species were lower in periurban areas than in urban areas. Element contents of *U. fastigiata* specimens from pristine areas were lower than those of periurban areas for all elements, except Br, Mn, and Ca. All concentration values were similar to or lower than the values reported in the literature.[101]

7.3.2 Brazil

Among the industrial centers of the country, the industrial complex of Cubatão in the state of São Paulo was called the most-polluted city in the world during the 1980s. Today, it is certainly the most extensively studied industrialized region of South America (see Section 7.2).

7.3.2.1 The Cubatão Region

A comparison of vegetation maps derived from aerial photographic inventories proved the strong degradation of the Atlantic Rain Forest ecosystem around the industrial complex between 1962 and 1977. In this period about 85% of the total area of 75 km^2, which in 1962 had been occupied by primary and secondary forests unaffected by pollution, presented changes in their vegetation structure that were attributed to air pollution impact. Until 1989, the continuation of the forest decline and the frequent occurrence of mudslides were followed by means of aerial photography.[60,102-104] The dramatic aggravation of the environmental situation in Cubatão induced the Brazilian scientific community to call for remedial measures and intense studies.[105] In the following years, the CETESB initiated several investigations that dealt with the pollution-induced changes of the vegetation structure, the sensitivity and resistance of native tree species to pollution impact, and methods for the recovery of the vegetation cover. Degradation started with the generalized death of emergent trees, followed by a dieback of the upper and middle tree layers. Epiphyte communities were also severely affected. During the initial phase of degradation, shrub and herb layers showed drastically increased density.[55,106,107] Phytosociological studies performed at an area where landslides had occurred 15 years before still showed the frequent occurrence of pioneer species in comparison to late secondary and climax species.[108] Preliminary results from several field surveys suggested that species of a limited number of families, such as Palmae, Araceae, Melastomataceae, Moraceae, Ulmaceae, Urticaceae, and Compositae may be relatively resistant to the local air pollution conditions,[109] whereas trees identified through anatomical characteristics of dead trunks belonging to the families Myrtaceae and Lauraceae had disappeared at the most polluted areas.[110] The high abundance of a small number of families, particularly the Melastomataceae, in the areas affected by the emissions from the steel and fertilizer factories of Cubatão has been confirmed by further studies.[60,107,111] A strong reduction of biodiversity of the tree stratum was found by comparing the polluted Mogi Valley and the nonpolluted Pilões Valley as well as the recent data and data obtained at the beginning of industrialization.[112] Epiphytes like bromeliads and orchids exuberant at nonpolluted areas have nearly disappeared from the polluted forest portions.[107] Negative pollution effects on the epiphytic bryophyte communities had also been reported.[113] The secondary succession of affected areas was considered severely disturbed including the premature death of young individuals even of the pioneer species that were supposed to be relatively resistant to air pollution.[60,107,111]

Estimates have shown that pollution had caused a reduction of the aboveground biomass at the polluted areas of Mogi Valley and Caminho do Mar by about 66 and 54%, respectively, and a reduction of root biomass at the same sites by about 78 and 67%, respectively.[107] The resulting increase of the shoot–root–ratio was considered to affect soil stability, hydrological conditions, and the nutrient retention in the ecosystem.

Other effects were an impairment of height and radial growth of trees, the death of higher individuals when reaching canopy height, and the stimulation of ramification at the bases of the trunks promoting a shift from arboreous to bushy vegetation formations. Remanescent trees and shrubs frequently exhibit visible injury symptoms such as strong interveinal chloroses and necroses, marginal necroses, and leaf deformations. All these effects are hindering the closure of the canopy and the normal stratification of the forest, thus promoting increased insolation and, as a consequence, interference in normal secondary succession and reduction of biodiversity.[60,107,111,114]

Studies at a higher and a less pollution-affected site at the Biological Reserve on the Plateau of Paranapiacaba (800 m above sealevel), at a distance of about 7 km from the fertilizer and steel industries of Cubatão, proved that air pollution had caused disturbances in the nutrient cycling of the Atlantic Forest ecosystem.[115] The nutrient residence time on the forest floor was higher and the nutrient use efficiency index was lower at the more affected site when compared with the less affected one, which may be attributed to a lower input of nutrients through litter production[116] and decomposition[117,118] and a higher litter accumulation on the forest floor.[119] Thus, contrary to undisturbed forests, contribution of litter fall to the total nutrient flux from the canopy to the soil was low compared with the contribution of precipitation.[120]

Comparison of one undisturbed Pilões Valley site with a severely pollution-affected Mogi Valley site near the fertilizer and steel factories at the coastal plain near Cubatão showed reduced rain interception rates at the polluted site compared with the reference site because of the dieback of the tree layer.[111] Litter fall was higher at the polluted site than at the reference area, in spite of the lower tree density. As a consequence of increased litter production and decreased decomposition rates, litter stocks on the forest floor were elevated. Chemical analyses of fresh litter revealed elevated contents of S, P, Fe, and Al. Studies on the pollution effects on vegetation and soil at different sites were also performed in an interdisciplinary research project executed in the Cubatão region between 1990 and 1996.[121] It was shown that at the pollution-affected area near the industries (Mogi Valley) and partly at Paranapiacaba the soil was strongly acidified, even in the deeper layers, due to the high atmospheric input and soil internal acid production through mineralization of organic matter and nitrification. As a consequence, nutrients are leached from the soil, thus making the soil extremely poor in cations like Mg and K. The rhizosphere of the trees is exposed to strongly elevated concentrations of Al and also F.[122] Contrary to other groups, the same authors found that litter production differed only slightly among the sites and element concentrations were considered not modified. Increased fluxes by litter fall could be found with respect to Fe, and — to a lower degree — to S, P, and N, whereas Mg, K, and Mn fluxes were lower at polluted sites than at the reference area. The authors concluded that compared with other regions of tropical and temperate zones, "The area under investigation represents one of the most polluted landscapes with forest vegetation."

Investigations of the standing vegetation at four sites with different pollution characteristics (Pilões Valley, Caminho do Mar, Mogi Valley, Paranapiacaba) performed by another working group of the same cooperation project confirmed the high pollution impact on forest portions more exposed to the industrial emissions. Particularly, foliar F concentrations of trees growing at Mogi Valley near to the F-emitting fertilizer factories were significantly elevated. Even at the Plateau of Paranapiacaba, Melastomataceae species exhibited increased F contents due to the F input into the soil in the past and its continuous cycling within the ecosystem.[114] Foliar S levels were increased at both of the most polluted sites — Caminho do Mar and Mogi Valley. An accumulation of Fe and Mn could also be observed in some areas. The high input of acidic pollutants as well as of nutrients such as N and P caused alterations in the nutritional balance of the trees, as, for example, a scarcity of cations like K and Ca.[60,123]

Besides biochemical changes, such as a reduction of cell sap pH and a diminished acid-buffering capacity of leaves,[123,124] several morphological and anatomical alterations in pollution-affected trees were described. Wood samples of *Cecropia glazioui*, a pioneer tree species considered relatively sensitive to air pollution, showed quantitative and qualitative changes in the xylem structure. Pollution caused a reduction of diameter and length of vessels, an increase of vessel frequency, and a reduction of the proportion of solitary vessels. These modifications as well as the observed clustering of elements were interpreted as means to provide greater safety in water transport, pointing to the existence of physiological water stress in pollution-affected trees.[125] Adult individuals of *Tibouchina pulchra*, a pioneer tree species apparently tolerant of air pollution, exhibited alterations of wood anatomy similar to those of *C. glazioui* and also a reduced complexity of vestures, frequently observed in wood from unfavorable environments.[126]

Active biomonitoring with bioindicator plants was performed at the Cubatão region in order to detect the main phytotoxic components of the complex air pollution mixture, to gain information on the spatial and temporal distribution of air pollution effects, and to distinguish between effects of air pollution and of soil contamination. In a pilot study, moss-bags (*Sphagnum recurvum*) were exposed at 24 experimental sites in the Cubatão region for 12 periods of 1 month each.[54,55] Chemical analyses of the plant material revealed a high deposition of heavy metals to the forest areas downwind from the emission sources, particularly of steel and fertilizer industries, and also to residential areas in the vicinity of the industrial plants, whereas pollution impact was low at sites protected from the prevailing wind direction and at greater distances from the complex. High accumulation rates of Al, Fe, Zn, and Ni were found, whereas the cadmium input proved to be low.

As part of the German–Brazilian research cooperation, an active monitoring network was installed and maintained at 12 sites including indicator plants sensitive to O_3, PAN, and HF and an accumulative indicator species.[123,124,127,128] Typical O_3-induced injury symptoms on leaves of the sensitive tobacco strain Bel W3 occurred at all the exposure sites. In accordance with results of measurements of atmospheric concentrations,[58] severe injury was observed at middle and higher altitudes near to the petrochemical complex of Caminho do Mar and at the Plateau of Paranapiacaba, whereas low damage levels were found at sites in the vicinity of emission sources.[123,128] Ozone-induced leaf lesions were also found in other species such as morning glory (*Ipomoea tricolor*) and bush bean.[60] Development of characteristic gray necrotic bands on leaves of *Urtica urens* exposed at the region near the petrochemical plants were attributed to the impact of PAN. Foliar symptoms in *Petunia* cultivars were also referred to as caused by organic compounds of photochemical smog.[128]

The impact of gaseous and particulate fluorides was assessed using sensitive and accumulative indicator plants. In comparative studies with several varieties, the *Gladiolus* cv 'White Friendship' and the daylily *Hemerocallis* cv. 'Red Moon' proved to be highly sensitive to HF under field conditions.[129,130] Routine exposure of these two cultivars in the monitoring network revealed the strong vegetation risk by fluorides emitted mainly from the superphosphate-producing industries at the bottom of Mogi Valley.[123,124,128,131] Chemical analyses of *Lolium multiflorum* cultures at the most-polluted site showed a 3-year average value of approximately 250 $\mu g\ g^{-1}$ dw and a monthly maximum value of 1200 $\mu g\ g^{-1}$ dw of F. The recommended threshold values for fodder and pasture grass suggested between 40 and 80 $\mu g\ g^{-1}$ dw maximum and 30 and 50 $\mu g\ g^{-1}$ dw on an annual average.[132,133] The S content of the grass cultures exposed at the polluted areas were up to 87% higher than at the reference site Pilões Valley, reaching values of 7.5 mg g^{-1} dw, whereas only a slight N accumulation in *Lolium* was found. In agreement with data from laser remote sensing, which showed very high SO_2 and NO_x concentrations at an altitude of 400 to 500 m asl in front of the mountain slopes,[58] highest foliar S and N concentrations were recorded at an elevation of about 450 m asl, proving the vegetation risk by sulfurous compounds in parts of the region. Corresponding to the results obtained using moss-bags,[54,55] a very strong accumulation of iron due to the emissions from the steelworks was detected at Mogi Valley, where foliar Fe concentrations were 6.5 times higher than at the reference area. Iron pollution was also found at the top of the mountains, at the beginning of the valley, and near the petrochemical complex.[60,123]

Saplings of *Tibouchina pulchra* were exposed to ambient air at eight sites of the Cubatão monitoring network for consecutive periods of 16 weeks each. Plants exposed at the most-polluted site Mogi Valley occasionally showed interveinal necroses. Increased leaf abscission and a reduced leaf area were observed at the same site. By SEM photomicrographs, a degeneration of the ornamentation of leaf surfaces and of stomata was demonstrated, with increasing intensity from

Pilões Valley to Paranapiacaba and finally to Mogi Valley. The erosion of epicuticular wax layers resulted in the disappearance of the striated aspect and obliteration and atrophy of stomata, which may make these leaves more susceptible to insect and parasite attack.[60,126] The changes in wood anatomy of these saplings were comparable with those described for adult individuals growing at the investigation area.

Alterations in the chemical composition of *T. pulchra* saplings confirmed the high burden of gaseous and particulate pollutants acting on the forest vegetation in the vicinity of the industrial complex. Particularly F, S, N, Fe, Mn, and Zn levels of plants exposed at the polluted sites of Caminho do Mar and Mogi Valley were significantly raised in comparison with reference samples. *T. pulchra* again proved to be a natural Al, Fe, and F accumulator that exhibits high background levels of these elements.[114,120,134] In general, the exposure experiments showed that *T. pulchra* might be a promising accumulative indicator of air pollution. Klumpp et al.[60,124] analyzed compounds of the endogenous antioxidative defense system of the same saplings, showing an increase of the peroxidase (POD) activity and a reduction of the ascorbic acid contents, respectively, at the sites where pollution impact was most severe. After 9 months of exposure at polluted sites (Mogi Valley, Caminho do Mar), *T. pulchra* saplings showed a reduction of photosynthetic capacity, transpiration and stomatal conductance and a stimulation of dark respiration when compared with plants exposed at the reference area Pilões Valley.[135]

Tropical fruit tree species, *Psidium guajava* (guava), *P. cattleyanum* (strawberry guava), and *Mangifera indica* (mango), were tested for their sensitivity to the air pollution conditions of the Cubatão region and their suitability as bioindicators. Exposure of the two *Psidium* species at monitoring sites near the fertilizer plants (Mogi Valley) and the oil refinery (Caminho do Mar) resulted in interveinal chloroses. Similar to the results obtained with other species, a strong S accumulation in leaves was found at these two sites, whereas highly elevated foliar F concentrations of more than 600 $\mu g\ g^{-1}$ dw only occurred at plants exposed at Mogi Valley. Comparing the two *Psidium* species, *P. guajava* was more sensitive and accumulated more toxic elements than *P. cattleyanum*. Mainly at Mogi Valley, but to a lesser extent also at Caminho do Mar, biochemical stress indicators showed alterations. POD activity was increased significantly, whereas ascorbic acid content diminished. At the same time, an increase of the content of soluble sulfhydryl compounds was stated. As a result of the pollution stress, growth characteristics of both species changed. At the most-polluted sites, shoot and particularly root growth decreased, leading to higher shoot–root–ratios, which may have important consequences for the mineral and water uptake of the plants as well as for their physical stability in the ground [60,134] and which is in good agreement with the findings based on estimates of the total biomass of the tree layer.[107]

There are few reports on the effects of soil contamination in the region, although the strong acidification and intoxication of the soils of the region are known. In an experiment, saplings of *T. pulchra* were cultivated in soils drawn from the same sites of the Cubatão region where the other studies were developed and exposed to ambient air at a location without significant F pollution for 48 weeks. Foliar F levels of plants growing in the acid soils from the heavily polluted Mogi Valley and, to a lesser degree, from the formerly F-polluted Paranapiacaba increased drastically during the first weeks, whereas F concentrations of plants grown in soil from the reference area rose only slowly. Strong F accumulation was followed by premature leaf loss.[60] Additional experiments were performed also by exposing *T. pulchra* saplings grown in contaminated and non-contaminated soils of the region to the ambient air of different sites (Pilões Valley, Caminho do Mar, Mogi Valley) in order to detect possible interactions of soil and air pollution. Both factors, polluted soil and polluted air, caused similar responses of the plants, such as accumulation of pollutants in leaves and changes of the antioxidative system as already described, the effect of air pollution being more pronounced than that of soil pollution at least during the limited experimental period. Effects of exposure site and of soil origin were additive rather than synergistic.[60,136] Germination tests gave the surprising results that soil from a medium-polluted area of Paranapiacaba prejudiced germination of seeds and development of seedlings, whereas soil taken from a severely damaged forest near the emission sources had no detrimental effect.[137]

Since the mid-1980s, strong efforts have been made by public and private initiative to recover degraded areas of the coastal mountain slopes. The initial planting of grass of the exotic genus *Brachiaria* was soon substituted by manual planting of native pioneer species of the Atlantic Forest that were considered relatively resistant to air pollution.[106,138] As a next step, aerial sowing of pelleted seeds and fern spores, was performed on the steep slopes using helicopters and airplanes.[106,139] The success of these activities, however, has been controversial from the beginning,[111,140,141] and a successful reconstitution of the vegetation cover without further drastic reduction of air pollutant emissions is questioned.[111,138]

7.3.2.2 Other Brazilian Regions

Although there exist a lot of industrial districts, particularly in the south and southeast areas of the country, and more recently in the northeastern states, some of them exceeding the Cubatão area with respect to their economic importance, there are no comprehensive studies on the environmental impact of industrial emissions.

In the early 1980s, the Instituto de Botânica São Paulo (IBt) initiated studies to prove that the strong degradation of the Atlantic Forest, remnants of the State Park "Fontes do Ipiranga" situated in the suburbs of São Paulo, was caused by dust emissions from an iron-sintering plant located nearby. Along a transect coincident with the prevailing wind direction chemical analyses of soil, leaves, and litter and qualitative analyses of the chemical composition of the dust deposited upon the leaf surfaces were done at areas in different distances to the emission source. Concentrations of S, Fe, and some heavy metals such as Zn, Cr, Pb, and Ni were quite elevated, reaching toxic levels and decreasing with increased distance.[142] Due to the high deposition of particulate matter, nutrient content and pH value of the soil were highest at 50 m from the stack and decreased along the transect.[143] Within the context of the bioindicator studies in the vicinity of Cubatão (see Section 7.3.2.1), one exposure site was also installed at the IBt area in São Paulo, very near to the forme investigation site.[142,143] As the smelter had been shut down previously, there were no accumulations of S, F, Fe, Al, and Zn in exposed *Lolium* cultures and *Tibouchina* saplings when compared with the results from the reference area Pilões Valley and normal background values cited in the literature; only foliar nitrogen concentrations were slightly enhanced. Severe injury, however, was regularly observed in *Nicotiana tabacum*, *Urtica urens*, and *Petunia* hybrids, bioindicator species sensitive to O_3, PAN, and other photochemical oxidants,[60,128] a result which had been expected as national O_3 standards are frequently exceeded and elevated atmospheric PAN concentrations were measured in this part of the city.[58]

In a preliminary study, bioindicator plants (*Gladiolus* sp., *Lolium multiflorum*) were exposed at 11 sites in the surroundings of the aluminum smelter of Ouro Preto (Minas Gerais), which is supposed to emit 42 tons of gaseous and 120 tons of particulate F per year. The leaves of *Gladiolus* developed tip and margin necroses typical of fluoride impact. Foliar F contents of *Lolium* cultures, however, remained relatively low, probably due to insufficient analytical conditions and exposure methods.[63,144] Leaf injury was observed at *Hemerocallis* sp. and *Dracaena* sp. plants growing in residential areas near to an aluminum smelter at Alumínio (São Paulo), whereas *Hibiscus rosa-sinensis* did not show leaf lesions in spite of foliar F levels of up to 1500 $\mu g\ g^{-1}$ dw.[62] Exposure of *Dracaena alba* and *Dioscorea* sp. cultivated in private gardens in the vicinity of the emission source resulted in foliar lesions, premature leaf loss, and accumulation of high quantities of fluoride in the leaf tissues. Foliar F concentrations directly correlated with measured deposition rates.

Based on the degree of foliar injury recorded near four different aluminum smelters in Brazil,[145] ranking of 230 native and exotic species according to their relative susceptibility to F was done. Several species of the Arecaceae and monocotyledonous species proved to be very sensitive. An enlarged list of native and cultivated Brazilian species similarly classified according to their sensitivity to fluorides was prepared.[64] Foliar symptoms observed in the vicinity of fluoride sources, particularly aluminum smelters and ceramic factories, in five Brazilian states were shown in a pictorial atlas together with the taxonomic description, mineral contents, and additional observations.[145] Several species of the families Lauraceae, Moraceae, Euphorbiaceae, Pinaceae, and Rosaceae are also considered susceptible. Different genera of a given family and species of a given genus were also classified as sensitive, moderately tolerant, and tolerant.

Since 1995, a biomonitoring network has been installed in the region of the petrochemical complex of Camaçari (Bahia), approximately 65 km from the city of Salvador. Lemon grass (*Cymbopogon citratus*) and coriander (*Coriandrum sativum*) cultivated in artificial substrate as well as moss-bags (*Sphagnum* sp.) were exposed to ambient air at eight sites for periods of 4 weeks (higher plants) and 8 weeks (moss-bags) and subsequently analyzed for their heavy metal content. First results showed elevated As levels at some sites and slightly elevated Cu contents at other sites. As accumulation was quite higher in *Sphagnum* than in coriander.[146,147] Naturally occurring vegetation growing near a copper smelter within the same industrial complex shows foliar symptoms that can probably be attributed to acute SO_2 damage.[148,149]

Several studies on the use of bioindicator plants for the monitoring of air quality in industrial areas have been conducted at Porto Alegre (Rio Grande do Sul).[150] At areas adjacent to coal-fired power plants and oil refineries, cultures of *Lolium multiflorum* were exposed and S accumulation in leaves determined permitting to outline zones of low, medium, and high accumulation rates around the emission source.[151] Photosynthetic capacity, the number of living cells and S accumulation in the three lichen species *Parmotrema tinctorum*, *Usnea subcomosa*, and *Telochistes exilis*, were affected by the SO_2 emissions from an oil refinery. *P. tinctorum* presented the best correlation between ambient SO_2 concentrations and physiological response.[151]

Studies on the environmental impact of the coal-fired power plant at Candiota (Rio Grande do Sul) revealed a significant accumulation of F in pasture grass at a distance of up to 8 km from the source, and the amount surpassed the common international limit values and caused various cases of fluorosis in cattle.[65,152] Repeated analyses in the following years showed reduced F content of pasture, reflecting the installation of new electrostatic filters. In the vicinity of another power plant, at Charqueadas (Rio Grande do Sul), where no technical improvements were made, there was no change of foliar F levels. Fluoride concentrations in honey samples were elevated at distances of several kilometers from both the power plants when compared with reference samples.[153] In the neighborhood of a textile mill at Teresópolis (State of Rio de Janeiro), *Lactuca sativa*, *Daucus carota*, and *Brassica rapa* cultivated at different distances to the factory showed a strong reduction of fresh and dry weight up to 90% when compared with a reference site in a rural area. Growth depression as well as chloroses and necroses on leaves of lettuce and turnip were attributed to SO_2 emissions.[154]

Spanish moss (*Tillandsia usneoides*) was used to monitor atmospheric mercury contamination inside a chlor-alkali plant in Rio de Janeiro. After 2 weeks, transplants exposed in the external area of the factory contained up to 35 $\mu g\ g^{-1}$ dw and those exposed at sites inside the factory up to 10, 400 $\mu g\ g^{-1}$ dw, whereas Hg level of control plants was 0.2 $\mu g\ g^{-1}$ dw. Thus, the exposure of *Tillandsia* transplants permitted detection of critical areas inside the industrial facilities.[155,156]

Knowledge of the effects of urban air pollution on plants is even more restricted, in part probably due to the difficulty in establishing defined cause–effect relationships in areas where a lot of mobile and stationary emission sources have to be considered. In Porto Alegre, epiphytic vegetation of *Melia azedarach* trees was mapped and sulfur and heavy metal concentrations of *Tillandsia aeranthos* and *T. recurvata* were analysed.[150] Clear differences in S accumulation of *Tillandsia* species between samples from the more-industrialized northern part of the city and the residential areas in the south were observed.[151] Studies of the morphological parameters of the plants growing in different areas showed that urban air pollution allowed a good development of the vegetative parts, but caused an inhibition of the reproductive organs. *T. recurvata* proved to be more sensitive than the other species due to poor development and disappearance from the most-polluted sites downtown.[157]

In the city of Curitiba (State of Paraná, southern Brazil), pigment contents in leaves of *Tabebuia alba* and *Pittosporum undulatum* trees growing in central and suburban parks were analyzed. Total chlorophyll concentrations were higher at polluted locations than at the less-polluted ones with a shift of the Chl *b*/Chl *a* ratio to higher values, and carotenoid contents were significantly reduced at the same sites.[158] Net photosynthesis of *Tabebuia chrysotricha* trees growing along main roads was about 50% lower than that of trees of the suburbs.[159] Studies with *Ligustrum lucidum* showed that trees of polluted sites downtown presented a reduction of net photosynthesis and leaf expansion and increased foliar Fe levels when compared with trees from the periphery, which was attributed to the emissions from the automobile traffic.[160] In *Spathodea campanulata* trees of urban areas of Brasília, the pH value of bark extracts was lower and the conductivity was higher in trees growing at polluted areas than in individuals from reference sites.[161]

A monitoring network with *Nicotiana tabacum* as O_3 indicator is currently under way in São Paulo.[162] In an industrialized district of São Bernardo do Campo, a densely populated city with several industries and intense traffic, *Lolium multiflorum* and *N. tabacum* were exposed as bioindicators of air pollution in a monitoring grid for a period of 1 year. Heavy metal and Al concentrations of *Lolium* leaves were elevated in comparison to normal background values. Exposure sites at the border of a highway and near metallurgical plants exhibited the highest Al and Pb levels, but also Cu and Zn concentrations were higher than at the other locations.[163]

7.3.2.3 Exposure Experiments under Controlled Conditions

Experiments under controlled conditions such as fumigation in closed or open-top chambers or treatments with artificial rain are an important tool for the diagnosis of air pollution damage, for the establishment of dose–effect relationships as a basis for risk prognoses, and for the development of biomonitoring methods. Few experiments of this kind have been done, some in Brazil with native or cultivated plants, some in foreign countries using Brazilian plant species.

The effects of simulated acid rain (SAR) with a chemical composition similar to that of Cubatão were investigated on soybean cultivated either in sand or in acid soil from Mogi Valley or in acidity-corrected soil.[164] Acid rain treatment (pH 5) reduced the amount of epicuticular wax and caused a decrease of chlorophyll content and of photosynthetic activity. These effects were most severe at plants cultivated in acid soil. In similar experiments, acid rain promoted an increase of POD activity and a reduction of nitrate reductase (NR) activity in leaves, whereas soil contamination alone induced a strong reduction of foliar NR activity and a stimulation of NR in the roots.[165]

Treatment of eight cultivars of soybean with simulated rain containing different amounts of F revealed a differential resistance of these cultivars considering the development of visible symptoms as the criterion of evaluation. The most sensitive cultivar 'Bossier' accumulated more F than the less sensitive cultivar 'Doko' due to differences in stomatal conductivity. Decline in chlorophyll content and increase of peroxidase activity were also more pronounced in the sensitive cultivar than in the resistant one.[166] On morphological and anatomical level, fluoride treatment with simulated rain resulted in necroses mainly to the younger leaves of soybean, the area near to the trichomes being the most sensitive part of the leaves. The development of visible injuries was preceded by cellular injuries.[167]

In a screening study, 10 tropical grass species were tested with respect to their respective sensitivity to F-contaminated simulated rain in comparison with the traditional accumulative bioindicator species *L. multiflorum*. Only *Panicum maximum* and *Paspalum notatum* developed symptoms such as marginal and apical necroses even at low foliar F concentrations, whereas F accumulation was highest in *Eragrostis curvula*, followed by *Andropogon gayana* and *Cenchrus ciliaris*, but did not result in typical injury symptoms. Considering the response to the pollutant and the handling of the plant material, *P. maximum* cv. 'Colonião' and *Chloris gayana* were selected the most promising sensitive and accumulative bioindicators.[168]

Epiphytic plant species have frequently been used in air pollution studies, as their mineral nutrition depends exclusively on the input of ions from dust or precipitation, thus permitting investigation of air pollution effects independently of possible interactions with soil conditions. Sulfur accumulation in individuals of *Tillandsia aeranthos* and *T. recurvata* fumigated with 400 μg SO_2 m^{-3} for up to 60 days was very low. After 25 days of fumigation, *T. aeranthos* exhibited visible injury symptoms such as folding of leaf tips, chloroses, and striping necroses and anatomical changes such as intracellular formation of crystals and modifications of the trichomes leading to an obstruction of the stomata, whereas *T. recurvata* did not show any response.[169,170] More realistic SO_2-concentrations (130 μg m^{-3} for 84 days) caused leaf chloroses and significantly elevated foliar S content in *T. aeranthos* leaves.[171]

Fumigation of *T. aeranthos* and *T. recurvata* with 150 and 300 μg HF m^{-3} did not cause visible leaf injuries nor structural alterations. Foliar F levels of both species, however, were strongly elevated reaching a value up to 700 μg g^{-1} dw.[170] Comparable results were obtained when both species were fumigated with 3 μg HF m^{-3} for 3 months. No typical visible symptoms, but very high foliar fluoride levels (up to 800 μg g^{-1} dw), were observed. Treatments with F-contaminated fog resulted in severe tip necroses of the leaves and F accumulation.[171]

Fumigation of different orchid species with 3 μg HF m^{-3} for 8 weeks did not cause any visible injury symptoms. Fluoride accumulation in plants differed widely, depending on plant tissue, developmental stage, and on the plant species. *Cattleya intermedia* and *Laelia purpurata* (both with CAM type of CO_2 fixation) presented the highest F concentrations in their bulbs, whereas in *Aspasia lunata* (C_3-type) F accumulation occurred mainly in the leaves.[78,172] Tropical plant species with CAM and C_3/CAM type of CO_2 fixation showed also a differential response to O_3 fumigation.[173] Comparative fumigation studies with several Brazilian *Gladiolus* cultivars demonstrated the differential susceptibility of these varieties to HF. Among the cultivars tested, cv. 'White Friendship' proved to be most sensitive to fluoride and most adequate for biomonitoring studies,[78,174] a result which was in good agreement with the experience of field experiments.[130]

Ipomoea species treated with 140 to 200 μg O_3 m^{-3} for 2 weeks developed foliar symptoms (water-soaked spots, stippling, necroses) that resembled those of O_3-sensitive bush bean varieties like Pinto or Blue Lake. *I. tricolor* cv. 'Scarlet O'Hara' and cv. 'Himmelblau' were particularly sensitive, whereas *I. cardinal* was relatively tolerant of O_3. Due to their rapid and climbing growth, however, there remained some doubt with respect to the suitability of these species for routine biomonitoring despite their high O_3 susceptibility.[171]

7.3.3 Colombia

In Colombia, the airborne heavy metal deposition in the highly industrialized Cauca Valley was examined using the epiphytes *T. recurvata* and *T. usneoides* as accumulative indicators.[46] Through analyses of the element contents of the plants collected in the whole investigation area, a regional deposition pattern could be established and the main emission sources were identified. The highest heavy metal concentrations (Pb, Cd, Zn, Cu, Ni, Mn, Fe) were found in the Cali-Yumbo industrial center, where, besides the industrial emissions, contributions from traffic sources (Pb, Zn) and from pesticide application (Cd, Cu) were responsible for elevated heavy metal levels in plants. The northern part of the industrial complex was characterized by the high emissions of Ca and Fe containing dusts from a cement factory with an epiphyte desert within a radius of about 1.5 km around the plant, whereas the southern part was dominated by Zn and Pb and locally by Cu, Cd, and Ni. The methodology even permitted determination of single emission sources.

Further studies in the Colombian cities of Medellín and Cali using *T. recurvata* as indicator plant, allowed characterization of the pollution situation with respect to heavy metals, pesticides, PCBs, and PAHs. By means of statistical analyses, 10 classes of independent pollution sources could be discriminated. Medellín was polluted mainly by Zn, Cd, Ni, Cr, and fluoranthene, whereas in Cali high levels of Pb, Cu, benzo-a-pyrene, and pesticides were detected. When compared with data from Europe and the U.S.A., particularly Zn, Cu, Cr, Ni, and DDT reached preoccupying levels in those cities.[175] Contrary to the results from controlled exposure of *Tillandsia* species to SO_2 and HF (see Section 7.3.2.3), these epiphytic species proved to be useful accumulative indicators for heavy metals and organic pollutants in the present study as well as in recent investigations in Costa Rica.[176]

7.3.4 Venezuela

Although there exists a remarkable knowledge of air pollution problems in Venezuela, and particularly in the capital Caracas, studies on pollution effects are scarce and deal mainly with the accumulation of heavy metals in plants. After 8 weeks of exposure at the Botanical Garden of Caracas, samples of *Tillandsia recurvata* contained 51 to 1075 μg Pb g^{-1} dw, depending on the distance to a busy road. Lead concentrations of *Tillandsia* samples collected at roads in downtown Caracas were between 13 and 1450 μg g^{-1} dw with an average value of 497 μg g^{-1} dw.[177] Up to 240 μg g^{-1} dw Pb were found in lichens collected at busy roads in Merida, the second city of the country. Correlation of Pb contents and traffic volume proved to be highly significant. The fruticose lichen species *Usnea barbata* accumulated more than the corticole species *Puntelia reddenda* and *Parmotrema tinctorum*, *P. austrosinensis*, and *P. hababiana*.[178] Analyses of bryophytes and epiphyllous organisms[179] suggested that rain forests of southwestern Venezuela are affected by long-range transport of heavy metals from industrial areas of Venezuela, Guyana, or Brazil. Lead accumulation in the grasslands around two Pb-recycling plants in northern Venezuela was investigated.[180] *Panicum maximum* contained up to 330 μg g^{-1} dw at sites near the industrial installations. Lead concentrations reached maximum values of 15,000 μg g^{-1} dw in the upper soil layer and 530 μg g^{-1} dw in mixed grass samples. By comparison with analytical results from a highway, it was concluded that Pb contamination was caused by the industrial emissions and not by automobile traffic.

Chemical analyses of natural populations of the epiphytic lichen *Parmotrema madagascariaceum* and transplant experiments with the same species were performed in two cloud forests at about 3 km north and 18 km southwest, respectively, of Caracas.[181] Lichens and tree bark of the phorophytes contained elevated levels of Pb, B, and Mn when compared with data from temperate regions. In lichen transplants, elevated mortality was observed and Pb and Zn concentrations increased within several months. Accumulation of heavy metals was attributed to the intense traffic in the city, the use of leaded gasoline, and urban refuse incineration. Considering the frequent occurrence of acid fog in these forests, the vegetation is at risk of being affected by heavy metals and probably gaseous air pollutants.

7.3.5 Other Countries

Bark of *Tilia cordata* was used as an indicator of air quality in nine cities of central and south Chile. Sulfur concentrations were generally about 1000 μg g^{-1} dw. In the two major cities, Valparaiso and Santiago, S content of 1790 and 3420 μg g^{-1} dw was measured, respectively. Sulfur content showed good correlation with conductivity of the bark extracts.[182] Vegetation damage caused by air and water pollution and soil contamination has been reported from the mining areas in the Central Andes.[70] Negative effects of the smoke emitted by metallurgical plants and tin mines at La Oroya in northern Peru on 700,000 ha of grazing lands were observed as early as 1933.[183] As a consequence of the severe damage to cattle and pasture grass, mitigation measures were implemented in the 1940s.[184] However, in the 1960s, highly elevated heavy metal and arsenic contents in pasture were again reported, As concentrations, e.g., were between 37 and 269 μg g^{-1} dw.[185] Highly elevated As and Cu concentrations and slightly increased Zn and Mn levels in vegetation growing at different distances to the copper mine Turmalina in northern Peru were attributed to uptake by the roots; foliar absorption from the atmosphere was considered less important.[72]

The aerospace center at Kourou in French Guyana is a particular emission source of air pollutants. Each launching of an Ariane V rocket is estimated to release approximately 55 tons of HCl and 91 tons of Al_2O_3 to the troposphere. The fallout which is being deposited to the vegetation in the surroundings is extremely acidic with a pH value near 1. Based on the results of experiments simulating the acid deposition, 25 native species were grouped in a ranking list according to their sensitivity to this treatment. Particularly, sun-adapted species such as *Cecropia* sp. and *Montichardia arborescens* were highly susceptible.[186]

7.4 CONCLUSIONS AND PERSPECTIVES

The database for an evaluation of the air pollution situation in South America is very restricted, as discussed earlier. In addition, the papers reviewed vary considerably with respect to their quality, ranging from very simple studies presented in regional congresses to interdisciplinary research published in international journals. the following conclusions, however, can be drawn:

- Control of air pollutant emissions frequently does not exist or is insufficient, resulting in severe environmental impact.
- Although physicochemical measurements of ambient air pollutant concentrations are done in some countries, their performance is still unsatisfactory. Limit and threshold values are normally adopted from the countries of the Northern Hemisphere, not taking into account the different technological and environmental conditions of the subcontinent.
- In spite of the serious environmental problems, studies on the effects of air pollutants on ecosystems or even single plant species are scarce.
- The consciousness that the main goal of environmental legislation should be to avoid negative effects on humans, fauna, flora, and materials rather than to observe limit values is not well developed. Consequently, the use of bioindicator plants for diagnosis of air pollution effects and monitoring of air quality is also not yet very common.

Against this background, what are the perspectives for the near future and what kind of recommendations can be given? All indications point to an intensification of environmental problems in South America rather than to an improvement unless measures will be taken to reduce air pollutant emissions. Despite their socioeconomic problems related to uncontrolled urbanization, insufficient systems of education and public health, and unequal income distribution, many South American countries are actually experiencing a rapid economic growth with countries like Argentina, Brazil, and Chile being among the world's most-promising markets. This development is favored by the trend to the globalization of the world economy, which promotes a rapid translocation of entire industries from one site to another, generally from industrialized countries to developing ones. Not only more favorable fiscal conditions, subsidies, and low wage rates, but also weaker environmental standards and less rigid control are important criteria for the establishing of new industrial plants.

Growing prosperity is accompanied by a dramatic increase of vehicular traffic. Due to the growth of the automobile fleet and the concomitant bad public transport systems, insufficient infrastructure, and unsatisfactory technical condition of the automobile transport, rapidly growing emissions of nitrogen oxides and hydrocarbons can be expected.

The intensification of agricultural productivity in order to satisfy the increasing demand for food, on the one hand, and agricultural practices such as biomass burning and ongoing deforestation, on the other, are processes that contribute to rising air pollution levels in remote areas. The same holds true for the extraction of ore deposits, particularly the mining of gold and silver that occurs in small, low-technology mines that are difficult to control and that have already contaminated vast areas of the Amazon. Summarizing the trends, we can foresee with increasing air pollution levels in the urban and industrial centers of South America in the near future, and as has happened in the industrialized countries to a growing degree, also at remote sites.

The result of the processes sketched above is a progressive destruction of the natural resources of air, water, soil, fauna, and flora, which is incompatible with a sustainable development as required by the Environment Summit Rio 1992. Health problems of the urban population will continue to grow. Today, millions of cases of respiratory illness and an elevated mortality in Latin American megacities are already attributed to air pollution impact with the implication of high additional costs for national budgets. Chronic and acute air pollution effects on plants are expected to cause severe vegetation damage and consequently soil erosion. The consequences of environmental pollution for the biodiversity of tropical ecosystems has not been studied in detail yet.[187] Finally, the potential impact of air pollution on agricultural productivity should not be neglected. Pollution-induced crop loss may even be more drastic in subtropical and tropical countries than forecasts on the basis of data from temperate latitudes would suggest.[188,189]

All these potential consequences of a supposed increase of air pollutant concentrations in South America and the expansion of the total air pollution–affected area certainly require a lot of measures to be taken in order to moderate the negative effects:

- A more efficient emission control will be necessary. A rigorous environmental legislation has to be introduced, and, even more important, already existing laws and regulations need to be brought into effect by powerful national environmental authorities.
- The already existing air quality–monitoring networks should be improved in order to produce more reliable data, and measurements of air pollution concentrations need to be extended to other severely polluted areas of South America.
- We suggest creating a data bank including as much information on air pollution impact on ecosystems and single plant species as possible. Based on this set of information, more intense studies on the effects on different organizational levels from the individual plant to populations, food webs, and the whole ecosystem should be initiated. Priority should be given to multidisciplinary projects.
- In addition to this, experimental studies including fumigation experiments with defined doses of air pollutants and experiments with filtered and nonfiltered air are needed in order to enlarge the knowledge of resistance vs. tolerance of tropical and subtropical plant species and to establish dose–response relationships. This would also facilitate a broader use of bioindication methods, preferentially with native plant species for diagnosis of air pollution effects and monitoring of air quality.
- Up to now, progress in this direction has been widely impaired because of missing connections between different research activities. An intensified scientific exchange between research groups on a national and international level is therefore a prerequisite for more efficient environmental research. Cooperative projects between South American and industrialized countries of the Northern Hemisphere are strongly recommended, as they will contribute to a more rapid increase of knowledge, transfer of knowledge, and formation of human resources. Such projects, however, should really be collaborative, and not only international,[187] in order to guarantee that the South American countries indeed profit by the cooperation. This chapter, in addition to summarizing the present knowledge of air pollution effects on plants and vegetation in South America, intends to stimulate these processes: exchange of information between research groups and initiation of international cooperation projects.

REFERENCES

1. Finkelman, J., Chemical safety and health in Latin America: an overview, *Sci. Total Environ.*, 188, (Suppl. 1), S3, 1996.
2. Kretzschmar, J. G., Particulate matter levels and trends in Mexico City, São Paulo, Buenos Aires, and Rio de Janeiro, *Atmos. Environ.*, 28, 3181, 1994.

3. Mage, D., Ozolins, G., Peterson, P., Webster, A., Orthofer, R., Vandeweerd, V., and Gwynne, M., Urban air pollution in megacities of the world, *Atmos. Environ.*, 30, 681, 1996.
4. Miguel, A. H., Environmental pollution research in South America, *Environ. Sci. Technol.*, 25, 590, 1991.
5. Arndt, U. and Schweizer, A., The use of bioindicators for environmental monitoring in tropical and subtropical countries, in *Biological Monitoring: Signals from the Environment*, Ellenberg, H., Ed., Vieweg & Sohn, Braunschweig, Germany, 1991, 199.
6. Agrawal, M. and Agrawal, S. B., Phytomonitoring of air pollution around a thermal power plant, *Atmos. Environ.*, 23, 763, 1989.
7. Agrawal, M., Singh, S. K., Singh, J., and Rao, D. N., Biomonitoring of air pollution around urban and industrial sites, *J. Environ. Biol.*, 1, 211, 1991.
8. De Bauer, L. I., Uso de plantas indicadoras de aeropolutos en la Ciudad de Mexico, *Agrociencia*, 9 (D), 139, 1972.
9. De Bauer, L. I. and Krupa, S. V., The Valley of Mexico: summary of observational studies on its air quality and effects on vegetation, *Environ. Pollut.*, 65, 109, 1990.
10. De Bauer, L. I., Hernàndez, T. T. and Manning, W. J., Ozone causes needle injury and tree decline in *Pinus hartwegii* at high altitudes in the mountains around Mexico City, *JAPCA*, 35, 838, 1985.
11. Hernández, T. T. and Pola, C. N. P., Effects of oxidant air pollution on *Pinus maximartinezii* Rzedowski in the México City region, *Environ. Pollut.*, 92, 79, 1996.
12. Pandey, J. and Pandey, U., Adaptational strategy of a tropical shrub *Carissa carandas* L. to urban air pollution, *Environ. Monit. Assess.*, 43, 255, 1996.
13. Rao, M. V. and Dubey, P. S., Explanations for the differential response of certain tropical tree species to SO_2 under field conditions, *Water Air Soil Pollut.*, 51, 297, 1990.
14. Rao, M. V. and Dubey, P. S., Occurrence of heavy metals in air and their accumulation by tropical plants growing around an industrial area, *Sci. Total Environ.*, 126, 1, 1992.
15. Tovar, D. C., Air pollution and forest decline near Mexico City, *Environ. Monit. Assess.*, 12, 49, 1989.
16. Rodhe, H., Cowling, E., Galbally, I., Galloway, J., and Herrera, R., Acidification and regional air pollution in the tropics, in *Acidification in Tropical Countries*, Rodhe, H. and Herrera, R., Eds., SCOPE 36, John Wiley & Sons, Chichester, U.K., 1988, chap. 1.
17. World Research Institute, International Institute for Environment and Development, *Internationaler Umweltatlas: Jahrbuch der Welt-Ressourcen*, Vol. 1, Ecomed, Landsberg, Germany, 1988.
18. Thomas, V., Pollution Control in São Paulo, Brazil: Costs, Benefits and Effects on Industrial Location, World Bank Staff Working Paper No. 501, 1981.
19. Moreira-Nordemann, L. M., Forti, M. C., Di Lascio, V. L., Espirito Santo, C. M. and Danelon, O. M., in *Acidification in Southeastern Brazil*, in Rodhe, H. and Herrera, R., Eds., SCOPE 36, John Wiley & Sons , Chichester, U.K., 1988, chap. 8.
20. CETESB Companhia de Tecnologia de Saneamento Ambiental, Relatório de qualidade do ar no estado de São Paulo 1994, Série Relatórios, São Paulo, Brazil, 1995.
21. Faiz, A., Gautam, S., and Burki, E., Air pollution from motor vehicles: issues and options for Latin American countries, *Sci. Total Environ.*, 169, 303, 1995.
22. Moretton, J., Guaschino, H., Amicone, C., Beletzky, V., Sánchez, M., Santoro, V. and Noto, B., *Contaminación del Aire en Argentina: Aspectos Generales, Legislación y Situación en Capital Federal y Provincia de Buenos Aires*, Ediciones Universo, Buenos Aires, 1996.
23. Yunus, M., Singh, N., and Iqbal, M., Global status of air pollution, in *Plant Response to Air Pollution*, Yunus, M. and Iqbal, M., Eds., John Wiley & Sons, Chichester, UK, 1996, chap. 1.
24. Massambani, O. and Andrade, F., Seasonal behaviour of tropospheric ozone in the São Paulo (Brazil) metropolitan area, *Atmos. Environ.*, 28, 3165, 1994.
25. Tanner, R. L., Miguel, A. H., Andrade, J. B., Gaffney, J. S., and Streit, G. E., Atmospheric chemistry of aldehydes: enhanced peroxyacetyl nitrate formation from ethanol-fueled vehicular emissions, *Environ. Sci. Technol.*, 22, 1026, 1988.
26. Grosjean, D., Miguel, A. H. and Tavares, T. M., Urban air pollution in Brazil: acetaldehyde and other carbonyls, *Atmos. Environ.*, 24B, 101, 1990.
27. Romo-Kröger, C. M., Elemental analysis of airborne particulates in Chile, *Environ. Pollut.*, 68, 161, 1990.

28. Rutllant, J. and Garreaud, R., Meteorological air pollution potential for Santiago, Chile: towards an objective episode forecasting, *Environ. Monit. Assess.*, 34, 223, 1995.
29. Wulf, K., Alarmfelder auf der Smog-Skala, *Akzente*, 3, 49, 1997.
30. WHO-UNEP, *Earthwatch (Global Environmental Monitoring System): Urban Air Pollution in Megacities of the World*, Blackwell Publishers, Oxford, U.K., 1992.
31. Aramendía, P. F., Fernández Prini, R., and Gordillo, G., Buenos Aires en Buenos Aires? *Cienc. Hoy*, 6, 55, 1995.
32. Olcese, L. E. and Toselli, B. M., Effects of meteorology and land use on ambient measurements of primary pollutants in Córdoba City, Argentina, *Meteorol. Atmos. Phys.*, 62, 241, 1997.
33. Branquinho, C. L. and Robinson, V. J., Some aspects of lead pollution in Rio de Janeiro, *Environ. Pollut.*, 10, 287, 1976.
34. Escalona, L. and Sanhueza, E., Elemental analysis of the total suspended matter in the air in downtown Caracas, *Atmos. Environ.*, 15, 61, 1981.
35. Trindade, H. A. and Pfeiffer, W. C., Relationship between ambient lead concentrations and lead in gasoline in Rio de Janeiro, Brazil, *Atmos. Environ.*, 16, 2749, 1982.
36. Van Grieken, R., Van't Dack, L., Dantas, C. C., and Amorim, W. M., Elemental constituents of atmospheric aerosols in Recife, North-East Brazil, *Environ. Pollut.* (Ser. B), 4, 143, 1982.
37. Lara, V., Bifano, C. and Sanhueza, E., Pb, Cd, Mn y Fe en las partículas respirables del centro de Caracas, *Acta Cient. Venez.*, 35, 369, 1984.
38. Caridi, A., Kreiner, A. J., Davidson, J., Debray, M., and Santos, D., Determination of atmospheric lead pollution of automotive origin, *Atmos. Environ.*, 23, 2855, 1989.
39. Ishizaki, C. and Sanhueza, E., International air quality standards and contaminants levels in downtown Caracas atmosphere, *Interciencia*, 4, 6, 1979.
40. Jaffé, R., Carrero, H., Cabrera, A., and Alvarado, J., Organic compounds and heavy metals in the atmosphere of the city of Caracas, Venezuela — I. Atmospheric particles, *Water Air Soil Pollut.*, 71, 293, 1993.
41. Jaffé, R., Cabrera, A., Carrero, H., and Alvarado, J., Organic compounds and heavy metals in the atmosphere of the city of Caracas, Venezuela – II. Atmospheric deposition, *Water Air Soil Pollut.*, 71, 315, 1993.
42. Gordon, C. A., Herrera, R. and Hutchinson, T. C., Studies of fog events at two cloud forests near Caracas, Venezuela – II. Chemistry of fog, *Atmos. Environ.*, 28, 323, 1994.
43. Böhm, G. M., Saldiva, P. H. N., Pasqualucci, C. A. G., Massad, E., Martins, M. A., Zin, W. A., Cardoso, W. V., Criado, P. M. P., Komatsuzaki, M., Sakae, R. S., Negri, E. M., Lemos, M., Capelozzi, V. M., Crestana, C., and Silva, R., Biological effects of air pollution in São Paulo and Cubatão, *Environ. Res.*, 49, 208, 1989.
44. Saldiva, P. H. N., Pope, C. A., Schwartz, J., Dockery, D. W., Lichtenfels, A. J., Salge, J. M., Barone, I., and Böhm, G. M., Air pollution and mortality in elderly people: a time-series study in São Paulo, Brazil, *Arch. Environ. Health*, 50, 159, 1995.
45. Ostro, B., Sanchez, J. M., Aranda, C., and Eskeland, G. S., Air pollution and mortality: results from a study of Santiago, Chile, *J. Exposure Anal. Environ. Epidemiol.*, 6, 97, 1996.
46. Schrimpff, E., Räumliche Verteilung von Schwermetallniederschlägen, angezeigt durch Epiphyten im Cauca-Tal/Kolumbien. Zur Schwermetallbelastung eines tropischen Raumes, *Fortschr. Ber. VDI-Z.* 15, 16, 1981.
47. Sanhueza, E., Africano, M., and Romero, J., Air pollution in tropical areas, *Sci. Total Environ.*, 23, 3, 1982.
48. Sanhueza, E., Cuenca, G., Gómez, M. J., Herrera, R., Ishizaki, C., Martí, I., and Paolini, J., Characterization of the Venezuelan environment and its potential for acidification, in *Acidification in Tropical Countries*, Rodhe, H., Herrera, R., Eds., SCOPE 36, John Wiley & Sons , Chichester, U.K., 1988, chap. 7.
49. Morales, J. A., Bifano, C., and Escalona, A., Rainwater chemistry at the Western Savannah region of the Lake Maracaibo Basin, Venezuela, *Water Air Soil Pollut.*, 85, 2325, 1995.
50. Catoggio, J. A., Succar, S. D., and Roca, A. E., Polynuclear aromatic hydrocarbon content of particulate matter suspended in the atmosphere of La Plata, Argentina, *Sci. Total Environ.*, 79, 43, 1989.

51. Alzuet, P. R., Gaspes, E., and Ronco, A. E., Mutagenicity of environmental samples from an industrialized area of the Rio de La Plata estuary using the *Salmonella*/microsomal assay, *Environ. Toxicol. Water Qual.*, 11, 231, 1996.
52. Ares, J. O., Fluoride cycling near a coastal emission source, *JAPCA*, 28, 344, 1978.
53. Ares, J. O., Villa, A., and Gayoso, A. M., Chemical and biological indicators of fluoride input in the marine environment near an industrial source (Argentina), *Arch. Environ. Contam. Toxicol.*, 12, 589, 1983.
54. Gutberlet, J., *Industrieproduktion und Umweltzerstörung im Wirtschaftsraum Cubatão/São Paulo (Brasilien)*, Tübinger Geographische Studien, 106, Universität Tübingen, Germany, 1991.
55. Gutberlet, J., *Cubatão: Desenvolvimento, Exclusão Social e Degradação Ambiental*, EDUSP, São Paulo, 1996.
56. Targa, H. J. and Klockow, D., Introduction, in *Air Pollution and Vegetation Damage in the Tropics – the Serra do Mar as an Example, Final Report 1990 – 1996*, Klockow, D., Targa, H. J., and Vautz, W., Eds., GKSS Geesthacht, Germany, 1997, chap. 1.
57. Dean, W., *A Ferro e Fogo: A História e a Devastação da Mata Atlántica Brasileira*, Editora Schwarcz, São Paulo, 1996.
58. Jaeschke, W., Chemistry Module, in *Air Pollution and Vegetation Damage in the Tropics — the Serra do Mar as an Example, Final Report 1990 – 1996*, Klockow, D., Targa, H. J., and Vautz, W., Eds., GKSS Geesthacht, Germany, 1997, chap. 3.
59. Alonso, C. D. and Godinho, R., A evolução da qualidade do ar em Cubatão, *Quím. Nova*, 15, 126, 1992.
60. Klumpp, A., Domingos, M., Klumpp, G., and Guderian, R., *Vegetation Module, in Air Pollution and Vegetation Damage in the Tropics — the Serra do Mar as an Example, Final Report 1990 – 1996,* Klockow, D., Targa, H. J., and Vautz, W., Eds., GKSS Geesthacht , Germany, 1997, chap. 5.
61. Neves, N. M. S., Air monitoring at Camaçari petrochemical complex, *Water Sci. Tech.*, 33, 9, 1996.
62. CETESB Companhia de Tecnologia de Saneamento Ambiental, Danos à Vegetação por Fluoretos Gasosos em Alumínio — SP, Informação Técnica No. 001/93 DPTE, São Paulo, Brazil, 1993.
63. Prado Filho, J. F., Uso de bioindicadores para monitoramento do ar, *Ambiente*, 7, 57, 1993.
64. Arndt, U., Flores, F. and Weinstein, L., *Efeitos do Flúor Sobre as Plantas: Diagnose de Danos na Vegetação do Brasil*, (in Portuguese, English, German), Editora da UFRGS, Porto Alegre, Brazil, 1995.
65. Martins, A. F. and Zanella, R., Estudo analítico-ambiental na região carboenergética de Candiota, Bagé (RS), *Ciênc. Cult.*, 42, 264, 1990.
66. Pfeiffer, W. C., Fiszman, M., De Lacerda, L. D., Van Weerelt, M., and Carbonell, N., Chromium in water, suspended particles, sediments and biota in the Irajá River estuary, *Environ. Pollut.*, 4, 193, 1982.
67. Karez, C. S., Magalhaes, V. F., Pfeiffer, W. C., and Amado Filho, G. M., Trace metal accumulation by algae in Sepetiba Bay, Brazil, *Environ. Pollut.*, 83, 351, 1994.
68. Moreira, J. C., Threats by heavy metals: human and environmental contamination in Brazil, *Sci. Total Environ.*, 188 (Suppl.1), S61, 1996.
69. Jordão, C. P., Pereira, J. C., Brune, W., Pereira, J. L., and Braathen, P. C., Heavy metal dispersion from industrial wastes in the Vale do Aço, Minas Gerais, Brazil, *Environ. Technol.*, 17, 489, 1996.
70. Ruthsatz, B. and Wey, H., Concept for a biological monitoring study, in *Biological Monitoring: Signals from the Environment*, Ellenberg, H., Ed., Vieweg & Sohn, Braunschweig, Germany, 1991, 75.
71. Romo-Kröger, C. M. and Llona, F., A case of atmospheric contamination at the slopes of the Los Andes Mountain range, *Atmos. Environ.*, 27A, 401, 1993.
72. Bech, J., Poschenrieder, C., Llugany, M., Barceló, J., Tume, P., Tobias, F. J., Barranzuela, J. L., and Vásquez, E. R., Arsenic and heavy metal contamination of soil and vegetation around a copper mine in northern Peru, *Sci. Total Environ.*, 203, 83, 1997.
73. Nriagu, J. O., Mercury pollution from the past mining of gold and silver in the Americas, *Sci. Total Environ.*, 149, 167, 1994.
74. Silva, A. P., Ferreira, N. L. S., Pádua, H. B., Veiga, M. M., Silva, G. C., Oliveira, E. F., Castro e Silva, E., and Ozaki, S. K., Mobilidade do mercúrio no Pantanal de Poconé, *Ambiente*, 7, 52, 1993.
75. Alho, C. J. R. and Vieira, L. M., Fish and wildlife resources in the Pantanal wetlands of Brazil and potential disturbances from the release of environmental contaminants, *Environ. Toxicol. Chem.*, 16, 71, 1997.
76. Wheatley, B. and Wyzga, R., Mercury as a global pollutant — human health issues, *Water Air Soil Pollut.*, 97, 1, 1997.

77. Lima, J. S., Distribution and Accumulation of Copper as Cupric Fungicide Residue in a Cocoa Agrarian Ecosystem in Bahia/Brazil, Ph. D. thesis, Universität Kassel, Germany, 1992.
78. Franz-Gerstein, C., Entwicklung und Erprobung von Bioindikatoren für die Tropen und Subtropen, in *Bioindikation — Neue Entwicklungen, Nomenklatur, Synökologische Aspekte*, Arndt, U., Fomin, A., and Lorenz, S., Eds., Verlag Heimbach, Ostfildern,Germany, 1996, 79.
79. Lacher Jr., T. E. and Kendall, R. J., Tropical ecotoxicology, *Environ. Toxicol. Chem.*, 16, 1, 1997.
80. Dardozzi, R. and Ramel, C., Chemical hazards in developing countries, Pontificiae Academiae Scientiarum Scripta Varia, *Sci. Total Environ.*, 188, (Suppl. 1), 1996.
81. Sandroni, S. and Anfossi, D., Historical data of surface ozone at tropical latitudes, *Sci. Total Environ.*, 148, 23, 1994.
82. Logan, J. A. and Kirchhoff, V. W. J. H., Seasonal variations of tropospheric ozone at Natal, Brazil, *J. Geophys. Res.*, 91, 7875, 1986.
83. Kirchhoff, V. W. J. H., Increasing concentrations of CO and O_3, rising deforestation rates and increasing tropospheric carbon monoxide and ozone in Amazonia, *Environ. Sci. Pollut. Res.*, 3, 210, 1996.
84. Lazutin, L., Bezerra, P. C., Fagnani, M. A., Pinto, H. S., Martin, I. M., Silva, E. L. P., Mello, M. G. S., Turtelli, A., Zhavkov, V., and Zullo, J., Jr., Surface ozone study in Campinas, São Paulo, Brazil, *Atmos. Environ.*, 30, 2729, 1996.
85. Greenberg, J. P., Zimmerman, P. R., Heidt, L., and Pollock, W., Hydrocarbon and carbon monoxide emissions from biomass burning in Brazil, *J. Geophys. Res.*, 89, 1350, 1984.
86. Crutzen, P. J., Delany, A. C., Greenberg, J., Haagenson, P., Heidt, L., Lueb, R., Pollock, W., Seiler, W., Wartburg, A., and Zimmerman, P., Tropospheric chemical composition measurements in Brazil during the dry season, *J. Atmos. Chem.*, 2, 233, 1985.
87. Crutzen, P. J. and Andreae, M. O., Biomass burning in the tropics: impact on atmospheric chemistry and biogeochemical cycles, *Science*, 250, 1669, 1990.
88. Ares J. O., Villa, A., and Mondadori, G., Air pollutant uptake by xerophytic vegetation: fluoride, *Environ. Exp. Bot.*, 20, 259, 1980.
89. Weißflog, L., Paladini, E., Gantuz, M., Puliafito, J. L., Puliafito, S., Wenzel, K. D., and Schüürmann, G., Immission patterns of airborne pollutants in Argentina and Germany — I. First results of a heavy metal biomonitoring, *Fresenius Environ. Bull.*, 3, 728, 1994.
90. Wenzel, K. D., Weißflog, L., Paladini, E., Gantuz, M., Guerreiro, P., Puliafito, C., and Schüürmann, G., Immission patterns of airborne pollutants in Argentina and Germany — II. Biomonitoring of organochlorine compounds and polycyclic aromatics, *Chemosphere*, 34, 2505, 1997.
91. Levin, A. G. and Pignata, M. L., *Ramalina ecklonii* (Spreng.) Mey. and Flot. as bioindicator of atmospheric pollution in Argentina, *Can. J. Bot.*, 73, 1196, 1995.
92. Carreras, H. A., Cañas, M. S., and Pignata, M. L., Differences in responses to urban air pollutants by *Ligustrum lucidum* Ait. and *Ligustrum lucidum* Ait. f. *tricolor* (Rehd.) Rehd, *Environ. Pollut.*, 93, 211, 1996.
93. Cañas, M. S., Carreras, H. A., Orellana, L., and Pignata, M. L., Correlation between environmental conditions and foliar chemical parameters in *Ligustrum lucidum* Ait. exposed to urban air pollutants, *J. Environ. Manage.*, 49, 167, 1997.
94. Pignata, M. L., Cañas, M. S., Carreras, H. A., and Orellana, L., Exploring chemical variables in *Ligustrum lucidum* Ait. f. *tricolor* (Rehd.) Rehd. in relation to air pollutants and environmental conditions, *Environ. Manage.*, 21, 793, 1997.
95. González, C. M. and Pignata, M. L., The influence of air pollution on soluble proteins, chlorophyll degradation, MDA, sulfur and heavy metals in a transplanted lichen, *Chem. Ecol.*, 9, 105, 1994.
96. González, C. M. and Pignata, M. L., Chemical response of the lichen *Punctelia subrudecta* (Nyl) Krog transplanted close to a power station in an urban-industrial environment, *Environ. Pollut.*, 97, 195, 1997.
97. González, C. M., Casanovas, S. S., and Pignata, M. L., Biomonitoring of air pollution in Córdoba, Argentina employing *Ramalina ecklonii* (Spreng.) Mey. and Flot, *Environ. Pollut.*, 91, 269, 1996.
98. Cañas, M. S., Orellana, L., and Pignata, M. L., Chemical response of the lichens *Parmotrema austrosinense* and *P. conferendum* transplanted to urban and non-polluted environments, *Ann. Bot. Fenn.*, 34, 27, 1997.

99. Cañas, M. S. and Pignata, M. L., Temporal variation of pigments and peroxidation products in the lichen *Parmotrema uruguense* (Kremplh.) Hale transplanted to urban and non-polluted environments, *Symbiosis*, 24, 147, 1998.
100. Carreras, H. A. and Pignata, M. L., Variación en la respuesta química de *Usnea sp.* a contaminantes atmosféricos de la Ciudad de Córdoba en diferentes épocas del año, XI Jornadas Científicas de la Sociedad de Biología de Córdoba, Córdoba, Resúmenes, 44, 1997.
101. Calvelo, S., Baccalá, N., Arribére, M., Ribeiro Guevara, S., and Bubach, D., Concentrations of biological relevant elements of foliose and fruticose lichens from Nahuel Huapi National Park (Patagonia), analyzed by INAA, *Sauteria*, 9, 87, 1998.
102. CETESB Companhia de Tecnologia de Saneamento Ambiental, Avaliação das degradações da Serra do Mar — Cubatão — SP, Report, São Paulo, Brazil, 1986.
103. CETESB Companhia de Tecnologia de Saneamento Ambiental, Mapeamento do uso do solo, das cicatrizes de escorregamentos e caracterização preliminar da estrutura da cobertura vegetal da Serra do Mar, em Cubatão — SP, Report, São Paulo, Brazil, 1987.
104. Bragança, C. F., Kono, E. C., Aguiar, L. S. J., and Santos, R. P., Avaliação da degradação da Serra do Mar, *Ambiente,* 1, 77, 1987.
105. Queiroz Neto, J. P., Monteleone Neto, R., and Marques, R. A., A dramática situação em Cubatão. *Ciênc. Cult.*, 35, 1164, 1983.
106. SMA Secretaria Do Meio Ambiente do Estado de São Paulo, *The Rain-Forest of the Serra do Mar: Degradation and Reconstitution*, Documents Series, São Paulo, Brazil, 1990.
107. Pompéia, S. L., Sucessão Secundária da Mata Atlântica em Áreas Afetadas Pela Poluição Atmosférica Cubatão, SP, Ph. D. thesis, Universidade de São Paulo, São Paulo, Brazil, 1997.
108. CETESB Companhia de Tecnologia de Saneamento Ambiental, Aspectos Fitosociológicos da Vegetação da Serra do Mar Degradada Pela Poluição Atmosférica de Cubatão - SP, Report, São Paulo, Brazil, 1989.
109. CETESB Companhia de Tecnologia de Saneamento Ambiental, Levantamento de Espécies Vegetais Resistentes e Tolerantes à Poluição Atmosférica do Polo Industrial de Cubatão, Report, São Paulo, Brazil, 1988.
110. CETESB Companhia de Tecnologia de Saneamento Ambiental, Espécies Arbóreas da Serra do Mar Sensíveis à Poluição Atmosférica do Polo Industrial de Cubatão — SP, Report, São Paulo, Brazil, 1988.
111. Leitão Filho, H. F., Pagano, S. N., Cesar, O., Timoni, J. L., and Rueda, J. J., *Ecologia da Mata Atlântica em Cubatão*, Editora UNESP/Editora UNICAMP, Campinas, Brazil, 1993.
112. Coutinho, L. M., Contribuição ao Conhecimento da Ecologia da Mata Pluvial Tropical, Boletim No. 257, Botânica No. 18, da Fac. Fil., Ciên., Let., Universidade de São Paulo, SP, Brazil, 1962.
113. Rebelo, C. F., Struffaldi-De Vuono, Y., and Domingos, M., Estudo ecológico de comunidades de briófitas epífitas da Reserva Biológica de Paranapiacaba, SP, em trechos de floresta sujeita à influência da poluição aérea, *Rev. Bras. Bot.*, 18, 1, 1995.
114. Klumpp, A., Klumpp, G., Domingos, M., and Silva, M. D., Fluoride impact on native tree species of the Atlantic Forest near Cubatão, Brazil, *Water Air Soil Pollut.*, 87, 57, 1996.
115. Struffaldi-De Vuono, Y., Lopes, M. I. M. S., and Domingos, M., Air pollution and effects on soil and vegetation of Serra do Mar, near Cubatão, São Paulo, Brazil, in *Air Pollution and Forest Decline, Proc. 14th Int. Meeting for Specialists in Air Pollution Effects on Forest Ecosystems*, Bucher, J. B. and Bucher-Wallin, I., Eds., Interlaken, Switzerland, 396, 1989.
116. Domingos, M., Poggiani, P., Struffaldi-De Vuono, Y., and Lopes, M. I. M. S., Produção de serapilheira na floresta da Reserva Biológica de Paranapiacaba, sujeita aos poluentes atmosféricos de Cubatão, SP, *Hoehnea*, 17, 47, 1990.
117. Struffaldi-De Vuono, Y., Domingos, M., and Lopes, M. I. M. S., Decomposicão da serapilheira e liberação de nutrientes na floresta da Reserva Biológica de Paranapiacaba, sujeita aos poluentes atmosféricos de Cubatão, São Paulo, Brazil, *Hoehnea*, 16, 179, 1989.
118. Moraes, R. M., Struffaldi-De Vuono, Y. and Domingos, M., Aspectos da decomposição da serapilheira em florestas tropicais preservada e sujeita à poluição atmosférica, no estado de São Paulo, Brasil, *Hoehnea*, 22, 91, 1995.
119. Lopes, M. I. M. S., Struffaldi-De Vuono, Y., and Domingos, M., Serapilheira acumulada na floresta da Reserva Biológica de Paranapiacaba, sujeita aos poluentes atmosféricos de Cubatão, SP, *Hoehnea*, 17, 59, 1990.

120. Domingos, M., Poggiani, F., Struffaldi-De Vuono, Y., and Lopes, M. I. M. S., Precipitação pluvial e fluxo de nutrientes na floresta da Reserva Biológica de Paranapiacaba, sujeita aos poluentes atmosféricos de Cubatão, SP, *Rev. Bras. Bot.*, 18, 119, 1995.
121. Klockow, D., Targa, H. J., and Vautz, W., *Air Pollution and Vegetation Damage in the Tropics – the Serra do Mar as an Example, Final Report 1990–1996*, GKSS Geesthacht, Germany, 1997.
122. Mayer, R. and Lopes, M. I. M. S., Soil Module, in *Air Pollution and Vegetation Damage in the Tropics — the Serra do Mar as an Example, Final Report 1990–1996*, Klockow, D., Targa, H. J. and Vautz, W., Eds., GKSS Geesthacht, Germany, 1997, chap. 4.
123. Domingos, M., Klumpp, A., and Klumpp, G., Air pollution impact on the Atlantic Forest at the Cubatão region, SP, Brazil, *Ciênc. Cult.*, 50, 230, 1998.
124. Klumpp, A., Klumpp, G., and Domingos, M., Diagnose von Immissionswirkungen in einem tropischen Belastungsgebiet mit Hilfe verschiedener Verfahren der Bioindikation, *Essener Ökol. Schr.*, 4, 167, 1994.
125. Alves, E. S., The effects of the pollution on wood of *Cecropia glazioui* (Cecropiaceae), IAWA J., 16, 69, 1995.
126. Mazzoni-Viveiros, S. C., Aspectos Estruturais de *Tibouchina pulchra* Cogn. (Melastomataceae) Sob o Impacto de Poluentes Atmosféricos Provenientes do Complexo Industrial de Cubatão, SP — Brasil, Ph. D. thesis, Universidade de São Paulo, São Paulo, Brazil, 1996.
127. Klumpp, A., Klumpp, G., and Domingos, M., Plants as bioindicators of air pollution at the Serra do Mar near the industrial complex of Cubatão, Brazil, *Environ. Pollut.*, 85, 109, 1994.
128. Klumpp, A., Klumpp, G., and Domingos, M., Bio-indication of air pollution in the tropics - the active monitoring programme near Cubatão (Brazil), *Gefahrstoffe — Reinhalt. Luft*, 56, 27, 1996.
129. Klumpp, G., Klumpp, A., Domingos, M., and Guderian, R., *Hemerocallis* as bioindicator of fluoride pollution in tropical countries, *Environ. Monit. Assess.*, 35, 27, 1995.
130. Klumpp, A., Modesto, I. F., Domingos, M., and Klumpp, G., Susceptibility of various *Gladiolus* cultivars to fluoride pollution and their suitability for bioindication, *Pesquis. Agropec. Bras.*, 32, 239, 1997.
131. Klumpp, A., Domingos, M., and Klumpp, G., Assessment of the vegetation risk by fluoride emissions from fertiliser industries at Cubatão, Brazil, *Sci. Total Environ.*, 192, 219, 1996.
132. Doley, D., Ambient air quality objectives for fluoride: an Australian perspective, *Clean Air*, 21, 55, 1987.
133. Van der Eerden, L. J., Fluoride content in grass as related to atmospheric fluoride concentrations: a simplified predictive model, *Agric. Ecosyst. Environ.*, 37, 257, 1991.
134. Klumpp, A., Domingos, M., Moraes, R. M., and Klumpp, G., Effects of complex air pollution on tree species of the Atlantic Rain Forest near Cubatão, Brazil, *Chemosphere*, 36, 989, 1998.
135. Pradella, D. Z. A., Estudos com *Tibouchina pulchra* Cogn. (Melastomataceae) — (manacá da serra) Submetida a Estresse por Poluição na Serra do Mar em Cubatão - SP — Brasil, Masters thesis, Universidade Federal de São Carlos, São Paulo, Brazil, 1997.
136. Domingos, M., Biomonitoramento do Potencial Fitotóxico da Poluição Aérea e da Contaminação do Solo na Região do Polo Industrial de Cubatão, São Paulo, Utilizando *Tibouchina pulchra* Cogn. como Espécie Indicadora, Ph.D. thesis, Universidade de São Paulo, São Paulo, 1998.
137. Silva Filho, N. L., Teixeira, N. T., and Silva, L. F. C., Influência de solos sujeitos à poluição de Cubatão na germinação de sementes com diferentes composições químicas, *Ecossistema*, 16, 111, 1991.
138. IBt Instituto de Botânica, Recomposição da Vegetação da Serra do Mar em Cubatão, Report, Série Pesquisa, São Paulo, 1989.
139. Pompéia, S. L., Pradella, D. Z. A., Martins, S. E., Dos Santos, R. C., and Diniz, K. M., A semeadura aérea na Serra do Mar em Cubatão, *Ambiente*, 3, 13, 1989.
140. Pompéia, S. L., Pradella, D. Z. A., Diniz, K. M., and Santos, R. P., Comportamento dos manacás-da-serra (*Tibouchina* sp.) semeados por via aérea em Cubatão, II Congresso Nacional sobre Essências Nativas, São Paulo, Brazil, Anais, 506, 1992.
141. Silva Filho, N. L., Composição da cobertura vegetal de um trecho degradado da Serra do Mar, Cubatão, SP, I. Estado da arte, II Congresso Nacional sobre Essências Nativas, São Paulo, Brazil, Anais, 971, 1992.
142. Struffaldi-De Vuono, Y., Lopes, M. I. M. S., and Domingos, M., Poluição atmosférica e elementos tóxicos na Reserva Biológica do Instituto de Botânica, São Paulo, Brazil, *Rev. Bras. Bot.*, 7, 149, 1984.

143. Struffaldi-De Vuono, Y., Lopes, M. I. M. S., and Domingos, M., Alterações provocadas pela poluição atmosférica na fertilidade do solo da Reserva Biológica do Instituto de Botânica, São Paulo, Brazil, *Rev. Bras. Bot.*, 11, 95, 1988.
144. Prado Filho, J. F., Plantas que detectam poluição, *Ciênc. Hoje*, 14, 18, 1992.
145. Weinstein, L. H. and Hansen, K. S., Relative susceptibilities of Brazilian vegetation to airborne fluoride, *Pesqui. Agropec. Bras.*, 23, 1125, 1988.
146. Lima, J. S., Carvalho Filho, D. M., Couto, E., Santana, D. L. and Souza, H. C., Comparação entre o coentro e o *Sphagnum* sp. como bioacumuladores do arsênio no Pólo Petroquímico de Camaçari — BA, *Rev. Bras. Ecol.*, 1, 91, 1997.
147. Lima, J. S., Carvalho Filho, D. M., Couto, E., Korn, M. A. G., Melo, M. H., and Gomes, R. C. T., Capim-santo (*Cymbopongon citratus*) como bioindicador de poluição atmosférica no Pólo Petroquímico de Camaçari — BA, *Rev. Bras. Ecol.*, 1, 95, 1997.
148. Arndt, U., personal communication, 1997.
149. Klumpp, A., personal observation, 1995.
150. Flores, F. E. V., O uso de plantas como bioindicadores de poluição no ambiente urbano-industrial: experiências em Porto Alegre, RS, Brasil, *Tübinger Geog. Stud.*, 96, 79, 1987.
151. Flores, F. E. V., Bioindicação vegetal no Centro de Ecologia da UFRGS, XLVI Congresso Nacional de Botânica, Ribeirão Preto, São Paulo, Brazil, Resumos, 374, 1995.
152. Fiedler, H., Martins, A. F. and Solari, J. A., Meio ambiente e complexos carboelétricos: o caso Candiota, *Ciênc. Hoje*, 12, 38, 1990.
153. Flores, E. M. M. and Martins, A. F., Use of pollution bioindicators for fluoride in the vicinity of coal thermoeletric power plants, *S. Braz. J. Chem.*, 1, 61, 1993.
154. Mendonça, B. R. and Silva, E. A. M., Efeito da poluição sobre bioindicadores vegetais, *Ambiente*, 5, 37, 1991.
155. Calasans, C. F. and Malm, O., Utilização de *Tillandsia usneoides* para avaliação de poluição atmosférica por mercúrio, in Portuguese and English, *Bromélia*, 1, 7, 1994.
156. Calasans, C. F. and Malm, O., Elemental mercury contamination survey in a chlor-alkali plant by the use of transplanted Spanish moss, *Tillandsia usneoides* (L.), *Sci. Total Environ.*, 208, 165, 1997.
157. Strehl, T. and Lobo, E. A., Analysis of the morphological characters of *Tillandsia aeranthos* (Loisel.) L.B. Smith and *T. recurvata* (L.) L. (Bromeliaceae) as bioindicators of the urban pollution in Porto Alegre City, Southern Brazil, *Aquilo Ser Bot.*, 27, 19, 1989.
158. Borges, M., Andrade, T. J., Jankowski, A., Ferreira, E. B., and Inoue, M. T., Pigmentos foliares em *Tabebuia alba* e *Pittosporum undulatum* como bioindicadores da poluição urbana, II Congresso Nacional sobre Essências Nativas, São Paulo, Brazil, Anais, 778, 1992.
159. Inoue, M. T., Wandembruck, A., and Mores, M., Plantas indicadores de poluição ambiental: uma abordagem metodológica exemplificada em *Tabebuia chrysotricha*, II Congresso Nacional sobre Essências Nativas, São Paulo, Brazil, Anais, 782, 1992.
160. Inoue, M. T. and Reissmann, C. B., Efeitos da poluição na fotossíntese, dimensões da folha, deposição de particulados e conteúdo de ferro e cobre em alfeneiro (*Ligustrum lucidum*) da arborização de Curitiba, PR, *Floresta*, 21, 3, 1993.
161. Bustamante, M. M. C. and Baumgarten, L. C., Possibilidades do uso de *Spathodea campanulata* Beuv. na bioindicação de poluentes atmosféricos em Brasília, DF, II Congresso de Ecologia do Brasil, Londrina, Brazil, Resumos, 593, 1994.
162. Pradella, D. Z. A., personal communication, 1997.
163. Silva, M. D. and Klumpp, A., unpublished data, 1997.
164. Alves, P. L. C. A., Oliva, M. A., Cambraia, J., and Sant'anna, R., Efeitos da chuva ácida simulada e de um solo de Cubatão (SP) sobre parâmetros relacionados com a fotossíntese e a transpiração de plantas de soja, *Rev. Bras. Fisiol. Veg.*, 2, 7, 1990.
165. Alves, P. L. C. A. and Oliva, M. A., Reações da soja à chuva ácida e solo contaminado, *Ambiente*, 7, 34, 1993.
166. Bustamante, M. M., Oliva, M. A., Sant'anna, R., and Lopes, N. F., Sensibilidade da soja ao flúor, *Rev. Bras. Fisiol. Veg.*, 5, 151, 1993.
167. Azevedo, A. A., Ação do Flúor, em Chuva Simulada, Sobre a Estrutura Foliar de *Glycine max* (L.) Merril, Ph.D. thesis, Universidade de São Paulo, São Paulo, Brazil, 1995.

168. Oliva, M. A., Figueiredo, J. G. and Souza, M. M., Bioindicação do flúor mediante gramíneas tropicais, in *Indicadores Ambientais*, Maia, N. B. and Martos, H., Eds., PUC São Paulo, Sorocaba, Brazil, 1997, 191.
169. Arndt, U. and Strehl, T., Begasungsexperimente mit SO_2 an Tillandsien zur Entwicklung eines Bioindikators, *Angew. Bot.*, 63, 43, 1989.
170. Strehl, T. and Arndt, U., Alterações apresentadas por *Tillandsia aeranthos* e *T. recurvata* (Bromeliaceae) expostas ao HF e SO_2, *Iheringia Ser. Bot.*, 39, 3, 1989.
171. Arndt, U., Franz-Gerstein, C., Hinger, D., Schachner, J., Scherrieble, T., Flores, F. and Strehl, T., Untersuchungen zur Eignung subtropischer Pflanzen als Bioindikatoren in Entwicklungsländern. *VDI-Ber.*, 901, 559, 1991.
172. Franz-Gerstein, C., Untersuchungen zur Eignung epiphytischer Orchideen als Bioindikatoren in tropischen und subtropischen Immissionsgebieten, Ph. D. thesis, Universität Hohenheim, Germany, 1992.
173. Grams, T., Einfluß von erhöhter bodennaher Ozonkonzentration auf Pflanzen mit Crassulaceen-Stoffwechsel (CAM), *Bielefelder Ökol. Beitr.*, 12, 99, 1998.
174. Franz-Gerstein, C., Fomin, A., Klumpp, A., Klumpp, G., Domingos, M., and Arndt, U., Untersuchungen zur Eignung verschiedener Gladiolensorten als Bioindikatoren für Fluorwasserstoff, *Verh. Ges. Ökol.*, 24, 293, 1995.
175. Schrimpff, E., Air pollution patterns in two cities of Colombia, S.A. according to trace substances content of an epiphyte (*Tillandsia recurvata* L.), *Water Air Soil Pollut.*, 21, 279, 1984.
176. Brighigna, L., Ravanelli, M., Minelli, A., and Ercoli, L., The use of an epiphyte (*Tillandsia caput-medusae Morren*) as bioindicator of air pollution in Costa Rica, *Sci. Total Environ.*, 198, 175, 1997.
177. Tugues, J. L., Las plantas como indicadoras de plomo en la atmósfera, *Acta Bot. Venez.*, 11, 107, 1976.
178. Burguera, J. L., Burguera, M., and Belandria, M. G., The amounts of lead in roadside soil and some lichen species and their correlation with motor vehicles traffic volume, in *Proc. Int. Heavy Metals in the Environment Conf.*,Vernet, J. P., Ed., CEP Consultants, Geneva, 1989, 460.
179. Montagnini, F., Neufeld, H. S., and Uhl, C., Heavy metal concentrations in some non-vascular plants in an Amazonian Rainforest, *Water Air Soil Pollut.*, 21, 317, 1984.
180. Wüstemann, M., Beispiele zur Problematik von Immissionswirkungen in tropischen und subtropischen Ländern — Untersuchungen in der Umgebung eines Bleiemittenten in Venezuela, Diploma thesis, Universität Hohenheim, Germany, 1983.
181. Gordon, C. A., Herrera, R., and Hutchinson, T. C., The use of a common epiphytic lichen as a bioindicator of atmospheric inputs to two Venezuelan cloud forests, *J. Trop. Ecol.*, 11, 1, 1995.
182. Godoy, R., Steubing, L., and Debus, R., Die Borke von *Tilia cordata* Mill. als Bioindikator für die Luftbelastung chilenischer Städte, *Verh. Ges. Ökol.*, 17, 567, 1989.
183. Millones, J. O., Patterns of land use and associated environmental problems of the Central Andes, *Mount. Res. Dev.*, 2, 49, 1982.
184. Vizcarra-Andreu, M. A., *La Atmósfera Contaminada y sus Relaciones con el Público*, Tecnosfera, Pacific Press, Lima, Peru, 1982.
185. Anonymus, Contaminación atmosférica en las comunidades de la región central del país, *Salud Ocupacional*, 16/17(18), Lima, Peru, 1972.
186. Garrec, J. P., Nolibos, I., Rose, C., Nourrisson, G., Laroussinie, D. and Caron, H., Sensibilité à l'acidité HCl des différentes écosystèmes végétaux situés aux abords du Centre Spatial Guyanais, Report INRA, Champenoux/Silvolab Kourou, 1996.
187. Lacher, T. E., Jr, and Goldstein, M. I., Tropical ecotoxicology: status and needs, *Environ. Toxicol. Chem.*, 16, 100, 1997.
188. Wahid, A., Maggs, R., Shamsi, S. R. A., Bell, J. N. B., and Ashmore, M. R., Air pollution and its impacts on wheat yield in the Pakistan Punjab, *Environ. Pollut.*, 88, 147, 1995.
189. Wahid, A., Maggs, R., Shamsi, S. R. A., Bell, J. N. B., and Ashmore, M. R., Effects of air pollution on rice yield in the Pakistan Punjab, *Environ. Pollut.*, 90, 323, 1995.

CHAPTER 8

Effects of Air Pollution on Plant Diversity

Madhoolika Agrawal and S. B. Agrawal

CONTENTS

8.1 INTRODUCTION

The rapid growth of industries, urbanization, and transportation and their injudicious planning without due regard for sustainable development are leading to serious environmental hazards. The destruction of ecological balance as a consequence has a significant unfavorable effect on the biosphere, and the entire life-support system is rapidly facing danger. The significant contribution of air pollution to the problems of ill health of human beings, loss of agricultural productivity, forest decline, etc. has been a cause of increasing public concern throughout the world. It has also become a topic of intense scientific, industrial, and government interest. The environment constitutes the largest portion of the biosphere and is one of the vital components on which the life depends. Biosphere contributes significantly to the chemical composition of the atmosphere and its properties, and, vice versa, changes in atmospheric composition have a strong feedback effect on ecosystem functioning and biodiversity. Air quality is critical to plant health.

Earth's atmosphere is finite and its capacity to cleanse itself is limited. Earth has evolved in an atmosphere of life-supporting chemical constituents and therefore has a limited ability for homeostasis and resilience against changes in air quality.[1] A wide array of air pollutants including particulates, liquids, and gases is being emitted both from natural and anthropogenic sources. After

1-56670-341-7/00/$0.00+$.50

emission, air pollutants are subjected to physical, chemical, and photochemical transformations, which ultimately decide the fate and atmospheric concentrations of the pollutants. The application of abatement technology, increased urbanization, population growth especially in developing countries, and the shift to large fossil-fueled electric generating plants have altered the nature of air quality problems and the focus of air pollution research during recent years. The problems of air pollution that were thought to be concentrated around large, uncontrolled point sources have been supplanted by those associated with regional-scale elevated concentrations of pollutants.

Air pollutants do not remain confined near the source of emission, but spread over far off distances, transcending natural and political boundaries depending upon topography and meteorological conditions, especially wind direction and speed and vertical and horizontal thermal gradients. The topic of the regional impact of air pollution on different ecosystems is one of the major ecological, scientific, and political issues of the modern industrialized world. Adverse effects of air pollutants on biota and ecosystems have been documented worldwide.[2-9] Since humans live in an extremely holocenotic environmental complex where all factors are interrelated, a slight change in any one factor may produce numerous and involved alterations in other factors. In the last three decades, environmental scientists have begun to appreciate the complex interactions of air pollutants with abiotic and biotic components of an ecosystem. Tingey et al.[10] have emphasized the consideration of ecological impacts of air pollution in developing ambient air quality standards.

8.2 CAUSES OF BIODIVERSITY LOSS

The continued functioning of ecosystems is dependent on the constituent species and their distribution, as well as on genetic variation within those species and the dynamics of the interactions that exist between different species and the physical environment. Ecosystems have characteristic biodiversity at different levels of organization depending upon the abiotic and biotic factors.[11] Both genetic and natural selection processes are responsible for the diversity of an area. Biodiversity originates at the gene level and extends through species, populations, communities, to ecosystems. Any kind of disturbance is a key factor in analyzing the dynamics of communities and its role in the functioning and maintenance of the ecosystem. In response to environmental stimuli, biodiversity constantly changes, leading to loss of genes, species, populations, and habitats.[11,12]

Humans now have the dominant influence on biodiversity. Humans have made changes in our landscapes due to agriculture, silviculture, road building, industrial development, urbanization, and pollution — the major causes of biodiversity loss at the global scale. The issue of biodiversity loss has raised an alarm throughout the world. The present species extinction rate has surpassed even the massive extinction reported during the geologic past. Habitat fragmentation and destruction, species overexploitation, introduction of exotic species, and release of toxic chemicals into air, water, and soil are implicated as major causes of loss of biodiversity.[13-15] The first two stressors attained immediate importance as they are directly related to species loss, but the release of toxic chemicals has a subtle effect on biological diversity because of the adverse effects of pollution on sensitive species. As plants live outdoors and are continuously exposed to air pollutants, they show the effects of pollutants in due course of time. An air pollutant may affect species differently in a community, where some species exhibit no response and some may be unfavorably affected. After several generations, the sensitive species may be progressively eliminated leading to species loss from the community. This response to low-intensity stressors for a prolonged duration is microevolution leading to a shift in genetic structure toward greater resistance.[16] Genetic diversity is found responsive to anthropogenic stresses.[17-18]

Alteration of the chemical status of the atmosphere has the potential of adversely affecting plant diversity at the local, regional, and global scale. Emissions of gases such as sulfur dioxide (SO_2), hydrogen fluoride (HF), chlorine (Cl_2), ammonia (NH_3), etc. and particulate containing toxic metals adversely affect the vegetation structure and function locally around large polluting sources.[3,8,9] Regional-scale impact on vegetation may be ascribed to secondary air pollutants formed as a result of different transformations of primary pollutants present in air masses traveling long distances. Airborne pollutants such as ozone (O_3) and peroxyacetyl nitrate (PAN) and acid depositions are the problems of regional scale affecting plant species directly and indirectly through interactions with other natural stresses such as water and nutrient deficiencies, temperature stress, and insect and pathogen infestations.[19,20] Global-level effects of changes in atmospheric composition on biodiversity are due to increases in persistent pesticides and trace metals, greenhouse gases, and stratospheric O_3-depleting gases, etc. The depletion of stratospheric O_3 with the consequence of increasing UV-B radiation will have direct effects on biodiversity and ecosystem functioning by impairment of photosynthesis, nitrogen metabolism, and orientation of marine phytoplankton,[21] which may affect dimethylsulfide (DMS) emission and thus affect cloud formation.[22] Increased UV-B not only affects primary producers, but the next tropic levels as well.[23]

Disruption of the carbon cycle due to increase in the concentration of CO_2 in atmosphere[24] and the nitrogen cycle due to human-induced nitrogen fixation leading to increase in nitrous oxide emissions and higher levels of N depositions may cause drastic changes in species diversity.[25] Plant species richness has been shown to reduce in short grass steppe, tall grass prairie, tundra, and deciduous forests upon N fertilization and depositions.[26,27] CO_2 enhancement experiments suggest changes in plant–plant interactions and alterations of competitive balances among species,[28,29] which might lead to a decrease in plant species diversity. Polley et al.[30] correlated the increase in woody C_3 *Prosopis glandulosa* in the C_4 grasslands in the southwestern U.S. to a 27% rise in CO_2 concentration. Vine growth in tropical forest has been found to be favored, leading to increase in tree mortality due to increased CO_2.[31] Increase of some insects, plant pathogens, and weeds has been implicated to the increase in CO_2 and consequent temperature.[32] These perturbations, commonly termed as *global change*, can reduce species diversity, which in turn will affect ecosystem functioning.[33] A change in climate faster than the migration of most species will reduce the suitable area for a large number of species, leading to drastic reduction in global species diversity. A big poleward shift of vegetation patterns is predicted under a double-CO_2 climate.[34]

The biosphere is a sink of atmospheric pollutants. Depositions of atmospheric S and N compounds of anthropogenic origin affect the ecosystem functioning in temperate and boreal forests and grasslands, as well as in lakes, and rivers, by acidifying the water and soil through N and S enrichment. These perturbations have been found to affect biodiversity severely.[35-40] Depositions of atmospheric pollutants in the biosphere is an important ecosystem function, highly dependent on biodiversity. There is a great deal of concern about the effects of air pollutants on biodiversity, but there has been relatively less known about the effects of air pollutants on species of natural and seminatural communities. The chemical composition of the atmosphere has changed on an unprecedented scale over the past decades. Air pollution thus constitutes a major evolutionary challenge for biodiversity in managed and natural ecosystems. This chapter is concerned with the effects of air pollutants on plant diversity at different levels of organization.

8.3 AIR POLLUTION AND PLANT RESPONSES

The plant response to air pollutants depends upon the chemical toxicity and exposure pattern of the pollutant and the sensitivity level of the species. Plant species differ in their perception and response to air pollutants due to differential expression under the influence of genetic and/or environmental factors leading to constitutive and inducible differences among the plants within and between the populations to the same air pollutant.[16,41] Pollutants may influence plant growth and development through multiple pathways and mechanisms over widely varying timescales. Effects may be direct due to deposition on plant surfaces, resulting in impaired physiology and metabolism of foliar surfaces, or indirect, as a consequence of altered nutrients or toxic metals availability or due to altered susceptibility to other stresses. Air pollutants in combination may induce additive, antagonistic, or snyergistic impacts on plants depending upon the specific pollutants, their concentration and exposure duration, and environmental conditions. Lower-level chronic exposures are generally more representative of toady's regional-scale air pollution situations.

Variability in species response has been suggested to be governed by a sequence of events, which either reduces penetration of pollutant in leaf interior (pollutant avoidance)[42] and/or enhances the ability of plant tissue to withstand the pollutant and its products after reaching the target (pollutant tolerance).[43,44] These sequences of events initially depend upon constitutive resource availability; thereafter regulation of gene expression and acceleration of secondary metabolite production may modify the response pattern.[45]

Air pollutants enter the plant body through stomatal pores in the leaves. Darrall[46] has reviewed the stomatal response of plants to a variety of air pollutants as of partial stomatal closure, wider opening, or no response depending upon pollutant concentration and exposure duration, environmental conditions, and plant species and their cultivars. Some plants are known to exhibit intrinsically higher stomatal conductance and tend to be more susceptible to damage than others.[47] Implications of stomatal responses to air pollutants also depend on responses of other critical physiological processes such as carbohydrate assimilation and partitioning.[48] Variability in inter- and intraspecific response to air pollutants with respect to photosynthate partitioning may alter the nature of competition within a plant community.[49,50] After entering into the leaf interior, air pollutants may be metabolized, sequestered, accumulated, or excreted. Detoxification of pollutants and their products at the cell surface and interior has been widely documented in the literature.[51-53]

8.4 IMPACT OF AIR POLLUTION ON PLANT DIVERSITY AT DIFFERENT LEVELS OF ORGANIZATION

Biological diversity, or in shortened form biodiversity, means the variability among living organisms inhabiting the Earth. In a broader perspective, biodiversity is ecological diversity referring to the number of species in a given area, the ecological roles of the species, the changes in composition across a region, and their grouping together with processes and interactions within and between systems.[54]

Three attributes of biodiversity are suggested: composition (number of alleles or species); structure (physical arrangement of biotic components and biomass distribution); and function (natural processes of energy flow and nutrient cycling).[55] It is well established that the stresses from regionally transported gaseous pollutants and acidic deposition can cause structural and functional changes in plant communities,[56,57] leading to subtle degradation of ecosystems.[58] Air pollutants can affect various levels of ecosystem organization, starting from an individual leaf to an entire ecosystem.[59,60] Although the species-level diversity is the most-studied element of biological diversity in relation to atmospheric pollutants, genetic- and community-level diversities are also very important.

8.4.1 Genetic Diversity

This element of biodiversity is the foundation of all other aspects of diversity. Loss of plant species of small population sizes can lead to reduced genetic diversity, lowered fitness, and increased extinction risk.[61] Intrinsic variations within populations serve as the immediate source of genetic diversity. Taylor and Murdy[62] demonstrated plant-to-plant variation of foliar injury in the herbaceous annual *Geranium carolinianum* L. due to SO_2 exposure. Even after a 30-year period, the populations continued to exhibit a high degree of variability. Bell et al.[63] have also shown variations across the taxa for most physiological and growth parameters in response to air pollution. Evolution of SO_2 resistance in grassland species in industrialized regions of the U.K. has been reported.[63] The ecological and biological issues affecting the genetic variations are physiological mechanisms of resistance, speed and frequency with which adaptation rises, and the role of microevolution.[16] Roose et al.[64] have shown that microevolution may have negative physiological costs resulting in the plants being less competitive under conditions of other stresses. Since plants follow dissimilar mechanisms of resistance to air pollutants and consequently have their own fitness costs, a shift in competitive interactions may favor the species with minimum cost of resistance. When the cost of resistance is very high and the rate of selection against resistant individuals is low, species replacement takes place.[65]

Dunn[66] had suggested long ago that ambient levels of O_3 in the Los Angeles basin were high enough to drive the selection of resistant genotypes of *Lupinus bicolor.* Recent studies on *Populus tremuloides*, *Trifolium repens*, and *Plantago major* have provided variations in genetic diversity among populations due to evolution of resistance to ambient O_3 under field conditions.[67-69] Inherent resistance of populations to O_3 is correlated with the O_3 level at the source of the plant material. Direct involvement of O_3 in this genetic diversity other than any environmental factors has been proved as the percentage of resistant individuals increased in populations exposed to elevated levels of O_3 over long periods of time.[70,71]

Genetic diversity has profound implications on community structure. Reduction in population size and restriction of exchange of alleles among populations due to air pollution effects will increase the chances of spatial isolation of the species. It has been hypothesized that the evolution of resistance due to air pollution contributes to the loss of genetic diversity within plant species.[72,73] Development of resistant populations of grasses and herbs growing in the vicinity of coal-fired plants has been reported.[9] A reduction in genetic variability of populations and the evolution of resistance and adaptation to gaseous air pollutants have been observed by Ayzaloo and Bell[75] and Coleman and Mooney.[76] Genetic diversity can be affected at the community level by mortality or reduced growth and reproduction of sensitive species which are subsequently replaced by more-resistant species. Parsons and Pitelka[77] were of the view that significant losses of genetic diversity may only occur due to population extinction or genetic drift at restricted areas experiencing intense pollution episodes.[62,78,79] The chronic level of pollution in recent years should not actually be a cause of great concern for genetic diversity.

8.4.2 Community Diversity

Plant community structure depends on various environmental factors. Air quality is an important factor which has changed substantially due to rapid industrialization and urbanization in the recent past. Air pollutants have not been normal constituents of the environment in which plants have evolved over the centuries. Thus, nature has not provided a continuous selection pressure to isolate genes for resistance. However, genetic modification and recombination may provide morphological and physiological systems that exclude and tolerate excesses of the pollutants.[80]

8.4.2.1 Methodological Approach for Studying Community Response

There are a variety of research approaches used for studying the community response to air pollution. Among various approaches, controlled environment growth chamber, greenhouse, and open-top chamber (OTC) studies have limitations for studying the community-level diversity response to air pollution.[81] Duchelle et al.[82] have used OTC to investigate the response of small samples of herbaceous communities to O_3. Open-air fumigation though an expensive methodology provided direct measures of community response on a long-term basis.[83,84] Armentano and Menges[85] used statistical approaches to relate symptoms to influencing factors under native conditions. Mathematical models to evaluate the long-term behavior of communities subjected to chronic air pollution were also tried.[86] The success of models depends upon the state of the art of pollution effect research based on fumigation studies.[87] Quasi-experimental approach depends upon the gradient of pollutant concentration in an area under natural conditions with other factors more or less uniform.[88] This approach can be successfully used in the field, where the response to air pollutants has clear-cut spatial and temporal variations and influencing factors are known through previous studies.[8,9] Long-term studies are required to elucidate the subtle effects of air pollution and its by-products at statistically acceptable levels.[89] Field studies of natural communities around pollution stress regimes are the most-used approach for studying changes in community diversity in response to air pollution.

8.4.2.2 Case Studies

The differential susceptibility of plant species to air pollutants may be reflected in the altered pattern of species composition and structure at the community level. Smith[90] classified the relationship between air pollutants and communities into three categories. Class I depicts the relationship where the vegetation acts only as a sink for the pollutants. Class II relationships occur when the pollutants have subtle deleterious effects on the plant community. Class III relationships represent acute morbidity or mortality of the members of the plant community. Murray[91] documented information for class II relationships, where changes in community structure were studied in an area around an aluminum smelter in New South Wales, Australia from immediate proximity of the source in a sequence of increasing distance. It was reported that species susceptible to air pollution stress are removed and others which are more tolerant become prevalent. It may be suggested that variation in air quality taxes the adaptive ability of organisms with the result that only those species which have already adapted to these conditions or those which can acclimatize to the new conditions may participate in community formation.

The impact of air pollution on plant diversity in a community may be viewed in terms of changes in ecological function and/or a loss of ecological integrity. Diversity is principally a mechanism that generates community stability. The more stable the community, the greater the species diversity. Air pollutants diminish the stability by increasing the mortality of the most-sensitive species. In a study to assess the effect of coal-fired power plant emission on the vegetation of a mixed oak forest with white pine, Rosenberg et al.[58] found that the important value index (IVI) of certain species and species diversity were inversely related to the distance from the emission source. An obvious community simplification around an iron smelter near Wawa, Ontario has been reported.[3] Wood and Nash[92] and Freedman and Hutchinson[93] have reported low-diversity communities with low standing crop productivity and nutrient status with a high proportion of weedy herbaceous species with high tolerance to stress around metal smelters.

A decline in the abundance of lichens has been suggested to be an early response to the unfavorable effect o atmospheric pollutants on an ecosystem.[94,95] Das[96] observed that along the Red Road of Calcutta, India with a traffic load of about 50 to 70 and sometimes 100 motor vehicles per minute, no lichen on tree trunks was found, whereas along Ballygunge circular road with 5 to 10 motor vehicles per minute, some foliose lichens were present. The diversity and abundance of lichens were found to be reduced in the presence of atmospheric S and N compounds.[97]

Photochemical oxidants are known to cause species-specific damage in agroecosystems and horticulture.[98] Diversity and functioning of microbial populations in the soil have been found to be unfavorably altered as a feed back mechanism to atmospheric depositions.[99-101] Microbial diversity is known to be sensitive to acid/alkaline depositions[102] and heavy metals.[103]

In a detailed study conducted around two coal-fired power plants situated in Sonbhadra District of Uttar Pradesh, India to analyze the impact of emission on structure of herbaceous communities, the number of individuals and species richness were found to increase with decreasing pollution load.[9] The species richness values ranged from 2.65 at a heavily polluted site to 9.31 at the less-polluted site during the rainy season. The IVI estimates indicated dominance of a few plant species, like *Cassia tora*, *Cyanodon dactylon*, and *Dichanthium annulatum* at the sites receiving higher pollution load. On the other hand, *Paspalidium flavidum*, *Phyllanthus simplex*, and *Rungia repens* were absent at heavily polluted sites and were more dominant at less-polluted sites. Some plant species were, however, found uniformly distributed. The successful survival of a species in the polluted area may be due to adequate biomass formation and their suitable structure, their ability to survive the lasting impact of pollutants, and their ability to reproduce under the pollution stress.[104] Thus the species having high IVI at heavily polluted sites were termed resistant and those with high IVI at distant sites receiving lower pollution load were kept in the sensitive category. Plant species having more or less the same IVI irrespective of pollution load were termed intermediate. The variations in sensitivity were linked with different strategies and life-forms of the species. The most sensitive species were annual herbs with hygromorphic or mesomorphic leaves. Grasses were found resistant except those that are pioneer species after forest clearing. The spread of pioneer grasses around a copper cliff smelter was found to be restricted to sites where residual soil toxicity is not excessive.[93]

Singh et al.[9] have shown that plant species with long reproductive phenophase better tolerated the power plant emissions compared with those with short reproductive phase. Plant species abundant at heavily polluted sites were found common in wastelands. A similar pattern was reported by Freedman and Hutchinson.[93] The herb layer of melick-beech forest community exposed to SO_2 (100 $\mu g\ m^{-3}$), NO_2 (100 $\mu g\ m^{-2}$), and O_3 (200 $\mu g\ m^{-3}$) singly and in combination for 4 h daily for 2 years showed undetectable (in some grasses to evergreen herbs) to severe reduction in leaf area (in most of the geophytes).[105] The elimination of sensitive species from the most-polluted site reduced the species diversity. The Shannon–Wiener's diversity index ranged from 2.62 to 4.24 between heavily polluted and least-polluted sites.[9] Guderian and Küppers[106] also observed that with decreasing SO_2 load the species diversity increased near an iron-ore roasting furnace in Biersdorf, Germany. McClenahen[59] has reported that species richness and Shannon–Wiener's diversity index reduced along a gradient of polluted air containing elevated concentrations of SO_2, HF, and particulate contaminants.

Narayan et al.[8] have observed reduction in number of woody species in response to aluminum factory emissions in Renukoot situated in the Sonbhadra District of Uttar Pradesh, India. Tree density and canopy cover gradually increased with increasing distance from the emission source. It has been suggested that emission has suppressed the growth of new colonizers; however, the individuals that were well established were flourishing well due to lack of interspecies competition. Species diversity of herbaceous and woody plants was inversely proportional to the pollution load. Negative influence of an aluminum smelter at Kitimat, British Columbia, Canada on the survival, growth, and regeneration of surrounding forest has also been reported.[107]

Diversity depends on both the number of kinds or classes of individuals present (species richness) and the distribution of abundance across the classes (evenness). The distribution of abundance reflects the niche division or the way in which species divide their functional role in the community.[108] The higher concentration of dominance (cd) at heavily polluted sites compared with less-polluted ones suggests that the number of individuals was not evenly distributed at all the study sites downwind of the coal-fired power plants[9] and the aluminum factory.[8] When the dominance was due to concentration of single species, cd values were higher, and, when several species contributed to the community formation, the index was low. It may be suggested that communities are heterogeneous at distantly situated sites, because ambient conditions favored the survival, growth, and regeneration of vegetation and also new arrivals. In contrast, at the sites closest to the emission source, subsequent to the loss of sensitive plant species, a niche was created that became available to opportunistic, more-tolerant species making them more abundant at the polluted sites. The inverse of the concentration of individuals is evenness. With the increase in evenness, species interactions are likely to increase.

The dominance–diversity curve is the best representation of both richness and the distribution of abundances.[108,109] Species sequences are arranged from left to right as most abundant to least abundant. The length of the sequence is richness, whereas the shape of the curve represents the distribution of abundance. It is known that undisturbed, stable, and equilibrium communities follow a lognormal pattern of disbribution.[110] Whittaker[108,111,112] indicated that a geometric series pattern of species abundance is often found in a harsh environment. The shape of the species sequence vs. IVI distribution curve slipped from the shape of longnormal distribution at sites receiving emissions from the aluminum factory at Renukoot, Uttar Pradesh, India.[8] The shifting of curve shape was more pronounced at the sites receiving higher pollution load. The slipping of curves toward geometric series was less obvious for herbaceous vegetation as compared with woody vegetation. The disturbances and inequilibrium of the communities caused by the factory emission resemble the shifting of communities backward through succession.

β diversity, which is the turnover of species along an environmental gradient, is useful in assessing the impact of air pollution on plant diversity at the community level.[113] β diversity was found to increase with increase in distance from an emission source.[8] The higher β diversity at places distant from emission source suggests a higher rate of turnover due to the presence of higher numbers of species.

Woodwell[2] described the results of a 7-year study on a late successional oak/pine forest and recognized five zones of damage from ionized radiation. He generalized that forest strata removed layer by layer starting from trees, tall shrubs, low shrubs, herbs, and finally mosses and lichen, as the intensity of pollution load increased. Mixed pine–spruce hardwood forests at Sudbury showed a variable response of the forested area to the smelter complex.[93] The study clearly observed a central devastated zone, followed by heavily damaged forest, then by areas with diminishing injury, and then by areas with no measurable effects at a distance of about 40 km from the source. The number of species increased with increasing distance form smelter complex. Understory cover and overstory basal area correlated negatively with pollution level. Shannon–Weiner's diversity index, however, did not correlate well with pollution load. The study further concluded that overstory vegetation was more responsive to pollution effects compared with understory.

Community responses in a boreal forest receiving emission from an oil refinery showed the understory to be more affected.[114] Species diversity increased due to invasion of weed species in areas denuded as a result of tree death. The study suggested that community attributes responded directly to the pollution gradient up to the severely damaged zone; thereafter, variability increased.

The work of McClenahen[59] on seven deciduous forest stands located along 50 km of the Ohio River Valley being affected by industrial air pollutants such as SO_2, HF, and Cl_2 showed decline in species diversity of different strata except the shrub layer near the sources. One interesting finding was that due to reduction in overstory tree density, other strata showed higher density at areas with higher pollution load. *Acer saccharum* was suggested to be replaced by tolerant species. Resistant species have been shown to increase at the expense of sensitive species when shift in competition took place.[115] A similar result was observed by Bowen[116] and Narayan et al.[8] It was reported that fluoride accumulation occurs maximally in tree followed by shrub and herb species, suggesting that the canopy of the overstory woody layer acted as a filter for the groundcover herbs.

In contrast, overstory trees were found least affected and ground vegetation more in oak forests in a Pennsylvania valley receiving emission from thermal power plants.[58] Species richness and the Shannon–Weiner diversity index increased with increasing distance from the source. Altered plant community composition and, at certain sites, a change in successional pattern have been reported in Mediterranean ecosystem types of Southern California experiencing chronic levels of photochemical oxidants.[117,118] Sensitive species were absent at polluted sites, thus changing the competitive relationship. Decline in foliar cover of coastal sage was found to be associated with regional oxidant level.[119] Native shrubs were replaced by non-invasive grasses, and therefore diversity increased in more-polluted areas compared to the least-polluted ones. A similar result was reported in a sage scrub area downwind of an oil refinery.[120] Higher density was, however, paralleled with reduced dominance. An ordination study found a greater effect at sites with lower foliar canopy.

The effects of O_3 on plant communities were generally studied on seminatural grasslands.[120-123] Ashmore and Ainsworth[124] found a decline in forb component with increasing O_3 level. It was further suggested that species composition might be affected adversely in parts of central and northern Europe in high-O_3 years.[125] Barbo et al.[126] found unfavorable effects on canopy structure, species richness, evenness, and diversity index resulting in a shift in vegetation dominance in an early successional forest community in the Shenandoah National Park, southern U.S. at ambient plus elevated O_3 levels.

Farrar et al.[127] correlated the scarcity of Scots pine (*Pinus sylvestris* L.) to SO_2 concentrations in the industrial Peninnes of England. Adverse effects of high concentrations of SO_2 on distribution of Scots pine have also been reported in the Ruhr area of Europe.[128,129] Materna[104] suggested that European silver fir (*Abies alba* Mill.) may be the most sensitive tree species showing severe losses in forest during the last decades. He further reported dieback of this species to start at a long-term concentration of 20 μg m^{-3} SO_2. The exposure to chronic level of pollutants also increased the vulnerability of species to other natural environmental factors such as frost damage. Materna[130] made a detailed study in a mountain forest where silver fir, Norway spruce, and common beech (*Fagus sylvatica* L.) associations were dominant. Silver fir formed the important species due to its deep rooting system and high biomass production. Because of its greater sensitivity, dieback of this species started first, followed by gradual dieback of Norway spruce leading to vulnerability of the beech forest having reduced reproductive capacity among the three species. Disappearance of Norway spruce in some regions of central Europe has been documented.[104] The species such as silver birch (*Betula verrucosa* Ehrh.) and mountain ash (*Sorbus aucuparia* L.) have become more dominant due to higher biomass production and reproductive ability under situations of air pollution. Since the degree of susceptibility of plants to air pollution is also influenced by natural ecological conditions, the injury to trees in forest is suggested to be an integrated effect of air pollution and other stress factors.[104,131-133]

Plant adaptations to other environmental stresses may modify the plant response to tolerate the pollution level. Winner et al.[134] and Reich[47] have suggested that physiological adaptations of high stomatal resistance under drought provide resistance to the plants by reducing gaseous reflux. Community response varied with interannual climate fluctuations in Montana prairie.[84]

8.5 SECONDARY EFFECTS OF PLANT DIVERSITY LOSS

Competitive relationships between plant species alter because of increased vitality or dieback of sensitive plant species.[8,9,135-137] Intraspecific competition was suggested to be a major factor determining relative abundance of sensitive and tolerant aspen clones in clean and polluted areas. Direct evidence of air pollution effects on competition comes from the study of Steubing and Fangmeier,[138] who exposed the understory of a beech forest in artificial chambers at an SO_2 concentration simulating the ambient peak level. Forb species showed reduction in leaf area and reproductive capacity, while grass species and the vine *Hedera helix* proved to be resistant. In clover fescue plots, Heagle et al.[139] observed resistance of clover species leading to good performance compared with fescue. Bennett and Runeckles[140] found that competitive ability of ryegrass increased compared with clover due to increased tillering by ryegrass under O_3 treatment. Armentano and Bennett[81] have clearly suggested that intraspecific competition is more sensitive to pollution than interspecific, but pollution can alter species competitive abilities. The effect is initially visible in clonal substitution followed by species.

Forest systems, while resilient to stresses, may be strongly controlled by a balance in competitive potential of a few key species. Responses of individual species and the developmental stage of the ecosystem at the time of stress are key determinants of competition response. Changes in relative growth rate between species is suggested to be a useful indicator of subtle changes in competitive potential.[60] Shifts in the relative interspecific competitive potential of yellow poplar and white oak in mixed deciduous forest stands in eastern Tennessee have been reported.[141] Competition can both compensate for and amplify the effects of air pollution.

Plant diversity loss is implicated in altered host–parasite relationships,[142] plant–pollinator relationships,[143] plant–pathogen relationships,[144] etc. A high degree of association between attach by bark beetles and prior oxidant damage was noted for ponderosa pine in Southern California.[145] The alterations of resource allocation to various organs due to air pollution may influence the ability of the plant to resist diseases.[144,146] The damage to coniferous forests in Norway by an insect, *Exteleia dodecellar*, was linked to the acid rain problem.[132] A similar case of insect damage to spruce was reported in middle Europe.[147] Alterations in honeybee (*Apis mellifera* L.) populations due to SO_2 and fluoride pollutants have been linked with detrimental effect on fruit yield.[143] Reduction in vigor of mycorrhiza of lobolly pine due to acid rain may have adverse effect on nutrient and water uptake and growth potential.[99]

The change in species composition of a natural ecosystem can affect the nutrient cycling adversely by changing the nutrient reserves in various parts of the soil profile.[104] The rate of decomposition also differ among plant species. Erosion of soils due to tree mortality in forested area may accelerate. The hydrological function of forests is another area of concern related to plant diversity loss.

8.6 CONCLUSIONS

Air pollution can influence plant species in diverse ways and thus affect ecosystems at various levels of organization. In the past few decades, dieback and decline of forests in the U.S. and Europe have been linked with air pollution levels in those areas. Early information on the effects of air pollution on plant diversity came from studies conducted near large point sources of pollution where increased mortality, declining vigor, and compositional changes in communities were reported downwind of industrial operations. But the application of abatement technology, strict air quality regulations, and the shift to tall stacks has altered the nature of air quality and prevailing concentrations of air pollutants in ambient air. Sulfur dioxide, HF, and heavy metals, the causes of severe damage during the past, are now of limited importance. The major focus at present has shifted to the impacts of regional elevated concentrations of oxidants, especially O_3, and acid rain. Thus the recent trend of change in air quality has increased the complexity of the studies assessing the effect of air pollution on plant diversity.

Loss in diversity has many unfavorable implications related to ecosystem functions such as energy flow and biogeochemical cycles. By virtue of perturbing energy fixation at the producer level, air pollution has the potential to perturb the energy flow in natural and managed ecosystems. Human activities are critical to the relationship between biological diversity and ecological processes. The current problems of air pollution and global climate change have clearly emphasized the need to understand the biodiversity response to these anthropogenic stress factors. Living things exist with specific ecological processes, and an unfavorable alteration in any of these has serious implications for the existence of each other. The values placed on biodiversity are strongly linked to the human influences on it and their underlying social and economic driving forces. This area requires a serious research effort throughout the world in view of the increased loss of biodiversity during the last few decades. Long-term monitoring programs are required at selected areas of long-term or shortterm pollution history, and data on structural and functional attributes of biotic components need to be collected periodically to evaluate changes in diversity, resiliency, and productivity.

The prevention of changes in ecosystem functioning and biodiversity that will affect the composition of the atmosphere is likely to be one of the most important policy issues of the future. Developing countries require a special effort in this direction, as the rate of biodiversity loss linked with other activities is already several magnitudes greater than developed countries. A global effort is warranted for such problems.

ACKNOWLEDGMENTS

The authors are thankful to the Ministry of Environment, Forest and Wildlife, Government of India, New Delhi, and the Council of Scientific and Industrial Research (Division of Scientific & Technical Personnel), New Delhi for financial assistance.

REFERENCES

1. Lovelock, J. D., *Gaia: A New Look at Life on Earth*, Oxford University Press, Oxford, U.K., 1987, 157.
2. Woodwell, G. M., Effects of pollution on the structure and physiology of ecosystems, *Science*, 168, 429, 1970.
3. Gordon, R. J. and Gorham, E., Ecological aspects of air pollution from an iron-sintering plant at Wawa, Ontario, *Can. J. Bot.*, 41, 1063, 1963.
4. Treshow, M., The impacts of air pollutants on plant populations, *Phytopathology*, 5, 1, 1968.
5. Miller, P. R., Taylor, O. C., and Wilson, R. G., Oxidant Air Pollution Effects on a Western Coniferous Forest Ecosystem, U.S. EPA Report No. 600/D-82-276, 1982.
6. Schreiber, R. K. and Newman, J. R., Acid precipitation effects on forest habitats: implications for wildlife, *Conserv. Biol*, 2, 249, 1988.
7. Moser, T. J., Barker, J. R., and Tingey, D. T., Eds., Ecological Exposure and Effects of Airborne Toxic Chemicals: An Overview, U.S. EPA Environmental Research laboratory, Corvallis, EPA Report No. 600/3-91/001, 1991.
8. Narayan, D., Agrawal, M., Pandey, J., and Singh, J., Changes in vegetation characteristics downwind of an aluminum factory in India, *Ann. Bot.*, 73, 557, 1994.
9. Singh, J., Agrawal, M., and Narayan, D., Effect of power plant emissions on plant community structure, *Ecotoxicology*, 3, 110, 1994.
10. Tingey, D. T., Hogsett, W. E., and Henderson, S., Definition of adverse effects for the purpose of establishing secondary national ambient air quality standards, *J. Environ. Qual.*, 19, 635, 1990.
11. Solbrig, O. T., The origin and function of biodiversity, *Environment*, 33. 17. 1991.
12. Wilson, E. O., Threats to biodiversity, *Sci. Am.*, 261, 108, 1989.
13. Barker, J. R., Henderson, S., Noss, R. F., and Tingey, D. T., Biodiversity and human impacts, in *Encyclopedia of Earth System Science*, Nierenberg, W., A., Ed., Academic Press, San Diego, 1991.
14. Barker, J. R. and Tingey, D. T., Eds., *Air Pollution Effects on Biodiversity*, Van Nostrand Reinhold, New York, 1992, 307.
15. Reid, W. V. and Miller, K. R., *Keeping Options Alive: The Scientific Basis for Conserving Biodiversity*, World Resources Institute, Washington, D.C., 1989.
16. Taylor, G. E. and Pitelka, L. F., Genetic diversity of plant populations and the role of air pollution, in *Air Pollution Effects on Biodiversity*, Barker, J. R. and Tingey, D. T., Eds., Van Nostrand Reinhold, New York, 1992, 111.
17. Bishop, J. A. and Cook, L. M., *Genetic Consequences of Man Made Changes*, Academic Press, New York, 1981.
18. Bradshaw, A. D. and McNeilly, T., *Evolution and Pollution*, Edward Arnold, London, 1981.
19. Smith, W. H., *Air Pollution and Forests*, Springer-Verlag, New York, 1990.
20. Davison, A. W., Barnes, J. D., and Renner, C. J., Interactions between air pollutants and cold stress, in *Air Pollution and Plant Metabolism*, Schulte-Hostede, S., Blank, L. W., Darrall, N. W., and Wellburn, A. R., Eds., Elsevier Applied Science, London, 1988.
21. Tevini, M., UV-B effects on terrestrial plants and aquatic organisms, *Prog. Bot.*, 55, 174, 1994.
22. Bates, T. S., Cline, J. D., Gammon, R. H., and Kelley-Hansen, S. R., Regional and seasonal variations in the flux of oceanic dimethylsulfide to the atmosphere, *J. Geophys. Res.*, 92, 2930, 1987.
23. Bothwell, M. L., Sherbot, D. M. J., and Pollock, C. M., Ecosystem response to solar ultraviolet-B radiation: influence of tropic level interactions, *Science*, 265, 97, 1994.
24. Keeling, C. D., *Atmospheric CO_2 Concentrations, Mauna Loa Observatory, Hawaii 1958–1986*, Carbon Dioxide Information Analysis Center, Oak Ridge, TN, 1986.
25. Vitousek, P. V., Beyond global warming: ecology and global change, *Ecology*, 75, 1861, 1994.

26. Lauenroth, W. K., Dodd, J. L., and Sims, P. L., The effect of water- and nitrogen-induced stresses on plant community structure in a semiarid grassland, *Oecologia*, 36, 211, 1978.
27. Schulze, E. D., Air pollution and forest decline in a spruce (*Picea abies*) forest, *Science*, 244, 776, 1989.
28. Mooney, H. A., Drake, B. G., Luxmoore, R. J., Oechel, W. C., and Pitelka, L. F., Predicting ecosystem responses to elevated CO_2 concentrations, *Bioscience*, 41, 96, 1991.
29. Morse, S. R. and Bazzaz, F. A., Elevated CO_2 and temperature alter recruitment and size hierarchies in C_3 and C_4 annuals, *Ecology*, 75, 966, 1994.
30. Polley, H. W., Johnson, H. D., and Mayeux, H. S., Increasing CO_2: comparative response of the C_4 grass *Schizachirium* and grassland invader *Prosopis*, *Ecology*, 75, 976, 1994.
31. Phillips, O. and Gentry, A. H., Increasing turnover through time in tropical forests, *Science*, 263, 954, 1994.
32. Pimentel, D., Brown, H., Vecchio, F., LaCapra, V., Hausman, S., Lee, O., Diaz, A., Williams, J., Cooper, S., and Newberger, E., Ethical issues concerning potential global climate change on food production, *J. Agric, Environ, Ethics*, 5, 113, 1992.
33. Ehrlich, P. R. and Mooney, H. A., Extinction, substitution, and ecosystem services, *Bioscience,* 33, 248, 1983.
34. Kramer, W. P. and Leemans, R., Assessing impacts of climate change on vegetation using climate classification systems, in *Vegetation Dynamics and Global Change*, Solomon, A. M. and Shugart, H. H., Eds., Chapman & Hall, New York, 1993, 190.
35. Lauenroth, W. K. and Preston, W. M., *The Effects of SO_2 on a Grassland: A Case Study in the Northern Great Plains of the United States*, Springer-Verlag, New York, 1984.
36. Schulze, E.-D., Lange, O. L., and Oren, R., *Forest Decline and Air Pollution: A Study of Spruce (Picea abies) on Acid Soils*, Springer-Verlag, Berlin, 1989.
37. Reuss, J. O. and Johnson, D. W., *Acid Deposition and the Acidification of Soils and Waters*, Springer-Verlag, New York, 1986.
38. Wellburn, A. R., Why are atmospheric oxides of nitrogen usually phytotoxic and not alternative fertilizers? *New Phytol.*, 115, 395, 1990.
39. Pearson, J. and Stewart, G. R., The deposition of atmospheric ammonia and its effects on plants, *New Phytol.*, 125, 283, 1993.
40. Rennenberg, H. and Polle, A., Metabolic consequences of atmospheric sulphur influx into plants, in *Proceedings of the 3rd International Symposium on Air Pollution and Plant Metabolism*, Wellburn, A. and Alsher, R., Eds., Chapman & Hall, London, 1994.
41. Tingey, D. T. and Anderson, C. P., The physiological basis of differential plant sensitivity to changes in atmospheric quality, in *Ecological Genetics and Air Pollution*, Taylor, G. E., Jr., Pitelka, L. F., and Clegg, M. T., Eds., Springer-Verlag, New York, 1991, 209.
42. Mansfield, T. A. and Freer-Smith, P. H., The role of stomata in resistance mechanisms, in *Gaseous Air Pollutants and Plant Metabolism*, Koziol, M. J. and Whatley, F. R., Eds., Butterworth, London, 1984, 131.
43. Malhotra, S.S. and Khan, A. A., Biochemical and physiological impact of major pollutants, in *Air Pollution and Plant Life*, Treshow, M., Ed., John Wiley & Sons, London, 1984, 113.
44. Hippeli, S. and Elstner, E. F., Mechanisms of oxygen activation during plant stress: biochemical effects of air pollutants, *J. Plant Physiol.*, 148, 249, 1996.
45. Barnes, J., Bender, J., Lyons, T., and Borland, A., Natural and man-made selection for air pollution resistance, *J. Exp. Bot.*, 1999.
46. Darrall, N. M., The effect of air pollutants on physiological processes in plants, *Plant Cell Environ.*, 12, 1, 1989.
47. Reich, P. B., Quantifying plant response to ozone: a unifying theory, *Tree Physiol.*, 3, 63, 1987.
48. Wolenden, J., Wookey, P. A., Luca, P. W., and Mansfield, T. A., Action of pollutants individually and in combination, in *Air Pollution Effects on Biodiversity*, Barker, J. R. and Tingey, D. T., Eds., Van Nostrand Reinhold, New York, 1992, 309.
49. Grime, J. P., Crick, J. C., and Rincon, J. E., The ecological significance of plasticity, in *Plasticity in Plants*, Jennings, D. H. and Trewavas, A. J., Eds., Cambridge University Press, London, 1986.
50. Mansfield, T. A., Factors determining root:shoot partitioning, in *Scientific Basis of Forest Decline Sympotomatology*, Cape, J. N. and Mathy, P., Eds., Commission of the European Communities, Brussels, 1988, 171.

51. Kangasjärvi, J., Talvinen, J., Utriainen, M., and Karjalainen, R., Plant defence systems induced by ozone, *Plant Cell Environ.*, 17, 783, 1994.
52. Polle, A. and Rennenberg, H., Significance of antioxidants in plant adaptation to environmental stress, in *Plant Adaptation to Environmental Stress*, Fowden, L., Mansfield, T. A., and Stoddart, J., Eds., Chapman & Hall, London, 1993, 263.
53. Iqbal, M., Abdin, M. A., Mahmooduzzafar, Yunus, J., and Agrawal, M., Resistance mechanisms in plants against air pollution, in *Plant Response to Air Pollution*, Yunus, M. and Iqbal M., Eds., John Wiley & Sons, Chichester, 1996, 195.
54. Heywood, V. H. and Baste, I., Introduction, in *Global Biodiversity Assessment*, Heywood, V. H., Ed., UNEP, Cambridge University Press, Cambridge, 1995, 1.
55. Noss, R. F., Indicators for monitoring biodiversity: a hierarchal model, *Conserv. Biol.*, 4, 255, 1990.
56. Lauenroth, W. K. and Milchunas, D. G., SO_2 effects on plant community function, in *Sulphur Dioxide and Vegetation. Physiology, Ecology and Policy Issues*, Winner, W. E., Mooney, H. A., and Goldstein, R. A., Eds., Stanford University Press, Stanford, CA, 1985, 593.
57. Kozlowski, T. T., SO_2 effect on plant community structure, in *Sulphur Dioxide and Vegetation. Physiology, Ecology and Policy Issues*, Winner, W. E., Mooney, H. A., and Goldstein, R. A., Eds., Stanford University Press, Stanford, CA, 1985, 431.
58. Rosenberg, C. R., Hutnik, R. J., and Davis, D. D., Forest composition at varying distances from a coal-burning power plant, *Environ. Pollut.*, 58, 307, 1979.
59. McClenahen, J. R., Community changes in a deciduous forest exposed to air pollution, *Can. J. For. Res.*, 8, 432, 1978.
60. McLaughlin, S. B., Effects of air pollution on forest, a critical review, *J. Air Pollut. Control Assoc.*, 35, 512, 1985.
61. Schnoewals-Cox, C. M., Chambers, S. M., MacBryde, B., and Lawrence Thomas, W., Eds., *Genetics and Conservation*, Benjamin/Cummings Publishing, Menlo Park, CA, 1983.
62. Taylor, G. E. and Murdy, W. H., Populations differentiations of an annual plant species, *Geranium carolinianum* L. in response to sulphur dioxide, *Bot. Gaz.*, 136, 212, 1975.
63. Bell, J. N. B., Ashmore, M. R., and Wilson, G. B., Ecological genetics and chemical modification of the atmosphere, in *Ecological Genetic and Air Pollution*, Taylor, G. E., Pitelka, L. F., and Clegg, M. T., Eds., Springer-Verlag, New York, 1991, 33.
64. Roose, M. L., Bradshaw, A. D., and Roberts, T. M., Evolution of resistance to gaseous air pollution, in *Effects of Gaseous Air Pollutants in Agriculture and Horticulture*, Unsworth, M. H. and Ormrod, D. P., Eds., Butterworth Science, London, 1982, 379.
65. Antonovics, J., Bradshaw, A. D., and Turner, R., Heavy metal tolerance in plants, *Adv. Ecol. Res.*, 7, 1, 1971.
66. Dunn, D. B., Some effects of air pollution on *Lupinus* in the Los Angeles area, *Ecology*, 40, 621, 1959.
67. Reiling, K. and Davison, A. W., Spatial variation in ozone resistance of British populations of *Plantago major* L., *New Phytol.*, 122, 699, 1992.
68. Berrang, P., Karnosky, D. F., and Bennett, J. P., Natural selection for ozone tolerance in *Populus tremuloides*: an evaluation of nationwide trends, *Can. J. For. Res.*, 21, 1091, 1991.
69. Macnair, M. R., Tansley Review No. 49. The genetics of metal tolerance in vascular plants, *New Phytol.*, 124, 541, 1993.
70. Heagle, A. S., McLaughlin, M. R., Miller, J. E, Joyner, R. I., and Spruill, S. E., Adaptation of a white clover population to ozone stress, *New Phytol.*, 119, 61, 1991.
71. Davison, A. W. and Reiling, K., A rapid change in the ozone resistance of *Plantago major* after summers with high ozone concentrations, *New Phytol.*, 131, 337, 1995.
72. Bergmann, F. and Scholz, F., Selection effects of air pollution in Norway spruce (*Picea abies*) populations, in *Genetic effects of Air Pollutants in Forest Tree Populations*, Scholz, F., Gregorius, H.-R., and Rudin, D., Eds., Springer-Verlag, New York, 1989, 143.
73. Karnosky, D. F., Ecological genetics and changes in atmospheric chemistry: the application of knowledge, in *Ecological Genetics and Air Pollution*, Taylor, G. E., Pitelka, L. F., and Clegg, M. R., Eds., Springer-Verlag, New York, 1991, 321.
74. Horsman, D. C., Robert, T. M., and Bradshaw, A. D, Evolution of sulfur dioxide tolerance in perennial ryegrass, *Nature*, 276, 493, 1978.

75. Ayzaloo, M. and Bell, J. N.B., Studies on the tolerance to sulfur dioxide of grass populations in polluted areas. I. Identification of tolerant populations, *New Phytol.*, 88, 203, 1981.
76. Coleman, J. S. and Mooney, H. A., Anthropogenic stress and natural selection: variability in radish biomass accumulation increases with increasing SO_2 dose, *Can. J. Bot.*, 68, 102, 1990.
77. Parsons, D. J. and Pitelka, L. F., Plant ecological genetics and air pollution stress: a commentary on implications for natural populations, in *Ecological Genetics and Air Pollution*, Taylor, G. D., Jr., Pitelka, L. F., and Clegg, J. R., Eds., Springer-Verlag, New York, 1991, 337.
78. Murdy, W. H., Effect of SO_2 on sexual reproduction in *Lepidium virginicum* L. originating from regions with different SO_2 concentration, *Bot. Gaz.,* 140, 299, 1979.
79. Heggestad, H. E. and Heck, W. W., Nature, extent, and variation of plant response to air pollutants, *Adv. Agron.*, 23, 111, 1971.
80. Reinert, R. A., Heggestad, H. E., and Heck, W. W., Response and genetic modification of plants for tolerance to air pollutants, in *Breeding Plants for Less Favourable Environments*, Christiansen, M. N., Ed., John Wiley & sons, Chichester, 1982, 259.
81. Armentano, T.V. and Bennett. J. P., Air pollution effects on the diversity and structure of communities, in *Air Pollution Effects on Biodiversity*, Barker, J. and Tingey, D. T., Eds., Van Nostrand Reinhold, New York, 1992, 159.
82. Duchelle, S. F., Skelley, J. M, Sharick, R. L., Chevone, B., Yang, Y. S., and Nielessen, J. E., Effects of ozone on the productivity of natural vegetation in a high meadow of the Shenandoah National Park of Virginia, *J. Environ. Manage.*, 17, 299, 1983.
83. Reich, P. B., Amundsen, R. G., and Lassoie, J. R., Reduction in soybean yield after exposure to O_3 and SO_2 using linear gradient exposure technique, *Water Air Soil Pollut.*, 17, 29, 1982.
84. Lauenroth, W. K., Milchunas, D. G., and Dodd, J. L., Responses of the vegetation, in *The Effects of SO_2 on a Grassland*, Lauenroth, W. K. and Preston, E. M., Eds., Springer-Verlag, New York, 1984, 97.
85. Armentano, T. V. and Menges, E. S., Air pollution-induced foliar injury to natural populations of jack and white pine in a chronically polluted environment, *Water Air Soil Pollut.*, 33, 395, 1987.
86. Shugart, H. H. and McLaughlin, S. B., Modelling SO_2 effects on plant growth and community dynamics, in *Sulfur Dioxide and Vegetation: Physiology, Ecology and Policy Issues*, Winner, W. E., Mooney, H. A., and Goldstein, R. A., Eds., Stanford University Press, Stanford, CA, 1985.
87. Dale, V. H. and Gardner, R. H., Assessing regional impact of growth declines using a forest succession model, *J. Environ. Manage.*, 24, 83, 1987.
88. Cook, T. D. and Campbell, D. T., *Quasi-Experimentation Design and Analysis Issues for Field Settings*, Houghton-Mifflin, Boston, 1979.
89. Likens, G. E., Some aspects of air pollution effects on terrestrial ecosystems and prospects for the future, *Ambio*, 18, 172, 1989.
90. Smith, W. H., Air pollution effect on the structure and function of the temperate forest ecosystem, *Environ. Pollut.*, 6, 111, 1974.
91. Murray, F., Effects of fluoride on plant communities around an aluminum smelter, *Environ. Pollut.*, 24, 45, 1981.
92. Wood, C. W. and Nash, T. N. III, Copper smelter effluents effects on Sonoran Desert vegetation, *Ecology*, 57, 1311, 1976.
93. Freedman, B. and Hutchinson, T.C., Long term effects of smelter pollution at Sudbury, Ontario, on forest community composition, *Can. J. bot.*, 58, 2123, 1980.
94. Le Blanc, F. and Rao, D. N., Effects of air pollutants on lichen and bryophytes, in *Response of Plants to Air Pollution*, Mudd, J.D. and Kozlowski, T. T., Eds., Academic Press, New York, 1975, 237.
95. Ferry, B. W., Baddeley, J.S., and Hawksworth, D. L., *Air Pollution and Lichens*, Athlone Press, London, 1973, 386.
96. Das, T. M., Lower and higher plant groups as indicators of air pollution, in *Symp. Biomonit. State Environ.*, INSA, New Delhi, 1985, 232.
97. Anderson, F. K. and Treshow, M., Response of lichens to atmospheric pollution, in *Air Pollution and Plant Life*, Treshow, M., Ed., John Wiley & Sons, New York, 1984, 259.
98. Guderian, R., Tingey, D. T., and Rabe, R., Effects of photochemical oxidants on plants, in *Pollution by Photochemical Oxidants*, Guderian, R., Ed., Springer-Verlag, Berlin, 1985, 129.

99. McCool, P. M., Effect of air pollution on mycorrhiza, in *Air Pollution and Plant Metabolism*, Schulte-Hostede, S., Darrall, N. M., Blank, L.W., and Wellburn, A. R., Eds., Elsevier Applied Sciences, London, 1988, 356.
100. Papen, H., Hellman, B., Papke, H., and Rennenberg, H., Emission of N-oxides from acid irrigated and limed soils of a coniferous forest in Bavaria, in *Biogeochemistry of Global Change, Radiatively Active Trace Gases*, Oremland, R. S., Ed., Chapman & Hall, London, 1993, 245.
101. Steudler, P. A., Bowden, R. D., Melillo, J. M., and Aber, J. D., Influence of nitrogen fertilization on methane uptake in temperate forest soils, *Nature*, 341, 314, 1989.
102. Adamson, E., and Seppelt, R. D., A comparison of airborne alkaline pollution damage in selected lichens and mosses at Casey Station, Wilkes Land, Antarctica, in *Antarctica Ecosystems*, Kerry, K. R. and Hempel, G., Eds., Springer-Verlag, New York, 1990, 347.
103. Zabowski, D., Zasoske, R. J., Little, W., and Ammirati, J., Metal content of fungal sporocarps from urban, rural and sludge-treated sites, *J. Environ. Qual.*, 19, 372, 1990.
104. Materna, J., Impact of atmospheric pollution on natural ecosystem, in *Air Pollution and Plant Life*, Treshow, M., Ed., John Wiley & Sons, New York, 1984, 397.
105. Fangmeier, A., Effects of open-top fumigations with SO_2, NO_2 and ozone on the native herb layer of a beech forest, *Environ. Exp. Bot.*, 29, 199, 1989.
106. Guderian, R. and Küppers, K., Responses of plant communities to air pollution, in *Effects of Air Pollutants on Mediterranean and Temperate Forest Ecosystems*, USDA Forest Service Gen. Tech. Rep., PSW-43, Berkeley, CA, 1980, 187.
107. Bunce, H. W. F., Fluoride emissions and forest survival, growth and regeneration, *Environ. Pollut.*, 35, 247, 1984.
108. Whittaker, R. H., *Communities and Ecosystems*, Macmillian, New York, 1975.
109. Wilson, J. B., Methods for fitting dominance-diversity curves, *J. Veg. Sci.*, 2, 35, 1991.
110. May, R. M., Patterns in multi-species communities, in *Theoretical Ecology: Principles and Applications*, May. R. M., Ed., Blackwell, London, 1981, 197.
111. Whittaker, R. H., Vegetation of the Siskiyou Mountains, Oregon and California, *Ecol. Monogr.*, 30, 279, 1960.
112. Whittaker, R. H., Evolution and measurement of species diversity, *Taxon*, 21, 213, 1972.
113. Cody, M.L., Diversity, rarity, and conservation in Mediterranean climate regions, in *Conservation Biology*, Soule, M. E., Ed., Sinauer Associates, Sunderland, MA, 1986, 122.
114. Winner, W. E. and Bewley, J. B., Contrasts between bryophyte and vascular plant synecological responses in a SO_2-stressed white spruce association, *Oecologia*, 35, 311, 1978.
115. Rapport, D. J., Regier, H.A., and Hutchinson, T. C., Ecosystem behaviour under stress, *Am. Nat.*, 125, 617, 1985.
116. Bowen, S. E., Spatial and temporal patterns in the fluoride content of vegetation around two aluminum smelters in the Hunter Valley, New South Wales, *Sci. Total Environ.*, 68, 97, 1988.
117. McBridge, J. R., Miller, P. R., and Laven, R. D, Effects of oxidant air pollutants on forest succession in the mixed conifer forest of southern California, in *Air Pollution Effects on Forest Ecosystems*, a symposium coordinated by the Acid Rain Foundation, St. Paul, MN, 1985.
118. O'Leary, J. F. and Westman, W.E., Regional disturbance effects on herb successional patterns in coastal sage scrub, *J. Biogeogr.*, 15, 775, 1988.
119. Westman, W.E., Oxidant effects on California coastal sage scrub, *Science*, 205, 1001, 1979.
120. Preston, K. P., Effects of sulphur dioxide pollution on a Californian sage scrub community, *Environ. Pollut.*, 51, 179, 1988.
121. Ashmore, M. R. and Davison, A. W., Towards a critical level of ozone for natural vegetation, in *Critical Levels for Ozone in Europe: Testing and Finalizing the Concepts*, Kärenlampi, F. and Skärby, L., Eds., UN Economic Commission for Europe convention on Long-Range Transboundary Air Pollution, Workshop report, University of Kuopio Printing Office, 1996, 58.
122. Bergmann, E., Bender, J., and Weigel, H., Effects of chronic ozone stress on growth and reproduction capacity of native herbaceous plants, in *Exceedance of Critical Loads and Levels*, Knoflacher, M. and Schnneider, S. G, Eds., Federal Ministry for Environment, Youth and Family, Vienna, 1996, 177.
123. Fürher, J., Shariat-Madari, H., Perier, R., Tschannen, W., and Grub, A., Effects of ozone on managed pasture: II Yield, species composition, canopy structure and forage quality, *Environ. Pollut.*, 86, 307, 1994.

124. Ashmore, M.R. and Ainsworth, N., The effects of cutting on the species composition of artificial grassland communities, *Funct. Ecol.*, 9, 708, 1995.
125. Ashmore, M. R., Thwaites, R. H., Ainsworth, N., Cousins, D.A., Power, S. A., and Morton, A. J., Effects of ozone on calcareous grassland communities, *Water Air Soil Pollut.*, 85, 1527, 1995.
126. Barbo, D. N., Chappelka, A. H., and Stolte, K. W., Ozone effects on productivity and diversity of an early successional forest community, in *Proc. of Eighth Biennial Southern Silvicultural Res. Council*, Nov. 1–3, Auburn, AL, 1994, 291.
127. Farrar, J. F., Relton, J., and Rutter, A. J., Sulphur dioxide and the scarcity of *Pinus sylvestris* in the industrial Pennines, *Environ. Pollut.*, 14, 63, 1977.
128. Knabe, W., Air quality criteria and their importance for forest, *Mitt. Forstl. Bundes Versuchsanst. Wien*, 92, 129, 1970.
129. Knabe, W., Kiefernwaldverbreitung und Schwefeldioxid Immisionen im Ruhrgebiet, *Staub*, 30, 32, 1970.
130. Materna, J., Relationship between SO_2 concentration and damage of forest trees in the region of the Slavkok forest, *Mitt. Forstl. Bundes Versuchsanst. Wien*, 92, 129, 1970.
131. Linzon, S. N., Effects of airborne sulphur pollutants on plants, in *Sulphur in the Environment*, Part II, Nriagu, J. O., Ed., John Wiley & Sons, New York, 1978, chap. 4, 110.
132. Huttunen, S., The integrative effects of airborne pollutants on boreal forest ecosystems, in *Effects of Airborne Pollution on Vegetation*, UN Economic Commission Europe, Warsaw, Poland, 180, 111.
133. Treshow, M., Pollution effects on plant distribution, *Environ. Conserv.*, 79, 279, 1980.
134. Winner, W. E., Koch, G. W., and Mooney, H. A., Ecology of SO_2 resistance IV. Predicting metabolic responses of fumigated shrubs and trees, *Oecologia*, 52, 16, 1982.
135. Miller, P. R., Ozone effects in the San Bernardino National Forest, in *Air Pollution and the Productivity of the Forest*, Davis, D. D., Millen, A. A., and Dochinger, L. S., Eds., Izaak Walton League, Arlington, 1983, 161.
136. Keller, T., Growth and premature leaf fall in American Aspen as bioindicators for ozone, *Environ. Pollut.*, 52, 183, 1988.
137. Barrang, P., Karnosky, D. F., and Bennett, J. P., Natural selection for ozone tolerance in *Populus tremuloides*: field verification, *Can. J. For. Res.*, 19, 519, 1989.
138. Steubing, L. and Fangmeier, A., SO_2 sensitivity of plant communities in a beech forest, *Environ. Pollut., 44, 297, 1987.*
139. Heagle, A. S., Rebbeck, J., Shafer, S. F., Blum, U., and Heck, W. W., Effects of long-term ozone exposure and soil moisture deficit on growth of a ladino clover–tall fescue pasture, *Phytopathology*, 79, 128, 1989.
140. Bennett, J. P. and Runeckles, V. C., Effects of low levels of ozone on growth of crimson clover and annual ryegrass, *Crop Sci.*, 17, 443, 1977.
141. Doyle, T. W., Competition and Growth Relationship in a Mixed-Aged Mixed Species Forest Communities, Ph. D. dissertation, University of Tennessee, Knoxville, TN, 1983.
142. Treshow, M., Interactions of air pollutants and plant disease, in *Responses of Plants to Air Pollution*, Mudd, J. B. and Kozlowski, T. T., Eds., Academic Press, New York, 1975, 307.
143. Carlson, C. E. and Dewey, J. E., Environmental pollution by fluorides, Flathead National Forest and Glacier National Park, Division of State and Private Forestry, U.S. Department of Agriculture Forest Service, Northern Region Headquarters, 10, 1971, 57.
144. Manion, P. D., *Tree Disease Concepts*, Prentice-Hall, Englewood Cliffs, NJ, 1981.
145. Stark, R. W., Miller, P. P., Cobb, F. W., Jr., Wood, D. L., and Parmeter, J. R., Jr., Incidence of bark beetle infestation on injured trees, *Hilgardia*, 39, 121, 1968.
146. Skelly, J. M., Photochemical oxidant impact on Mediterranean and temperate forest ecosystems: real and potential effects, in *Effects of Air Pollutants on Mediterranean and Temperate Forest Ecosystems*, Miller, P. R., Ed., U.S. Department of Agriculture Forest Service Rep. PSW-43, 38, 180.
147. Edmunds, G. G. and Alstad D. A., Effects of air pollutants on insect populations, *Annu. Rev. Entomol.*, 27, 369, 1982.

CHAPTER 9

Effects of Tropospheric Ozone on Woody Plants

Katrien Bortier, Reinhart Ceulemans, and Ludwig de Temmerman

CONTENTS

9.1 INTRODUCTION

The same industrial development that has resulted in an increase in the concentration of atmospheric carbon dioxide (CO_2) has also led to an increase of tropospheric ozone (O_3) concentrations. Analyses of historical measurements suggest that surface O_3 concentrations at mid to high latitiudes have more than doubled during the last century.[1,2] Elevated levels are found in urban areas, but also occur in rural and remote regions by transport of O_3 and its precursors. Photochemical production of O_3 from nitrogen oxides and volatile organic compounds appears to be the major source of O_3 in the troposphere. Approximately 80% of NO_x emission is due to anthropogenic activities in the Northern Hemisphere.[1]

1-56670-341-7/00/$0.00+$.50

Tropospheric O_3 has long been defined as a phytotoxin.[3] Much plant physiological research has been conducted in recent years on the reaction of woody plant species to tropospheric O_3. Ozone has been suggested to cause the greatest amount of damage to vegetation as compared with any gaseous pollutant[4-6] and its relative importance may still increase because of the decline in the occurrence of other air pollutants.[7] Furthermore, O_3 is considered as the most widespread of all atmospheric air pollutants.[8] Many studies show that ambient O_3 concentrations are potentially high enough to cause significant reductions in growth and yield of agricultural crops and trees.[6,9-11] In this chapter, the overall responses and reactions of woody plants to tropospheric O_3 levels are discussed at different hierarchical levels of organization based on an extensive, recent literature review. Furthermore, the variation in response to O_3 among different genera and the interactions with other biotic and abiotic factors are documented.

9.2 VARIATIONS IN CONCENTRATIONS AND DOSES

Concentrations of O_3 of anthropogenic origin vary significantly with time and geographic location. Pronounced diurnal cycles are found at low elevation sites, showing low concentrations at night and highest concentrations for a few hours after midday.[1] High elevation sites show little diurnal variation and ozone concentrations increase with altitude.[12] Over Europe average daytime O_3 concentrations range from 35 to 55 ppb in summer, depending on the location. Peak concentrations usually occur episodically in spring and summer, but annual profiles vary from year to year. Ozone concentrations have increased between 1 and 2 % per year during the last two decades.[13]

One of the difficulties in interpreting and comparing vegetation responses to O_3 is the wide range of methods used to describe O_3 exposure. Concentration, duration, shape of exposure, and frequency of peaks are all important in determining plant response.[5,14] Examples of exposure indices are the 7- or 8-h seasonal mean, the sum of all hourly mean concentrations using no threshold (SUM0, total dose), the AOT 30, 40,... (equal to the sum of all hourly mean concentrations above a threshold of 30, 40,... ppb over a defined time period), and the SUM06, 08... (equal to the sum of all hourly average concentrations above 0.06, 0.08,... ppm).[1] The seasonal average concentration does not strongly correlate with plant injury,[14] as it does not capture the important exposure components related to plant response. It does not, for example, allow discrimination between an O_3 profile with episodic peaks and one without peaks but an equal total dose, which has been observed to cause less detrimental effects. Exposure indices that differentially weigh O_3 concentrations and that incorporate exposure duration have been recommended, as peak concentrations are more damaging than low concentrations where O_3 effects are cumulative.[5] They have been found to perform better than cumulative measures or mean values that use all concentrations above zero.[10,14,15] Like low concentrations, high O_3 concentrations (> 90 ppb) are now thought to be less damaging than moderately high O_3 concentrations, because they often coincide with periods of low O_3 flux and low stomatal uptake (i.e., dry and hot conditions).[5,15]

For Europe the proposed exposure index is the AOT40 (accumulated exposure over a threshold of 40 ppb), calculated as the sum of the differences between the hourly ozone concentrations and 40 ppb for each hour when the concentration exceeds 40 ppb over a defined period of time.[10,16] It is a cumulative index that does not weigh O_3 concentrations, but removes concentrations below a threshold of 40 ppb, as lower concentrations are considered of minor importance in causing adverse biological responses. However, Kärenlampi and Skärby[10] emphasized that 40 ppb should not be regarded as an absolute threshold for plant damage, as some biological responses may occur at lower concentrations. The critical level for trees provisionally is set at an AOT40 of 10 ppm.h, calculated over a 6-month growing season for daylight hours.[10] As an indication, for example, for Belgium the AOT40 of 10 ppm.h was exceeded in 1989, 1990, and 1995 (De Temmerman, unpublished results). Researchers in Europe have agreed to use this concept as a way to identify areas at risk and to try to relate O_3 profiles to the biological responses of plants.[17]

9.3 METHODOLOGIES AND EXPOSURE TECHNIQUES

Several experimental methods have been used to study the effects of O_3 on trees (Figure 9.1, Table 9.1). For a comprehensive review of techniques and methodologies the reader is referred to Saxe[18] and Manning and Krupa.[19] The most widely used approach for O_3 experiments is the classic open-top chamber (OTC) technique used for both young seedlings in pots and for taller trees.[11,20] This allows studies in the field with a range of concentrations of O_3 above ambient and offers a more accurate reflection of natural environmental conditions than laboratory exposure systems. There remain, however, problems in translating results obtained from OTCs to field conditions due to differences in microclimate. The growth environment within OTCs will differ in terms of temperature, relative humidity, and wind speed, and these need to be monitored both within and outside the chambers during experiments.[21] Other commonly used field chambers are greenhouses (including solardomes) and branch chambers. Branch chambers or branch bags[22-24] are one of the viable methods of carrying out O_3 impact studies on mature tree species. However, branches are not wholly autonomous units and both export carbohydrates to other parts of the tree and import water, nutrients, and physiological growth regulators.[25] Laboratory exposure systems offer the advantage of being highly controllable, allowing perfectly reproducible studies. However, the conditions are less realistic and laboratory-grown plants may differ physiologically and morphologically from field-grown plants.[26] Examples of laboratory exposure systems are phytotrons and continuously stirred tank reactors (CSTRs). The most realistic exposure is, of course, a chamberless approach in the field and several strategies have been used. Effects of O_3 in the field can be estimated by measuring plants at different locations (field plots) with different O_3 exposures.[27-29] Continuous monitoring of O_3 concentrations and environmental variables are essential. Also open air fumigation systems have been developed and used in some studies.[26,30-32] Alternative approaches include cultivar comparisons, the use of indicator plants, and the use of protective chemicals. The most-encountered problem in the field is the lack of absolute control plants.[19] As every method has its advantages and disadvantages, the choice of the methodology has to depend on the objectives set and processes looked upon.

Because of the difficulties associated with fumigation of mature trees, the number of studies with mature trees is limited and most experiments have been carried out with seedlings or saplings for periods of up to 5 years (Table 9.1). The question remains how to translate these results to the longer term and to scale up from seedling to mature tree.[6,33] Recently, some studies reported on comparisons between the response of seedlings and mature trees to O_3, and there is increasing evidence that the sensitivities of seedlings and mature trees differ. However, no consistent patterns have been discovered. Studies with northern red oak (*Quercus rubra* L.)[34-38] have shown higher sensitivities of mature trees as compared with seedlings, while the opposite was found for giant sequoia (*Sequoiadendrum giganteum* Bucholz.)[39] and black cherry (*Prunus serotina* Ehrh.),[15] with seedlings showing higher sensitivities.

9.4 PLANT RESPONSES TO TROPOSPHERIC OZONE

Table 9.1 summarizes O_3 treatment studies reported in the literature since 1990 on woody plants only, but makes no claims to completeness. Although the table does not report results or relate O_3 doses to effects, it reviews the parameters, tree species, conditions, and methodologies that have been examined in recent years. In order to compare different studies and to establish critical values for O_3 damage, it is crucial to correctly report O_3 doses used. In some studies measured or calculated O_3 uptake has been used instead of ambient exposure, because of its higher physiological relevance.[28,40,41] This issue has been discussed in detail by Reich.[9]

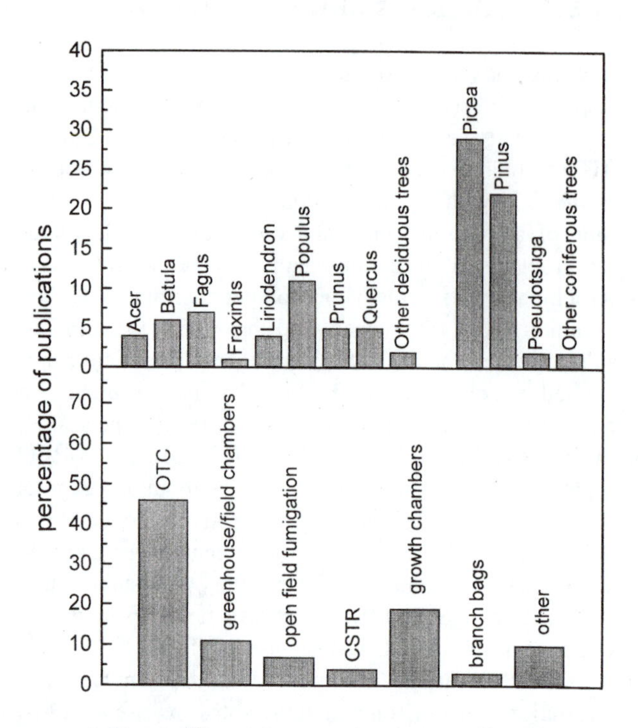

Figure 9.1 Relative representation of different tree species (top panel) and use of different experimental fumigation techniques (bottom panel) in the recent open literature (1990 to 1997) summarized in Table 9.1. Number of reports or publications are shown as percentage of the total number of studies. Tree species were divided in deciduous and coniferous species. OTC = open top chamber, CSTR = continuously stirred tank reactor.

In the following sections O_3 uptake and its effects on trees at different heirarchical levels of organization are discussed (Figure 9.2). The effects are being reviewed on biochemistry, ultrastructure, physiology, growth, allocation patterns, and belowground processes.

9.4.1 Ozone Uptake: Role of Stomates

Reported O_3 exposures are usually ambient doses, i.e., the concentration of O_3 in ambient air integrated over the duration of exposure.[5] Plants can vary in their response to a particular O_3 dose because of, among other reasons, differences in O_3 uptake, or absorbed dose, reflecting differences in leaf conductance.[9] Plant injury is most closely related to the fraction of O_3 entering the plant, i.e., the O_3 flux.[15,42]

The main route of entry of O_3 into the leaf is via the stomates (see Figure 9.2). Stomatal uptake is estimated to be four orders of magnitude larger than uptake through fractures in the leaf cuticle.[33] Factors that influence stomatal aperture also affect plant response to O_3. Ozone can injure cuticles, but most injury takes place after entry through the stomates. The flux of O_3 from the troposphere into the plant depends on different resistances at various levels, i.e., an aerodynamic resistance depending on atmospheric turbulence, a boundary layer resistance caused by the layer of laminar air adjacent to the leaf, the stomatal resistance exerted by the stomatal pores and an internal resistance in the plant leaf.[33] High O_3 concentrations generally occur in spring and summer under atmospheric conditions of high temperature, high irradiance, and low wind speed. Under these conditions boundary layer and stomatal resistance are (generally) high and thus limit O_3 uptake into the leaf.[5]

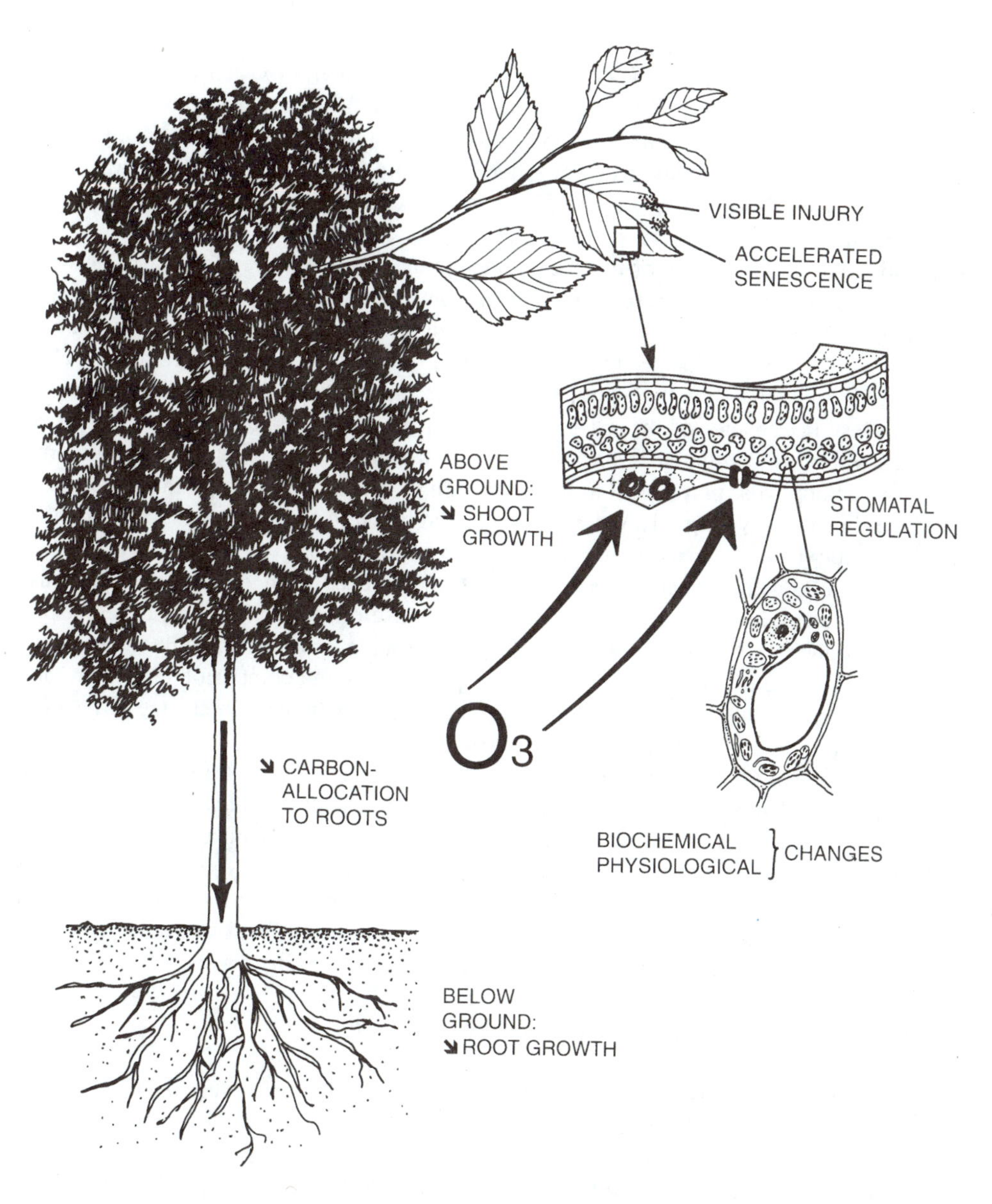

Figure 9.2 Simplified schematic flow diagram of the impact of tropospheric O_3 on different functional processes and tree structures at various hierarchical levels of organization. The contents of the chapter are presented according to this overall schematic representation.

9.4.2 Biochemical Plant Responses at the Cell Level

After entering the plant through the stomates, O_3 dissolves in the water phase of the apoplast and can react with unsaturated fatty acids in the plasmalemma with formation of aldehydes (i.e., malondialdehyde) and H_2O_2.[43] Moreover O_3 reacts in the substomatal cavity with hydrocarbons such as ethylene and some terpenes.[44,45] These oxidative processes (ozonolysis) also produce H_2O_2 in a humid environment. In contrast to O_3, H_2O_2 dissolves very well in the water phase and can be transported through the membranes and circulated in the plant. Hydrogen peroxide is one of the most important O_3 derived oxidants causing plant damage. Further reactions of H_2O_2 lead to other reactive oxygen species such as superoxide and hydroxyl radicals which can initiate lipid peroxidation, a chain reaction destroying the membranes.[46]

The initial site of injury caused by O_3 and/or O_3 generated reactive oxygen species is the plasma membrane,[47] resulting in changes in permeability, fluidity, and ionic and metabolic disturbances within the cell.[48] The plasma membranes are protected by antioxidants such as the hydrophilous ascorbate (vitamin C), the lipophilous α-tocopherol (vitamin E) present in the membranes and by enzymes such as superoxide dismutase (SOD) and peroxidases. Their level generally increased after O_3 exposure.[49-52]

In addition to the destruction of the plasma membranes, O_3 exposure induces oxidative stress similar to that caused by other stresses, i.e., high and low light intensities, drought and cold stress.[43] Hydroxyl radicals formed by the iron-catalyzed reaction of superoxide and H_2O_2 are extremely toxic.[53] These reactive compounds react on the sites where iron is available, mostly close to membranes. Destruction of the thylakoid membranes in the chloroplasts has a direct influence on the light-harvesting capacity of the leaves.[43] However, the cells are provided with a number of antioxidants which are continuously formed and regenerated.[49,54] Also carotenoids, and more specifically the xanthophylls, play a role in the protection of the thylakoid membranes.[55] In general, studies have shown that younger leaves are less affected by O_3 than older leaves, possibly because of increased antioxidant concentrations and activity.[56]

Many metabolic pathways are altered by O_3. Ozone-induced changes in protein patterns can be mediated by altered gene expression.[47] Ozone can cause different effects on different protein pools.[57] For instance, it is known to decrease the concentration of Rubisco, while the synthesis of enzymes involved in antioxidation may increase. Lipid content and composition,[58,59] nutrient concentrations,[20,60] and carbon partitioning[57,61,62] into different compounds can change in response to O_3 exposure, but it is difficult to find consistent patterns.

Under influence of O_3 exposure, plants can increase the formation of stress ethylene.[63,64] Stress ethylene formation can be caused by reaction of superoxide with methional, but probably more important is the enzymatic pathway where ethylene is formed enzymatically out of S-adenosyl methionine (SAM). Ethylene formation coincides with the polyamine biosynthesis as they are using the same precursor, SAM. Plants with the ability to increase polyamine biosynthesis can reduce ethylene production and consequently reduce the negative effects. Stress ethylene production is considered the major cause of enhanced senescence and early leaf abscission.[65]

9.4.3 Ultrastructural Changes at the Subcellular Level

Typical O_3 induced ultrastructural changes reported are reduction in chloroplast size; disintegration of thylakoids and cellular organelles[66,67]; increased size and number of plastoglobuli[59,68,69] and extrusion from the chloroplast[70]; reduced or increased amounts of starch[66-68]; thickened cell walls[67]; appearance of exudates on the walls of mesophyll cells; and mesophyll collapse[66,67] (see Table 9.1). Damaged cells often are located near stomata or stomatal cavities.[71] Changes in plastoglobuli are likely due to early senescence, and are related to changes in lipid composition and translocation of materials from thylakoids.[70,72] Starch accumulation observed along the veins of leaves[67,73] may be the result of disturbed phloem loading (see also Section 9.4.5).

9.4.4 From Cell to Leaf Level: Functional/Physiological Plant Responses

Photosynthesis seems to be the process most affected by O_3 exposure.[18] Ozone-induced reductions in net photosynthesis have been observed for a variety of hardwood and coniferous tree species (see reviews by Pye;[6] Reich;[9] Saxe;[18,74] Figure 9.3). Various factors can alter the photosynthetic performance of a plant, and the degree of response to O_3 can be influenced by environmental conditions (humidity, light, etc.), biological factors (leaf age, diseases), or genotype.[14]

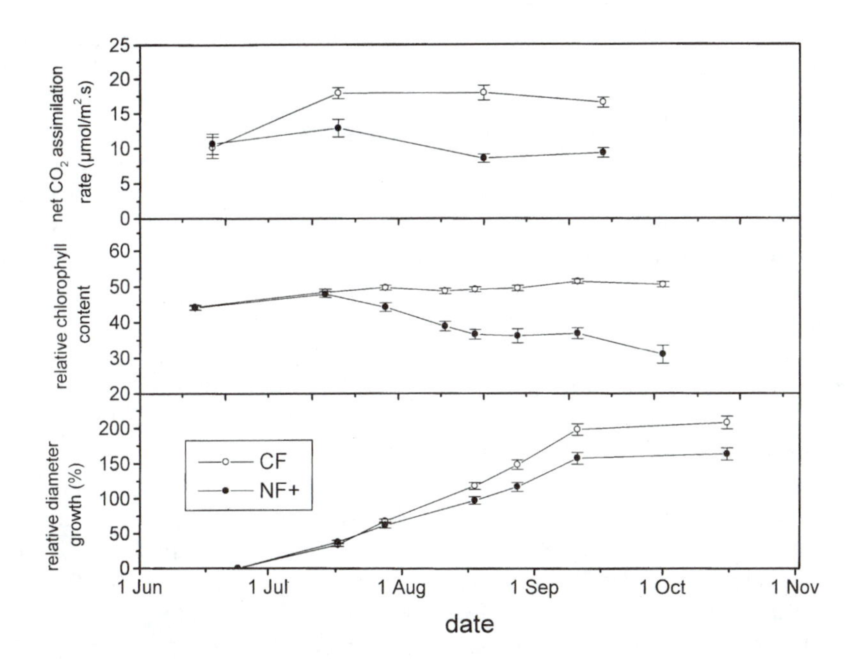

Figure 9.3 Evolution of net CO_2 assimilation rate (top panel), relative chlorophyll content (middle panel), and relative stem diameter growth (bottom panel) of black poplar (*Populus nigra* cv. Wolterson) grown in OTCs under two different ozone concentrations. The control chambers (CF) were supplied with charcoal-filtered air while the ozone-treated chambers (NF+) received nonfiltered air to which tropospheric ozone was added (AOT40 at the end of the growing season: 8883 ppb.h). Mean values of 15 replications + standard error have been shown. (Bortier et al., unpublished data.)

The reductions in net photosynthetic rate, light saturated net photosynthesis, and carboxylation efficiency by O_3 have been related to reduced levels and activity of Rubisco, impaired electron transport, light harvesting, and regeneration of ribulose biphosphate[56,75-77] The effects of O_3 on photosynthesis are generally interpreted as direct effects, rather than indirect ones, through effects on the stomata.[18,47,56,70] Stomatal conductance decreased for several species after O_3 exposure,[18,78] and this could be a response to increased internal CO_2 concentrations[9] due to the inhibition of photosynthesis, or it may be simply a symptom of damage to the guard cells.[56] Oscillations in gas exchange have also been measured in response to O_3, and this is attributed to loss of stomatal control and an uncoupling of the relationship between photosynthesis and stomatal conductance.[56,60,79,80] This results in a disturbed water balance, reduced ability to control water loss, and higher sensitivity to drought. Accordingly, reduced water use efficiency has been reported in response to O_3.[62,80,81]

Ozone-induced inhibition of Rubisco activity is usually associated with decreases in Rubisco concentration. The loss in activity may be caused by effects of the secondary oxidants on protein structure, and this can result in degradation and loss of protein.[82] Another hypothesis for the reduction of Rubisco quantity is a decline in synthesis, related to a decline in the messenger RNA for Rubisco. The largest effects are observed in the older leaves. Losses in Rubisco will reduce carboxylation capacity, but can also lead to symptoms of early senescence.[82,83]

A loss in canopy carbon gain will not only result from loss of photosynthetic capacity by individual leaves, but to a greater or lesser extent by a decrease of photosynthesizing leaf area caused by accelerated leaf shedding.[56-84]

For dark respiration, increases as well as decreases and no effects in response to O_3 have been reported.[18,75,78] The resulting effect depends on how severely leaves are injured: a slight injury will result in increased respiration reflecting a stimulated metabolic activity for detoxification as well as repair of membranes and cellular constituents.[57] When the injury is very severe, respiration will decrease, because metabolic activities will cease or because the substrate becomes limiting due to a reduction in photosynthesis.[36]

9.4.5 Leaf Level: Visible Injury and Leaf Senescence

The appearance of characteristic lesions on the leaves, chlorosis (see Figure 9.3), bleaching, and accelerated abscission of leaves have long been known to be associated with elevated O_3 levels.[21] These signs of damage by O_3 are observed especially on older and mature leaves.

Ozone-induced visible injury frequently is used to assess forest damage,[85] although the extent of foliar injury does not necessarily correlate with physiological damage or reductions in growth.[6] Injured cells may be compensated for by uninjured ones,[15] and early leaf abscission can be compensated for by an increased leaf production.[26,56,67,86] Biochemical and physiological injury may occur before the appearance of any visible symptoms.[15] Acute O_3 stress will generally result in visible symptoms, but low concentrations over long periods may lead to hidden damage without the appearance of visible foliar injury.[6]

In many tree species O_3 stimulates senescence processes.[15,80,87,88] Natural senescence is characterized by a controlled degradation of cellular and leaf functions during which cellular constituents are remobilized before abscission.[89] Lippert et al.[83] demonstrated high nitrogen (N) losses for beech after O_3 exposure due to inhibited N-translocation before leaf drop, which differs from natural autumnal senescence where N is withdrawn from the leaves. Similar results were reported for birch (*Betula pendula*).[80] Ozone-induced degradation of leaves should therefore not be confounded with natural senescence, as it seems more like an unregulated degradation than an accelerated natural senescence.[80,85]

Chlorosis is not a primary result of O_3 exposure, but a secondary effect due to impaired photosynthetic capacity.[47] When chlorophyll molecules are arranged structurally in the thylakoids, they are very resistant to direct oxidation,[90] and chlorosis more likely is associated with accelerated senescence than with direct effects of O_3 or its oxidative products.

For several poplar hybrids exposed to O_3, visible effects on stems have been reported.[63] Where leaves were shed, lesions or intumescences appeared on the stems, resulting in bark cracking and the exposure of soft cortical tissue. It has been hypothesized that ethylene is possibly responsible for the induction of these stem lesions.

There have been very few reports on the effects of O_3 on overall tree structure. In birch (*Betula pendula*)[73] and European aspen (*Populus tremula*), [91] the crown structure was altered significantly after exposure to elevated O_3 levels while in interamerican poplar (*Populus*) hybrids, a lowered branchiness has been observed.[92]

9.4.6 Carbon-Allocation and Below-Ground Responses: Tree Level

Ozone causes reduction in carbon uptake by reduction of photosynthesis and of photosynthetic leaf area. The subsequent translocation of carbon to different plant organs and to different pools can also be altered by O_3.[93-96] In many tree species carbon retention is increased in the leaves and consequently carbon allocation to the roots is reduced.[68,97,98] Very often, the shoot/root ratio increases following O_3 exposure, due to higher reductions in root growth than in shoot growth.[99] This higher retention of C in the leaves may be explained by higher carbon demands for repair of damaged foliage, by reduced assimilate transport in the phloem, or by decreased phloem loading.[91,93,100] In loblolly pine needles (*Pinus taeda*) a decreased partitioning of assimilated carbon into starch and protein, and an increased partitioning into organic acids, lignin plus structural carbohydrates, and lipids plus pigments has been reported after O_3 exposure.[57] This shift in partitioning from storage compounds to soluble carbohydrates and carbon compounds involved in repair might be a compensatory response to maintain photosynthetic rates. Effects on the amounts of foliar starch are sometimes contrasting, since both increases[101] and decreases[93,94,102] have been reported. Further research is needed to elucidate the mechanisms underlying the effects of O_3 on carbohydrate metabolism.[95,103] In O_3-exposed birch (*Betula pendula*) accumulation of starch along the leaf vein network has been observed, suggesting problems with the mechanism of starch–sucrose turnover and phloem loading.[66]

Reduced root growth can alter the functioning of rhizosphere organisms and could make trees more susceptible to drought or nutrient deficiency.[86] Andersen et al.[104] reported lower carbohydrate levels in new roots of ponderosa pine seedlings (*Pinus ponderosa*) after O_3 exposure, which may result in reduced plant growth over time. In addition, O_3 exposure during one year resulted in less new root growth in the year following exposure (carryover effect).

9.4.7 Tree Level: Growth

As growth is the integration of several physiological and biochemical processes, O_3 induced physiological and biochemical changes can ultimately lead to growth reductions (see Figure 9.3).[72] But despite the evidence that growth of trees can be reduced by O_3, it has not yet been possible to quantify the role of O_3 in forest tree health and forest production.[29] Growth response to O_3 depends on many factors such as species, developmental stage, exposure profiles, climatic variables, biotic and abiotic factors; attempts have been made to compile data and to evaluate thresholds and critical levels.[9,14,105] Modeling has also become an essential tool in the study of air pollution effects on forests. Models of forest trees and stands can be useful in predicting how forests will respond to existing or projected O_3 scenarios.[47]

Relating disturbances in physiological processes to reductions in growth is very complex. Decreases in photosynthesis may reduce growth, but not necessarily. Plants have different ways to compensate for stress. For instance, trees with indeterminate growth, such as poplars, can sometimes compensate for losses in photosynthetic capacity and leaf loss by increased photosynthetic rates of the retained foliage[106] as well as by increased production or area of new leaves.[56,67,86] Increases in the synthesis of Rubisco in young leaves of hybrid poplar correlate well with reductions in Rubisco content of leaves experiencing accelerated senescence.[82] In poplar, mature leaves near the stem base export most of the fixed carbon to the lower stem and the roots, so premature abscission of mature leaves causes reductions in root growth. In *Populus tremuloides*, the loss of mature leaves can somewhat be compensated for by an increased carbon allocation to the roots by recent mature leaves.[93]

Both direct and indirect effects of O_3 can alter tree growth. Direct effects that have been described in this chapter include, e.g., changes in photosynthesis, stomatal physiology, photosynthate allocation, and visible symptoms of injury. But, additionally, changes in the physiological state of the tree can change its response to other stresses. Examples of these indirect effects are changes in drought resistance, winter hardiness, and susceptibility to pathogens.[107] The ability of the tree to withstand these stresses will determine its survival.[47]

9.5 INTERACTIONS AND GENOTYPIC DIFFERENCES

9.5.1 Interactions Between Ozone and Other Abiotic and Biotic Factors

A number of stresses generally affect the growth and physiology of a tree. The response of a tree to O_3 can be altered by various biotic and abiotic factors, and vice versa. Abiotic factors that can interact with O_3 include temperature, moisture, solar radiation, nutrition, CO_2, and other air pollutants.[108] Episodes of elevated O_3 generally coincide with periods of minimal rainfall and drought, and research has been done on the interactive effects between these two stresses (see Table 9.1). Drought can protect plants from ozone damage, because stomatal closure in response to drought limits O_3 uptake.[109,110] However, in some studies more-pronounced responses for plants experiencing water deficit were reported.[100,111] Ozone can alter the response of the tree to drought by causing disturbances in stomatal functioning, thereby reducing stomatal control of water loss.[112] As the rising concentration of atmospheric CO_2 is one of the major environmental concerns, the interactions between O_3 and CO_2 have been investigated for a number of tree species (see Table 9.1). A common hypothesis is that elevated CO_2 may reduce O_3 damage, because in most tree species elevated levels of CO_2 reduce stomatal conductance,[113] which will result in a reduced O_3 flux into the leaf. This hypothesis is not always supported and studies that have examined the interactive effects of CO_2 and O_3 have shown a variety of effects. In a number of studies, elevated CO_2 resulted in a protection against ozone damage.[75,114] In others no protective effect was observed and even an increased plant sensitivity to O_3 was reported.[76,115]

Interaction studies involving O_3 and other gaseous pollutants or acid rain have shown variable responses ranging from antagonistic to additive and synergistic effects. The responses also depend on other factors like genotype, age, soil type, and environmental conditions.[108] Furthermore, the role of nutrition in O_3 tolerance is uncertain. Some studies indicate that plants with optimal N are more sensitive than N-deficient plants,[116,117] while others have found that well-fertilized plants cope better with O_3 stress.[101]

Ozone can alter the response of trees to biotic stresses. Damage due to O_3 may change their tolerance or resistance to insect herbivores and plant pathogens.[108] In general, O_3 exposure has shown to increase palatability, increase herbivorous consumption, and enhance insect performance.[103] In addition to increased susceptibility to invasion by plant pathogens, inhibitory effects of O_3 against microorganisms and fungi have been reported as well. Mycorrhizae have been shown to offer beneficial effects in ameliorating O_3 stress, while O_3 can have negative effects on mycorrhizal development. A decrease in photosynthesis and carbon allocation to the roots would imply less carbohydrates available for the mycorrhizae.[4,108]

9.5.2 Genotypic and Species Differences

Tree species exhibit a wide range of sensitivity to O_3 (see Table 9.1). A common generalization is that coniferous species seem less sensitive to O_3 than broad-leaved trees. Reich[9] hypothesized that differences between species can be explained by differences in leaf conductance and associated O_3 uptake. Conifers generally have lower leaf conductances, and therefore lower O_3 uptake rates. For the same reason, fast-growing species are usually more sensitive than slow-growing species.[6] Plants could also vary in their sensitivity to O_3 due to differences in detoxification and repair capacities.[5]

Reports in the literature have ranked tree species according to relative sensitivity. The criteria for ranking include measures of growth or visible injury; however, rankings based on visible injury do not correlate with rankings according to growth measures.[9,108] In addition to the used criteria, plant material (source, age, genotype) and methodology play an important role in comparing results for different species (see Table 9.1). Within a species, a wide genotypic variation in sensitivity can occur.[11,30,68,118,119] Family and geographic differences may cause a species being ranked as tolerant by some and sensitive by others. Different stages in the life cycle can be more or less sensitive, and evolutionary adaptation to O_3 can change the sensitivity of a species over time.[118,120]

9.6 CONCLUSIONS

Tropospheric O_3 has profound negative impacts on the growth, development, and productivity of many plants and vegetations, including trees and forests. Significant effects of O_3 have been observed on a wide range of characterstics such as early leaf senescence, decreased photosynthetic assimilation, altered stomatal behavior, decreased growth and productivity, and reduced carbon allocation to roots. Although related species or genera may show very different responses to O_3 and there may be large differences in sensitivity between different cultivars or clones of the same species, the initial mechanism of O_3 induced stress on plants is uniform. A better understanding of the effects of O_3 and O_3-derived oxidants is necessary for a more-detailed insight into the impact of O_3 on plant growth and development. As rising tropospheric O_3 levels are likely to be a continuing problem, overall growth and yield of trees and forests may be increasingly affected. In particular, the responses of trees to increased tropospheric O_3 levels in combination with other environmental changes will play a very important role in determining growth, development, survival, and abundance of individual plants as well as plant communities in the future.

ACKNOWLEDGMENTS

The authors acknowledge the support from the Ministry of the Flemish Community, Forests and Green Areas Division (research contracts B&G/13/1995 and B&G/14/1995) during the preparation of this manuscript.

REFERENCES

1. Stockwell, W. R., Kramm, G., Scheel, H.-E., Mohnen, V. A., and Seiler, W., Ozone formation, destruction and exposure in Europe and the United States, in *Forest decline and Ozone*, Sandermann, H., Wellburn, A. R., and Heath, R. L., Eds., Springer-Verlag, Berlin, 1997, 1.
2. Volz, A. and Kley, D., Evaluation of the Montsouris series of ozone measurements made in the nineteenth century, *Nature*, 332, 240, 1988.
3. Richards, B. L., Middleton, J. T., and Hewitt, W. B., Air pollution with relation to agronomic crops: V. Oxidant stipple of grape, *Agron. J.*, 50, 559, 1958.
4. Spence, R. D., Rykiel, E. J., Jr., and Sharpe, P. J. H., Ozone alters carbon allocation in loblolly pine: assessment with carbon-11 labeling, *Environ. Pollut.*, 64, 93, 1990.
5. Musselman, R. C., McCool, P. M., and Lefohn, A. S., Ozone descriptors for an air quality standard to protect vegetation, *J. Air Waste Manage. Assoc.*, 44, 1383, 1994.
6. Pye, J. M., Impact of ozone on the growth and yield of trees: a review, *J. Environ. Qual.*, 17, 347, 1988.
7. Tenga, A. Z., Hale, B. and Omrod, D. P., Growth responses of young cuttings of *Populus deltoides*× *nigra* to ozone in controlled environments, *Can. J. For. Res.*, 26, 649, 1993.
8. Simini, M., Skelly, J. M., Davis, D. D., Savage, J. E., and Comrie, A. C., Sensitivity of four hardwood species to ambient ozone in north central Pennsylvania, *Can. J. For. Res.*, 22, 1789, 1992.
9. Reich, P. B., Quantifying plant response to ozone: a unifying theory, *Tree Physiol.*, 3, 63, 1987.
10. Kärenlampi, L. and Skärby, L., Eds., *Critical Levels for Ozone in Europe: Testing and Finalising the Concepts*, Kuopio University, Kuopio, 1996.
11. Karnosky, D. F., Gagnon, Z. E., Dickson, R. E., Coleman, M. D., Lee, E. H., and Isebrands, J. G., Changes in growth, leaf abscission, and biomass associated with seasonal tropospheric ozone exposures of *Populus tremuloides* clones and seedlings, *Can. J. For. Res.*, 26, 23, 1996.
12. Mansfield, T. A. and Pearson, M., Physiological basis of stress imposed by ozone pollution, in *Plant Adaptation to Environmental Stress*, Fowden, L., Mansfield, T. A., and Stoddart, J., Eds., Chapman & Hall, London, 1993.
13. Hough, A. M. and Derwent, R. G., Changes in the global concentration of tropospheric ozone due to human activities, *Nature*, 344, 645, 1990.

14. Lefohn, A. S., Ozone standards and their relevance for protecting vegetation, in *Surface-Level Ozone Exposures and Their Effects on Vegetation*, Lefohn A. S., Ed., Lewis Publishers, Chelsea, 1992, 271.
15. Fredericksen, T. S., Skelly, J. M., Steiner, K. C., Kolb, T. E., and Kouterick, K. B., Size-mediated foliar response to ozone in black cherry trees, *Environ. Pollut.*, 91, 53, 1996.
16. Führer, J. and Achermann, B., Eds., Critical Levels for Ozone, UN-ECE workshop report, Liebefeld-Bern, 1994.
17. Mortensen, L., Bastrup-Birk, A., and Ro-Poulsen, H., Critical levels of O_3 for wood production of European beech (*Fagus sylvatica* L.), *Water Air Soil Pollut.*, 85, 1349, 1995.
18. Saxe, H., Photosynthesis and stomatal responses to polluted air, and the use of physiological and biochemical responses for early detection and diagnostic tools, *Adv. Bot. Res.*, 18, 1, 1991.
19. Manning, W. J. and Krupa, S. V., Experimental methodology for studying the effects of ozone on crops and trees, in *Surface-Level Ozone Exposures and Their Effects on Vegetation*, Lefohn A. S., Ed., Lewis Publishers, Chelsea, 1992, 93.
20. Samuelson, L. J., Kelly, J. M., Mays, P. A., and Edwards G. S., Growth and nutrition of *Quercus rubra* L. seedlings and mature trees after three seasons of ozone exposure, *Environ. Pollut.*, 91, 317, 1996.
21. Schmieden, U. and Wild, A., The contribution of ozone to forest decline, *Physiol. Plant.*, 94, 371, 1995.
22. Schaap, W., Use of Branch and Whole Tree Exposure Systems to Evaluate Ozone Impacts on Forest Trees, Ph.D. thesis, University of Washington, Seattle, 1992.
23. Maier-Maercker, U., Experiments on the water balance of individual attached twigs of *Picea abies* (L.) Karst. in pure and ozone-enriched air, *Trees Struct. Func.*, 11, 229, 1997.
24. Momen, B., Anderson, P. D., Helms, J. A., and Houpis, J. L. J., Acid rain and ozone effects on gas exchange of *Pinus ponderosa*: a comparison between trees and seedlings, *Int. J. Plant Sci.*, 158, 617, 1997.
25. Sprugel, D. G., Hinckley, T. M., and Schaap, W., The theory and practice of branch autonomy, *Annu. Rev. Ecol. Syst.*, 22, 662, 1991.
26. Pääkkönen, E., Vahala, J., Holopainen, T., Karjalainen, R., and Kärenlampi, L., Growth responses and related biochemical and ultrastructural changes of the photosynthetic apparatus in birch (*Betula pendula*) saplings exposed to low concentrations of ozone, *Tree Physiol.*, 16, 597, 1996.
27. Bartholomay, G. A., Eckert, R. T., and Smith, K. T., Reductions in tree-ring widths of white pine following ozone exposure at Acadia National Park, Maine, USA, *Can. J. For. Res.*, 27, 361, 1997.
28. Wieser, G. and Havranek, W. M., Environmental control of ozone uptake in *Larix decidua* Mill.: a comparison between different altitudes, *Tree Physiol.*, 15, 253, 1995.
29. McLaughlin, S. B. and Downing, D. J., Interactive effects of ambient ozone and climate measured on growth of mature forest trees, *Nature*, 374, 252, 1996.
30. Pääkkönen, E., Holopainen, T., and Kärenlampi, L., Variation in ozone sensitivity among clones of *Betula pendula* and *Betula pubescens*, *Environ. Pollut.*, 95, 37, 1997.
31. Holland, M. R., Mueller, P. W., Rutter, A. J., and Shaw, P. J. A., Growth of coniferous trees exposed to SO_2 and O_3 using an open-air fumigation system, *Plant Cell Environ.*, 18, 227, 1995.
32. Holopainen, J. K., Kainulainen, P., and Oksanen, J., Effects of gaseous air pollutants on aphid performance on Scots pine and Norway spruce seedlings, *Water Air Soil Pollut.*, 85, 1431, 1995.
33. Guderian, R., *Air Pollution by Photochemical Oxidants*, Guderian, R., Ed., Ecological Studies, 52, Springer, Berlin, 1985.
34. Edwards, G. S., Wullschleger, S. D. and Kelly, J. M., Growth and physiology of northern red oak: preliminary comparisons of mature tree and seedling responses to ozone, *Environ. Pollut.*, 83, 215, 1994.
35. Kelly, J. M., Samuelson, L., Edwards, G., Hanson, P., Kelting, D., Mays, A., and Wullschleger, S., Are seeddlings reasonable surrogates for trees? An analysis of ozone impacts on *Quercus rubra*, *Water Air Soil Pollut.*, 85, 1317, 1995.
36. Wullschleger, S. D., Hanson, P.J., and Edwards, G. S., Growth and maintenance respiration in leaves of northern red oak seedlings and mature trees after 3 years of ozone exposure, *Plant Cell Environ.*, 19, 577, 1996.
37. Hanson, P. J., Samuelson, L. J., Wullschleger, S. D., Tabberer, T. A., and Edwards, G. S., Seasonal patterns of light-saturated photosynthesis and leaf conductance for mature and seedling *Quercus rubra* L. foliage: differential sensitivity to ozone exposure, *Tree Physiol.*, 14, 1351, 1994.

38. Samuelson, L. J. and Edwards, G. S., A comparison of sensitivity to ozone in seedlings and trees of *Quercus rubra* L., *New Phytol.*, 125, 373, 1993.
39. Grulke, N. E. and Miller, P. R., Changes in gas exchange characteristics during the life span of giant seqoia: implications for response to current and future concentrations of atmospheric ozone, *Tree Physiol.*, 14, 659, 1994.
40. Samuelson, L. J. and Kelly, J. M., Ozone uptake in *Prunus serotina*, *Acer rubrum*, and *Quercus rubra* forest trees of different sizes, *New Phytol.*, 136, 255, 1997.
41. Weber, J. A., Scott Clark, C. and Hogsett, W. E., Analysis of the relationships among O_3 uptake, conductance, and photosynthesis in needles of *Pinus ponderosa*, *Tree Physiol.*, 13, 157, 1993.
42. Sandermann, H., Jr., Wellburn, A. R., and Heath, R. L., Forest decline and ozone: synopsis, in *Forest Decline and Ozone*, Sandermann, H., Wellburn, A. R., and Heath, R. L., Eds., Springer-Verlag, Berlin, 1997, 369.
43. Hippelli, S. and Elstner, E. F., Mechanisms of oxygen activation during plant stress: biochemical effects of air pollutants, *J. Plant Physiol.*, 148, 249, 1996.
44. Elstner, E. F., Osswald, W., and Youngman, R. J., Basic mechanisms of pigment bleaching and loss of structural resistance in spruce (*Picea abies*) needles: advances in phytomedical diagnostics, *Experientia*, 41, 591, 1985.
45. Hewitt, C. N., Kok, G. L., and Fall, R., Hydroperoxides in plants exposed to ozone mediate air pollution damage to alkene emitters, *Nature*, 344, 56, 1990.
46. Winston, G. W., Physiochemical basis for free radical formation in cells: production and defenses, in *Stress Responses in Plants: Adaptation and Acclimation Mechanisms*, Alscher, R. G. and Cumming, J. R., Eds., Wiley-Liss, New York, 1990, 57.
47. Heath, R. L. and Taylor, G. E. Jr., Physiological processes and plant responses to ozone exposure, in *Forest Decline and Ozone*, Sandermann, H., Wellburn, A. R., and Heath, R. L., Eds., Springer-Verlag, Berlin, 1997, 317.
48. Heath, R. L., Alterations of plant metabolism by ozone exposure, in *Plant Responses to the Gaseous Environment*, Alscher, R. G. and Wellburn, A. R., Eds., Chapman & Hall, London, 1994, 121.
49. Sen Gupta, A., Alscher, R. G., and McCune, D., Response of photosynthesis and cellular antioxidants to ozone in *Populus* leaves, *Plant Physiol.*, 96, 650, 1991.
50. Kangasjärvi, J., Talvinen, J., Utriainen, M., and Karjalainen, R., Plant defence systems induced by ozone, *Plant Cell Environ.*, 17, 783, 1994.
51. Bermadinger, E., Guttenberger, H., Grill, D., Krupa, S. V., and Arndt, U., Physiology of young Norway spruce, *Environ. Pollut.*, 68, 319, 1990.
52. Wellburn, F. A. M., Lau, K. K., Millin, P. M. K., and Wellburn, A. R., Drought and air pollution affect nitrogen cycling and free radical scavenging in *Pinus halepensis* (Mill.), *J. Exp. Bot.*, 47, 1361, 1996.
53. Fridovich, I., Biological effects of the superoxide radical, *Arch. Biochem. Biophys.*, 247, 1, 1986.
54. Bowler, C., Van Montagu, M., and Inzé D., Superoxide dismutase and stress tolerance, *Annu. Rev. Plant Physiol. Plant Mol. Biol.*, 43, 83, 1992.
55. Young, A. and Britton, G., Carotenoids and stress, in Stress Responses in *Plants: Adaptation and Acclimation Mechanisms*, Alscher, R. G. and Cumming, J. R., Eds., Wiley-Liss, New york, 1990, 87.
56. Clark, C. S., Weber, J. A., Lee, E. H., and Hogsett W. E., Reductions in gas exchange of *Populus tremuloides* caused by leaf aging and ozone exposure, *Can. J. For. Res.*, 26, 1384, 1996.
57. Friend, A. L. and Tomlinson, P. T., Mild ozone exposure alters ^{14}C dynamics in foliage of *Pinus taeda* L., *Tree Physiol.*, 11, 215, 1992.
58. Wingsle, G., Mattson, A., Ekblad, A., Hallgren, J.-E., and Selstam, E., Activities of glutathione reductase and superoxide dismutase in relation to changes of lipids and pigments due to ozone in seedlings of *Pinus sylvestris* L., *Plant Sci.*, 82, 167, 1992.
59. Antonnen, S., Herranen, J., Peura, P., and Kärenlampi, L., Fatty acids and ultrastructure of ozone-exposed aleppo pine (*Pinus halepensis* Mill.) needles, *Environ. Pollut.*, 87, 235, 1995.
60. Tjoelker, M. G., Volin, J. C., Oleksyn, J., and Reich, P. B., Interaction of ozone pollution and light effects on photosynthesis in a forest canopy experiment, *Plant Cell Environ.*, 18, 895, 1995.
61. Peace, E. A., Lea, P. J., and Darall, N. M., The effect of open-air fumigation with SO_2 and O_3 on carbohydrate metabolism in Scots pine (*Pinus sylvestris*) and Norway spruce (*Picea abies*), *Plant Cell Environ.*, 18, 277, 1995.

62. Maurer, S. and Matyssek, R., Nutrition and the ozone sensitivity of birch (*Betula pendula*). 2. Carbon balance, water-use efficiency and nutritional status of the whole plant, *Trees Struct. Func.*, 12, 11, 1997.
63. Kargiolaki, H., Osborne, D. J., and Thompson, F. B., Leaf abscission and stem lesions (intumescences) on polar clones after SO_2 and O_3 fumigation: a link with ethylene release? *J. Exp. Bot.*, 42, 1189, 1991.
64. Ballach, H.-J., Niederée, C., Wittig, R., and Woltering, E. J., Reactions of cloned polars to air pollution, *Environ. Sci. Pollut. Res.*, 2, 201, 1995.
65. Mehlhorn, H., O'Shea, J. M., and Wellburn, A. R., Atmospheric ozone interacts with stress ethylene formation by plants to cause visible plant injury, *J. Exp. Bot.*, 42, 17, 1991.
66. Günthardt-Goerg, M. S., Matyssek, R., Scheidegger, C., and Keller, T., Differentiation and structural decline in the leaves and bark of birch (*Betula pendula*) under low ozone concentrations, *Trees*, 7, 104, 1993.
67. Günthardt-Goerg, M. S., Schmutz, P., Matyssek, R., and Bucher, J. B., Leaf and stem structure of poplar (*Populus × euramericana*) as influenced by O_3, NO_2, their combination, and different soil N supplies, *Can. J. For. Res.*, 26, 649, 1996.
68. Wulff, A., Antonnen, S., Heller, W., Sandermann, H., Jr., and Kärenlampi, L., Ozone-sensitivity of Scots pine and Norway spruce from northern and local origin to long-term open-field fumigation in central Finland, *Environ. Exp. Bot.*, 36, 209, 1996.
69. Selldén, G., Sutinen, S., and Skärby, L., Controlled ozone exposures and field observations in Fennoscandia, in *Forest Decline and Ozone*, Sandermann, H., Wellburn, A. R., and Heath, R. L., Eds., Springer-Verlag, Berlin, 1997, 317.
70. Mikkelsen, T. N. and Heide-Jorgensen, H. S., Acceleration of leaf senescence in *Fagus sylvatica* L. by low levels of tropospheric ozone demonstrated by leaf colour, chlorophyll fluorescence and chloroplast ultrastructure, *Trees*, 10, 145, 1996.
71. Antonnen, S., Sutinen, M. L., and Heagle, A. S., Ultrastructure and some plasma membrane characteristics of ozone-exposed loblolly pine needles, *Physiol. Plant.*, 98, 309, 1996.
72. Pääkkönen, E., Holopainen, T., and Kärenlampi, L., Differences in growth, leaf senescence and injury, and stomatal density in birch (*Betula pendula* Roth) in relation to ambient levels of ozone in Finland, *Environ. Pollut.*, 96, 117, 1997.
73. Matyssek, R., Günthardt-Goerg, M. S., Saurer, M., and Keller, T., Seasonal growth, $\delta^{13}C$ in leaves and stem, and phloem structure of birch (*Betula pendula*) under low ozone concentrations, *Trees*, 6, 69, 1992.
74. Saxe, H., Physiological and biochemical tools in diagnosis of forest decline and air pollution injury to plants, in *Plant Response to Air Pollution*, Yunus, M. and Iqbal, M., Eds., John Wiley & Sons, Chichester, U.K., 1996, 449.
75. Kellomäki, S. and Wang, K.-Y., Effects of elevated O_3 and CO_2 concentrations on photosynthesis and stomatal conductance in Scots pine, *Plant Cell Environ.*, 20, 995, 1997.
76. Kull, O., Sober, A., Coleman, M. D., Dickson, R. E., Isebrands, J. G., Gagnon, Z., and Karnosky, D. F., Photosynthetic responses of aspen clones to simultaneous exposures of ozone and CO_2 , *Can. J. For. Res.*, 26, 639, 1996.
77. Reichenauer, T., Bolhar-Nordenkampf, H. R., Ehrlich, U., Soja, G., Postl, W. F., and Halbwachs, F., The influence of ambient and elevated ozone concentrations on photosynthesis in *Populus nigra*, *Plant Cell Environ.*, 20, 1061, 1997.
78. Coleman, M. D., Isebrands, J. G., Dickson, R. E., and Karnosky, D. F., Photosynthetic productivity of aspen clones varying in sensitivity to troposheric ozone, *Tree Physiol.*, 15, 585, 1995.
79. Flagler, R. B., Lock, J. E., and Elsik, C. G., Leaf-level and whole-plant gas exchange characteristics of shortleaf pine exposed to ozone and simulated acid rain, *Tree Physiol.*, 14, 361, 1994.
80. Matyssek, R., Günthardt-Goerg, M. S., Keller, T., and Scheidegger, C., Impairment of gas exchange and structure in birch leaves (*Betula pendula*) caused by low ozone concentrations, *Trees*, 5, 5, 1991.
81. Shan, Y., Feng, Z., Izuta, T., Aoki, M., and Totsuka, T., The individual and combined effects of ozone and simulated acid rain on growth, gas exchange rate and water-use efficiency of *Pinus armandi* Franch., *Environ. Pollut.*, 91, 355, 1996.
82. Pell, E. J., Landry, L. G., Eckardt, N. A., and Glick, R. E., Air pollution and Rubisco: effects and implications, in *Plant Responses to the Gaseous Environment*, Alscher, R. G. and Wellburn, A. R., Chapman & Hall, London, 1994,239.

83. Lippert, M., Steiner, K., Payer, H. D., Simons, S., Langebartels, C., and Sandermann, H., Jr;, Assessing the impact of ozone on photosynthesis of European beech (*Fagus sylvatica* L.) in environmental chambers, *Trees*, 10, 268, 1996.
84. Matyssek, R., Günthardt-Goerg, M. S., Landolt, W., and Keller, T., Whole-plant growth and leaf formation in ozonated hybrid poplar (Populus × euramericana), *Environ. Pollut.*, 81, 207, 1993.
85. Patterson, M. T. and Rundel, P. W., Stand characteristics of ozone-stressed populations of *Pinus jeffreyi* (Pinaceae): extent, development, and physiological consequences of visible injury, *Am. J. Bot.*, 82, 150, 1995.
86. Woodbury, P. B., Laurence, J. A., and Hudler, G. W., Chronic ozone exposure alters the growth of leaves, stems and roots of hybrid *Populus*, *Environ. Pollut.*, 85, 103, 1994.
87. Reich, P. B. and Lassoie, J. P., Influence of low concentrations of ozone on growth, biomass partitioning and leaf senescence in young hybrid poplar plants, *Environ. Pollut.*, 39, 39, 1985.
88. Shelburne, V. B., Reardon, J. C., and Paynter, V. A., The effects of acid rain and ozone on biomass and leaf area parameters of shortleaf pine (*Pinus echinata* Mill.), *Tree Physiol.*, 12, 163, 1993.
89. Addicott, F. T., *Abscission*, University of California Press, Los Angeles, 1982.
90. Heath, R. L., Alteration of chlorophyll in plants upon air pollutant exposure, in *Biological Markers of Air-Pollution Stress and Damage in Forests*, Woodwell, G. M., Cook, E. R., Cowling, E. B., Johnson, A. H., Kimmerer, T. W., Matson, P. A., McLaughlin, S. S., Ruynal, D. J., Swank, W. T., Waring, R. H., Winner, W. E., and Woodman, J. N., Eds., U.S. National Academy Press, Washington, D.C., 347, 1989.
91. Matyssek, R., Keller, T., and Koike, T., Branch growth and leaf gas exchange of *Populus tremula* exposed to low ozone concentrations throughout two growing seasons, *Environ. Pollut.*, 79, 1, 1993.
92. Gardner, S. D. L., The Effects of Elevated CO_2 and Tropospheric O_3 on the Growth and Development of Hybrid Poplar, Ph.D. Thesis, University of Sussex, Brighton, U.K., 1996.
93. Coleman, M. D., Dickson, R. E., Isebrands, J. G., and Karnosky, D. F., Carbon allocation and partitioning in aspen clones varying in sensitivity to tropospheric ozone, *Tree Physiol.*, 15, 593, 1995.
94. Samuelson, L. J. and Kelly, J. M., Carbon partitioning and allocation in northern red oak seedlings and mature trees in response to ozone, *Tree Physiol.*, 16, 853, 1996.
95. Smeulders, S. M., Gorissen, A., Joosten, N. N., and Van Veen, J. A., Effects of short-term ozone exposure on the carbon economy of mature and juvenile Douglas firs (*Pseudotsuga menziesii* (Mirb.) Franco), *New Phytol.*, 129, 45, 1995.
96. Fialho, R. C. and Bücker, J., Changes in levels of foliar carbohydrates and myo-inositol before premature leaf senescence of *Populus nigra* induced by a mixture of O_3 and SO_2, *Can. J. Bot.*, 74, 965, 1996.
97. Coleman, M. D., Dickson, R. E., Isebrands, J. G., and Karnosky, D. F., Root growth and physiology of potted and field-grown trembling aspen exposed to tropospheric ozone, *Tree Physiol.*, 16, 145, 1996.
98. Andersen, C. P., Wilson, R., Plocher, M., and Hogsett, W. E., Carry-over effects of ozone on root growth and carbohydrate concentrations of ponderosa pine seedlings, *Tree Physiol.*, 17, 805, 1997.
99. Cooley, D. R. and Manning, W. J., The impact of ozone on assimilate partitioning in plants: a review, *Environ. Pollut.*, 47, 95, 1987.
100. Gorissen, A., Joosten, N. N., Smeulders, S. M., and Van Veen, J. A., Effects of short-term ozone exposure and soil water availability on the carbon economy of juvenile Douglas-fir, *Tree Physiol.*, 14, 647, 1994.
101. Landolt, W., Günthardt-Goerg, M. S., Pfenniger, I., Einig, W., Hampp, R., Maurer, S., and Matyssek, R., Effect of fertilization on ozone-induced changes in the metabolism of birch (*Betula pendula*) leaves, *New Phytol.*, 137, 389, 1997.
102. Friend, A. L., Tomlinson, P. T., Dickson, R. E., O'Neill, E. G., Edwards, N. T., and Taylor, G. E., Biochemical composition of loblolly pine reflects pollutant exposure, *Tree Physiol.*, 11, 35, 1992.
103. Wellburn, A. R., Barnes, J. D., Lucas, P. W., Mcleod, A. R., and Mansfield, T. A., Controlled O_3 exposures and field observations of O_3 effects in the U.K., in *Forest Decline and Ozone*, Sandermann, H., Wellburn, A. R., and Heath, R. L., Eds., Springer-Verlag, Berlin, 1997, 317.
104. Andersen, C. P., Hogsett, W. E., Wessling, R., and Plocher, M., Ozone decreases spring root growth and root carbohydrate content in ponderosa pine the year following exposure, *Can. J. For. Res.*, 21, 1288, 1991.

105. Skärby, L. and Karlsson, P. E., Critical levels for ozone to protect forest trees — best available knowledge from the Nordic countries and the rest of Europe, in *Critical Levels for Ozone in Europe: Testing and Finalising the Concepts*, Kärenlampi, L. and Skärby, L.,Eds., Kuopio University, Kuopio, 1996, 36.
106. Temple, P. J. and Miller, P. R., Foliar ozone injury and radial growth of ponderosa pine, *Can. J. For. Res.*, 24, 1877, 1994.
107. Pfirrmann, T., Barnes, J. D., Steiner, K., Schramel, P., Busch, U., Kuchenhoff, H., and Payer, H. D., Effects of elevated CO_2, O_3 and K deficiency on Norway spruce (*Picea abies*): nutrient supply, content and leaching, *New Phytol.*, 134, 267, 1996.
108. Chappelka, A. H. and Chevone, B. I., Tree responses to ozone, in *Surface-Level Ozone Exposures and Their Effects on Vegetation*, Lefohn A. S., Ed., Lewis Publishers, Chelsea, 1992, 271.
109. Dobson, M. C., Taylor, G., and Freer-Smith, P. H., The control of ozone uptake by *Picea abies* (L.) Karst. and *P. sitchensis* (Bong.) Carr. during drought and interacting effects on shoot water relations, *New Phytol.*, 116, 465, 1990.
110. Karlsson, P. E., Medin, E. L., Wickstro, H., Sellden, G., Wallin, G., Ottoson, S., and Skärby, L., Ozone and drought stress — interactive effects on the growth and physiology of Norway spruce (*Picea abies* (L.) Karst.), *Water Air Soil Pollut.*, 85, 1325, 1995.
111. Pearson, M. and Mansfield, T. A., Effects of exposure to ozone and water stress on the following season's growth of beech (*Fagus sylvatica*), *New Phytol.*, 126, 511, 1994.
112. Pearson, M. and Mansfield, T. A., Interacting effects of ozone and water stress on the stomatal resistance of beech (*Fagus sylvatica* L.), *New Phytol.*, 123, 351, 1993.
113. Ceulemans, R. and Mousseau, M., Effects of elevated CO_2 on woody plants, *New Phytol.*, 127, 425, 1994.
114. Schwanz, P., Haberle, K. H., and Polle, A., Interactive effects of elevated CO_2, ozone and drought stress on the activities of antioxidative enzymes in needles of Norway spruce trees (*Picea abies* (L.) Karsten) grown with luxurious N-supply, *J. Plant Physiol.*, 148, 351, 1996.
115. Lippert, M., Steiner, K., Pfirrmann, T., and Payer, H. D., Assessing the impact of elevated O_3 and CO_2 on gas exchange characteristics of differently K supplied clonal Norway spruce trees during exposure and the following season, *Trees Struct. Func.*, 11, 306, 1997.
116. Tjoelker, M. G. and Luxmoore, R. J., Soil nitrogen and chronic ozone stress influence physiology, growth and nutrient status of *Pinus taeda* L. and *Liriodendron tulipifera* L. seedlings, *New Phytol.*, 119, 69, 1991.
117. Pell, E. J., Sinn, J. P., and Johansen, C. V., Nitrogen supply as a limiting factor determining the sensitivity of *Populus tremuloides* Michx. to ozone stress, *New Phytol.*, 130, 437, 1995.
118. Karnosky, D. F. and Witter, J. A., Effects of genotype on the response of *Populus tremuloides* Michx. to ozone and nitrogen deposition, *Water Air Soil Pollut.*, 62, 189, 1992.
119. Schafer, S. R., Reinert, R. A., Eason, G., and Spruill, S. E., Analysis of ozone concentration — biomass response relationships among open-pollinated families of loblolly pine, *Can. J. For. Res.*, 23, 706, 1993.
120. Berrang, P., Karnosky, D. F., and Bennett, J. P., Natural selection for ozone tolerance in *Populus tremuloides*: an evaluation of nationwide trends, *Can. J. For. Res.*, 21, 1091, 1991.
121. Davis, D. D. and Skelly, J. M., Growth response of four species of eastern hardwood tree seedlings exposed to ozone, acidic precipitation, and sulfur dioxide, *J. Air Waste Manage. Assoc.*, 42, 309, 1992.
122. Samuelson, L. J., Ozone-exposure responses of black cherry and red maple seedlings, *Environ. Exp. Bot.*, 34, 355, 1994.
123. Boerner, R. E. J. and Rebbeck, J., Decomposition and nitrogen release from leaves of three hardwood species grown under elevated O_3 and/or CO_2, *Plant Soil*, 170, 149, 1995.
124. Lovett, G. M. and Hubbell, J. G., Effects of ozone and acid mist on foliar leaching from eastern white pine and sugar maple, *Can. J. For. Res.*, 21, 794, 1991.
125. Rebbeck, J., Chronic ozone effects on three north-eastern hardwood species: growth and biomass, *Can. J. For. Res.*, 26, 1788, 1996.
126. Rebbeck, J. and Loats, K. V., Ozone effects on seedling sugar maple (*Acer saccharum*) and yellow-poplar (*Liriodendron tulipifera*): gas exchange, *Can. J. For. Res.*, 27, 1595, 1997.
127. Günthardt-Goerg, M. S., Different responses to ozone of tobacco, poplar, birch, and alder, *J. Plant Physiol.*, 148, 207, 1996.

128. Matyssek, R., Günthardt-Goerg, M. S., Maurer, S., and Keller, T., Nighttime exposure reduces whole-plant production in *Betula pendula*, *Tree Physiol.*, 15, 159, 1995.
129. Maurer, S., Matyssek, R., Günthardt-Goerg, M. S., Landolt, W., and Einig, W., Nutrition and the ozone sensitivity of birch (*Betula pendula*). 1. Responses at the leaf level, *Trees Struct. Func.*, 12, 1, 1997.
130. Tuomainen, J., Pellinen, R., Roy, S., Kiiskinen, M., Eloranta, T., Karjalainen, R., and Kangasjärvi, J., Ozone affects birch (*Betula pendula* Roth) phenylpropanoid, polyamine and active oxygen detoxifying pathways at biochemical and gene expression level, *J. Plant Physiol.*, 148, 179, 1996.
131. Ainsworth, N. and Ashmore, M. R., Assessment of ozone effects on beech *(Fagus sylvatica)* by injection of a protectant chemical, *For. Ecol. Manage.*, 51, 129, 1992.
132. Arndt, U., Billen, N., Seufert, G., Ludwig, W., Borkhart, K., and Ohnesorge, B., Visible injury responses, *Environ. Pollut.*, 68, 355, 1990.
133. Braun, S. and Fluckiger, W., Effects of ambient ozone on seedlings of *Fagus sylvatica* L. and *Picea abies* (L.) Karst., *New Phytol.*, 129, 33, 1995.
134. Krause, G. H. M. and Höckel, F.-E., Long-term effects of ozone on *Fagus sylvatica* L. — an open-top chamber exposure study, *Water Air Soil Pollut.*, 85, 1337, 1995.
135. Luwe, M., Antioxidants in the apoplast and symplast of beech (*Fagus sylvatica* L.) leaves: seasonal variations and responses to changing ozone concentrations in air, *Plant Cell Environ.*, 19, 321, 1996.
136. Polle, A. and Morawe, B., Seasonal changes of the antioxidative systems in foliar buds and leaves of field-grown beech trees (*Fagus sylvatica* L.) in a stressful climate, *Bot. Acta*, 108, 314, 1995.
137. Siefermann-Harms, D., Krupa, S. V., and Arndt, U., Chlorophyll, carotenoids and the activity of the xanthophyll cycle, *Environ Pollut.*, 68, 293, 1990.
138. Wollmer, H., Kottke, I., Krupa, S. V. and Arndt, U., Fine root studies *in situ* and in the laboratory, *Environ. Pollut.*, 68, 383, 1990.
139. Reiner, S., Wiltshire, J. J. J., Wright, C. J., and Colls, J. J., The impact of ozone and drought on the water relations of ash trees (*Fraxinus excelsior* L.), *J. Plant Physiol.*, 148, 166, 1996.
140. Wiltshire, J. J. J., Wright, C. J., Colls, J. J., Craigon, J., and Unsworth, M. H., Some foliar characteristics of ash trees (*Fraxinus excelsior*) exposed to ozone episodes, *New Phytol.*, 134, 623, 1996.
141. Cannon, W. N., Jr., Roberts, B. R., and Barger, J. H., Growth and physiological response of water stressed yellow-poplar seedlings exposed to chronic ozone fumigation and ethylenediurea, *For. Ecol. Manage.*, 61, 61, 1993.
142. Sury, R. von, Flückiger, W., and Von Sury, R., Effects of air pollution and water stress on leaf blight and twig cankers of London planes (*Platanus acerifolia* (Ait.) Willd.) caused by *Apiognomonia veneta* (Sacc. & Speg.) Hohn, *New Phytol.*, 118, 397, 1991.
143. Ainsworth, N., Fumagalli, I., Giorcelli, A., Mignanego, L., Schenone, G., and Vietto, L., Assessment of EDU stem injections as a technique to investigate the response of trees to ambient ozone in field conditions, *Agric. Ecosyst. Environ.*, 59, 33, 1996.
144. Ballach, H.-J., Oppenheimer, S., and Mooi, J., Reactions of cloned poplars to air pollution: premature leaf loss and investigations of the nitrogen metabolism, *Z. Naturforsch.*, 47, 109, 1992.
145. Ballach, H.-J., Mooi, J., and Wittig, R., Premature aging in *Populus nigra* L. after exposure to air pollutants, *Angew. Bot.*, 66, 14, 1992.
146. Ballach, H.-J., Niederee, C., Wittig, R., and Woltering, E. J., Reactions of cloned poplars to air pollution: ozone-induced increase of stress ethylene and possible antisenescence strategies, *Environ. Sci. Pollut. Res.*, 2, 1, 1995.
147. Frost, D. L., Taylor, G., and Davies, W. J., Biophysics of leaf growth of hybrid poplar: impact of ozone, *New Phytol.*, 118, 407, 1991.
148. Taylor, G. and Frost, D. L., Impact of gaseous air pollution on leaf growth of hybrid poplar, *For. Ecol. Manage.*, 51, 151, 1992.
149. Coleman, M. D., Dickson, R. E., Isebrands, J. G., and Karnosky, D. F., Photosynthetic productivity of aspen clones varying in ozone sensitivity, *Agric. Ricerca*, 15, 18, 1993.
150. Karnosky, D. F., Gagnon, Z. E., Reed, D. D., and Witter, J. A., Growth and biomass allocation of symptomatic and asymptomatic *Populus tremuloides* clones in response to seasonal ozone exposures, *Can. J. For. Res.*, 22, 1785, 1992.
151. Chappelka, A., Renfro, J., Somers, G., and Nash, B., Evaluation of ozone injury on foliage of black cherry (*Prunus serotina*) and tall milkweed (*Asclepias exaltata*) in Great Smoky Mountains National Park, *Environ. Pollut.*, 95, 13, 1997.

152. Fredericksen, T. S., Joyce, B. J., Skelly, J. M., Steiner, K. C., Kolb, T. E., Kouterick, K. B., Savage, J. E., and Snyder, K. R., Physiology, morphology, and ozone uptake of leaves of black cherry seedlings, saplings, and canopy trees, *Environ. Pollut.*, 89, 273, 1995.
153. Long, R. P. and Davis, D. D., Black cherry growth response to ambient ozone and EDU, *Environ. Pollut.*, 70, 241, 1991.
154. Skelly, J. M., Savage, J. E., deBauer, M. D., and Alvarado, D., Observations of ozone-induced foliar injury on black cherry (*Prunus serotina*, var. Capuli) within the Desierto de Los Leones National Park, Mexico City, *Environ. Pollut.*, 95, 155, 1997.
155. Farage, P. K., The effect of ozone fumigation over one season on photosynthetic processes of *Quercus robur* seedlings, *New Phytol.*, 134, 279, 1996.
156. Kelting, D. L., Burger, J. A., and Edwards, G. S., The effects of ozone on the root dynamics of seedlings and mature red oak (*Quercus rubra* L.), *For. Ecol. Manage.*, 79, 197, 1995.
157. Samuelson, L. J., The role of microclimate in determining the sensitivity of *Quercus rubra* L. to ozone, *New Phytol.*, 128, 235, 1994.
158. Bender, J., Manderscheid, R., Jager, H. J., Krupa, S. V., and Arndt, U., Analyses of enzyme activities and other metabolic criteria after five years of fumigation, *Environ. Pollut.*, 68, 331, 1990.
159. Schweizer, B., Arndt, U., Krupa, S. V., and Arndt, U., CO_2/H_2O gas exchange parameters of one- and two-year-old needles of spruce and fir, *Environ. Pollut.*, 68, 275, 1990.
160. Wieser, G. and Havranek, W. M., Evaluation of ozone impact on mature spruce and larch in the field, *J. Plant Physiol.*, 148, 189, 1996.
161. Blaschke, H., Krupa, S. V,. and Arndt, U., Mycorrhizal populations and fine root development on Norway spruce exposed to controlled doses of gaseous pollutants and simulated acidic rain treatments, *Environ. Pollut.*, 68, 409, 1990.
162. Blaschke, H., Weiss, M., Blank, L. W., and Lutz, C., Impact of ozone, acid mist and soil characteristics on growth and development of fine roots and ectomycorrhiza of young clonal Norway spruce, *Environ. Pollut.*, 64, 255, 1990.
163. Dohmen, G. P., Koppers, A., Langebartels, C., Blank, L. W., and Lutz, C., Biochemical response of Norway spruce (*Picea abies* (L.) Karst.) towards 14-month exposure to ozone and acid mist: effects of amino acid, glutathione and polyamine titers, *Environ. Pollut.*, 64, 375, 1990.
164. Ebel, B., Rosenkranz, J., Schiffgens, A., Blank, L. W., and Lutz, C., Cytological observations on spruce needles after prolonged treatment with ozone and acid mist, *Environ. Pollut.*, 64, 323, 1990.
165. Ekeberg, D., Jablonska, A. M., and Ogner, G., Phytol as a possible indicator of ozone stress by *Picea abies*, *Environ. Pollut.*, 89, 55, 1995.
166. Fuhrer, G., Dunkl, M., Knoppik, D., Selinger, H., Blank, L. W., Payer, H. D., Lange, O. L., Blank, L. W., and Lutz, C., Effects of low-level long-term ozone fumigation and acid mist on photosynthesis and stomata of clonal Norway spruce (*Picea abies* (L.) Karst.), *Environ. Pollut.*, 64, 279, 1990.
167. Hampp, R., Einig, W., Egger, B., Krupa, S. V., and Arndt, U., Energy and redox status, and carbon allocation in one- to three-year-old spruce needles, *Environ. Pollut.*, 68, 305, 1990.
168. Heller, W., Rosemann, D., Osswald, W. F., Benz, B., Schonwitz, R., Lohwasser, K., Kloos, M., Sandermann, H., Jr., Blank, L. W., and Lutz, C., Biochemical response of Norway spruce (*Picea abies* (L.) Karst.) towards 14-month exposure to ozone and acid mist: Part I — Effects on polyphenol and monoterpene metabolism, *Environ. Pollut.*, 64, 353, 1990.
169. Holopainen, J. K., Kainulainen, P., and Oksanen, J., Growth and reproduction of aphids and levels of free amino acids in Scots pine and Norway spruce in an open-air fumigation with ozone, *Global Change Biol.*, 3, 139, 1997.
170. Karlsson, P. E., Medin, E. L., Wallin, G., Sellden, G., and Skärby, L., Effects of ozone and drought stress on the physiology and growth of two clones of Norway spruce (*Picea abies*), *New Phytol.*, 136, 265, 1997.
171. Kronfuss, G., Wieser, G., Havranek, W. M., and Polle, A., Effects of ozone and mild drought stress on total and apoplastic guaiacol peroxidase and lipid peroxidation in current-year needles of young Norway spruce (*Picea abies* L. Karst.), *J. Plant Physiol.*, 148, 203, 1996.
172. Lutz, C., Heinzmann, U., Gulz, P. G., Blank, L. W., and Lutz, C., Surface structures and epicuticular wax composition of spruce needles after long-term treatment with ozone and acid mist, *Environ. Pollut.*, 64, 313, 1990.

173. Lux, D., Leonardi, S., Muller, J., Wiemken, A., and Flückiger, W., Effects of ambient ozone concentrations on contents of non-structural carbohydrates in young *Picea abies* and *Fagus sylvatica*, *New Phytol.*, 137, 399, 1997.
174. Magel, E., Holl, W., Ziegler, H., Blank, L. W., and Lutz, C., Alteration of physiological parameters in needles of cloned spruce trees (*Picea abies* (L.) Karst.) by ozone and acid mist, *Environ. Pollut.*, 64, 337, 1990.
175. Mikkelsen, T. N., Dodell, B., and Lütz, C., Changes in pigment concentration and composition in Norway spruce induced by long-term exposure to low levels of ozone, *Environ. Pollut.*, 87, 197, 1995.
176. Nast, W., Mortensen, L., Fischer, K., and Fitting, I., Effects of air pollutants on the growth and antioxidative system of Norway spruce exposed in open-top chambers, *Environ. Pollut.*, 80, 85, 1993.
177. Payer, H. D., Pfirrmann, T., Kloos, M., Blank, L. W., and Lutz, C., Clone and soil effects on the growth of young Norway spruce during 14 months exposure to ozone plus acid mist, *Environ. Pollut.*, 64, 209, 1990.
178. Pfirrmann, T., Runkel, K. H., Schramel, P., Eisenmann, T., Blank, L. W., and Lutz, C., Mineral and nutrient supply, content and leaching in Norway spruce for 14 months to ozone and acid mist, *Environ. Pollut.*, 64, 229, 1990.
179. Polle, A. and Rennenberg, H., Superoxide dismutase activity in needles of Scots pine and Norway spruce under field and chamber conditions: lack of ozone effects, *New Phytol.*, 117, 335, 1991.
180. Schmitt, R., Sandermann, H., Jr., Blank, L. W., and Lutz, C., Biochemical response of Norway spruce (*Picea abies* (L.) Karst.) towards 14-month exposure to ozone and acid mist: Part II — Effects on protein biosynthesis, *Environ. Pollut.*, 64, 367, 1990.
181. Schneider, P., Horn, K., Lauterbach, R., Hock, B., Blank, L. W., and Lutz, C., Influence of ozone and acid mist on the contents of gibberellic acid (GA3) in spruce needles *(Picea abies* (L.) Karst), *Environ. Pollut.*, 64, 347, 1990.
182. Schneiderbauer, A., Back, E., Sandermann, H., Jr., and Ernst, D., Ozone induction of extensin mRNA in Scots pine, Norway spruce and European beech, *New Phytol.*, 130, 225, 1995.
183. Senser, M., Kloos, M., Lutz, C., Blank, L. W., and Lutz, C., Influence of soil substrate and ozone plus acid mist on the pigment content and composition of needles from young Norway spruce trees, *Environ. Pollut.*, 64, 295, 1990.
184. Shaw, P. J. A. and McLeod, A. R., The effects of SO_2 and O_3 on the foliar nutrition of Scots pine, Norway spruce and Sitka spruce in the liphook open-air fumigation experiment, *Plant Cell Environ.*, 18, 237, 1995.
185. Wedler, M., Weikert, R. M., and Lippert, M., Photosynthetic performance, chloroplast pigments and mineral content of Norway spruce (*Picea abies* (L.) Karst.) exposed to SO_2 and O_3 in an open-air fumigation experiment, *Plant Cell Environ.*, 18, 263, 1995.
186. Edwards, G. S., Sherman, R. E., and Kelly, J. M., Red spruce and loblolly pine nutritional responses to acidic precipitation and ozone, *Environ. Pollut.*, 89, 9, 1995.
187. Laurence, J. A., Amundson, R. G., Kohut, R. J., and Weinstein, D. A., Growth and water use of red spruce (*Picea rubens* Sarg) exposed to ozone and simulated acidic precipitation for four growing seasons, *For. Sci.*, 42, 355, 1997.
188. Vann, D. R., Strimbeck, G. R., and Johnson, A. H., Effects of mist acidity and ambient ozone removal on montane red spruce, *Tree Physiol.*, 15, 639, 1995.
189. Bambridge, L., Harmer, R., and Macleod, R., Root and shoot growth, assimilate partitioning and cell proliferation in roots of Sitka spruce (*Picea sitchensis*) grown in filtered and unfiltered chambers, *Environ. Pollut.*, 92, 343, 1996.
190. Shan, Y., Feng, Z., Izuta, T., Aoki, M., and Totsuka, T., The individual and combined effects of ozone and simulated acid rain on chlorophyll contents, carbon allocation and biomass accumulation of armand pine seedlings, *Water Air Soil Pollut.*, 85, 1399, 1995.
191. Shan, Y. F., Izuta, T., Aoki, M., and Totsuka, T., Effects of O_3 and soil acidification, alone and in combination, on growth, gas exchange rate and chlorophyll content of red pine seedlings, *Water Air Soil Pollut.*, 97, 355, 1997.
192. Paynter, V. A., Reardon, J. C., and Shelburne, V. B., Changing carbohydrate profiles in shortleaf pine (*Pinus echinata*) after prolonged exposure to acid rain and ozone, *Can. J. For. Res.*, 22, 1556, 1992.
193. Byres, D. P., Dean, T. J., and Johnson, J. D., Long-term effects of ozone and simulated acid rain on the foliage dynamics of slash pine (*Pinus elliottii* var. Elliottii Engelm.), *New Phytol.*, 120, 61, 1992.

194. Dean, T. J. and Johnson, J. D., Growth response of young slash pine trees to simulated acid rain and ozone stress, *Can. J. For. Res.*, 22, 839, 1992.
195. Johnson, J. D., Byres, D. P., and Dean, T. J., Diurnal water relations and gas exchange of two slash pine (*Pinus elliottii*) families exposed to chronic ozone levels and acidic rain, *New Phytol.*, 131, 381, 1995.
196. Diaz, G., Barrantes, O., Honrubia, M. and Gracia, C., Effect of ozone and sulphur dioxide on mycorrhizae of *Pinus halepensis* miller, *Ann. Sci. For.*, 53, 849, 1996.
197. Gerant, D., Podor, M., Grieu, P., Afif, D., Cornu, S., Morabito, D., Banvoy, J., Robin, C., and Dizengremel, P., Carbon metabolism enzyme activities and carbon partitioning in *Pinus halepensis* Mil. exposed to mild drought and ozone, *J. Plant Physiol.*, 148, 142, 1996.
198. Wellburn, F. A. M. and Wellburn, A. R., Atmospheric ozone affects carbohydrate allocation and winter hardiness of *Pinus halepensis* (Mill.), *J. Exp. Bot.*, 45, 607, 1994.
199. Beyers, J. L., Riechers, G. H., and Temple, P. J., Effects of long-term ozone exposure and drought on the photosynthetic capacity of ponderosa pine (*Pinus ponderosa* Laws.), *New Phytol.*, 122, 81, 1992.
200. Clark, C. S., Weber, J. A., Lee, E. H., and Hogsett, W. E., Accentuation of gas exchange gradients in flushes of ponderosa pine exposed to ozone, *Tree Physiol.*, 15, 181, 1995.
201. Grulke, N. E. and Lee, E. H., Assessing visible ozone-induced foliar injury in ponderosa pine, *Can. J. For. Res.*, 27, 1658, 1997.
202. Momen, B., Helms, J. A., and Criddle, R. S., Foliar metabolic heat rate of seedlings and mature trees of *Pinus ponderosa* exposed to acid rain and ozone, *Plant Cell Environ.*, 19, 747, 1996.
203. Momen, B. and Helms, J. A., Osmotic adjustment induced by elevated ozone: interactive effects of acid rain and ozone on water relations of field-grown seedlings and mature trees of *Pinus ponderosa*, *Tree Physiol.*, 15, 799, 1996.
204. Scagel, C. F. and Andersen, C. P., Seasonal changes in root and soil respiration of ozone-exposed ponderosa pine (*Pinus ponderosa*) grown in different substrates, *New Phytol.*, 136, 627, 1997.
205. Takemoto, B. K., Bytnerowicz, A., Dawson, P. J., Morrison, C. L., and Temple, P. J., Effects of ozone on *Pinus ponderosa* seedlings: comparison of responses in the first and second growing seasons of exposure, *Can. J. For. Res.*, 27, 23, 1997.
206. Temple, P. J., Riechers, G. H., and Miller, P. R., Foliar injury responses of ponderosa pine seedlings to ozone, wet and dry acidic deposition, and drought, *Environ. Exp. Bot.*, 32, 101, 1992.
207. Temple, P. J., Riechers, G. H., Miller, P. R., and Lennox, R. W., Growth responses of ponderosa pine to long-term exposure to ozone, wet and dry acidic deposition, and drought, *Can. J. For. Res.*, 23, 59, 1993.
208. Antonnen, S. and Kärenlampi, L., Slightly elevated ozone exposure causes cell structural changes in needles and roots of Scots pine, *Trees*, 10, 207, 1996.
209. Meinander, O., Somersalo, S., Holopainen, T., and Strasser, R. J., Scots pines after exposure to elevated ozone and carbon dioxide probed by reflectance spectra and chlorophyll a fluorescence transients, *J. Plant Physiol.*, 148, 229, 1996.
210. Pérez-Soba M., Dueck, T. A., Puppi, G., and Kuiper, P. J. C., Interactions of elevated CO_2, NH_3 and O_3 on mycorrhizal infection, gas exchange and N metabolism in saplings of Scots pine, *Plant Soil*, 176, 107, 1995.
211. Skeffington, R. A. and Sutherland, P. M., The effects of SO_2 and O_3 fumigation on acid deposition and foliar leaching in the Liphook forest fumigation experiment, *Plant Cell Environ.*, 18, 247, 1995.
212. Adams, M. B., Edwards, N. T., Taylor, G. E., and Skaggs, B. L., Whole-plant ^{14}C-photosynthate allocation in *Pinus taeda*: seasonal patterns at ambient and elevated ozone levels, *Can. J. For. Res.*, 20, 152, 1990.
213. Baker, T. R. and Allen, H. L., Ozone effects on nutrient resorption in loblolly pine, *Can. J. For. Res.*, 26, 1634, 1996.
214. Booker, F. L., Antonnen, S., and Heagle, A. S., Catechin, proanthocyanidin and lignin contents of loblolly pine (*Pinus taeda*) needles after chronic exposure to ozone, *New Phytol.*, 132, 483, 1996.
215. Edwards, N. T., Taylor, G. E., Adams, M. B., Simmons, G. L., and Kelly, J. M., Ozone, acidic rain and soil magnesium effects on growth and foliar pigments of *Pinus taeda* L., *Tree Physiol.*, 6, 95, 1990.
216. Horton, S. J., Reinert, R. A., and Heck, W. W., Effects of ozone on three open-pollinated families of *Pinus taeda* L. grown in two substrates, *Environ. Pollut.*, 65, 279, 1990.

217. Manderscheid, R., Jager, H. J., and Kress, L. W., Effects of ozone on foliar nitrogen metabolism of *Pinus taeda* L. and implications for carbohydrate metabolism, *New Phytol.*, 121, 623, 1992.
218. Meier, S., Grand, L. F., Schoeneberger, M. M., Reinert, R. A., and Bruck, R. I., Growth, ectomycorrhizae and nonstructural carbohydrates of loblolly pine seedlings exposed to ozone and soil water deficit, *Environ. Pollut.*, 64, 11, 1990.
219. Reddy, G. B., Reinert, R. A., and Eason, G., Effect of acid rain and ozone on soil and secondary needle nutrients of loblolly pine, *Dev. Plant Soil Sci.*, 45, 24, 1991.
220. Reinert, R. A., Shafer, S. R., Eason, G., Schoeneberger, M. M., and Horton, S. J., Responses of loblolly pine to ozone and simulated acidic rain, *Can. J. For. Res.*, 26, 1715, 1996.
221. Richardson, C. J., Sasek, T. W., and Fendick, E. A., Implications of physiological responses to chronic air pollution for forest decline in the southeastern United States, *Environ. Toxicol. Chem.*, 11, 1105, 1992.
222. Stow, T. K., Allen, H. L., and Kress, L. W., Ozone impacts on seasonal foliage dynamics of young loblolly pine, *For. Sci.*, 38, 102, 1992.

Table 9.1 Literature survey (over the period 1990 to 1997) on the influence of tropospheric ozone on deciduous and coniferous woody plants (species name, age or size of trees at the start of the experiment, system and duration of exposure, interactions with other environmental factors, exposure treatments and parameters that were examined are summarized.)

Species Name[a]	Age at Start of Treatment[b]	Exposure System[c]	Interaction[d]	Exposure Duration[e]	Exposure[f]	Effects Measured[g]	Ref.
Deciduous trees							
Acer rubrum	2 y	CSTR	Acid rain		40, 80 ppb (7 h/d, 5 d/w)	Visible injury, biomass	121
	1 y	OTC	—		sub-A, A, 2-A	Visible injury, height, S/R, A, g	122
	Seedling, sapling, mature	Field	—	—	A	O_3-uptake	40
	1–3 y	OTC	—	3 GS	0.4-A, 0.6-A, 0.95-A, A	Leaf loss, radial growth, height, LA, visible injury, soil nutrients, ph	8
Acer saccharum	Seedling	OTC	—	2 GS	CF, A, 2-A	Decomposition and N release	123
	Mature	Field	Acid mist	5 h	25, 70, 140 ppb	Foliar leaching	124
	1 y	OTC	—	2 GS	CF, A, 0.5-A, 1.5-A, 2-A	Biomass, height, radial growth, LA, leaf production, node number, branch number, bud break, visible injury	125
	Seedling	OTC	—	1–2 GS	Sub-A→1.7-A	A, g, respiration, biomass, S/R	126
	35 y	Open air fum. of branches	Irradiance	3 m	A, 2-A	A, g, WUE, dark respiration, chlorophyll, N	60
Alnus glutinosa		Field/GC	—	1 GS	CF, 75 ppb (12 h/d, 12 h/n, 24 h)	Visible injury, leaf abscission, ultrastructure	127
Betula pendula	3 cm	GH	—	1 GS	CF, 50, 75, 100 ppb (24 h/d)	Visible injury, LA, leaf thickness, leaf production, ultrastructure	66
		Field/field fum.ch.	—	1 GS	CF, 75 ppb (12 h/d, 12 h/n, 24 h)	Visible injury, leaf abscission, ultrastructure	67
	Cuttings	Field fum. Ch.	Nutrients	1 GS	CF, 90/40 ppb (d/n)	Carbohydrate pools, enzyme activities	101
	3 cm	Field fum. Ch.	—	5.5 m	0, 50, 75, 100 ppb	A, A/c_i , g, dark respiration, WUE, N (leaves), stomatal density, aperture, mesophyll ultrastructure	80
	3 cm	Field cum. Ch.	—	4.5 m	0, 50, 75, 100 ppb	LA, radial growth, S/R, visible injury, A, E, ^{13}C, S, P, cations, ultrastructure starch, phloem	73
	Cuttings	Field fum. Ch.	—	20 w	CF, 75 ppb (12 h/d, 12 h/n, 24 h)	Biomass, S/R, g, visible injury, leaf abscission	128
	Cuttings	Field fum. Ch.	Nutrients	1 GS	CF, 90/40 ppb (d/n)	Stomatal density, leaf abscission, visible injury, leaf turnover, WUE, A, A/c_i, Rubisco activity, chlorophyll fluorescence, N	129

			Nutrients	1 GS	CF, 90/40 ppb (d/n)	Leaf turnover, C allocation, biomass, WUE, nutrient status	62
	Sapling	Open air fum.	—	1 GS	A, 1.7-A	LA, biomass, Rubisco, chlorophyll, A, E, WUE, ultrastructure chloroplast	26
	Sapling	Field	—	2 GS	A	Biomass, height, leaf senescence, visible injury, stomatal density, ultrastructure chloroplast	72
	Sapling	Open air fum.	—	2 GS	A, 1.6-A/ 1.7-A	Leaf senescence, visible injury, stomatal density, ultrastructure chloroplast, leaf production, height, LA	30
	Seedling		—	8 h	150 ppb	Visible injury, plasma membrane damage, antioxidative enzymes, mRNA (gene expression)	130
Fagus sylvatica		OTC	EDU	4 m	30/15 ppb (d/n), id. + 80/30 ppb (2 w, d/n)	Radial growth, height, biomass, leaf loss, leaf conductance, A	131
	Seedling	OTC	SO_2, acid prec.	5 y	CF, 25-90 ppb	Visible injury, diseases	132
	Seedling	OTC	Altitude	up to 3 y	CF, A	Biomass, shoot length, S/R, visible injury, leaf abscission, water loss, frost hardiness, carbohydrates	133
	6 y	OTC	—	4 GS	CF, A	LA, radial growth, shoot length, leaf abscission, bud swelling, visible injury	134
	3 y	GC	—	6 m	0.2-A, A, 1.5-A, 2-A	A, A/c_i, g, E, respiration, chlorophyll fluorescence, visible injury	83
	30 y	field	—		A	Antioxidative system: ascorbate and glutathione	135
	3 y	OTC	SO_2, NO_2	4 GS	CF, A, A + 30 ppb (8 h/d)	Visible injury, chlorophyll fluorescence, ultrastructure	70
	3 y	OTC	—	3 GS	CF, A, A + 30 ppb (8 h/d)	Radial growth	17
	3 y	GH	Drought	128 d	CF, fluctuating episodes (6 h/d)	g, water potential	112
		field	—	2-3 GS		Antioxidative system, chlorophyll, proteins	136
	Seedling	OTC	SO_2, acid prec.	5 y	CF, 25-90 ppb	Pigments	137
	9-12 y	OTC	SO_2, acid prec.	5 y	CF, 25-90 ppb	Fine root production and mycorrhizae	138
Fraxinus excelsior	2 y	OTC	Drought	1 m (22 d)	0, 150 ppb (8 h/d)	g, radial growth, ring width, cell number/ring width	139
	Young	OTC	—	3 GS	CF, 150 ppb (8 h/d, 25 d/GS)	LA, leaf production and abscission, stomatal density	140
Liriodendron tulipifera	Seedling	OTC	—	2 GS	CF, A, 2-A	Decomposition and N release	123
	1 y	CSTR	Drought, EDU	12 w	CF, 150 ppb (6 h/d)	A, dark respiration, g, E, ethylene, S/R, height, biomass, LA	141
	2 y	CSTR	Acid rain		40, 80 ppb (7 h/d, 5 d/w)	Visible injury, biomass	121
	1 y	OTC	—	2 GS	CF, A, 0.5-A, 1.5-A, 2-A	Biomass, height, radial growth, LA, leaf production, node number, branch number, bud break, visible injury	125
	Seedling	OTC	—	1-2 GS	sub-A → 1.7-A	A, g, respiration, biomass, S/R	126

Table 9.1 continued

Species Name[a]	Age at Start of Treatment[b]	Exposure System[c]	Interaction[d]	Exposure Duration[e]	Exposure[f]	Effects Measured[g]	Ref.
	1-3 y	OTC	—	3 GS	0.4-A, 0.6-A, 0.95-A, A	Leaf loss, radial growth, height, LA,visible injury, soil: nutrients, pH	8
Platanus acerifolia	2 y	semi-OTC	drought	1 y	CF, A	Leaf blight and twig cankers, shoot growth, bud burst	142
Populus deltoides	1-2 y	GC	SO_2	112 d	CF, 70-80 ppb	Leaf abscission, LA, stem intumescences, visible injury, ethylene	63
P. deltoides × *P. cv. caudina*	Cuttings	GH	—	3 h	CF, 180 ppb	A, g, chlorophyll, cellular antioxidants	49
P. deltoides × *P. maximowiczii*	40-60 cm	GH/field	EDU	10 d/2y	85 ppb (8 h/d)/A	Visible injury, chlorophyll, height, radial growth, biomass, A, g	143
P. deltoides × *nigra*	Cuttings	CSTR	—	up to 4 w (2,4,6,8 exposures)	0-160 ppb	Height, radial growth, leaf production and abscission, LA, biomass, leaf greenness	7
	Cuttings	OTC	—	98, 112 d	0.5-A, A, 2-A (8-12 h/d)	Biomass, leaf production and abscission, radial growth, internode length, height, LA, S/R	86
P. × *euramericana*	40-60 cm	GH/field	EDU	10 d/2y	85 ppb (8 h/d)/A	Visible injury, chlorophyll, height, radial growth, biomass, A, g	143
		Field/GC	—	1 GS	CF, 75 ppb (12 h/d, 12 h/n, 24 h)	Visible injury, leaf abscission, ultrastructure	127
	Cuttings	GC	NO_2, N	12 w	CF, 2-A	Visible injury, biomass, LA, leaf abscission, radial growth, ultrastructure, C-uptake	67
	5 cm	GH	—	143 d	CF, A, 50 ppb, 100 ppb	Biomass, S/R, LA, leaf length, leaf loss, stomatal density, ultrastructure, mesophyll, S, P, N, starch, cations	84
P. maximowiczii × *P. nigra*	6 w	OTC	NO,NO_2,SO_2	6 w	4, 23 ppb	Amino acids, N, proteins, starch, ultrastructure chloroplasts	144
P. nigra	Cuttings	-OTC	NO,NO_2,SO_2	- 2 × 6 w	1-37 ppb	Leaf abscission, A, ethylene, pigments, amino acids, N, starch, EFE, proteins, ultrastructure	145
		-GC		- 18 d			
	6 w	OTC	NO,NO_2,SO_2	6 w	4, 23 ppb	Amino acids, N, proteins, starch, ultrastructure chloroplasts	144
	Cuttings	GC	—	22 d	CF, 36 ppb	Ethylene, chlorophyll, visible injury, leaf abscission	146
	Cuttings	OTC	SO_2	32 d	CF, 52/18 ppb (8/16 h/d)	Carbohydrates, myo-inositol, pigments, visible injury	96
	1-2 y	GC	SO_2	112 d	CF, 70-80 ppb	Leaf abscission, LA, stem intumescences, visible injury, ethylene	63
	Cuttings	OTC	—	1 GS	CF, A, A + 50 ppb	A, A/c_i, E, g, chlorophyll fluorescence, pigments	77
P. nigra × *P. deltoides*	Cuttings	GC	—	- 61 d	- CF, 50 ppb (16 h/d)	LA, leaf abscission, biomass, visible injury, biophysics of cell expansion	147
				- 42 d	-id. + 3 peak episodes (120 ppb)		

	Cuttings	- OTC - GC	—	- 70 d - 60 d	- CF, A - 50 ppb	LA, leaf loss, biophysics of cell expansion	148
P. nigra × *P. maximowiczii*	6 w	OTC	NO,NO_2,SO_2	6 w	4, 23 ppb	Amino acids, N, proteins, starch, ultrastructure chloroplasts	144
Populus tremula	13 months	Field fum. ch.	—	2 × 6 m	CF, A, 50, 100 ppb	Branch and stem diameter, branch length, biomass, leaf loss, LA, A, A/c_i, g, WUE, N, S, P, cations, chlorophyll	91
P. tremuloides	40-70 cm	GH	Climate	6 h	150 ppb	Leaf production, visible injury	120
	8 weeks	GC	—	33 d	CF, peak 100, 200, 100/200 ppb	A, A/c_i, g, LA, leaf production, length internodes, leaf abscission	56
	Cuttings	OTC	—	2 GS	CF, 0.5-A, A, 1.5-A, 2-A	A, leaf abscission, biomass	149
	Cuttings	OTC	—	1 GS	CF, CF + 2-A	A, S/R, C-translocation, C-partitioning	93
	6, 14, 25 leaves	OTC	—	3 m	CF, A, 2-A, 100 ppb (6 h/d, 4 d/w)	A, respiration, g, LA, biomass	78
	Cuttings	- GC - OT	— — — —	- 12 w - 3 × 3 m	- A, A + 150 ppb (8 h/d) - CF, sim. A, 2-sim. A	Root respiration, soil CO_2 efflux, root length, biomass	97
	4, 20 cm	OTC	—	2 GS	CF, A, 80 ppb (6 h/d, 3 d/w)	Height, radial growth, biomass, visible injury	150
	Cuttings	OTC	N	97 d	CF, A, 80 ppb (6 h/d, 3 d/w)	Height, radial growth, biomass, visible injury	118
	Cuttings, seedlings	OTC	—	2 GS	CF, 0.5-, 1-, 1.5-, 2-A	Height, radial growth, biomass, leaf abscission, visible injury, bud break, first flush	11
	2-3 y	OTC	CO_2	2 GS	CF, 100 ppb (6 h/d, 5 d/w), 2-A	LA, biomass, A, A/c_i, g, N, chlorophyll	76
	Seedling	OTC	N	76 d	39 ppb, 73 ppb (8 h/d)	LA, biomass, leaf loss, S/R, N (plant components, soil)	117
Prunus serotina	Seedling	OTC	—	2 GS	CF, A, 2-A	Decomposition and N-release	123
	Seedling - sapling	Field			A	Visible injury	151
	2 y	CSTR	Acid rain		40, 80 ppb (7 h/d, 5 d/w)	Visible injury, biomass	121
	Seedling, 3-7-21 m	Field	—	—	A	A, g, water potential, stomatal size, density, leaf thickness, LA, LMA	152
	3 y/ 3-7-21 m	field	—	—	A	Visible injury, A, g	15
	Seedling	field	EDU	4 y	A	Height, radial growth, biomass	153
	1 y	OTC	—	2 GS	CF, A, 0.5-A, 1.5-A, 2-A	Biomass, height, radial growth, LA, leaf production, node number, branch number, bud break, visible injury	125
	1 y	OTC	—		sub-A, A, 2-A	Visible injury, height, S/R, A, g	122
	Seedling, sapling, mature	Field	—	—	A	O_3 uptake	40
	Mature	Field	—	—	A	Visible injury	154
	1-3 y	OTC	—	3 GS	0.4-A, 0.6-A, 0.95-A, A	Leaf loss, radial growth, height, LA, visible injury, soil: nutrients, pH	8
Quercus robur	Sapling	GH	—	1 GS	20, 80 ppb	Chlorophyll fluorescence, g, A, A/c_i	155
	2 y	CSTR	Acid rain		40, 80 ppb (7 h/d, 5 d/w)	Visible injury, biomass	121

Table 9.1 continued

Species Name[a]	Age at Start of Treatment[b]	Exposure System[c]	Interaction[d]	Exposure Duration[e]	Exposure[f]	Effects Measured[g]	Ref.
	6 months + mature	OTC, field	—	2 m + 7 m	Sub-A, A, 2-A	Height, radial growth, LA, biomass, leaf loss, branch length, visible injury, A, g, respiration, E, below-ground CO_2-efflux	34
	2 y/30 y	OTC	—	2 GS	Sub-A, A, 2-A	LA, LMA, A, g, leaf water potential, stomatal density and size	37
	4 y/32 y	OTC	—	3 GS	Sub-A, A, 2-A	Height, radial growth, biomass, S/R, fine root production and turnover, A, A/c_i, chlorophyll fluorescence, respiration, water potential, g, N	35
	2 y/30 y	OTC	—	3 GS	Sub-A, A, 2-A	Root dynamics	156
	1 y/31 y	OTC	Microclimate	1 GS	Sub-A, A, 2-A	Biomass, N, A, g, respiration	157
	2 y/30 y	OTC	—	1 GS	Sub-A, A, 2-A (24 h/d)	Radial growth, leaf abscission, leaf expansion, height, biomass, A, A/c_i, g, chlorophyll fluorescence, water potential, visible injury	38
	2 y/30 y	OTC	—	1 GS	Sub-A, A, 2-A (24 h/d)	Height, radial growth, LA, biomass, leaf abscission, S/R, nutrients	20
	Seedling, mature	OTC	—	3 GS	Sub-A, A, 2-A	C-partitioning and allocation	94
	Seedling, sapling, mature	Field	—	—	A	O_3 uptake	40
	1-3 y	OTC	—	3 GS	0.4-A, 0.6-A, 0.95-A, A	Leaf loss, radial growth, height, LA, visible injury, soil: nutrients, pH	8
	4 y/32 y	OTC	—	3 GS	CF, A, 2-A	Leaf expansion, LMA, respiration, N	36
Coniferous trees							
Abies alba	Seedling	OTC	SO_2, acid prec.	5 y	CF, 25-90 ppb	Visible injury, diseases	132
	Seedling	OTC	SO_2, acid prec.	5 y	CF, 25-90 ppb	Enzymatic activities, protein and N content	158
	Seedling	OTC	SO_2, acid prec.	5 y	CF, 25-90 ppb	A, E	159
	Seedling	OTC	SO_2, acid prec.	5 y	CF, 25-90 ppb	Pigments	137
	9-12 y	OTC	SO_2, acid prec.	5 y	CF, 25-90 ppb	Fine root production and mycorrhizae	138
Larix decidua		field	—		A (2 altitudes)	g, O_3 uptake, leaf and soil water potential	28
	Mature	Branch ch.	—	1-2 GS		g, O_3 uptake	160
Picea abies	Seedling	OTC	SO_2, acid rain	5 y	CF, 25-90 ppb	Visible injury, diseases	132
	Seedling	OTC	SO_2, acid prec.	5 y	CF, 25-90 ppb	Enzymatic activities, protein and N content	158
	Seedling	OTC	SO_2	5 y	CF, 25-90 ppb	Thiols, ascorbic acid, glutathione-reductase, pigments	51
	Seedling	OTC	SO_2, acid prec.	5 y	CF, 25-90 ppb	Mycorrhizal populations and fine root development	161

3 y	GC	acid mist, soil pH	2 GS	25 ppb, 50 ppb (+ peaks)	Growth and development of fine roots and ectomycorrhizae	162
Seedling	OTC	Altitude	up to 3 y	CF, A	Visible injury, leaf abscission, biomass, shoot length, S/R, frost hardiness, carbohydrates	133
1 y	Plant cuvette	Drought	2h, 3 h	CF, 80 ppb	A, g, E, WUE, water relations, biomass	109
3 y	GC	Acid mist, soil pH	2 GS	25 ppb, 50 ppb + peaks	Amino acids and polyamines	163
3 y	GC	Acid mist, soil pH	2 GS	25 ppb, 50 ppb + peaks	Ultrastructure	164
	OTC	—	1 GS	CF, 100, 200 ppb	Phytol	165
3 y	GC	Acid mist, soil pH	2 GS	25 ppb, 50 ppb + peaks	A	166
13 y	OTC	SO_2	5 y	CF, 25-90 ppb	Energy and redox status, carbohydrate metabolism, C allocation	167
3 y	GC	Acid mist, soil pH	2 GS	25 ppb, 50 ppb + peaks	Enzyme activities, monoterpenes, polyphenoles	168
2 y	Open air fum.	SO_2	3 y	A, 1.3-A	H, radial growth, cone production	31
Seedling	Open air fum.	N	4 y	A, 1.2/1.7-A	Aphids	32
Seedling	Open air fum.	N	4 y	A, 1.2/1.7-A	Growth and reproduction of aphids, free amino acids	169
	OTC	Drought	3 GS	CF, 1.5-A	Needle water potential, biomass, needle conductance, radial growth	110
Cuttings	OTC	Drought	4 m	CF, A, A + 30 ppb (7 h/d)	Biomass, dark respiration, A, water potential	170
Seedling		Drought	17 w	100 ppb	Guaiacol peroxidase activity and lipid peroxidation	171
4 y	GC	CO_2, K	1 GS	20, 80 ppb	A, A/c_i, N, P, chlorophyll + delayed effects	115
3 y	GC	Acid mist, soil pH	2 GS	25 ppb, 50 ppb + peaks	Surface structure and epicuticular wax composition	172
Up to 2 y	OTC	—	up to 1 y	CF, A	Pools of nonstructural carbohydrates	173
3 y	GC	Acid mist, soil pH	2 GS	25, 50 ppb + peaks	Carbohydrates, energy status (adeninenucleotides)	174
18 y	branch ch.	Drought		CF, A	Water potential, E, xylem sap flow rates	23
Graftings	OTC	—	3 × 7 m	CF, A + 30 ppb (8 h/d)	Pigments, N (needles)	175
2 y	OTC	NO_2, SO_2	2 GS	CF, A, A + 30 ppb (8 h/d)	Antioxidative enzymes, ascorbic acid, glutathione, radial growth, bud number, needle length, shoot length	176
3 y	GC	Acid mist, soil pH	2 GS	25, 50 ppb + peaks	Radial growth, bud break, visible injury	177
2 y	Open air fum.	—	3 GS	A, 1.3-A	Carbohydrate metabolism	61
3 y	GC	Acid mist, soil pH	2 GS	25, 50 ppb + peaks	Mineral and nutrient content and leaching	178
5 y	GC	CO_2	1 GS	20, 75 ppb	Nutrient content and leaching	107
Young, mature	GC, field	—	h → 20 w	10-600 ppb, A	SOD activity	179
3 y	GC	Acid mist, soil pH	2 GS	25, 50 ppb + peaks	Protein biosynthesis	180

Table 9.1 continued

Species Name[a]	Age at Start of Treatment[b]	Exposure System[c]	Interaction[d]	Exposure Duration[e]	Exposure[f]	Effects Measured[g]	Ref.
	3 y	GC	Acid mist, soil pH	2 GS	25, 50 ppb + peaks	Gibberellic acid	181
	Seedling, sapling		—	1h→several weeks	20-200 ppb	Induction of extensin mRNA	182
	Seedling	OTC	SO_2, acid prec.	5 y	CF, 25-90 ppb	A, E	159
	3 y	GC	Acid mist, soil pH	2 GS	25, 50 ppb + peaks	Pigment content and composition	183
	5 y	GC	CO_2, drought	1 GS	20, 80 ppb	Activities of antioxidative enzymes and soluble protein contents	114
	2 y	Open air fum.	—	5 y	A, 1.3-A	Foliar nutrition (P, cations, N, S)	184
	Seedling	OTC	SO_2, acid prec.	5 y	CF, 25-90 ppb	Pigments	137
	2 y	Open air fum.	SO_2	3 y	A, 1.3-A	Pigments, A, A/c_i, Al, mineral content	185
	Mature	Branch ch.	—	1-2 GS		G, O_3 uptake	160
	9-12 y	OTC	SO_2, acid prec.	5 y	CF, 25-90 ppb	Fine root production and mycorrhizae	138
	2-4 y	Open air fum.	—	1-3 GS	A, 1.4-A	Visible injury, biomass, h, S/R, N, P, K, Ca, Mg, chlorophyll, fatty acids, phenolic compounds, ultrastructure	68
Picea rubens	10 y	OTC	Acid rain	4 GS	CF, NF, 1.5-A	Soil nutrients, foliar nutrients (N, P, K, Ca, Mg, S)	186
	Saplings	OTC	Acid prec.	3-4 GS	0.5-A →2-A	Biomass, water use	187
	Mature	Branch ch.	Acid mist	1 GS	CF, A	Pigments, shoot length, starch, biomass, needle and twig number	188
Picea sitchensis	Cuttings	OTC	—	1 GS	CF, A	Biomass, root length, nonstructural carbohydrates, cell proliferation in root apices	189
	2 y	Open air fum.	SO_2	3 y	A, 1.3-A	Height, radial growth, cone production	31
	2 y	Open air fum.	—	5 y	A, 1.3-A	Foliar nutrition (P, cations, N, S)	184
Pinus armandi	3.7 cm	GC	Acid rain	97 d	CF, 300 ppb (8 h/d, 6 d/w)	Chlorophyll, C allocation, biomass	190
	3.7 cm	GC	Acid rain	97 d	CF, 300 ppb (8 h/d, 6 d/w)	A, WUE, respiration, E, root respiration, biomass, visible injury, chlorophyll	81
Pinus densiflora	1 y		pH soil	1 GS	CF, 150 ppb (8 h/d, 6d/w)	Biomass, A, E, dark respiration, chlorophyll, WUE	191
Pinus echinata	13 months	OTC	Acid rain	29 m	CF, 0.9-A, 1.7-A, 2.5-A	A, g, E, WUE	79
	Seedling	OTC	Acid rain	2 y	CF, A, 1.7-A, 2.5-A (9-12 h/d)	Carbohydrate contents	192
	6 months	OTC	Acid rain	28 m	CF, A, 1.7-A, 2.5-A (9-12 h/d)	Biomass, LA, leaf length, leaf loss, N, P, K, SO_4 (soil, leaves)	88
Pinus elliottii	Seedling	OTC	Acid rain	28 m	CF, A, 2-A, 3-A	LA, bud number, fascicle survival, leaf litterfall, biomass	193
	Seedling	OTC	Acid rain	28 m	CF, A, 2-A, 3-A	Radial growth, height, mean unit leaf rate	194
	Seedling	OTC	Acid rain	22 m	CF, NF, 2-A, 3-A	Xylem water potential, A, leaf conductance	195
Pinus halepensis	15-18 months	GC	—	2-16 d	150-600 ppb (7-12 h/d)	Ultrastructure, visible injury, fatty acids	59
	Seedling		SO_2	1 y	CF, 50 ppb	Biomass, mycorrhizae	196

	3 y	GH	Drought	3 m	CF, 100 ppb (14 h/d)	C-partitioning and metabolism, Rubisco-activity, NAD malic enzyme activity, chlorophyll, proteins	197
	2 y	Solardomes	—	2 GS	25-120 ppb	Radial growth, height, biomass, ultrastructure, carbon allocation, winter hardiness	198
	2 y	Solardomes	Drought	2 GS	25-120 ppb	Nitrogen cycling and free radical scavenging	52
Pinus jeffreyi		Field	—			Leaf loss, visible injury, A, chlorophyll fluorescence	85
Pinus ponderosa	2 y	OTC	—	1 GS	CF, simulated A (low, elevated)	Carbohydrate storage (in roots) and root growth	104
	2 y	OTC	—	2 GS	CF, simulated A (low, elevated)	Roots: biomass, carbohydrates	98
	3 y	OTC	Drought	3 y	CF, A, 1.5-A	A	199
	2 y	OTC	—	2 GS		g, A, A/c_i	200
	40 y	Field	—	—		Visible injury	201
	Seedling, mature		Acid rain	15 m	A, 2-A	Respiration rates	202
	Seedling	Branch ch.	Acid rain	15 m	CF, A, 2-A	A, g, WUE	24
	19 months	GH	—	70 d	CF, diurnal patterns + peak 100 ppb/200 ppb (1 h)	A, g, O_3 uptake, A/c_i, AI, leaf loss	41
	26 months, mature	Branch cuvettes	Acid rain	14 m	A, 2-A (9-16 h/d)	Water relations	203
	3 y	OTC	Nutrients	2 GS		Root and soil respiration	204
	Seedling	OTC	Acid deposition, drought	2 GS	CF, A, 2-A	Visible injury, chlorophyll fluorescence, A, growth	205
	2 y	OTC	Acid deposition, drought	3 GS	CF, A, 1.5-A	Visible injury	206
	2 y	OTC	Acid deposition, drought	3 y	CF, A, A + 1.5-A	Biomass, radial growth, needle loss	207
	2 y	OTC	Drought	3 GS	CF, A, 1.5-A	Leaf loss, radial growth, visible injury	106
Pinus strobus		Field	—	—	A	Ring width	27
	Mature	Field	Acid mist	5 h	25, 70, 140 ppb	Foliar leaching	124
Pinus sylvestris	Seedling	Open air fum.	—	2 GS	A, 1.2-1.5-A	Ultrastructure of needles and roots	208
	2 y	Open air fum.	SO_2	3 y	A, 1.3-A	Height, radial growth, cone production	31
	Seedling	Open air fum.	N	4 y	A, 1.2/1.7-A	Aphids	32
	Seedling	Open air fum.	N	4 y	A, 1.2/1.7-A	Growth and reproduction of aphids, levels of free amino acids	169
	25-30 y	OTC	CO_2	2 GS	CF, A, 35-80 ppb (12 h/d)	A, A/c_i, AI, g, respiration, N	75
	Sapling	OTC	CO_2	3 m	CF, A, 2-A	Chlorophyll fluorescence, reflectance spectra	209
	2 y	Open air fum.	—	3 GS	A, 1.3-A	Carbohydrate metabolism	61
	4 y	GC	NH_3, CO_2	11 w	CF, 55 ppb (9 h/d), 20 ppb (15 h/d)	LA, biomass, root length and branching, mycorrhizal infection, A, g, E, N, C (needles)	210
	Young, mature	GC, field	—	h $\rightarrow$ 20 w		SOD-activity	179
	Seedling, sapling		—	1h $\rightarrow$ several weeks	20-200 ppb	Induction of extensin mRNA	182
	2 y	Open air fum.	—	5 y	A, 1.3-A	Foliar nutrition (P, cations, N, S)	184
	2 y	Open air fum.	SO_2	4 y	A, 1.3-A	Acid deposition and foliar leaching	211
	Seedling		—	5 d/10 d	300 ppb/75 ppb (8 h/d)	Pigments, lipids, antioxidative enzymes	58

Table 9.1 continued

Species Name[a]	Age at Start of Treatment[b]	Exposure System[c]	Interaction[d]	Exposure Duration[e]	Exposure[f]	Effects Measured[g]	Ref.
	2-4 y	Open air fum.	—	1-3 GS	A, 1.4-A	Visible injury, biomass, height, S/R, N, P, K, Ca, Mg, chlorophyll, fatty acids, phenolic compounds, ultrastructure	68
Pinus taeda	Seedling	OTC	—	1 GS	A, 2-A	C allocation	212
		OTC	—	1 GS	CF, NF, 1.5-A, 2-A (12 h/d)	Visible injury, needle growth, ultrastructure, plasma membrane characteristics	71
		OTC	—	3 y	0.5-A → 3-A	Fascicle abscission, nutrient resorption	213
	3 y	OTC	—	140 d	CF, A,1.5-A, 2-A (12 h/d)	Phenolic compounds	214
	1 y	OTC	Mg, acid rain	6 m	sub-A, A, 2-A	Biomass, height, radial growth, pigments	215
	Seedling	OTC	Acid rain	3 GS	CF, A, 2-A	Soil nutrients, foliar nutrients (N, P, K, Ca, Mg, S)	186
	1 y	OTC	Mg, acid rain	3 y, 3_6 m	CF, A, 2-A	Biochemical composition of plant components	102
	1 y	OTC	—	3 GS	CF, A, 2-A	C translocation and partitioning	57
	Seedling	CSTR	—	8 w	0, 160, 320 ppb (6 h/d, 4 d/w)	Height, radial growth, needle N, visible injury, biomass	216
		OTC	—	2 GS	0.2-A →3-A	Foliar N metabolism, chlorophyll, protein, amino acids, enzyme activities, root radial growth rate	217
	Mature	Field	—	6 y	A	Radial growth	29
	Seedling	CSTR	Drought	6, 12 w	0, 50, 100, 150 ppb (5 h/d, 5 d/w)	Visible injury, biomass, nonstructural carbohydrates, ectomycorrhizae	218
	Seedling	GH	Acid rain	11 w	0, 80, 160, 240, 320 ppb	Nutrient content and leaching	219
	Seedling	GH	Acid rain	11 w	0, 80, 160, 240, 320 ppb	Height, biomass, radial growth	220
	Seedling	OTC	Acid prec.		0.5-A → 3-A	A, needle abscission	221
	3-4 months	CSTR, GH	—	12 w	0, 80, 160, 240, 320 ppb (6 h/d, 4 d/w)	Biomass	119
	6 months	CSTR	—	12 w	CF, 120 ppb (7 h/d, 5 d/w)	A, C allocation, visible injury, biomass, number + length of branches, needles, needle area	4
			Acid rain		CF, A, 1.3-A, 1.7-A, 2-A, 3-A	Leaf production and abscission	222
	1 y	OTC	N	1 GS	sub-A, A, A + 60 ppb (24 h)	Height, radial growth, biomass, WUE, A, g, nutrients	116
Pseudotsuga menziesii	3 y	GC	Drought	9 d	0, 50, 250 ppb	A, root-soil respiration, biomass partitioning and translocation, starch	100
	3, 25 y	GC, branch bags	—	9 d	0, 100, 200 ppb (8 h/d)	Biomass, visible injury, C partitioning and translocation, starch, root/soil respiration, C uptake	95

1. Species name: Pop. = *Populus*. 2. Age at start of treatment: m = meter. 3. Exposure system: branch ch. = branch chambers; CSTR = continuously stirred tank reactor; field fum. ch. = field fumigation chamber; GC = growth chamber; GH = greenhouse; OTC = open top chamber; open air fum. = open air fumigation. 4. Interaction: acid prec. = acid precipitation; EDU = ethylenediurea. 5. Exposure duration: GS = growing season; w = weeks. 6. Exposure concentration: A = ambient concentration; CF = charcoal filtered air. 7. Effects measured: A = net CO_2 assimilation rate; E = evaporation rate; g = stomatal conductance; LA = leaf area; LMA = leaf mass per area; S/R = shoot/root ratio; WUE = water use efficiency.

CHAPTER 10

Extracellular Antioxidants: A Protective Screen Against Ozone?

Tom Lyons, Matthias Plöchl, Enikö Turcsányi, and Jeremy Barnes

CONTENTS

10.1 INTRODUCTION

In many parts of the industrialized world, tropospheric concentrations of ozone (O_3) are recognized to be high enough to cause significant crop loss,[1-3] and to drive subtle changes in the composition and diversity of natural ecosystems.[4-7] However, it has taken three decades of research to begin to unravel the principal mechanisms governing the phytotoxicity of this ubiquitous pollutant.

Although not a radical species itself, O_3 is a strong oxidant (redox potential = +2.07 V). Following uptake into the leaf interior, the gas is believed to react with constituents of the extraprotoplasmic matrix (the apoplast) to yield additional reactive oxygen species (ROS), and subsequently, if the pollutant (and/or its reactive products) escapes interception, with components of the

1-56670-341-7/00/$0.00+$.50

plasma membrane and cytosol.[8-11] Several contemporary reviews are available that discuss the importance of intracellular antioxidant systems in combating the oxidative stress induced by O_3 once the extracellular defenses have been breached, as well as the way in which plants may sense and respond (i.e., acclimate) to the oxidative stress induced by O_3 — phenomena that share similarities with other plant pathologies.[12-19] In this chapter, we focus on building a better understanding of the possibly fundamental role played by constituents of the apoplast in scavenging O_3, and/or its reactive products, at the first site of their interaction with plant tissue — the apoplast. The role of extracellular reactions in screening the plasma membrane from O_3-induced oxidation is discussed within the context of the various physical, chemical and biological impedances that affect the diffusion of O_3 to its initial site of action — the plasma membrane (Figure 10.1).

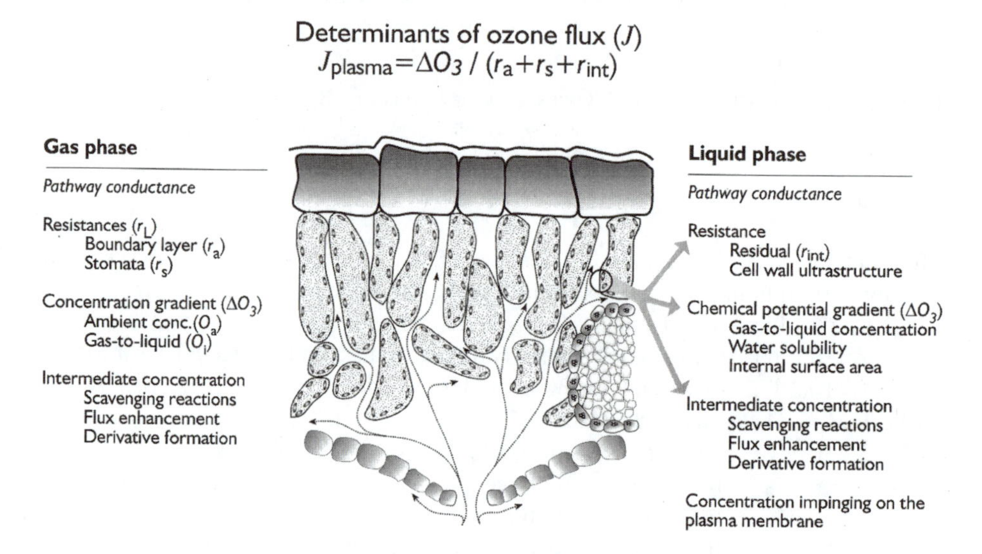

Figure 10.1 The model employed is based on a diffusion–reaction network as envisaged originally by Gaastra,[20] founded on the principle that the flux of O_3 impinging on the plasma membrane (J_{plasma}) is driven by the difference in O_3 concentrations between the bulk air (O_a) and leaf interior (O_i), i.e., ΔO_3, and is inversely proportional to the sum of the resistances to gas flow encountered as O_3 diffuses to its main target in the gas phase (r_a and r_s) and subsequently in the liquid-phase (r_{int}). The approach adopted is basically similar to other treatments of pollutant flux,[21-24] with the exception that much greater emphasis is placed on the residual resistance to O_3 flux and the factors that contribute to it. (Modified from Tingey and Taylor.[21])

10.2 FACTORS DETERMINING OZONE FLUX TO THE PLASMA MEMBRANE

10.2.1 Gas-Phase Conductance

10.2.1.1 Diffusion of Ozone Through Stomata and Intercellular Spaces

The extent of O_3 uptake is determined by several factors that operate at different scales of resolution.[5,24] However, stomatal conductance is by far the single most important variable governing the uptake of the pollutant at the leaf level.[25] It is no surprise to find, therefore, that intra- and interspecific variations in the impacts of O_3 often correlate with intrinsic differences in stomatal conductance; plants with higher rates of stomatal uptake exhibiting greater effects.[26-29] Moreover, since O_3 exposure generally results in a decline in stomatal aperture, plants that display the most rapid stomatal closure in response to O_3 are often reported to be the most "resistant" in population-level studies.[28] However, stomatal closure is not a universal response to O_3 exposure,[30] and Reiling and Davison's[31] detailed investigations on *Plantago major* L. suggest that patterns of stomatal response may be considerably more complex than is often credited. Indeed, O_3-induced declines in stomatal aperture may be of limited protective value,[32-34] since stomatal closure is commonly a downstream consequence of damage to the photosynthetic apparatus.[35-37]

Factors such as the volume of leaf intercellular air spaces and exposed mesophyll cell surface area may play an important role in determining the variable O_3 responses of different taxa,[21] as well as that of leaves at contrasting stages of development.[38] There is strong evidence linking O_3-sensitivity with differences in leaf intercellular air space volume.[39-43] Furthermore, thinner leaves are generally found to be more sensitive to O_3 than thicker leaves[41,42,44] — presumably because of differences in the length of the gas-phase diffusion pathway.[45] Recent experiments on *Betula* spp. also suggest that leaves exhibiting higher stomatal density may be more resistant to O_3.[41,42,46] The reason for this is not entirely clear. However, it has been suggested that this trait could lead to a more even distribution of O_3 uptake across the leaf, possibly facilitating greater interception and detoxification of the pollutant.[47]

10.2.1.2 Reaction with Biogenic Hydrocarbons

Gas-phase reactions not only affect the O_3 concentration gradient between the bulk air and leaf interior (i.e., the driving force for O_3 uptake into the leaf interior; ΔO_3), but may also lead to the production of potentially damaging ROS. Since the mid-1970s, much evidence has been provided to link gaseous emissions of the plant hormone ethylene,[15,48-54] and possibly other biogenic non-methane hydrocarbons (e.g., isoprene, propene, etc.), [55-57] with the degree of visible foliar injury induced by exposure to phytotoxic O_3 concentrations. The implication is that these emissions are in some way directly involved in the development of O_3 injury (or, to be more specific, O_3-dependent lesion development) — a view supported by research on a range of O_3-sensitive/resistant plant-pairings in which ethylene emission was demonstrated to be the only factor consistently associated with O_3 sensitivity.[58]

Until recently, the only major hypothesis proposed to explain the possible role of ethylene in O_3- dependent lesion formation was the formation of ROS from the chemical reaction between O_3 and biogenic hydrocarbons[49,55,56,59]; the existence of O_3–hydrocarbon reactions having been demonstrated *in vivo* through the detection of hydroxy hydroperoxides in isoprene-emitting plants under controlled levels of O_3.[59,60] However, the significance of such reactions has been questioned, first because the rates of the reactions involved are so slow[61] and, second, because experiments in which plants have been exposed to various combinations of O_3 and isoprene/propene have revealed little evidence of synergistic effects.[62] Alternative suggestions that the damage initiated by these compounds may result from the inhibition of peroxidase activity,[56] allowing the build up of hydrogen peroxide (H_2O_2) and subsequent accumulation of hydroxyl radicals ($OH^\cdot$), has also been challenged.[63] On the basis of Polle and Junkermann's [64] findings, unrealistically high concentrations (> 200 μM) of hydroxymethyl hydroperoxide would be required to inhibit peroxidase activity to the extent required to promote damage. Alternative explanations have thus been sought for the link between ethylene emissions and O_3 damage, recent experiments performed on *Lycopersicon esculentum* Mill. suggesting a direct role for O_3-driven ethylene emissions (via the "normal" biosynthetic pathway, i.e., 1-aminocyclopropane-1-carboxylic acid), in the induction of a hypersensitive-style response[65] and in the spread of cell death.[66] Findings consistent with Bae and colleagues[53] experiments using 2,5-norbornadiene to inhibit ethylene action,[53] where a strong reduction in the extent of visible O_3 injury was reported, without effects on O_3-induced ethylene emissions. Other factors have also been suggested to play a part, such as the production of cyanide during the biosynthesis of ethylene[67] and the close coupling between ethylene and polyamine biosynthesis (Section 10.2.2.1.4).

10.2.2 Liquid-Phase Conductance

The flux of O_3 to the leaf interior ($J_{internal}$) is, in practice, most often calculated by simple corollary with the pathlength and combined gas-phase resistances (aerodynamic resistance, r_a, and stomatal resistance, r_s) to the diffusion of effluxing water molecules.[5,24,68] It is important, however, to recognize that there is no *a priori* relationship between the gaseous flux of O_3 to the leaf interior and that impinging on the plasma membrane (J_{plasma}), since the various physical and chemical processes that constitute r_{int} (the internal resistance to O_3 flux) play a decisive role in determining J_{plasma}. Cell wall thickness, for instance, strongly influences the length of the diffusion pathway for O_3 and modifies the likelihood of interaction with antioxidative constituents in the apoplast (Section 10.2.2.1). On the other hand, the density of the cell wall (i.e., degree of cross-linking, suberization, or lignification) would be expected to influence the tortuosity of the diffusion pathway for O_3.[69] Such effects may contribute to the apparent O_3-suseptibility of ferns, which tend to have less suberized mesophyll cell walls,[70] and also to the contrasting effects of O_3 on different cell types within the same leaf.[40] It is also worthy of note that exposure to environmentally-relevant concentrations of O_3 may induce "acclimatory" changes in leaf anatomy, e.g., increases in mesophyll cell wall density/thickness.[71,72] This proposition is supported by the finding that the pollutant has been shown to induce the activity and expression of the monolignol biosynthesis enzyme, cinnamyl alcohol dehydrogenase (EC 1.1.1.195),[73-75] and to stimulate the incorporation of secondary compounds into cell wall components.[76]

10.2.2.1 Interactions with Constituents of the Apoplast

Once inside the leaf, the majority of the O_3 is usually assumed to be absorbed into the aqueous matrix which overlies the surface of the cells lining the substomatal cavity. Indeed, the hydrated leaf cell wall forms a near-perfect sink for O_3, and it is generally accepted that the intercellular space O_3 concentration is close to zero.[77] There are, however, two common misconceptions.

Ozone reacts readily with the water contained in the apoplast under physiological conditions to yield significant quantities of ROS. Ozone is not particularly soluble in water. At 20°C the gas is 33% and 0.7% as soluble as carbon dioxide and sulfur dioxide, respectively.[78] Moreover, once the pollutant has entered into solution the decomposition of O_3 is slow as the apoplast is weakly acidic[79] — Heath[9] calculated that the rate of loss of O_3 in water would be expected to decrease from 0.21% min^{-1} at pH 9.0 to 0.015% min^{-1} at pH 7.0. Although there are several reports of the production of ROS (including $OH^{\bullet}$, superoxide [$O_2^{\bullet -}$], and peroxyl radicals) through the reaction of O_3 with water,[80-82] questions remain whether rates of production are significant in a biological context.[10,12,56]

In contrast, O_3 is known to react readily with numerous solutes in the apoplastic fluid,[83] several of which are known to have the potential to react rapidly with O_3 at rates many orders of magnitude greater than that with water (Table 10.1). Indeed, O_3 uptake into the aqueous matrix of the cell wall is probably mediated by reactive absorption.[84,88] Since many of the products of the interaction between O_3 and constituents of the apoplast are ostensibly harmless, it is feasible that the sacrificial oxidation of certain compounds may serve as an important first line of defense against O_3.[79,89]

The plasma membrane is the initial target for O_3. There is little doubt that the majority of the damage resulting from O_3 stems from the oxidation of plasma membrane components.[9-11,63] However, there is growing recognition that it is constituents of the apoplast that form the initial target for O_3.[9,63,68,90] Table 10.1 shows *in vitro* rate constants for the reaction of O_3 with a number of biological molecules in aqueous solution; target hierarchies dependent not only on reaction rates but also on concentration.[91,92] It is important, however, to recognize that considerable caution is required when attempting to extrapolate from biochemical data to the real world. This situation is most clearly illustrated by Jakob and Heber's[93] recent findings that the extent of protection against

O_3 afforded by apoplastic ascorbate (ASC) observed during *in vitro* experiments may not be reciprocated *in vivo*.

Table 10.1 *In Vitro* Rate Constants for the Reaction of O_3 with Biomolecules in Aqueous Solution

Biomolecule		Rate constant ($M^{-1}s^{-1}$)	Ref.
Ascorbate,	pH 6.0–7.0	4.8×10^7	84
Glutathione,	pH 7.0	2.5×10^6	84
α-Tocopherol,	pH 7.0	1.0×10^6	85
Ureate,	pH 1.9–6.1	1.4×10^6	86
Cysteine,	pH 7.0	4.4×10^6	84
Methionine,	pH 7.0	4.0×10^6	84
Tryptophan,	pH 7.9	7.0×10^6	87
Methyl oleate,	pH 2.0–6.8	8.7×10^5	86
Methyl linoleate,	pH 4.2–5.7	1.1×10^6	86

10.2.2.1.1. Ascorbate — The most abundant solute in the leaf apoplast is ASC (concentration ranging from 10 to 4000 μM),[68,71,90,94-105] a compound recognized to serve essential antioxidant and metabolic functions in both plants and animals.[106-112] Among its many roles, there is strong evidence that ASC is involved in mediating O_3-tolerance. Recent work on a semi-dominant monogenic mutant of *Arabidopsis thaliana* L. (*soz1*; since renamed *vtc1* by Conklin and co-workers[113]) has categorically shown that at least one of the genes involved in determining O_3-resistance is connected with the biosynthesis of ASC,[114] a finding substantiated by historical evidence of a positive relationship between leaf ASC content and O_3 resistance.[115-119]

Although the pool of ASC located in the apoplast represents only a small fraction (~1%) of that in the bulk leaf, there is evidence to suggest that the size and rate of turnover of this pool may be sufficient to afford a significant degree of protection against O_3 and/or its toxic reaction products.[68,90,98,103] Ascorbate is readily oxidized by O_3, and several other ROS (including $OH^{\cdot}$, $O_2^{\cdot-}$, H_2O_2, and singlet oxygen [$^1O_2$]), to yield dehydroascorbic acid (DHA) (Figure 10.2). This compound must be reduced to regenerate ASC, or is rapidly and irreversibly hydrolyzed to yield 2,3-diketogulonic acid (DKG) and subsequently an array of degradation products, including oxalate.[106,107,111,120-123] Since DHA cannot be reduced efficiently in the apoplast, it is believed to be returned to the cytosol for regeneration.[90,94,97] This view is supported by the presence of a carrier-mediated system on the plasma membrane for the transport of ASC/DHA,[124-127] which shows a higher affinity for DHA than for ASC[127] (Figure 10.3). A plasma membrane–bound monodehydroascorbate (MDHA) reductase (EC 1.1.5.4) also exists, which may facilitate the fast regeneration of ASC from MDHA in the apoplast.[128-131] However, there is disagreement over the importance of this pathway in relation to the regeneration of ASC following its reaction with O_3 in the apoplast, some researchers consider that the O_3–ASC reaction does not yield significant amounts of MDHA given the weakly acidic environment of the cell wall.[85,132]

Evidence that extracellular ASC may play an important role in screening sensitive intracellular targets from the oxidizing action of O_3 (and/or its reaction products) arises from a variety of sources. *Ex vivo* experiments on isolated biological fluids (e.g., apoplastic washing fluid, AWF,[133] respiratory tract lining fluid, RTLF,[89,134] and blood plasma,[88,91]) indicate that ASC consumption is dependent upon the concentration and the length of exposure to O_3.[134] Using blood plasma as a surrogate for RTLF, medical researchers have demonstrated that ASC is oxidized faster than lipids and proteins; low-level oxidation of proteins and lipids restricted to cases where there is complete depletion of soluble antioxidants.[88,91] *In vitro* studies, using model solutions of ASC, ureate, glutathione, and albumin have revealed similar findings, consumption of antioxidants generally following the reactive hierarchy ureate > ascorbate > glutathione, with no evidence of oxidative modification of the

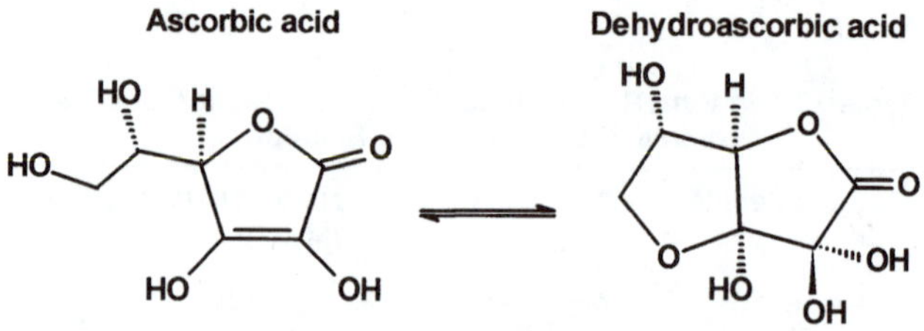

Figure 10.2 L-Ascorbic acid and its oxidized form dehydro-L-ascorbic acid. The reversible oxidation and the acidic properties (pK_a 4.2) of ascorbic acid are associated with the enediol structure at C-2 and C-3. At physiological pH, ascorbic acid exists mostly as the anion, ascorbate. Dehydroascorbic acid, which exists as a bicyclic monomeric hydrate in aqueous solution, does not have acidic characteristics, despite its trivial name. If not reduced back to ascorbate, dehydroascorbic acid irreversibly hydrolyses to form 2,3-diketo-L-gulonic acid, which can break down further to yield ~50 compounds.

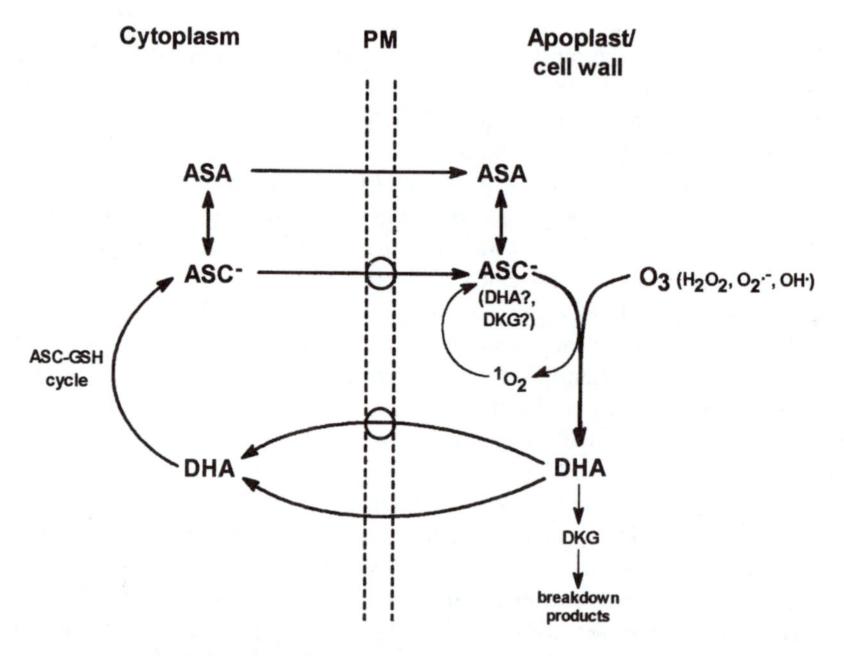

Figure 10.3 Proposed scheme by which O_3 (and its reactive derivatives: H_2O_2, $O_2^{\cdot-}$, OH·) may be intercepted in the apoplast by ASC. Ascorbate movement into the apoplast is controlled by a transporter in the plasma membrane (PM) and/or diffusion through the membrane as the neutral species, ascorbic acid (ASA). Following oxidation (producing $^1O_2$, which may react with additional ASC),[84] the resulting DHA may return (via transport and/or diffusion) to the cytoplasm for regeneration via the ASC–GSH cycle, or undergo hydrolysis to yield DKG and an array of subsequent breakdown products — it is possible that DHA and DKG may also function as antioxidants.[123]

protein.[92] Analogous studies undertaken on AWF isolated from leaves of *Sedum album* L., using $^1O_2$ generation to determine the reaction of O_3 with apoplastic constituents, provides evidence of a direct reaction between O_3 and apoplastic ASC.[133] Consistent with this assertion, a reversible and dose-dependent decline in the ASC content of the leaf apoplast has been found *in vivo* following controlled exposure to environmentally relevant O_3 concentrations.[68,90,97,98] Figure 10.4 illustrates the consumption of ASC in the leaf apoplast of *Vicia faba* L. plants during an 8-h exposure to 150 nmol mol^{-1} O_3 in the absence of significant changes in the intracellular pool of ASC (intracellular measurements made on leaf tissue remaining after the isolation of AWF) — data consistent with the oxidation of apoplastic ASC in the absence of intracellular oxidative stress[103] and lending support to other findings which suggest that the pool of ASC located in leaf cell walls may serve as an inducible forward-defensive screen against oxidative stress.[13,68]

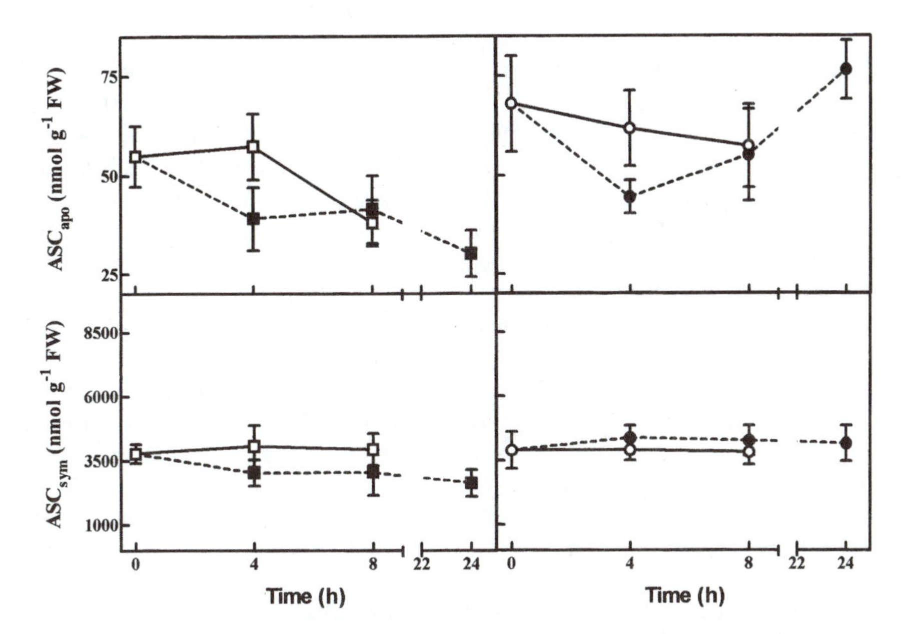

Figure 10.4 Impacts of O_3 on the ASC content of apoplastic washing fluid (ASC_{apo}) and residual leaf extracts (ASC_{sym}) prepared from the youngest fully expanded leaf borne on 28-d-old plants of *Vicia faba* L. Symbols: squares represent plants grown in charcoal/Purafil® filtered air (CFA); circles represent plants grown in CFA plus 75 nmol mol^{-1} O_3 for 7 h d^{-1}; filled symbols represent plants exposed to CFA plus 150 nmol mol^{-1} O_3 for 8 h and subsequently left to recover in CFA; open symbols represent control plants exposed to CFA. Data represent the mean (±SE) of measurements made on between six and eight independent plants exposed in duplicate chambers. (Redrawn from Turcsányi et al.[103])

10.2.2.1.2 Peroxidases — Peroxidases catalyse the H_2O_2-dependent oxidation of a range of substrates (SH_2):

$$SH_2 + H_2O_2 \rightarrow S + 2H_2O$$

The apoplast is known to contain both ascorbate-specific (EC 1.11.1.11) and nonspecific (EC 1.11.1.7) peroxidases, the activity of which is generally found to be stimulated by environmentally relevant O_3 concentrations,[90,100,101,135-139] although Kronfuss and colleagues report decreased activity in AWF isolated from needles of *Picea abies* [L.] Karst. following O_3 exposure.[140] The importance of peroxidases in scavenging O_3 reaction products before they interact with plasma membrane constituents is unclear. Primary detoxification potential would only be realized if O_3 uptake leads to the production of H_2O_2 in the apoplast. Additional functions of peroxidases in cell walls include the polymerization of lignin precursors[141] and the cross-linking of phenolics with proteins and polysaccharides;[142] processes leading to stiffening and increases in the density of the cell wall. Since apoplastic peroxidase activity is generally increased by O_3,[101,135-138] prior exposure to the pollutant might be expected to modify the diffusion pathway for O_3 in a manner consistent with the observed "acclimatory" changes in leaf anatomy (Section 10.2.2). However, further research is required to explore the link between O_3-induced changes in peroxidase activity, biophysical properties of cell walls, and O_3 penetration.

10.2.2.1.3. Superoxide Dismutases — Superoxide dismutases (SODs; EC 1.15.1.1) catalyze the dismutation $O_2^{\cdot -}$ to H_2O_2:

$$O_2^{\cdot -} + O_2^{\cdot -} + 2\ H^+ \rightarrow H_2O_2 + O_2$$

Apoplastic washing fluid isolated from foliage of *Avena sativa* L., *Hordeum vulgare* L., *Nicotiana tabacum* L., *Picea abies*, *Pinus sylvestris* L., and *Plantago major* has been shown to display SOD-like activity,[101,104,105,137,139,143] cyanide inhibition suggesting that the observed activity is due to the presence of Cu/ZnSOD isoforms.[101,143] Indeed, "cytosolic" Cu/ZnSOD has been immunolocated in the leaf cell walls of *Spinacia oleracea* L.[144] There are, however, conflicting reports of the effects of O_3 on apoplastic SOD activity. On the one hand, Castillo and co-workers[137] found that apoplastic SOD activity was stimulated by O_3 in *Picea abies* and suggested a forward-defensive role for the enzyme against the pollutant, while on the other, Lyons and colleagues[101] report a decline in Cu/ZnSOD activity in *Plantago major* in response to controlled exposure to 70 nmol mol^{-1} O_3 for 7 h d^{-1}. Clearly, further work is required, as there can be no doubt that SOD is a highly efficient scavenger of $O_2^{\cdot -}$[145] and could play an important role in scavenging the $O_2^{\cdot -}$ generated by the breakdown of O_3 in the apoplast under environmentally relevant conditions.[10] It has recently been proposed that apoplastic SOD may play an essential role in the lignification of the cell wall by supplying H_2O_2 to peroxidase and preventing peroxidase inactivation by $O_2^{\cdot -}$.[146] Thus, impacts of the pollutant on apoplastic SOD activity may affect lignification processes and, consequently, J_{plasma} (Section 10.2.2.).

10.2.2.1.4 Polyamines and Phenolics — Polyamines are essential for normal cellular function,[147,148] and conjugated forms (via an amide bond to hydroxycinnamic acids) are known to be associated with plant cell walls.[50] In some species, the conjugated polyamine pool can exceed that of the free polyamines.[149] However, the function of these conjugated forms is poorly understood. There is evidence that certain conjugates (e.g., caffeoly-putrescine) may play a direct role in the scavenging of ROS,[150] as well as acting as substrates for peroxidases.[149] Indeed, there are data which support a role for these hydroxycinnamic acid amides in protection against O_3. For example, apoplastic concentrations of caffeoly-putrescine have been shown to increase markedly in response to O_3-exposure in *Nicotiana tabacum* Bel-B, an O_3-tolerant cultivar.[50] Moreover, experimental inhibition of arginine decarboxylase (EC 4.1.1.19), a key enzyme in plant polyamine biosynthesis[151] which is induced by O_3 exposure,[50,152] has been demonstrated to increase the extent of visible O_3 injury.[153] In addition, catabolic enzymes for polyamines are present in the apoplast.[154-156] Interestingly, the activity of diamine oxidase (EC 1.4.3.6) has been shown[138] to decrease in response to O_3 exposure in AWF isolated from leaves of *Phaseolus vulgaris* L., an effect that may lead to increased apoplastic polyamine concentrations. It should also be recognized that the biosynthesis of polyamines is linked with ethylene formation through the shared intermediate, *S*-adenosyl methionine.[157] Consequently, it is possible that the relationship between polyamine levels and O_3-resistance may be associated, at least in part, with the repression of "stress" ethylene emissions.[15,67]

AWF isolated from leaves of *Petroselinum crispum* L. has also been shown to contain apiin, a flavone glycoside, as well as furanocoumarins (such as xanthotoxin and begapten), compounds that accumulate in response to O_3-exposure.[75,96] Although there is a dearth of available information, it seems feasible that extracellular polyamine conjugates, and as yet unidentified phenolic compounds, may play a significant role in scavenging ROS in the region of the cell wall.

10.2.2.1.5 Glutathione — The tripeptide glutathione (GSH; γ-glutamyl-cysteinylglycine) is the major storage and transport form of reduced sulfur in plants.[158] It is also involved in detoxification of xenobiotics mediated by glutathione *S*-transferases (EC 2.5.1.18),[159] the destruction of H_2O_2 via the ASC–GSH cycle[160] and possibly GSH peroxidases (EC 1.11.1.0),[161] as well as in the direct scavenging of ROS.[84,86,88,91,92] Through its antioxidant action, GSH is reversibly converted to the dimer GSSG, which can be recycled by GSH reductase (EC 1.6.4.2) at the expense of NADPH.[160] Indirect evidence that the leaf apoplast contains GSH has been provided by the presence of a specific GSH/GSSG transporter at the plasma membrane of *Vicia faba* protoplasts[162] and the existence of GSH-handling enzymes in the apoplast of *Glycine max* L. hypocotyls.[163] Although the rate constant for the reaction between O_3 and GSH is high (Table 10.1), research has revealed that AWF contains only small quantities (3 to 6.5 μM) of GSH[94,104,105] and many authors suggest that levels are below the limits of detection.[97,101,103] Together with the available medical literature,[88,91,92] this suggests that the apoplastic GSH pool may only play a minor role in the detoxification of O_3. Further work is, however, required to verify this conclusion.

10.3 SIMULATING OZONE DETOXIFICATION IN THE APOPLAST

Provisional estimates of the consumption of O_3 through its reaction with apoplastic constituents have focused, because of the complexities and the unknown, on modeling the extent of *potential* O_3 scavenging achieved through the direct reaction of the pollutant with apoplastic ASC.[61,132] The most-advanced model available at the present time (*SODA*, *S*imulated *O*zone *D*etoxification in the *A*poplast) has been developed during recent collaboration between the authors.[103,164,165] This model attempts to simulate the various physical and chemical factors that influence the uptake and diffusion of O_3 to its initial site of action — the plasma membrane — in both the gas- and liquid-phase (see Figure 10.1). Recent model updates endeavor to account for the modification in the pathlength for diffusion introduced by the tortuosity of the cell wall,[69] the rate of regeneration of DHA in the cytosol (based upon the reaction with GSH according to the data presented by Winkler and colleagues[166]) the distribution of the neutral species (ascorbic acid) between compartments of different pH,[167] and also the transport kinetics of ASC/DHA across the plasma membrane.[125,127] Much effort has been expended in attempting to validate and refine the existing model. Figure 10.5 illustrates the close agreement between predicted (modeled) and measured concentrations of ASC in the apoplast of fully parameterized plant tissue that can be achieved using *SODA*. Measured input parameters include the external O_3 concentration, stomatal conductance, relative compartment volumes, exposed mesophyll cell surface area, bulk leaf apoplast pH, and cell wall thickness. Assumptions that must be made are: (1) ASC is homogeneously distributed in the leaf apoplast; (2) stomatal uptake of O_3 is homogeneous across the leaf; (3) O_3 reacts directly, and solely, with apoplastic ASC *in vivo*, at a rate similar to that observed *in vitro* (Table 10.1), where between 1 and 3 mol of ASC are oxidized for every mole of O_3 consumed.[84,88]

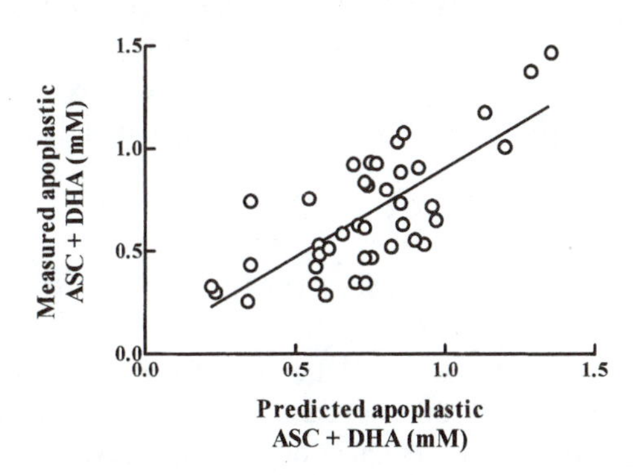

Figure 10.5 The relationship between measured and predicted (modeled) concentrations of ASC + DHA in the leaf apoplast of *Vicia faba* L. The solid line represents the least-square linear regression, $y = 0.861x + 0.042$; $r^2 = 0.546$, $P < 0.001$.

Applying SODA in a manner similar to the way in which Chameides[61] applied his model, it is possible to illustrate how "realistic" variations in leaf ascorbate concentration, stomatal conductance, cell wall thickness and O_3–ASC reaction stoichiometry *potentially* influence the amount of O_3 consumed by apoplastic ASC. The simulations shown in Figure 10.6 illustrate that, under standardized conditions (details provided in figure legend), (a) apoplastic ASC concentrations within the range 26 to 232 μM could intercept 10 to 60% of the incoming O_3 before the pollutant reaches the plasma membrane; (b) differences in g_{H_2O} have a large impact on the flux of O_3 impinging on the plasma membrane; (c) the amount of O_3 detoxified in the apoplast is extremely sensitive to variations in cell wall thickness; and (d) uncertainties surrounding the stoichiometry of the O_3–ASC reaction *in vivo*[84,88] may not have such an important influence on O_3 interception as might be intuitively suspected.

Analyses of experimental data using *SODA*, and also the model described by Chameides,[61] suggests that the size and the rate of turnover of the small pool of ASC in the leaf apoplast is *potentially* high enough to afford a significant degree of protection against environmentally relevant concentrations of O_3 (40 to 70% detoxification) in studies on *Phaseolus vulgaris*[168] and *Vicia faba*.[103] However, data for *Hordeum vulgare* and *Triticum aestivum*,[71] *Populus* hybrids,[102] and *Raphanus sativus* L.[169] suggest a minor role for apoplastic ASC in the interception of O_3 (<15% detoxification). The latter view is supported by the recent findings of Jakob and Heber.[93] These authors infiltrated leaves of *Spinacia oleracea* with solutions containing 10 mM ASC and an oxidation-sensitive fluorescent dye prior to fumigation ($\geq$65 nmol mol^{-1} O_3), while conducting parallel experiments *in vitro*. Ascorbate was demonstrated to prevent O_3-induced oxidation of the dye *in vitro*, but similar effects were not reciprocated *in vivo*. The authors suggested that the surprising lack of agreement between the data could be due to heterogeneity across the leaf in terms of ASC distribution and/or O_3 uptake — a very real possibility since O_3 would be expected to react primarily in the apoplast of cells in close proximity to the substomatal cavity.[79] This hypothesis is currently being tested by the authors.

10.4 CONCLUSIONS

There has been much recent discussion about the role played by constituents of the leaf apoplast in the provision of a first line of defense against O_3.[10,61,135] Studies involving the controlled fumigation of simple biochemical solutions,[84,86,92,170-172] as well as ASC-containing fluids from both plants and animals,[88,91,133] suggest that extracellular ASC concentrations may play a vital role in the interception and detoxification of O_3. Based on current knowledge, ascorbate tops the hierarchical series of first-line reaction targets in the leaf apoplast and there can be little doubt that this compound occupies a key position in protection against O_3-induced oxidative stress.[114] However, unexpected findings during recent *in vivo* studies highlight the need for further research to foster a better understanding of the function of extracellular antioxidants and their role in the detoxification of O_3.

ACKNOWLEDGMENTS

Experimental work performed in the Air Pollution Laboratory at Newcastle University presented in this chapter was financed through the provision of funds by The Royal Society, Newcastle University Equipment Fund, the Swales Foundation, the Overseas Development Agency, and the European Union. We are grateful to Prof. Heino Moldau (Tartu University, Estonia) for helpful comments on an earlier draft of this manuscript. T.L. and E.T. are indebted to Newcastle University and the Royal Society's East-West Exchange Scheme, respectively, for the provision of postdoctoral fellowships. M.P. appreciates the support provided by O. Sire, P. Ramge, and F.-W. Badeck during the early stages of the development of SODA, work that formed part of the BIATEX project financed by the German Ministry of Science and Technology. J.B. is personally indebted to the Royal Society for financial support.

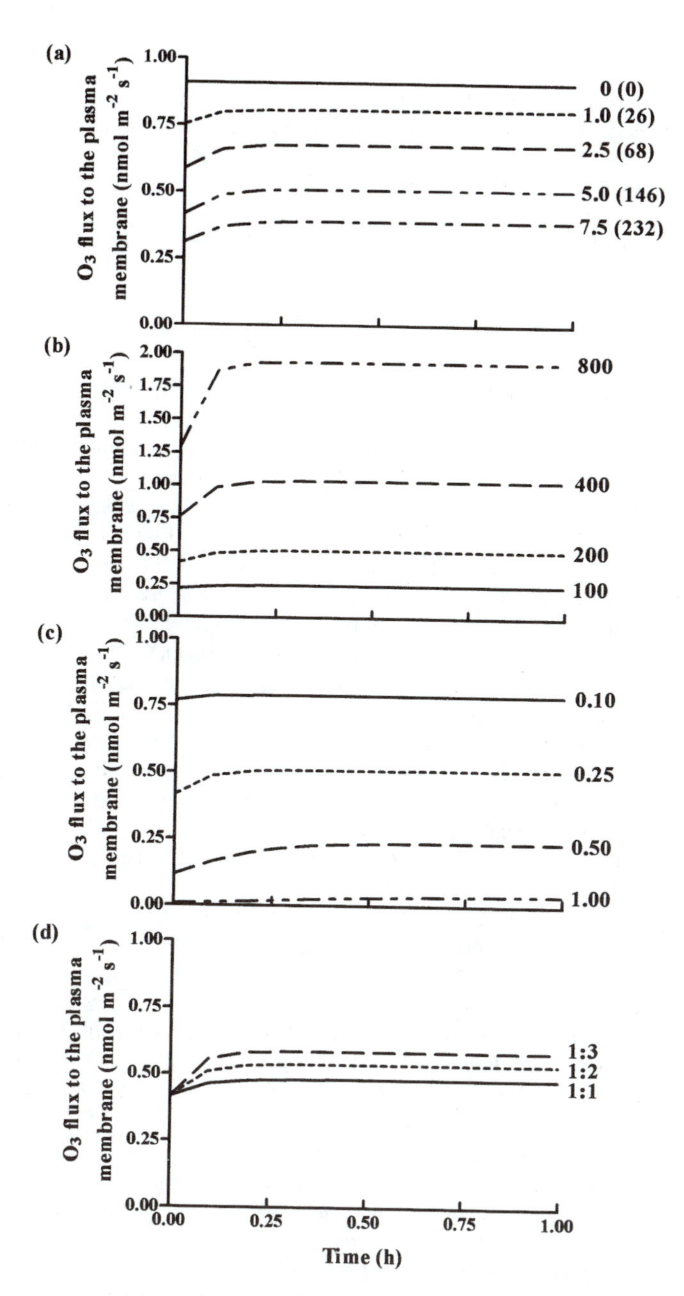

Figure 10.6 Simulated ozone (O_3) detoxification in the leaf apoplast. Manipulation of single variables: (a) bulk leaf ASC concentration 0 to 7.5 m*M*, steady-state apoplastic ASC concentrations (μ*M*) shown in parentheses; (b) stomatal conductance (g_{H_2O}) 100 to 800 mmol m^{-2} s^{-1}; (c) mesophyll cell wall thickness 0.10 to 1.00 μm; (d) O_3–ASC reaction stoichiometry 1:1–3. Additional inputs (standardized): leaf ASC concentration 5.0 m*M*; cell wall thickness 0.25 μm; stomatal conductance (g_{H_2O}) 200 mmol m^{-2} s^{-1}; O_3–ASC reaction stoichiometry 1:1.5 employing a second-order rate constant of 4.8×10^7 M^{-1} s^{-1}; apoplast pH 5.7; O_3 concentration 70 nmol mol^{-1}.

REFERENCES

1. Olszyk, D. M., Cabrera, H., and Thompson, C. R., California statewide assessment of the effects of ozone on crop productivity, *J. Air Pollut. Control Assoc.*, 38, 928, 1988.
2. UN-ECE, *Critical Levels for Ozone in Europe: Testing and Finalising the Concepts*, UN-ECE workshop report, Kärenlampi, L. and Skärby, L., Eds., Department of Ecology and Environmental Science, University of Kuopio, Kuopio, 1996, 363.
3. Turcsányi, E., Cardoso-Vilhena, J., Daymond, J., Gillespie, C., Balaguer, L., Ollerenshaw, J., and Barnes, J., Impacts of tropospheric ozone: past, present and likely future, in *Climate Change and Plants*, Singh, S. N., Ed., Springer-Verlag, Berlin, 1999.
4. Fuhrer, J., Skärby, L., and Ashmore, M. R., Critical levels for ozone effects on vegetation in Europe, *Environ. Pollut.*, 97, 91, 1997.
5. Barnes, J. D., Balaguer, L., Manrique, E., and Davison, A., Resistance to air pollutants: from cell to community, in *Handbook of Functional Plant Ecology*, Pugnaire, F. I. and Valladares, F., Eds., Marcel Dekker, New York, 1999, 735.
6. Barnes, J. D., Bender, J., Lyons, T., and Borland, A., Natural and man-made selection for air pollution resistance, *J. Exp. Bot.*, 1999, in press.
7. Davison, A. W. and Barnes, J. D., Effects of ozone on wild plants, *New Phytol.*, 139, 135, 1998.
8. Heath, R. L., Initial events in injury to plants by air pollutants, *Annu. Rev. Plant Physiol.*, 31, 395, 1980.
9. Heath, R. L., The biochemistry of ozone attack on the plasma membrane of plant cells, *Rec. Adv. Phytochem.*, 21, 29, 1987.
10. Heath, R. L., Biochemical mechanisms of pollutant stress, in *Assessment of Crop Loss from Air Pollutants*, Heck, W. W., Taylor, O. C., and Tingey, D. T., Eds., Elsevier, New York, 1988, 259.
11. Heath, R. L., Alterations of plant metabolism by ozone exposure, in *Plant Responses to the Gaseous Environment*, Alscher, R. G. and Wellburn, A. R., Eds., Chapman & Hall, London, 1994, 121.
12. Runeckles, V. C. and Chevone, B. I., Crop responses to ozone, in *Surface-Level Ozone Exposures and Their Effects on Vegetation*, Lefohn, A. S., Ed., Lewis Publishers, Chelsea, 1992, 189.
13. Polle, A. and Rennenberg, H., Significance of antioxidants in plant adaptation to environmental stress, in *Plant Adaptation to Environmental Stress*, Fowden, L., Mansfield, T. A. and Stoddart, J., Eds., Chapman & Hall, London, 1993, 263.
14. Harris, M. J. and Bailey-Serres, J. N., Ozone effects on gene expression and molecular approaches to breeding for air pollution resistance, in *Stress-Induced Gene Expression in Plants*, Basra, A. S., Ed., Harwood Academic Publishers, Chur, 1994, 185.
15. Kangasjärvi, J., Talvinen, J., Utriainen, M., and Kerjalainen, R., Plant defence systems induced by ozone, *Plant Cell Environ.*, 17, 783, 1994.
16. Schraudner, M., Langebartels, C., and Sandermann, H., Changes in the biochemical status of plant cells induced by the environmental pollutant ozone, *Physiol. Plant.*, 100, 274, 1997.
17. Sharma, Y. K. and Davis, K. R., The effects of ozone on antioxidant responses in plants, *Free Radical Biol. Med.*, 23, 480, 1997
18. Pell, E. J., Schlagnhaufer, C. D., and Arteca, R. N., Ozone-induced oxidative stress: mechanisms of action and reaction, *Physiol. Plant.*, 100, 264, 1997.
19. Sandermann, H., Ernst, D., Heller, W., and Langerbartels, C., Ozone: an abiotic elicitor of plant defence reactions, *Trends Plant Sci.*, 3, 47, 1998.
20. Gaastra, P., Photosynthesis of crop plants as influenced by light, carbon dioxide, temperature, and stomatal diffusive resistance, *Meded. Landbouwhogesch. Wageningen*, 59, 1, 1959.
21. Tingey, D. T. and Taylor, G. E., Variation in plant response to ozone: a conceptual model of physiological events, in *Effects of Gaseous Air Pollutants in Agriculture and Horticulture*, Unsworth, M. H. and Ormrod, D. P., Eds., Butterworth, London, 1982, 113.
22. Unsworth, M. H., Biscoe, P. V., and Black, V., Analysis of gas exchange between plants and polluted atmospheres, in *Effects of Air Pollutants on Plants*, Mansfield, T. A., Ed., Cambridge University Press, New York, 1976, 5.
23. O'Dell, R. A., Taheri, M., and Kabel, R. L., A model for uptake of pollutants by vegetation, *J. Air Pollut. Control Assoc.*, 27, 1104, 1977.

24. Runeckles, V. C., Uptake of ozone by vegetation, in *Surface-Level Ozone Exposures and Their Effects on Vegetation*, Lefohn, A. S., Ed., Lewis Publishers, Chelsea, 1992, 157.
25. Kerstiens, G. and Lendzian, K. J., Interactions between ozone and plant cuticles. 1. Ozone deposition and permeability, *New Phytol.*, 112, 13, 1989.
26. Mansfield, T. A. and Freer-Smith, P. H., The role of stomata in resistance mechanisms, in *Gaseous Air Pollutants and Plant Metabolism*, Koziol, M. J. and Whatley, F. J., Eds., Butterworths, London, 1984, 131.
27. Reich, P. B., Quantifying plant response to ozone: a unifying theory, *Tree Physiol.*, 3, 63, 1987.
28. Winner, E. W., Coleman, J. S., Gillespie, C., Mooney, H. A., and Pell, E. J., Consequences of evolving resistance to air pollution, in *Ecological Genetics and Air Pollution*, Taylor, G. E., Pitelka, L. F. and Clegg, M. T., Eds., Springer-Verlag, New York, 1991, 177.
29. Nebel, B. and Fuhrer, J., Inter- and intraspecific differences in ozone sensitivity in semi-natural plant communities, *Angew. Bot.*, 68, 116, 1994.
30. Sanders, G. E., Colls, J. J., Clark, A. G., Galaup, S., Bonte, J., and Cantuel, J., *Phaseolus vulgaris* and ozone: results from open-top chamber experiments in France and England, *Agric. Ecosys. Environ.*, 38, 31, 1992.
31. Reiling, K. and Davison, A. W., Effects of ozone on stomatal conductance and photosynthesis in populations of *Plantago major* L., *New Phytol.*, 129, 587, 1995.
32. Taylor, G. E., Plant and leaf resistance to gaseous air pollution stress, *New Phytol.*, 80, 523, 1978.
33. Guzy, M. R. and Heath, R. L., Responses to ozone of varieties of common bean (*Phaseolus vulgaris* L.), *New Phytol.*, 124, 617, 1993.
34. Robinson, M. F., Heath, J., and Mansfield, T. A., Disturbances in stomatal behaviour caused by air pollutants, *J. Exp. Bot.*, 49, 461, 1998.
35. Atkinson, C. J., Robe, S. V. and Winner, W. E., The relationship between changes in photosynthesis and growth for radish plants fumigated with SO_2 and O_3, *New Phytol.*, 110, 173, 1988.
36. Farage, P. K., Long, S. P., Lechner, E. G. and Baker, N. R., The sequence of change within the photosynthetic apparatus of wheat following short-term exposure to ozone, *Plant Physiol.*, 95, 529, 1991.
37. Farage, P. K. and Long, S. P., An *in vivo* analysis of photosynthesis during short-term O_3 exposure in three contrasting species, *Photosynth. Res.*, 43, 11, 1995.
38. Taylor, G. E., Hanson, P. J. and Baldochi, D. B., Pollutant deposition to individual leaves and plant canopies: sites of regulation and relationship to injury, in *Assessment of Crop Loss from Air Pollutants*, Heck, W. W., Taylor, O. C., and Tingey, D. T., Eds., Elsevier, New York, 1988, 227.
39. Evans, L. S. and Miller, P. R., Comparative needle anatomy and relative ozone sensitivity of four pine species, *Can. J. Bot.*, 50, 1067, 1972.
40. Evans, L. S. and Ting, I. P., Ozone sensitivity of leaves: relationship to leaf water content, gas transfer resistance, and anatomical characteristics, *Am. J. Bot.*, 61, 592, 1974.
41. Pääkkönen, E., Holopainen, T., and Kärenlampi, L., Ageing-related anatomical and ultrastructural changes in leaves of birch (*Betula pendula* Roth.) clones as affected by low ozone exposure, *Ann. Bot.*, 75, 285, 1995.
42. Pääkkönen, E., Holopainen, T., and Kärenlampi, L., Variation in ozone sensitivity among clones of *Betula pendula* and *Betula pubescens*, *Environ. Pollut.* 95, 37, 1997.
43. Wiese, C. B. and Pell, E. J., Influence of ozone on transgenic tobacco plants expressing reduced quantities of Rubisco, *Plant Cell Environ.*, 20, 1283, 1997.
44. Bennett, J. P., Rassat, P., Berrang, P., and Karnosky, D. F., Relationships between leaf anatomy and ozone sensitivity of *Fraxinus pennsylvanica* Marsh. and *Prunus serotina* Ehrh., *Environ. Pollut.*, 32, 33, 1992.
45. Chappelka, A. H. and Samuelson, L. J., Ambient ozone effects on forest trees of the eastern United States: a review, *New Phytol.*, 139, 91, 1998.
46. Pääkkönen, E., Paasisalo, S., Holopainen, T., and Kärenlampi, L., Growth and stomatal responses of birch (*Betula pendula* Roth.) clones to ozone in open-air and field fumigations, *New Phytol.*, 125, 615, 1993.
47. Pääkkönen, E., Metsärinne, S., Holopainen, T., and Kärenlampi, L., The ozone sensitivity of birch (*Betula pendula* Roth.) in relation to the developmental stage of leaves, *New Phytol.*, 132, 145, 1995.

48. Tingey, D. T., Standley, C., and Field, R. W., Stress ethylene evolution: a measure of ozone effects on plants, *Atmos. Environ.*, 10, 969, 1976.
49. Mehlhorn, H. and Wellburn, A. R., Stress ethylene formation determines plant sensitivity to ozone, *Nature*, 327, 417, 1987.
50. Langebartels, C., Kerner, K., Leonardi, S., Schrauder, M., Trost, M., Heller, W., and Sandermann, H., Biochemical plant responses to ozone. I. Differential induction of polyamine and ethylene biosynthesis in tobacco, *Plant Physiol.*, 95, 882, 1991.
51. Mehlhorn, H., O'Shea, J. M., and Wellburn, A. R., Atmospheric ozone interacts with stress ethylene formation by plants to cause visible plant injury, *J. Exp. Bot.*, 42, 17, 1991.
52. Wenzel, A. A., Schlautmann, H., Jones, C. A., Küppers, K., and Mehlhorn, H., Aminoethoxyvinylglycine, cobalt and ascorbic acid all reduce ozone toxicity in mung beans by inhibition of ethylene biosynthesis, *Physiol. Plant.*, 93, 286, 1995.
53. Bae, G. Y., Nakajima, N., Ishizuka, K., and Kondo, N., The role in ozone phytotoxicity of the evolution of ethylene upon induction of 1-aminocyclopropane-1-carboxylic acid synthase by ozone fumigation in tomato plants, *Plant Cell Physiol.*, 37, 129, 1996.
54. Sandermann, H., Ozone and plant health, *Annu. Rev. Phytopathol.*, 34, 347, 1996.
55. Elstner, E. F., Osswald, W., and Youngman, R. J., Basic mechanisms of pigment bleaching and loss of structural resistance in spruce (*Picea abies*) needles: advances in phytomedical diagnostics, *Experientia*, 41, 591, 1985.
56. Salter, L. and Hewitt, C. N., Ozone-hydrocarbon interactions in plants, *Phytochemistry,* 31, 4045, 1992.
57. Steinbrecher, R., Karbach, I., Mits, S., and Ziegler, H., Isoprene emission from Norway spruce (*Picea abies* L.) and behaviour in the atmosphere of a major spruce stand, in *Proceedings of the EUROTRAC Symposium 1992*, Borrell, P. M., Borrell, P., Cvitas, T., and Seiler, W., Eds., SPB Publishing BV, The Hague, 1993, 286.
58. Wellburn, F. A. M. and Wellburn, A. R., Variable patterns of antioxidant protection but similar ethene emission differences in several ozone-sensitive and ozone-tolerant plant selections, *Plant Cell Environ.*, 19, 754, 1996.
59. Hewitt, C. N., Kok, G. L., and Fall, R., Hydroperoxides in plants exposed to ozone mediate air pollution damage to alkene emitters, *Nature*, 344, 56, 1990.
60. Hewitt, C. N. and Kok, G. L., Formation and occurrence of organic hydroperoxides in the troposphere: laboratory and field observations, *J. Atmos. Chem.*, 12, 181, 1991.
61. Chameides, W. L., The chemistry of ozone deposition to plant leaves: the role of ascorbic acid, *Environ. Sci. Technol.*, 23, 595, 1989.
62. Stokes, N. J., Terry, G. M., and Hewitt, C.N., The impact of ozone, isoprene and propene on antioxidant levels in two leaf classes of velvet bean (*Mucuna pruriens* L.), *J. Exp. Bot.*, 49,115, 1998.
63. Heath, R. L. and Taylor, G. E., Physiological processes and plant responses to ozone exposure, in *Forest Decline and Ozone*, Ecological Studies Vol. 127, Sandermann, H., Wellburn, A. R., and Heath, R. L., Eds., Springer-Verlag, Berlin, 1997, 317.
64. Polle, A. and Junkermann, W., Inhibition of apoplastic and symplastic peroxidase activity from Norway spruce by the photo-oxidant hydroxymethyl hydroperoxide, *Plant Physiol.*, 104, 617, 1994.
65. Tuomainen, J., Betz, C., Kangasjärvi, J., Ernst, D., Yin Z.-H., Langerbartels, C., and Sandermann, H., Ozone induction of ethylene emission in tomato plants: regulation by differential accumulation of transcripts for biosynthetic enzymes, *Plant J.*, 12, 1151, 1997.
66. Greenberg, J. T., Guo, A., Klessig, D. F., and Ausubel, F. M., Programmed cell death in plants: a pathogen-triggered response activated coordinately with multiple defense functions, *Cell*, 77, 551, 1994.
67. Morgan, P. W. and Drew, M. C., Ethylene and plant responses to stress, *Physiol. Plant.*, 100, 620, 1997.
68. Polle, A., Wieser, G., and Havranek, W. M., Quantification of ozone influx and apoplastic ascorbate content in needles of Norway spruce trees (*Picea abies* L., Karst.) at high altitude, *Plant Cell Environ.*, 18, 681, 1995.
69. Nobel, P. S., *Biochemical Plant Physiology and Ecology,* W. H. Freeman and Co., San Francisco, 1983, 608.
70. Glater, R. A. B., Smog damage to ferns in the Los Angeles area, *Phytopathology,* 46, 696, 1956.

71. Kollist, H., Moldau, H., Mortensen, L., Rasmussen, S. K., and Jørgensen, L. B., Ascorbate levels and ozone decay in the cell walls of barley and wheat seedlings, in *Plant Peroxidases: Biochemistry and Physiology*, Obinger C., Burner, U., Ebermann, R., Penel, C., and Greppin, H., Eds., IV International Symposium, Vienna, 1996, 363.
72. Günthardt-Georg, M. S., McQuattie, C. J., Scheidegger, C., Rhiner, C., and Matyssek, R., Ozone-induced cytochemical and ultrastructural changes in leaf mesophyll cell walls, *Can. J. Forest Res.*, 27, 453, 1997.
73. Galliano, H., Cabané, M., Eckerskorn, C., Lottspeich, F., Sandermann, H., and Ernst, D., Molecular cloning, sequence analysis and elicitor-induced ozone-induced accumulation of cinnamyl alcohol dehydrogenase from Norway spruce (*Picea abies* L.). *Plant Mol. Biol.*, 23, 145, 1993.
74. Galliano, H., Heller, W., and Sandermann, H., Ozone induction and purification of spruce cinnamyl alcohol dehydrogenase, *Phytochemistry*, 32, 557, 1993.
75. Eckey-Kaltenbach, H., Ernst, D., Heller, W., and Sandermann, H., Biochemical plant responses to ozone IV. Cross-induction of defensive pathways in parsley plants, *Plant Physiol.*, 104, 67, 1994.
76. Langebartels, C., Ernst, D., Heller, W., Lütz, C., Payer, H.-D., and Sandermann, H., Ozone responses of trees: results from controlled chamber exposures at the GSF phytotron, in *Forest Decline and Ozone*, Ecological Studies Vol. 127, Sandermann, H., Wellburn, A. R., and Heath, R. L., Eds., Springer-Verlag, Berlin, 1997, 163.
77. Laisk, A., Kull, O., and Moldau, H., Ozone concentration in leaf intercellular air spaces is close to zero, *Plant Physiol.*, 90, 1163, 1989.
78. Hill, A. C., Vegetation: a sink for atmospheric pollutants, *J. Air Pollut. Control Assoc.*, 21, 341, 1971.
79. Dietz, K.-J., Functions and responses of the leaf apoplast under stress, in *Progress in Botany*, Vol. 58, Springer-Verlag, Berlin, 1997, 221.
80. Staehelin, J. and Hoigné, J., Decomposition of ozone in water: rate of initiation of hydroxide ions and hydrogen peroxide, *Environ. Sci. Technol.*, 16, 676, 1982.
81. Grimes, H. D., Perkins, K. K., and Boss, W. F., Ozone degrades into hydroxyl radicals under physiological conditions, *Plant Physiol.*, 72, 1016, 1983.
82. Byvoet, P., Balis, J. U., Shelley, S. A., Montgomery, M. R., and Barber, M. J., Detection of hydroxyl radicals upon interaction of ozone with aqueous-media or extracellular surfactant — the role of trace iron, *Arch. Biochem. Biophys.*, 319, 464, 1995.
83. Staehelin, J. and Hoigné, J., Decomposition of ozone in water in the presence of organic solutes acting as promoters and inhibitors of radical chain reactions, *Environ. Sci. Technol.*, 19, 1206, 1985.
84. Kanofsky, J. R. and Sima, P. D., Reactive absorption of ozone by aqueous biomolecule solutions: implications for the role of sulfhydryl compounds as targets for ozone, *Arch. Biochem. Biophys.*, 316, 52, 1995.
85. Giamalva, D. H., Church, D. F., and Pryor, W.A., Kinetics of ozonation. 4. Reactions of ozone with α-tocopherol and oleate and linoleate esters in carbon tetrachloride and in aqueous micellular solvents, *J. Am. Chem. Soc.*, 108, 6646, 1986.
86. Giamalva, D. H., Church, D. F., and Pryor, W. A., A comparison of the rates of ozonation of biological antioxidants and oleate and linoleate esters, *Biochem. Biophys. Res. Commun.*, 133, 773, 1985.
87. Pryor, W. A., Giamalva, D. H., and Church, D. F., Kinetics of ozonation. II. Amino acids and model compounds in water and comparison to rates in non-polar solvents, *J. Am. Chem. Soc.*, 106, 7074, 1984.
88. Van der Vliet, A., O'Neill, C. A., Eiserich, J. P., and Cross, C. E., Oxidative damage to extracellular fluids by ozone and possible protective effects of thiols, *Arch. Biochem. Biophys.*, 321, 43, 1995.
89. Kelly, F. J., Mudway, I., Krishna, M. T., and Holgate, S. T., The free radical basis of air pollution: focus on ozone, *Respir. Med.*, 89, 647, 1995.
90. Castillo, F. J. and Greppin, H., Extracellular ascorbic acid and enzyme activities related to ascorbic acid metabolism in *Sedum album* L. leaves after ozone exposure, *Environ. Exp. Bot.*, 28, 231, 1988.
91. Cross, C. E., Motchnik, P. A., Bruener, B. A., Jones, D. A., Kaur, H., Ames, B. N., and Halliwell, B., Oxidative damage to plasma constituents by ozone. *FEBS Lett.*, 298, 269, 1992.
92. Mudway, I. S. and Kelly, F. J., Modeling the interaction of ozone with pulmonary epithelial lining fluid antioxidants, *Toxicol. Appl. Pharmacol.*, 148, 91, 1998.
93. Jakob, B. and Heber, U., Apoplastic ascorbate does not prevent the oxidation of fluorescent amphiphilic dyes by ambient and elevated concentrations of ozone in leaves, *Plant Cell Physiol.*, 39, 313, 1998.

94. Polle, A., Chakrabarti, K., Schurmann, W., and Rennenberg, H., Composition and properties of hydrogen peroxide decomposing systems in extracellular and total extracts from needles of Norway spruce (*Picea abies* L., Karst.), *Plant Physiol.*, 94, 312, 1990.
95. Takahama, U. and Oniki, T., Regulation of peroxidase-dependent oxidation of phenolics in the apoplast of spinach leaves by ascorbate, *Plant Cell Physiol.*, 33, 379, 1992.
96. Eckey-Kaltenbach, H., Heller, W., Sonnenbichler, J., Zetl, I., Schäfer, W., Ernst, D., and Sandermann, H., Oxidative stress and plant secondary metabolism: 60″-*O*-malonyapinn in parsley, *Phytochemistry*, 34, 687, 1993.
97. Luwe, M. W. F., Takahama, U., and Heber U., Role of ascorbate in detoxifying ozone in the apoplast of spinach (*Spinacia oleracea* L.) leaves, *Plant Physiol.*, 101, 969, 1993.
98. Luwe, M. and Heber, U., Ozone detoxification in the apoplasm and symplasm of spinach, bean and beech leaves at ambient and elevated concentrations of ozone in air, *Planta*, 197, 448, 1995.
99. Luwe, M., Antioxidants in the apoplast and symplast of beech (*Fagus sylvatica* L.) leaves: seasonal variations and responses to changing ozone concentrations, *Plant Cell Environ.*, 19, 321, 1996.
100. Ranieri, A., D'Urso, G., Nali, C., Lorenzini, G., and Soldatini, G. F., Ozone stimulates apoplastic antioxidant systems in pumpkin leaves, *Physiol. Plant.*, 97, 381, 1996.
101. Lyons, T., Ollerenshaw, J., and Barnes J., Impacts of ozone on *Plantago major*: apoplastic and symplastic antioxidant status, *New Phytol.*, 141, 253, 1999.
102. Ranieri, A., Castagna, A., Padu, E., Moldau, H., Rahi, M., and Soldatini, G. F., The decay of O_3 through direct reaction with cell wall ascorbate is not sufficient to explain the different degrees of O_3-sensitivity in two poplar clones, *J. Plant Physiol.*, 154, 250, 1999.
103. Turcsányi, E., Lyons, T., Plöchl, M., and Barnes, J., Does ascorbate in the mesophyll cell walls form the first line of defence against ozone? 1. Testing the concept using broad bean (*Vicia faba* L.), *J. Exp. Bot.*, communicated, 1999.
104. Vanacker, H., Carver, T. L. W., and Foyer, C. H., Pathogen-induced changes in the antioxidant status of the apoplast in barley leaves, *Plant Physiol.*, 117, 1103, 1998.
105. Vanacker, H., Foyer, C. H., and Carver, T. L., Changes in apoplastic antioxidants induced by powdery mildew attack in oat genotypes with race non-specific resistance, *Planta*, 208, 444, 1999.
106. Loewus, F. A., Ascorbic acid: metabolism, biosynthesis, function, in *The Biochemistry of Plants*, Vol. 3, Priess, J., Ed., Academic Press, New York, 1980, 77.
107. Loewus, F. A., Ascorbic acid and its metabolic products, in T*he Biochemistry of Plants*, Vol. 14, Priess, J., Ed., Academic Press, New York, 1988, 85.
108. Davies, M. B., Austin, J. and Partridge, D. A., Vitamin C: its chemistry and biochemistry, the Royal Society of Chemistry, Cambridge, 1991, 154.
109. Moser, U. and Bendich, A., Vitamin C, in *Handbook of the Vitamins*, 2nd edition, Machlin, L. J., Ed., Marcel Dekker, New York, 1991, 195.
110. Arrigoni, O., Ascorbate system in plant development, *J. Bioenerg. Biomembr.*, 26, 407, 1994.
111. Smirnoff, N., The function and metabolism of ascorbic acid in plants, *Ann. Bot.*, 78, 661, 1996.
112. Smirnoff, N. and Pallanca, J. E., Ascorbate metabolism in relation to oxidative stress, *Biochem. Soc. Trans.*, 24, 472, 1996.
113. Conklin, P. L., Pallanca, J. E., Last, R. L., and Smirnoff, N., l-Ascorbic acid metabolism in the ascorbate-deficient arabidopsis mutant *vtc1*, *Plant Physiol.*, 115, 1277, 1997.
114. Conklin, P. L., Williams, E. H., and Last, R. L., Environmental stress sensitivity of an ascorbic acid-deficient *Arabidopsis* mutant, *Proc. Natl. Acad. Sci. U.S.A.*, 93, 9970, 1996.
115. Freebairn, H. T.,The prevention of air pollution damage to plants by the use of vitamin C sprays, *J. Air Pollut. Control Assoc.*, 10, 314, 1960.
116. Freebairn, H. T. and Taylor, O. C., Prevention of plant damage from airborne oxidising agents, *Proc. Am. Soc. Hortic. Sci.*, 76, 693, 1960.
117. Menser, H. A., Response of plants to air pollutants. III. A relation between ascorbic acid levels and ozone susceptibility of light-preconditioned tobacco plants, *Plant Physiol.*, 39, 564, 1964.
118. Lee, E. H., Jersey, J. A., Gifford, C., and Bennett, J., Differential ozone tolerance in soybean and snapbeans: analysis of ascorbic acid in O_3-susceptible and O_3-resistant cultivars by high-performance liquid chromatography, *Environ. Exp. Bot.*, 24, 331, 1984.
119. Mächler, F., Wasescha, M. R., Krieg, F., and Oertli, J. J., Damage by ozone and protection by ascorbic acid in barley leaves, *J. Plant Physiol.*, 147, 469, 1995.

120. Kurata, T., Miyake, N., and Otsuka, Y., Formation of L-threonolactone and oxalic acid in the autoxidation reaction of L-ascorbic acid — possible involvement of singlet oxygen, *Biosci. Biotech. Biochem.*, 60, 1212, 1996.
121. Deutsch, J. C., Ascorbic acid and dehydroascorbic acid interconversion without net oxidation or reduction, *Anal. Biochem.*, 247, 58, 1997.
122. Miyake, N., Otsuka, Y., and Kurata, T., Autoxidation reaction mechanism for L-ascorbic acid in methanol without metal ion catalysis, *Biosci. Biotech. Biochem.*, 61, 2069, 1997.
123. Deutsch, J. C., Ascorbic acid oxidation by hydrogen peroxide, *Anal. Biochem.*, 255, 1, 1998.
124. Rautenkranz, A. A. F., Li, L., Mächler, F., Märtinoia, E., and Oertli, J. J., Transport of ascorbic and dehydroascorbic acids across protoplast and vacuole membranes isolated from barley (*Hordeum vulgare* L. cv Gerbel) leaves, *Plant Physiol.*, 106, 187, 1994.
125. Foyer, C. H. and Lelandais, M., A comparison of the relative rates of transport of ascorbate and glutathione across the thylakoid, chloroplast and plasmalemma membranes of pea leaf mesophyll cells, *J. Plant Physiol.*, 148, 391, 1996.
126. Horemans, N., Asard, H., and Caubergs, R. J., Transport of ascorbate into plasma membrane vesicles of *Phaseolus vulgaris* L., *Protoplasma*, 194, 177, 1996.
127. Horemans, N., Asard, H., and Caubergs, R. J., The ascorbate carrier of higher plant plasma membranes preferentially translocates the fully oxidised (dehydroascorbate) molecule, *Plant Physiol.*, 114, 1247, 1997.
128. Horemans, N., Asard, H., and Caubergs, R. J., The role of ascorbate free radical as an electron acceptor to the cytochrome *b*-mediated *trans*-plasma membrane electron transport in higher plants, *Plant Physiol.*, 104, 1455, 1994.
129. Navas, P., Villalba, J. M., and Córdoba, F., Ascorbate function at the plasma membrane, *Biochim. Biophys. Acta*, 1197, 1, 1994.
130. Asard, H., Horemans, N., and Caubergs, R. J., Involvement of ascorbic acid and a b-type cytochrome in plant plasma membrane redox reactions, *Protoplasma*, 184, 36, 1995.
131. Bérczi, A. and Möller, I. M., NADH-monodehydroascorbate oxidoreductase is one of the redox enzymes in spinach leaf plasma membranes, *Plant Physiol.*, 116, 1029, 1998.
132. Plöchl, M., Ramge, P., Badeck, F.-W., and Kohlmaier, G. H., Modelling of deposition and detoxification of ozone in plant leaves, in *Proceedings of the EUROTRAC Symposium 1992*, Borrell, P. M., Borrell, P., Cvitas, T., and Seiler, W., Eds., SPB Publishing BV, The Hague, 1993, 748.
133. Kanofsky, J. R. and Sima, P. D., Singlet oxygen production from the reaction of ozone with plant leaves, *J. Biol. Chem.*, 270, 7850, 1995.
134. Kelly, F. J., Respiratory tract lining fluid antioxidants: impact of ozone, Respir. Med., 88, 818, 1994.
135. Castillo, F. J., Penel, C., and Greppin, H., Peroxidase release induced by ozone in *Sedum album* leaves, *Plant Physiol.*, 74, 846, 1984.
136. Castillo, F. J. and Greppin, H., Balance between anionic and cationic extracellular activities in *Sedum album* leaves after ozone exposure. Analysis by high-performance liquid chromatography, *Physiol. Plant.*, 68, 201, 1986.
137. Castillo, F. J., Miller, P. R., and Greppin, H., Extracellular biochemical markers of photochemical oxidant air pollution damage to Norway spruce, *Experientia*, 43, 111, 1987.
138. Peters, J. L., Castillo, F. J., and Heath, R. L., Alteration of extracellular enzymes in pinto bean leaves upon exposure to air pollutants, ozone and sulfur dioxide, *Plant Physiol.*, 89, 159, 1989.
139. Batini, P., Ederli, L., Pasqualini, S., Antonielli, M., and Valenti, V., Effects of ethylenediurea and ozone in detoxificant ascorbic-ascorbate peroxidase system in tobacco plants, *Plant Physiol.* Biochem., 33, 717, 1995.
140. Kronfuss, G., Wieser, G., Havranek, W. M., and Polle, A., Effects of ozone and mild drought stress on total and apoplastic guaiacol peroxidase and lipid peroxidation in current-year needles of young Norway spuce (*Picea abies* L., Karst.), *J. Plant Physiol.*, 148, 203, 1996.
141. Polle, A., Otter, T., and Seifert, F., Apoplastic peroxidases and lignification in needles of Norway spruce (*Picea abies* L.), *Plant Physiol.*, 106, 53, 1994.
142. Fry, S. C., Cross-linking of matrix polymers in the growing cell walls of angiosperms, *Annu. Rev. Plant Physiol.*, 37, 165, 1986.
143. Streller, S. and Wingsle, G., *Pinus sylvestris* L. needles contain extracellular CuZn superoxide dismutase, *Planta*, 192, 195, 1994.

144. Ogawa, K., Kanematsu, S., and Asada, K., Intra- and extra-cellular localization of "cytosolic" CuZn-superoxide dismutase in spinach leaf and hypocotyl, *Plant Cell Physiol.*, 37, 790, 1996.
145. Klug, D., Rabani, J., and Fridovitch, I., A direct demonstration of the catalytic action of superoxide dismutase through the use of pulse radiolysis, *J. Biol. Chem.*, 247, 4839, 1972.
146. Ogawa, K., Kanematsu, S., and Asada, K., Generation of superoxide anion and localization of CuZn-superoxide dismutase in the vascular tissue of spinach hypocotyls: their association with lignification, *Plant Cell Physiol.*, 38, 1118, 1997.
147. Tabor, C. W. and Tabor, H., Polyamines, *Annu. Rev. Biochem.*, 53, 749, 1984.
148. Smith, T. A., Polyamines, *Annu. Rev. Plant Physiol.*, 36, 117, 1985.
149. Martin-Tanguy, J., The occurrence and possible function of hydroxycinnamoyl acid amides in plants, *Plant Growth Reg.*, 3, 381, 1985.
150. Bors, W., Langebartels, C., Michel, C., and Sandermann, H., Polyamines as radical scavengers and protectants against ozone damage, *Phytochemistry*, 28, 1589, 1989.
151. Burtin, D. and Michael, A. J., Overexpression of arginine decarboxylase in transgenic plants, *Biochem. J.*, 325, 331, 1997.
152. Rowland-Bamford, A. J., Borland, A. M., and Lea, P. J., The effect of air pollution on polyamine metabolism, *Environ. Pollut.*, 53, 410, 1988.
153. Rowland-Bamford, A. J., Borland, A. M., Lea, P. J., and Mansfield, T. A., The role of arginine decarboxylase in modulating the sensitivity of barley to ozone, *Environ. Pollut.*, 61, 95, 1989.
154. Federico, R. and Angelini, R., Occurrence of diamine oxidase in the apoplast of pea epicotyls, *Planta*, 167, 300, 1986.
155. Angelini, R., Manes, F., and Federico, R., Spatial and functional correlation between diamine oxidase and peroxidase activities and their dependence upon de-etiolation and wounding in chickpea stems, *Planta,* 182, 89, 1990.
156. Federico, R., Angelini, R., Cona, A., and Niglio, A., Polyamine oxidase bound to cell walls from *Zea mays* seedlings, *Phytochemistry*, 31, 2955, 1992.
157. Abeles, F. B., Ethylene in *Plant Biology*, Academic Press, New York, 1973, 302.
158. Rennenberg, H., Glutathione metabolism and possible biological roles in higher plants, *Phytochemistry*, 21, 2771, 1982.
159. Marrs, K. A., The functions and regulation of glutathione *S*-transferases in plants, *Annu. Rev. Plant Physiol. Plant Mol. Biol.*, 47, 127, 1996.
160. Foyer, C. H. and Halliwell, B., Presence of glutathione and glutathione reductase in chloroplasts: a proposed role in ascorbic acid metabolism, *Planta*, 133, 21, 1976.
161. Eshdat, Y., Holland, D., Faltin, Z., and Ben-hayyim, G., Plant glutathione peroxidases, *Physiol. Plant.*, 100, 234, 1997.
162. Jamaï, A., Tommasini, R., Martinoia, E., and Delrot, S., Characterisation of glutathione uptake in broad bean leaf protoplasts, *Plant Physiol.*, 111, 1145, 1996.
163. Flury, T., Wagner, E., and Kreuz, K., An inducible glutathione *S*-transferase in soybean hypocotyl is localized in the apoplast, *Plant Physiol.*, 112, 1185, 1996.
164. Plöchl, M., Die Bedeutung apoplasticher Ascorbinsäure für die Aufnahme und Reduktion von ozon und Stickstoffdioxid in Pflanzenblättern: Eine Dynamische Modellierung der Involvierten Metabolischen Prozesse in System Blatt. Ph.D. Thesis, Faculty of Chemistry, Johann Wolfgang Goethe-University Frankfurt-Main, 1992.
165. Plöchl, M., Lyons, T., Ollerenshaw, J. H., and Barnes, J. D., Simulating ozone detoxification in the leaf apoplast through the direct reaction with ascorbate, *Planta*, 1999, communicated.
166. Winkler, B. S., Orselli, S. M., and Rex, T. S., The redox couple between glutathione and ascorbic acid: a chemical and physiological perspective, *Free Radical Biol. Med.*, 17, 333, 1994.
167. Slovik, S., Baier, M., and Hartung, W., Compartmental distribution and redistribution of abscisic acid in intact leaves. I. Mathematical formulation, *Planta*, 187, 14, 1992.
168. Moldau, H., Padu, E., and Bichele, I., Quantification of ozone decay and requirement for ascorbate in *Phaseolus vulgaris* L. mesophyll cell walls, *Phyton Ann. Bot.*, 37, 175, 1997.
169. Raddison, J. and Barnes, J., unpublished.
170. Kanofsky, J. R. and Sima, P., Singlet oxygen production from the reactions of ozone with biological molecules, *J. Biol. Chem.*, 266, 9039, 1991.

171. Kanofsky, J. R. and Sima, P. D., Singlet-oxygen generation at the gas-liquid interfaces: a significant artifact in the measurement of singlet-oxygen yields from ozone-biomolecule reactions, *Photochem. Photobiol.*, 58, 335, 1993.
172. O'Neill, C. A., Van der Vliet, A., Hu, M.-L., Kaur, H., Cross, C. E., Louie, S., and Halliwell, B., Oxidation of biological molecules by ozone: the effect of pH, *J. Lab. Clin. Med.*, 122, 497, 1993.

CHAPTER 11

Early Detection, Mechanisms of Tolerance, and Amelioration of Ozone Stress in Crop Plants

Edward H. Lee

CONTENTS

11.1 INTRODUCTION

Air pollution poses a worldwide threat to human health and the environment.[1-5] There is evidence that air quality is worsening in nearly all developing countries.[2] Air pollution is exacerbated as

1-56670-341-7/00/$0.00+$.50

countries industrialize because of increase in population density, greater use of motor vehicles, rapid economic development, and higher levels of energy consumption.[2]

Ozone (O_3) is a major atmospheric pollutant and perhaps one of the most important in terms of its agricultural impact. Ozone levels reach their highest concentration during the summer months. Ozone levels tend to peak around midday and become depleted at night because of reduced photochemical activity.[3,5] Mean O_3 levels of 0.04 to 0.06 μmol mol^{-1} and peak episodes of 0.1 to 0.2 μmol mol^{-1}, which are known to be phytotoxic to plants, are frequently observed.[6]

Although initial air pollutant research was largely descriptive in nature and focused on acute exposures, current emphasis has been to use both acute[5-12] and chronic[9] exposures and to determine (1) the combined effects of O_3 and other air pollutants on crop yield and quality;[8] (2) the interactive effects of O_3, UV-B radiation, and CO_2, as well as other environmental factors on plant response;[7] and (3) the basis for cultivar and species differences in sensitivity to O_3 exposure.[10]

Plants vary in their sensitivity to O_3 and other air pollutants.[13,14] The severity of leaf injury produced by a pollutant is determined not only by concentration, duration of exposure, and genetic makeup, but also by many environmental variables including light, soil moisture, temperature, and nutrition.[15-18] Plants grown under optimum conditions, such as high soil moisture and high relative humidity, are generally more sensitive to air pollutants than those under less favorable conditions.[7,16]

Recently, biochemical and molecular effects of O_3 in plants have received increased attention.[11,19-28] Biochemical changes are not only sensitive indicators, but also may reveal mechanisms of toxicity and offer an improved understanding of structure–function relationships.[29-31] Physiological, biochemical, and molecular measurements made in several laboratories and in open-top field chambers have contributed greatly to an understanding of the mechanisms of O_3 toxicity in crop plants.

The objective of this chapter is to provide the reader with an overview of early indicators, mechanisms of tolerance, and amelioration of O_3 stress in crop plants. The first part of the review briefly surveys known physiological and biochemical responses of crop plants to O_3 stress. The second part describes some physiological and biochemical indicators of O_3 stress with emphasis on nondestructive techniques currently being used to detect incipient O_3 stress. The third part discusses mechanisms of O_3 tolerance, with emphasis on detoxification. The last part summarizes information on the use of chemical protectants and carbon dioxide (CO_2)-enriched atmospheres to ameliorate O_3 damage in crop plants.

11.2 OZONE STRESS AND PLANT RESPONSES

Air pollution is not a new problem in agriculture. For decades, farmers in congested or industrialized areas have experienced air pollution damage to their crops, often without realizing it. Ozone primarily injures the leaf mesophyll tissues causing leaf chlorosis and/or a characteristic flecking of the upper leaf surface. Important physiological processes such as photosynthesis (Pn), respiration, carbon allocation, and stomatal function are often adversely affected by O_3.[17,19,32-34] Estimates of yield losses in major crop plants, resulting from ambient O_3 stress at a seasonal 7-h d^{-1} mean O_3 concentration of 0.06 μmol mol^{-1} [16,35] generally range from 5 to 20%.[3,6,7,16] This translates into economic losses in the U.S. of about \$4 to \$5 billion, or about 5.6% of the gross values of farm commodities.[35]

11.2.1 Effects of Ambient O_3 on Plants

Numerous books and articles have been published on the effects of ambient O_3 on agricultural crops. Irving[36] has summarized results of the National Acid Precipitation Assessment Program. Results from the National Crop Loss Assessment Network (NCLAN) and other studies provide an excellent overview of this subject.[3,4,37] Krupa and Kickert[7] have discussed in detail the atmospheric factors that are conducive to the transfer of O_3 into the plant canopy and subsequent uptake by the plant. In general, C_3 species such as soybean, cotton, peanuts, spinach, radish, bean, potato, and tomato are more sensitive to chronic O_3 stress than C_4 species such as sorghum and corn.[36,38] Oat appears to be sensitive and wheat is intermediate in response to O_3.[38]

Ambient levels of O_3 have been shown to cause foliar injury and reduce productivity of a number of crops.[3-7] Typical effects of elevated O_3 pollution include a decrease in plant growth,[8-10] an alteration in plant metabolism,[11-12] and a decrease in crop yield.[5-7]

11.2.2 Effects on Photosynthesis

Pn is the driving force in plant productivity. The ability to maintain CO_2 fixation and nitrate assimilation under environmental stress is essential to the maintenance of plant growth and production.[39] Plant exposure to chronic O_3 has been shown to significantly reduce Pn in agricultural crops.[40-43] Such plants also display a decrease in stomatal conductance (g_s) in response to chronic O_3 exposure.[44-46] This reduction in net Pn has been attributed to a decrease in g_s[45,47] and reduced activity of the photosynthetic enzyme, ribulose 1,5-bisphosphate carboxylase/oxygenase (Rubisco).[48,49] Damage to chloroplast membranes also contributes to a decrease in Rubisco activity.[50]

The response of stomatal and photosynthetic processes to air pollutants depends on the duration of exposure and concentration of the pollutant. Ozone sensitivity is also related to the developmental age of the leaf.[51] Many experiments have indicated that leaves are most sensitive to O_3 when they have just reached full size. Photosynthesis is inhibited to a greater extent in newly expanded or mature foliage than in young expanding leaves.[52] Fully expanded leaves showed the strongest decrease in the quantity and activity of Rubisco in O_3-fumigated potato[48,50] and protein mRNAs in *Arabidopsis*.[24] Chronic doses of O_3 exposure of soybean cultivars during mid to late growth stages generally caused a greater decrease in yield than exposure during early growth stages.[9,40] Ozone inhibition of growth and yield in crop plants occurs concomitantly with reductions in net Pn.[40] Effects on dry matter partitioning and Pn responses to O_3 suggest changes in carbohydrate translocation patterns within the plant. Pausch et al.[53] reported that O_3 significantly affected carbon and nitrogen translocation in soybean. An O_3-induced decrease in carbon allocation to the roots and leaves may have serious consequences in plant productivity. Pell et al.[54] indicated that O_3 stress induces shifts in C allocation in favor of shoot growth. Plant physiological strategies for coping with stress can also lead to shifts in shoot/root ratio; these include accelerated senescence of injured leaves in favor of production of new foliage, altered leaf area, and decreased rates of Pn and nitrogen and carbon allocation.

Decreased Pn is also associated with reductions in chlorophyll content, soluble protein, adenylate, and reduced carboxylation efficiency, resulting in reduced CO_2 uptake.[55] Such changes in CO_2 uptake would be expected to alter assimilate partitioning and may be involved in reduced yield responses.[40,56] But reduction in plant biomass and or yield without reduced Pn has also been reported.[57] This apparent lack of Pn response may also be partly related to differences in species or cultivar sensitivity, as well as growth stage. However, there is evidence that species with higher g_s have the greatest negative response to O_3 in terms of growth and yield.[58]

Ozone stress induces several specific changes at the level of the primary photochemical reaction of Pn.[59-61] If damage is severe, visible symptoms may be evident on the leaves or stems. Decreased photosynthetic rate may not be visually evident but can result in subtle changes in biomass production that may be recognized only at harvest or when ameliorative measures no longer help. Depression of Pn rates is, therefore, an important criterion in the evaluation of O_3 and other environmental stress conditions on plants.

11.2.3 Plant Sensitivity and Visual Foliar Injury

Sensitivity of plants to O_3 has been determined by observing changes in visible foliar symptoms, such as necrotic and chlorotic lesions following air pollutant exposure. Such visible foliar symptoms have been reported for many annual crop species including soybean,[45,62-64] corn,[41,42] wheat,[43,64-66] rice,[67-70] tobacco,[66,71] tomatoes,[8,72] and others,[73,74] as well as tree species.[75,76] Visible changes attributable to acute O_3 injury also include flecking of the upper leaf surface of fully expanded leaves.[77] However, the production of visible injury is not linearly related to the amount of O_3, or the time of exposure.[78]

Leaf damage from chronic O_3 exposure is often not very intense in the case of chronic injury; few symptoms may appear on the leaf surface. However, if exposure is prolonged, chlorotic lesions may gradually coalesce on the leaf surface, giving the injured leaf a bronzed or yellowed appearance. Bronzing or yellowing may be accompanied by premature defoliation. Reduced growth and/or reduced yield may also result from long-term exposure of vegetation to low levels of O_3. In the past, leaf injury has been determined by visual scoring procedures. However, these methods require a well-trained person to obtain consistent data. Data on visible injury scores tend to be somewhat subjective and are often not very consistent.

Visual injury caused by air pollution is often accompanied by reductions in the total chlorophyll and carotenoid content of many plant species.[67,79,80] Damage to foliage and/or reduced leaf photosynthetic pigment content in response to O_3 exposure can coincide with reduced plant growth and/or yield.[81]

11.3 PHYSIOLOGICAL AND BIOCHEMICAL INDICATORS OF OZONE STRESS

Numerous techniques are available for identifying or measuring response of plants to O_3, such as physiological and biochemical changes.[82-84] Many of these changes in stomatal conductance, Pn rate, water potential, membrane permeability, leaf pigments, proteins, lipids or other macromolecule, and metabolite pools can be used as indicators for the assessment of O_3 effects and other environmental stress effects on vegetation.[19,21,33,84,85] The detrimental effects of O_3 on plants are well documented, but a clear understanding of the biochemistry of these processes is still lacking.[26] Kainulainen et al.[86] indicated that rising levels of atmospheric O_3 may disturb primary carbohydrate, amino acid, phenylpropanoid, and polyamine metabolism of plants.[87-89] However, chlorophyll content has been used more often than other parameters as a criterion for assessing leaf injury.[90] Many air pollutants decrease chlorophyll content of leaves. Miller et al.[91] showed that O_3 treatments of soybean cultivars (ranging from 14 to 83 nmol mol^{-1} 12 h d^{-1} seasonal means) consistently induced visible injury and suppressed net carbon exchange and water use efficiency.

Recently, molecular biology studies focusing on stress gene expression have been conducted to characterize antioxidant defenses against O_3 stress in a wide range of plants.[27,92-95]

11.3.1 Physiological Indicators

Historically, attention has been focused on the quantitative effects of air pollution on plant yields. Recently, however, there has been a shift in interest toward the qualitative aspects of O_3 injury with emphasis on physiological alterations.[96] Examples of air pollutant-induced physiological changes in the leaves include reduced Pn rate, increased respiration rate, membrane lipid peroxidation, enhanced rate of senescence, and reduced transpiration (due to stomatal closure).[11,12,19,97]

These physiological changes can be detected in O_3-treated leaf tissues, using any one of several stress-detection techniques available developed. These include spectral reflectance, fluorescence, and videography.[98-100] Spectral reflectance is a good indicator of pigment composition and surface architecture of the leaves.[41,98,99] Ozone-stressed leaves usually have a lower reflectance and lower

Photosystem II (PSII) activity than nonstressed leaves of the same chronological age.[42] However, physiological leaf age is difficult to determine. Ozone stress influences the water content of the leaves. Consequently, O_3-damaged leaves develop necrotic areas that dehydrate rapidly. Kraft et al.[99] reported that leaves from noninjured and O_3-injured plants of white clover (*Trifolium repens* cv. Karina) differed significantly in chlorophyll reflectance. These investigators also found that O_3 deceased near-infrared (NIR) reflectance in leaves of white clover leaves, even before visible injury was apparent. However, these changes in NIR reflectance were not correlated with physiological or biochemical alterations or growth changes due to O_3 stress.[99] Nevertheless, reflectance and photosynthetic measurements have been found useful in detecting stress effects in leaves of crop plants. One of the best indicators of O_3-induced changes in leaf optical properties is a difference in reflectance at 550 nm caused by variations in leaf pigment concentration. High reflectance is associated with low total chlorophyll concentration, and, conversely, low reflectance is correlated with high chlorophyll concentration.[100]

11.3.2 Early Stress Detection

Chlorophyll fluorescence transients (CFTs) have been commonly used as a diagnostic tool to detect early stress conditions in plants and are measured during the dark–light transition of photosynthetically active leaves.[98,101-103] CFTs are sensitive indicators of photosynthetic energy conversion and have been important in the assessment of quantum yield and other photosynthetic parameters.[101] Such measurements are important in learning how general the stress response is in plants and if there are differences in sensitivity and resistance for different stressors. For the past two decades, we have used measurements of CFTs as a rapid, nondestructive technique for measuring incipient O_3 stress and for evaluating differences in stress tolerance in a number of cultivars of snap bean and soybean.[60] Our results indicate that the pattern of CFTs and the rate of fluorescence induction transients are useful in characterizing basic responses to O_3 stress. Two phases of fluorescence induction kinetics have been used to describe the functioning of the photosynthetic apparatus: (1) a rapid phase characterized by a fast fluorescence rise to the maximum fluorescence (F_m) which is completed in less than 0.5 s and (2) a slow phase characterized by a slow decrease in fluorescence (F_d) to a steady-state level (F_s), where Pn and fluorescence are in equilibrium, which takes 3 to 5 min. The rise in fluorescence from F_o (i.e., fluorescence independent of photochemical events) to F_m has been termed variable fluorescence (F_v); the ratio F_v to F_m is taken as an indicator of the functioning of the photosynthetic apparatus, and is commonly termed *photochemical efficiency.*[101] Any disturbance in the transfer of electrons between PSII and PS I reduces the ability of plants to convert absorbed light energy into chemical energy (ATP and NADPH).[101]

Under O_3 stress conditions, when the photosynthetic quantum conversion is disturbed or inhibited, chlorophyll fluorescence (Chl-Fl) remains high and the decrease from F_m to F_s is slow or does not occur at all.[103] Barnes et al.[104] and Grandjean and Fuhrer[105] observed that wheat plants exposed to chronic O_3 fumigation showed reduced photochemical efficiency indicating that O_3 impaired PSII activity. Grandjean and Fuhrer[105] suggested that reduced photosynthetic rates caused by O_3 were due to metabolic changes rather than direct damage to the photosynthetic apparatus. A similar conclusion was drawn by Farage et al.[55] who exposed wheat plants to acute O_3 fumigation. Schreiber et al.[59] observed that the degree of change in CFTs was greater for bean leaves following long exposures to low O_3 concentration than during brief exposures to high ones.

Recently, fluorescence imaging has been shown to be a promising tool to detect incipient plant stress.[103,106-107] In conjunction with the National Aeronautics and Space Administration (NASA) and the U.S. Department of Agriculture (USDA) Remote Sensing Laboratory at the Beltsville Agricultural Research Center (BARC), a high-resolution UV-induced fluorescence imaging system was developed, which images tissues in four fluorescence bands: blue, green, red, and far red. Fluorescence images in these bands and the corresponding ratio images blue:red and blue:far red are particularly sensitive to environmental stress.[103]

We have recently established the feasibility of using chlorophyll *a* fluorescence and fluorescence imaging to provide a rapid, nondestructive technique to detect incipient effects of O_3 stress. Cultivars of O_3-sensitive ('Forrest') and O_3-tolerant ('Essex') soybean plants were grown outdoors during the growing season in 3-m-diameter open-top chambers under increased O_3 and CO_2 conditions. Within 48 h after O_3 exposure, leaves of plants grown under elevated CO_2 showed significantly higher red (F680) and green (F550) intensities than those from plants grown under ambient CO_2; this was true for both cultivars. The blue:green and blue:far-red fluorescence ratios, which increased under O_3 stress, were higher in O_3-sensitive 'Forrest' than O_3-tolerant 'Essex' plants.

Buschmann et al.[106] have suggested that changes in the fluorescence ratio of the red band near 690 nm and the far-red band near 735 nm may be taken as an indicator of long-term stress effects on the chlorophyll content of the leaves. The inverse relationship between the intensity of Chl-Fl and the rate of Pn (Kautsky effect) can be used for detecting short-term damage to plants, which affects photosynthetic activity, but does not yet decrease the chlorophyll content of the leaf.[106] Lang et al.[107] demonstrated that fluorescence imaging of leaves in blue, green, red, and far-red emission bands is an excellent tool for detection of early stages of water and temperature stress in plants and is superior to the hitherto applied spectral point data measurement.

It is likely that the application of tools such as gas exchange, CFTs, and fluorescence imaging, will play an increasingly important role in the early detection of environmental stress from O_3 pollution and in distinguishing cultivar differences in sensitivity to O_3 exposure.[103,107]

11.3.3 Biochemical Indicators

Numerous reports have been published on the biochemical changes that occur at the cellular and molecular levels following O_3 exposure.[11,12,19,24,27,84,85,93,108] These include, for example, changes in the activities of antioxidative enzymes such as superoxide dismutase (SOD), peroxidase (POX), and glutathione reductase (GR) that remove the active oxygen species formed in the cell during normal metabolism and may or may not change under O_3 stress. A more extensive discussion of these changes is given in the next section.

A variety of biochemical techniques has been used to detect incipient O_3 stress in plant tissues; these include determination of ion leakage of cell membranes,[109,110] lipid peroxidation products,[111] free radical metabolites,[112] levels of secondary metabolites and antioxidants,[113] ethylene,[50,114] and polyamines,[88,115,116] accumulation of salicylic acid,[94,117] chlorophyll loss,[118] and measurement of antioxidant gene expression.[119-121] None of these tests is very specific for O_3; any oxidative stress imposed on a plant may produce similar results. Recently, Pell et al.[27] have proposed several models to explain O_3-induced injury to plant foliage; these included the induction of ethylene, chitinases, β-1,3-glucanase, antioxidants, and suppression of photosynthetic genes.

11.4 MECHANISMS OF OZONE TOLERANCE IN PLANTS

For a number of years, research on the mechanism of action of photochemical oxidants on plant leaves has targeted cell membranes, because of their vulnerability to attack on sulfhydryl or lipid linkages.[19] If membrane permeability is not restored, injured cells may lose water and become necrotic. Current research has focused on the effects of O_3 stress on leaf Pn, biochemical, and molecular responses in both agricultural crops and forest species.

Under conditions where no visible symptoms were observed, linear decreases in net Pn and cellular metabolites or antioxidants were observed with increased O_3 stress for a wide range of plant species.[28,121-125] The use of contrasting genotypes (cultivars or near-isolines) that differ in response to a given pollution exposure (sensitive vs. resistant) has provided invaluable information on the mechanism of plant response to air pollutants.[14,18] This approach has value whether the level of understanding is molecular, physiological, or whole plant.

The responses of plants to air pollutants are undoubtedly influenced by many environmental factors, as well as by plant genotype. There is evidence, for example, that O_3 stress decreases the tolerance of soybeans to soil moisture stress.[126,127] Soil moisture stress and O_3 stress both have important consequences on physiological processes such as Pn, respiration, carbon allocation, and stomatal function.[17,19,34] Ozone can react with unsaturated lipids to initiate lipid peroxidation in plant tissues. Early studies implicated vitamin C and E and the glutathione-dependent oxidant defenses as being critical to ameliorating O_3 toxicity caused by reactive oxygen species.[30,128,129]

To gain a better understanding of the processes regulating levels of antioxidant proteins within a particular experimental system and differential response to air pollutant stress, studies are being conducted to investigate regulation at the gene level.[130] Such studies on antioxidant gene expression will improve our understanding of the mechanisms of plant tolerance to O_3 stress.

Despite vigorous efforts to determine the basis for O_3 tolerance in crop plants, there is still no consensus on a primary mechanism. The literature contains voluminous data regarding O_3-induced metabolic and molecular changes in plants, but no consistent relationships have been established.[26] Various mechanisms of O_3 tolerance have been proposed to explain the basis for genotype differences in O_3 tolerance in plants. These include avoidance by stomatal closure, detoxification of free radicals through the action of natural scavengers affecting gene expression, and induction of antioxidant defense genes at the cellular level.

11.4.1 Avoidance Mechanism

Genetic differences in plant sensitivity to O_3 have been reported to be associated with differences in stomatal frequency or stomatal response.[131] Stomata provide the primary passage of air pollutants into the leaf and thus play an important role in O_3 injury to plants.[46] The rates of O_3 uptake are affected by leaf surface area, boundary layer, stomatal resistance, and leaf metabolism.[46,57] Stomatal closure due to air pollutants has often been correlated with reduction in Pn. One of the possible reasons for the reduced rates of CO_2 fixation associated with O_3 exposure is decreased g_s.

Fluxes of K^+ regulate stomatal movement[46] and O_3 is known to increase membrane permeability to K^+ ions.[19] Stomatal function is normally regulated by changes in the endogenous level of growth regulators such as abscisic acid (ABA). Adedipe and Ormrod,[132] and Downton et al.[133] have suggested that ABA-induced modification of stomatal function is responsible for O_3 resistance in some plants. Stomatal responses to O_3 depend on the levels of O_3 exposure, moisture conditions under which the plants are grown, and on the species and genotype. Butler and Tibbitts[131] compared g_s in two *Phaseolus vulgaris* genotypes having differential O_3 sensitivity and found that the more O_3-tolerant cultivar closed its stomata in response to O_3 exposure while the O_3 sensitive cultivar had small increases in g_s. Engle and Gabelman[134] observed rapid stomatal closure and complete prevention of injury in O_3 tolerant cultivars of *Allium cepa* when plants were exposed to elevated O_3. Similarly, in tobacco plants, the stomata of an O_3-resistant cultivar closed more rapidly and to a greater extent when exposed to O_3 than the stomata of a sensitive cultivar.[135-137] However, there are examples where differences in O_3 tolerance cannot be linked to changes in g_s.[136]

11.4.2 Endogenous Defense Mechanism for Scavenging Free Radicals and Coping With Oxidative Stress

It is well established that many oxidative stresses such as O_3, nitrogen oxides (NO_2), and sulfur dioxide (SO_2) damage plant tissues, at least in part, because they generate active oxygen and other free radicals.[138] Free radicals, such as superoxide (O^-_2) and hydroxyl ($\cdot OH$) are very reactive chemical species and can readily lead to uncontrolled reactions, which may result in cross-linking of DNA, proteins, and lipids to each other or one another, or oxidatively damage functional groups of these important biological macromolecules, causing molecular damage and cellular injury that lead to accelerated aging and senescence.[129]

Free radicals and related oxidants as agents of tissue damage have long been studied by animal scientists.[30,128] The field of antioxidants and oxyradicals in biology was stimulated by the discovery that SOD was responsible for scavenging free radicals.[139] Oxyradicals can cause serious damage by attacking almost every type of molecule found in living cells including DNA, proteins, and membrane lipids.[140-144] Damage to membrane lipids can be especially serious because once initiated, peroxidation proceeds by a spontaneous chain reaction which quickly destroys membrane integrity.[145,146] To minimize oxygen free radical damage, cells are equipped with enzymatic defense systems designed to eliminate these potentially harmful levels of hydrogen peroxide (H_2O_2) and oxyradicals.

Both enzymatic and nonenzymatic systems have been identified in higher plants for detoxifying oxyradicals and for providing cellular protection against free radical–induced cell damage. Enzymatic systems include SOD, catalase (CAT), GR, glutathione peroxidase (GSH-Px), and ascorbate peroxidase (APX).[28,92,95] Non-enzymatic systems include vitamin C, vitamin E, selenium, carotenoids, and flavonoids.[147-151] SOD catalyzes the dismutation of two superoxide radicals to produce H_2O_2 and dioxygen.[123]

An important feature of antioxidant defense mechanisms is the interaction series of redox-based antioxidant cycles and non-redox-based antioxidants which act additively and synergistically in protection.[129] However, the endogenous biological oxidants and free radicals formed during metabolism are not easy to measure directly, since most of these chemical species are highly reactive and short lived.[119]

Enhancing the ability of plants to detoxify damaging O_3 effectively, free radical products would eventually result in less O_3 injury than would occur in the plants with low antioxidant capacity. One of the most efficient detoxification mechanisms is the ascorbate/glutathione cycle, in which H_2O_2 is scavenged.[152-157] This cycle appears essential in photosynthetic tissues.[157] Since this scavenging pathway is particularly relevant under severe stress that leads to oxidative damage, many studies have been conducted to determine whether or not O_3 tolerance involves changes in the function of the ascorbate/glutathione cycle. The capacity of acclimation and deployment of free radical scavenging systems in the presence of air pollutants varies both between and within species.

Because H_2O_2 also has the potential to cause cellular damage, a second enzyme, CAT, reacts with two H_2O_2 molecules to form dioxygen and water. The combined action of SOD and CAT efficiently eliminates H_2O_2 and O^-_2 radicals, and indirectly protects against the more reactive $\cdot$OH, which results from interactions between H_2O_2 and O_2.[152] Although CAT is widespread, it is not efficient at removing low concentrations of peroxides because it has a low affinity for H_2O_2 and is largely restricted to peroxisome.[141] This problem is overcome by the selenium-containing enzyme GSH-Px and APX. Hydrogen peroxide scavenging in higher plants is usually accomplished by the enzyme APX.

Glutathione peroxidase is absent in most plants; however, it has recently been reported that some higher plants as well as algae contain a GSH-Px with properties resembling those of the animal GSH-Px.[153] This enzyme catalyzes the peroxidation of reduced GSH, thus forming the oxidized disulfide form of GHS (GSSG).[148] The supply of GSH is regenerated in an NADPH-dependent reaction catalyzed by GR. This system is critical in the protection of cell organelles and other proteins from peroxide and oxyradical damage. Glutathione has been reported to cause a massive and selective induction of plant defense genes.[154]

Wellburn and Wellburn[95] compared the free radical–scavenging ability in air pollution-tolerant and -sensitive cultivars. Rao et al. [158] and Rao and Dubey[159] showed that exposure of *Arabidopsis thaliana* to O_3 could enhance SOD, POX, GR, and APX activities. Under oxidative stress such as exposure to O_3, SO_2, low temperature, heat shock, or drought, the level of reduced GSH in foliar tissue has been shown to increase above control levels.[108,147,160,161]

Various studies have shown that differences in antioxidant enzyme activities between genotypes are often correlated with differences in O_3 tolerance. For example, Guri[108] determined that O_3-insensitive cultivars of *Phaseolus vulgaris* had twice as much GR activity as sensitive cultivars and more reduced form of GSH after O_3 exposure. Glutathione reductase also appeared to be involved in O_3 tolerance of *Nicotiana tabacum* cultivars after exposure to 300 nmol mol^{-1} O_3 for 24 h.[155] Mehlhorn et al.[162] reported a twofold increase in levels of APX and GR in peas which had been exposed for 3 weeks to either O_3 alone at 0.15 µmol mol^{-1} or to mixtures of SO_2, O_3, and NO_x at individual concentration of 0.05 µmol mol^{-1} each.

Ozone tolerance in common bean genotypes was also associated with antioxidant accumulation.[136] These increases may be the expression of an enzyme induction process. Lee et al.[137] studied the relationship between foliar O_3 tolerance and leaf ascorbic acid concentrations in O_3-susceptible (O_3-S) and O_3-resistant (O_3-R) cultivars of soybean cv. 'Hark' (O_3-S) and 'Hood' (O_3-R), and in snap bean cv. 'BBL-290' (O_3-S) and 'Astro' (O_3-R). Young trifoliate leaves were highly tolerant to O_3 and had proportionally higher ascorbic acid concentrations than newly expanded leaves. A threshold concentration of approximately 1000 µg ascorbic acid per gram of leaf fresh weight was required for good O_3 protection. Ascorbic acid is an antioxidant and a free radical scavenger which at physiological levels can protect against lipid peroxidation and leaf damage. Ozone stress also induced production and accumulation of ascorbic acid in O_3-treated leaves.[137] Plant tolerances to other oxidative stresses such as drought, low temperature stress, paraquat, SO_2, UV radiation, and the aging process also involve the ascorbate-glutathione cycle and antioxidant enzymes (e.g., GR and APX).[156,162-165]

The development of gene-transfer techniques and the use of transgenic plants have provided new approaches toward understanding the functional role(s) of antioxidative enzymes and plant tolerance under oxidative stress.[11,92,154,156,157] Willekens et al.[166] have studied the expression of antioxidant genes in response to near ambient conditions of oxidative stress such as O_3, SO_2, and UV-B in *Nicotiana plumbaginifolia* Viv. The genes analyzed encode four different SODs, three CATs (Cat1, Cat2, and Cat3), the cytosolic APX (cyt APX), and GSH-Px. Their data showed that all of these oxidative stresses had similar effects on mRNA accumulation of antioxidant genes in tobacco plants. Transgenic tobacco plants with enhanced cytosolic SOD and GR activities also have been shown to have higher tolerance to paraquat and SO_2 stresses than control plants.[156] Their results indicated that GR and SODs play an essential role in the tolerance of plants to paraquat-induced photooxidative and SO_2 stresses. Overproduction of Mn-SOD in chloroplasts also protected *N. tabacum* plants against O_3 injury.[122]

In recent years, transgenic plants that overexpressed SODs have been generated for several species, and the resulting plants were then subjected to various forms of oxidative stress to determine whether or not overexpression resulted in increased oxytolerance. Unfortunately, to date, transgenic experiments involving overexpression of antioxidative enzymes in response to O_3 stress have had mixed results.[95] Studies with Cu, Zn-SOD indicated that high copper, zinc-SOD contents in the cytosol are more important for O_3 tolerance than the enhanced activity of chloroplastic Cu-Zn-SOD isoform.[120,121] However, there is lack of evidence regarding genetic variation in CAT content providing differential tolerance to any oxidative stress. Transgenic plants with elevated APX levels have not been produced and the role of APX in O_3 tolerance has not been completely resolved. Overproduction of APX in the tobacco (*N. tabacum* cv. Bel W3) chloroplast did not provide protection against O_3 stress.[167] Antioxidant defenses are both highly parallel (different antioxidants play similar roles) and cooperative (e.g., the enzymes SOD plus CAT operate in tandem to decompose singlet oxygen to water). Consequently, measurement of only a single antioxidant does not have much relevance to the overall defenses of the cell.[25] Understanding the bases of the differential responses of the different antioxidative enzymes to oxidative or O_3 stress is needed to help clarify the specific functions of multiple forms of these enzymes.

11.4.3 Detection of Free Radicals and Antioxidants

Despite the short lives of free radicals and the extreme difficulty in measuring these highly reactive species at the cellular level, sensitive and specific measurement techniques have been developed to detect free radicals and their biological oxidation products at sites of tissue injury. One such method is the use of electron spin (ESR) or paramagnetic resonance (EPR) to detect free radical metabolites such as O^{-}_{2} and ·OH radicals *in vitro* and *in vivo* by spin trapping (e.g., Tiron and dimethylpyrolline-*N*-oxide, DMPO).[168,169]

Recently developed high-field nuclear magnetic resonance (NMR) spectrometers have been used for rapid and simultaneous determination of free radical components in extracts of tissue samples.[170] This approach has provided much useful information about reactive radical or H_2O_2-scavenging activities of antioxidants.

In addition, several biochemical methods have been used for indirect detection of enzymatic activities in tissues. These include measurement of antioxidants and scavenging enzymes,[171,172] detection of steady-state levels of antioxidant mRNA,[24,173] and use of high-performance liquid chromatographic techniques to detect ·OH radical in tissues.[119] These techniques, which are described in detail by Kaur and Halliwell,[174] involve generation, trapping, and detection of ·OH radical adduct products, Lee et al.[147] employed these techniques and used salicylate to trap ·OH radicals to form specific products to test the hypothesis that ethylenediurea (EDU) protects against O_3 damage by scavenging hydroxyl free radicals.

Much is known about the role of free radicals in the aging process in animal systems, and correlations have been made between free radical–scavenging enzyme levels and detoxification processes in these organisms.[148] However, knowledge concerning the role of antioxidants in protecting plant cells against atmospheric stress (e.g., O_3 and SO_2) associated with growth reduction and yield loss is limited.

11.4.4 Variation in Plant Tolerance in Relation to Genetic and Environmental Factors

Numerous screening tests for O_3 injury have demonstrated that cultivars of most agronomic and horticultural crops differ considerably in O_3 resistance.[14] Some of this variation is due to genetic factors, but much is in response to differences in environmental conditions both prior to and during fumigation.[4,17] Variation in tolerance also occurs with the geographic area from which plant populations of a species are derived.[175] For example, seedlings of green ash (*Fraxinus pennsylvanica*) grown from seeds collected in the eastern U.S. were more tolerant to O_3 than those collected from plants in Manitoba. Differences in biochemical makeup of the plant also cause differences in O_3 sensitivity. For example, Robinson and Rowland[176] found cultivar differences in key photosynthetic enzymes in spinach; they obtained a positive correlation between tolerance to O_3 and elevated leaf levels of glucose, glucose-6-phosphate, and respiratory substrates (pyruvate and malate). Ozone can increase the level of soluble sugars and carbohydrates in the leaves of treated plants.[177] Because sugars can act as scavengers of free radicals,[178] and as a precursor of an antioxidant (e.g., ascorbic acid),[179] it is possible that an increase or difference in levels of antioxidant can partly overcome phytotoxic effects of free radicals produced by O_3.

11.5 AMELIORATION OF OZONE-INDUCED OXIDATIVE STRESS

11.5.1 Use of Exogenous Application of Growth Regulators and Other Chemicals as Antioxidants, Antiozonants, and Antitranspirants

Interest in the use of exogenous application of plant growth regulators, growth retardants, fungicides, vitamins, and other chemicals as antioxidants, antiozonants, and antitranspirants for chemically protecting plants against oxidative stress imposed by air pollutants and other stress factors has grown dramatically.[180-193] Lee et al.[180,182] were among the first to propose that a hormonal mechanism might be involved in phytoprotection against air pollutant injury in crop plants. They postulated (1) that the susceptibility to air pollutants may be related to the concentration of endogenous plant hormones (such as GA_3,[180]) and antioxidants in plant tissues[137] and (2) that exogenous application of growth regulators and antioxidants or antiozonants might ameliorate air pollution injury by inducing detoxification mechanisms. Research has also been conducted to determine the role of avoidance mechanisms involving the use of antitranspirants to induce stomatal closure and thereby prevent the uptake of O_3 and other air pollutants.

The foliar application or soil drench of several chemical compounds to crop plants provides varying degrees of protection against air pollution injury; these include EDU;[181-187] piperonyl butoxide (butox);[187] ascorbic acid;[151] cytokinins;[145] abscisic acid;[133,189] growth retardants such as ancymidol, chlormequat, and (2-chloroethyl) trimethyl ammonium chloride;[190] triazoles such as paclobutrazol[180] and sumagic (S-3307).[191] These active ingredients are mixed with water containing a surfactant prior to foliar spraying or soil treatment.

Soil application of five inhibitors of gibberellin (GA) biosynthesis, namely, paclobutrazol, chlormequat, ancymidole, tetcyclasis, and flurprimidol was found to offer protection against SO_2 pollution in snap bean and soybean. The compounds EDU, ascorbic acid, and butox were found to provide varying degrees of protection against O_3 stress,

Although some of these compounds are believed to provide protection by virtue of their antioxidant abilities, others such as EDU are effective as antiozonants but do not appear to act as antioxidants.[147] Instead EDU appears to offer indirect protection by retention of chlorophyll and maintenance of GR and reduced GSH levels during O_3 exposure.[147] The exact mechanism by which these chemicals induce resistance to air pollutants remains to be elucidated.

11.5.2 Use of Elevated Carbon Dioxide to Ameliorate Ozone Stress

In studies in progress in USDA facilities at Beltsville, MD,[40-43] in Raleigh, NC,[194] and at several other laboratories,[195-196] investigators are determining the interactive effects of low-level CO_2 enrichment (i.e., 10% above ambient) and O_3 stress in crop plants. These studies are being conducted to test the hypothesis that elevated CO_2 will ameliorate the effects of air pollution stress. By using open-top chambers, soybean, winter wheat, corn, and cotton plants were grown in field plots at Beltsville and exposed to season-long ambient (~350 μmol mol^{-1}) and elevated (500 μmol mol^{-1}) CO_2.[40-43] In some of the experiments, soil moisture stress was also imposed. Typically, three O_3 regimes were used: charcoal filtered, nonfiltered, and nonfiltered air plus 40 nmol mol^{-1} O_3.

Comparative physiological and biochemical measurements were made on a number of parameters including shoot weight, yield, foliar gas exchange rates (e.g., Pn, g_s), photorespiration, electron transport system, and polyamine metabolism.[40-43,64,116] Comparative measurements of Pn, g_s, and shoot weight made in field-grown soybean grown under the above O_3 regimes revealed that elevated CO_2 ameliorated the deleterious effects of increased O_3. Rudorff et al.[41-43] also showed that CO_2 enrichment counteracted the negative effects of ambient O_3 (i.e., 55 ± 5 nmol mol^{-1} O_3) stress on growth and productivity of soybeans. However, the effects of CO_2 enrichment were reversed by increasing the ambient O_3 concentration by 40 nmol mol^{-1} O_3. Results also showed that increasing O_3 at ambient CO_2 increased the concentration of polyamines in soybean leaves, but at elevated CO_2, polyamine accumulation under O_3 fumigation was blocked.[116]

The mechanism of action responsible for reversing the stimulation by CO_2 is likely a combination of reduced g_s and reduced Rubisco activity.[195] Barnes et al.[195] reported that elevated CO_2 reduced the adverse effects of O_3 pollution in wheat, a C_3 species, and concluded that this protection was due at least in part to a reduction in stomatal conductance and reduced O_3 uptake by the boundary layer of the mesophyll cells. The decline in O_3 flux under CO_2-enriched conditions was associated with a reduced effect of O_3 on leaf photosynthetic capacity.

Although numerous studies have been conducted on the interactive effects of CO_2 and O_3, the specific mechanism by which elevated CO_2 ameliorates O_3-induced oxidative stress remains to be determined. It is also not clear why, in contrast to O_3, elevated CO_2 has failed to ameliorate air pollution injury from SO_2 in certain studies.[64] The question of what might happen at higher CO_2

level, e.g., 650 µmol mol^{-1}, is unclear. Lee and his associates speculate that this increase may cause partial stomatal closure which in turn may reduce the damaging effects caused by current levels of ambient O_3. Several laboratories have started to conduct experiments to determine whether or not changes in the antioxidant status in plant tissues play a role to play in the enhanced tolerance of CO_2-enriched leaves to O_3. Recent results showed that leaf antioxidant enzymes (e.g., GR, SOD, CAT, and SOD) are involved in conferring CO_2-induced tolerance to O_3 stress in both soybean cultivars 'Essex' and 'Forrest,' grown in open-top chambers in the field. Exposure of plants to elevated CO_2 (500 µmol mol^{-1} CO_2) during elevated ozonation (60 nmol O_3 mol^{-1} air) caused 'Essex' to be more tolerant to O_3 than 'Forrest,' in part because the activity per unit leaf area of antioxidant enzymes (GR, CAT, and SOD) was greater in leaves of 'Essex' than in those of 'Forrest.' Both high light and elevated carbohydrate levels favored the maintenance of high total (reduced + oxidized) ascorbic acid in leaves as well as an increased ratio of reduced to oxidized form of the compounds.[31] These findings may be useful in developing plants with improved stress tolerance in the coming century.

11.6 CONCLUSIONS

Plant resistance to air pollution stress is complex and involves a wide array of responses ranging from the molecular and cellular level to the whole-plant level. Genotype differences in response to O_3 and other air pollutants in crop plants are related to (1) stomatal behavior of the leaf surface and (2) the free radical–scavenging ability of endogenous antioxidant compounds in the leaf mesophyll cells. The primary enzymatic defense system that has evolved in plants is the ability to eliminate potentially harmful levels of free radicals. The biochemical basis for detoxification of free radicals has been shown to involve several key enzymatic and nonenzymatic antioxidant systems. Recent studies indicate that certain growth regulators and other chemicals as well as elevated CO_2 levels also are effective in ameliorating O_3 stress. Considerable improvements in diagnostic techniques have been developed for detecting O_3 stress. Among the most promising of these are use of fluorescence image analysis and chlorophyll fluorescence. This review chapter should be useful to researchers involved in studying air pollution effects in plants and to policy makers interested in the impacts of global change.

REFERENCES

1. McKee, D. J., *Tropospheric Ozone: Human Health and Agricultural Impacts*, CRC Press, Boca Raton, FL, 1994, 333 pp.
2. Yunus, M. and Iqbal, M., Eds., *Plant Response to Air Pollution*, John Wiley & Sons, New York, 1996, 545 pp.
3. Heggestad, H. E. and Bennett. J. H., Impact of atmospheric pollution on agriculture, in *Air Pollution and Plant Life*, Treshow, M., Ed., John Wiley & Sons, New York 1984, 357.
4. Heck, W. W., Taylor, O. C., and Tingey, D. T., Eds., *Assessment of Crop Loss From Air Pollutants*, Proc. of an International Conference Raleigh, NC. Oct. 25–29, 1987, Elsevier, London, 1988, 552 pp.
5. Krupa, S. V. and Manning, W. J., Atmospheric ozone: formation and effects on vegetation, *Environ. Pollut.*, 50, 101, 1988.
6. Heagle, A. S., Ozone and crop yields, *Annu. Rev. Phytopathol.*, 27, 397, 1989.
7. Krupa. S. V. and Kickert, R. N., Considerations for establishing relationships between ambient ozone (O_3) and adverse crop response, *Environ. Rev.*, 5, 55, 1997.
8. Deveaou, J. L., Ormrod, D. P., Allen, O. B., and Berkerson, D. W., Growth and foliar injury responses of maize, soybean and tomato seedlings exposed to mixtures of ozone and sulphur dioxide, *Agric. Ecosyst. Environ.*, 19, 223, 1987.

9. Heagle, A. S., Miller, J. E., Rawlings, J. O., and Vozzo, S. F., Effects of growth stage on soybean response to chronic ozone exposure, *J. Environ. Qual.*, 20, 562, 1991.
10. Miller, J. E., Pursley, W. A., Vozzo, S. F., and Heagle, A. S., Response of net carbon exchange rate of soybean to ozone at different stages of growth and its relation to yield, *J. Environ. Qual.*, 20, 571, 1991.
11. Sharma, Y. K. and Davis, K. R., Ozone-induced expression of stress-related genes in *Arabidopsis thaliana*, *Plant Physiol.*, 105, 1089, 1994.
12. Heath, R. L., Alterations of plant metabolism by ozone exposure, in *Plant Responses to the Gaseous Environment*, Alscher, R. G. and Wellburn, A. R., Eds., Chapman & Hall, London, 1994, 121.
13. Ormrod, D. P., Air pollution and seed growth and development, in *Plant Response to Air Pollution*, Yunus, M. and Iqbal, M., Eds., John Wiley & Sons, New York, 1996, 425.
14. Reinert, R. A., Heggestad, H. E., and Heck, W. W., Response and genetic modification of plants for tolerance to air pollutants, in *Breeding Plants for Less Favorable Environments*, Christiansen, M. N. and Lewis, C. F., Eds., John Wiley & Sons, New York, 1982, 259.
15. Leone, L. A. and Brennan, E., Modification of sulfur dioxide injury to tobacco and tomato by varying nitrogen and sulfur nutrition, *J. Air Pollut. Contr. Assoc.*, 22, 544, 1972.
16. Heck, W. W., Adams, R. M., Cure, W. W., Heagle, A. S., Heggestad, H. E., Kohut, R. J., Kress, L. W., Rawlings, J. O., and Taylor, O. C., A reassessment of crop loss from ozone, *Environ. Sci. Technol.*, 17, 573, 1983.
17. Darrall, N. M., The effect of air pollutants on physiological processes in plants, *Plant Cell Environ.*, 12, 1, 1989.
18. Heck, W. W., Future directions in air pollution research, in *Effects of Air Pollution on Agriculture and Horticulture*, Unsworth, M. H. and Ormrod, D. P., Eds., Proc. of the 32nd Univ. of Nottingham School of Agri. Sci., Butterworth, London, 1982, 419.
19. Heath, R. L., Biochemical mechanisms of pollutant stress, in *Assessment of Crop Loss from Air Pollutants*, Heck, W. W., Taylor, O. C., and Tingey, D. T., Eds., Elsevier, London, 1988, 259.
20. Shaaltiel, Y., Glazer, A., Bocion, P. F., and Gressel, J., Cross tolerance to herbicidal and environmental oxidants of plant biotypes tolerant to paraquat, sulfur dioxide, and ozone, *Pest. Biochem. Physiol.*, 31, 13, 1988.
21. Pell, E. J. and Reddy, G. N., Oxidative stress and its role in air pollution toxicity, *Curr. Top. Plant Physiol.*, 6, 67, 1991.
22. Alscher, R. G. and Wellburn, A. R., Eds., *Plant Responses to the Gaseous Environment*, Chapman Hall, London, 1994, 395 pp.
23. Eckey-Kaltenbach, H., Ernst, D., Heller, W., and Sandermann, H., Jr., Biochemical plant responses to ozone. IV. Cross-induction of defensive pathways in parsley (*Petroselinum crispum* L.) plants, *Plant Physiol.*, 104, 67, 1994.
24. Conklin, P. L. and Last, R. L., Differential accumulation of antioxidant mRNAs in *Arabidopsis thaliana* exposed to ozone, *Plant Physiol.*, 109, 203, 1995.
25. Beckman, K. B. and Ames, A. N., Oxidants, antioxidants, and aging, in *Oxidative Stress and the Molecular Biology of Antioxidant Defenses*, Scandalios, J. G., Ed., Cold Spring Harbor Laboratory Press, Cold Spring Harbor, 1997, 201.
26. Mudd, J. B., Biochemical basis for the toxicity of ozone, in *Plant Response to Air Pollution*, Yunus, M. and Iqbal, M., Eds., John Wiley & Sons, New York, 1996, 267.
27. Pell, E. J. Schlagnhaufer, C. D., and Arteca, R. N., Ozone-induced oxidative stress: mechanism of action and reaction, *Physiol. Plant.*, 100, 264, 1997.
28. Scandalios, J. G., Ed., *Oxidative Stress and the Molecular Biology of Antioxidant Defenses*, Cold Spring Harbor Laboratory Press, Cold Spring Harbor, 1997, 839 pp.
29. Mustafa, M. G. and Lee, S. D., Biological effects of environmental pollutants: methods for assessing biochemical changes, in *Assessing Toxic Effects of Environmental Pollutants*, Lee, S. D. and Mudd, J. B., Eds., Cambridge University Press, London, 1979, 60.
30. Chow, C. K., Ed., *Cellular Antioxidant Defense Mechanisms*, CRC Press, Boca Raton, FL, 3 Vols, 1988.
31. Chernikova, T., Ozone effects on growth, physiological characteristics and antioxidant enzymes in soybean cultivars exposed to ambient and elevated carbon dioxide, Ph.D. dissertation, University of Maryland, College Park, 1997, 217 pp.
32. Larson, R. A., Plant defenses against oxidative stress, *Arch. Insect Biochem. Physiol.*, 29, 175, 1995.

33. Saxe, H., Photosynthesis and stomatal responses to polluted air, and the use of physiological and biochemical responses for early detection and diagnostic tools, *Adv. Bot. Res.*, 18, 1, 1991.
34. Baker, N. R., Ed., *Photosynthesis and the Environment*, National Research Council, Research Press, Canada, 1997, 491 pp.
35. Adams, R. M., Glyer, D. J., and McCarl, B. A., The NCLAN economic assessment--approach, findings, and implications, in *Assessment of Crop Loss From Air Pollutants*, Heck, W. W. and Taylor, O. C., and Tingey, D. T., Eds., Elsevier, London, 1988, 473.
36. Irving, P. M., Effects on agricultural crops, in *NAPAP Interim Assessment, Vol. IV: Effects of Acidic Deposition*, The NADAP, Washington, D.C., 1988.
37. Dempster, J. P. and Manning, W. J., Special Issue: Response of crops to air pollutants, *Environ. Pollut.*, 53, 1, 1988.
38. Heath, R. L., Ozone, in *Responses of Plants to Air Pollution*, Mudd, J. B. and Kozlowski, T. T., Eds., Academic Press, New York, 1975, 23.
39. Lawlor, D. W., The effects of water deficit on photosynthesis, in *Environmental and Plant Metabolism*, Smirnoff, N., Ed., Bios Scientific Publ., U.K., 1995, 129.
40. Mulchi, C. L., Slaughter, L., Saleem, M., Lee, E. H., Pausch, R., and Rowland, R., Growth and physiological characteristics of soybean in open-top chambers in response to ozone and increased atmospheric CO_2, *Agric. Ecosyst. Environ.*, 38, 107, 1992.
41. Rudorff, B. F. T., Mulchi, C. L., Daughtry, C. S. T., and Lee, E. H., Growth, radiation use efficiency, and canopy reflectance of wheat and corn grown under elevated ozone and carbon dioxide atmospheres, *Remote Sensing Environ.*, 55, 163, 1996.
42. Rudorff, B. F. T., Mulchi, C. L., Lee, E. H., Rowland, R. A., and Pausch, R., Effects of enhanced O_3 and CO_2 enrichment on plant characteristics in wheat and corn, *Environ. Pollut.*, 94, 53, 1996.
43. Rudorff, B. F. T., Mulchi, C. L., Lee, E. H., Rowland, R. A., and Pausch, R., Photosynthetic characteristics in wheat exposed to elevated O_3 and CO_2, *Crop Sci.*, 36, 1247, 1996.
44. Aben, J. M., Janssen-Jurkovicova, M., and Adema, E. H., Effects of low-level ozone exposure under ambient conditions on photosynthesis and stomatal control of *Vicia faba* L., *Plant Cell Environ.*, 13, 436, 1990.
45. Mulchi, C. L., Lee, E. H., Tuthill, K., and Olinick, E. V., Influence of ozone stress on growth processes, yields and grain quality characteristics among soybean cultivars, *Environ. Pollut.*, 53, 151, 1988.
46. Mansfield, T. A. and Person, M., Disturbances in stomatal behavior in plants exposed to air pollution, in *Plant Response to Air Pollution*, Yunus, M. and Iqbal, M., Eds., John Wiley & Sons, New York, 1996, 179.
47. Slaughter, L. H., Mulchi, C. L., and Lee, E. H., Wheat kernel growth characteristics during exposure to chronic ozone pollution, *Environ. Pollut.*, 81, 73, 1993.
48. Eckardt, N. A. and Pell, E. J., O_3-induced degradation of Rubisco protein and loss of Rubisco m-RNA in relation to leaf age in *Solanum tuberosum* L., *New Phytol.*, 127, 741, 1994.
49. Rao, M. V., Paliyath, G., and Ormrod, D. P., Differential response of photosynthetic pigments, rubisco activity and Rubisco protein of *Arabidopsis thaliana* exposed to UV-B and ozone, *Phytochem. Photobiol.*, 62, 727, 1995.
50. Glick, R. E., Schlagnhaufer, C. D., Arteca, R. N., and Pell, E. J., Ozone-induced ethylene emission accelerates the loss of ribulose-1,5-bisphosphate carboxylase-oxygenase and nuclear-encoded mRNA in senescing potato leaves, *Plant Physiol.*, 109, 891, 1995.
51. Lee, E. H., Plant resistance mechanisms to air pollutants: rhythms in ascorbic acid production during growth under ozone stress, *Chronobiol. Intern.*, 8, 93, 1991.
52. Paakkonen, E., Metsarinne, S., Holopainen, T., and Karenlampi, L., The ozone sensitivity of birch (*Betula pendula*) in relation to the developmental stage of leaves, *New Phytol.*, 132, 145, 1995.
53. Pausch, R. C., Mulchi, C. L., Lee, E. H., Forseth, I. N., and Slaughter, L. H., Use of ^{13}C and ^{14}N isotopes to investigate O_3 effects of C and N metabolism in soybeans, Part I. C fixation and translocation, *Agric. Ecosyst. Environ.*, 59, 69, 1996.
54. Pell, E. J., Temple, P. J., Friend, A. L., Mooney, H. A., and Winner, W. E., Compensation as a plant response to ozone and associated stresses: an analysis of ROPIS experiments, *J. Environ. Qual.*, 23, 429, 1994.
55. Farage, P., Long, S., Lechner, E., and Baker, N., The sequence of change within the photosynthetic apparatus of wheat following short term exposure to ozone, *Plant Physiol.*, 95, 529, 1991.

56. Miller, J. E., Effects of ozone and sulfur dioxide stress on growth and carbon allocation in plants, in *Photochemical Effects of Environmental Compounds*, Saunders, J. A., Ed., Plenum Press, New York, 1987, 55.
57. Taylor, G. E. Jr., Norby, R. J., McLaughlin, S. B., Johnson, A. M., and Turner, R. S., Carbon dioxide assimilation and growth of red spruce *(Picea rubens* Sarg.) seedlings in response to ozone, precipitation chemistry, and soil type, *Oecologia*, 70, 163, 1986.
58. Reich, P. B. and Amundson, R. G., Ambient levels of ozone reduce net photosynthesis in tree and crop species, *Science*, 230, 566, 1985.
59. Schreiber, U., Vidaver, W., Runeckles, V., and Rosen, P., Chlorophyll fluorescence assay for ozone injury in intact plants, *Plant Physiol.*, 61, 80, 1978.
60. Lee, E. H., Chlorophyll fluorescence as an indicator to detect differential tolerance of snap bean cultivars in response to O_3 stress, *Taiwania*, 36, 220, 1991.
61. Owens, T. G., *In vivo* chlorophyll fluorescence as a probe of photosynthetic physiology, in *Plant Responses to the Gaseous Environment*, Alscher, R. G. and Wellburn, A. R., Eds., Chapman Hall, London, 1994, 195.
62. Heagle, A. S., Heck, W. W., Rawlings, J. O., and Philbeck, R. B., Effects of chronic doses of ozone and sulfur dioxide on injury and yield of soybeans in open-top field chambers, *Crop Sci.*, 23, 1184, 1983.
63. Heagle, A. S., Flagler, R. B., Paterson, R. P., Lesser, W. V., and Heck, W. W., Injury and yield response of soybean to chronic doses of ozone and soil moisture deficit, *Crop Sci.*, 27, 1016, 1987.
64. Lee, E. H., Pausch, R. C., Rowland, R. A., Mulchi, C. L., and Rudorff, B. F. T., Responses of field-grown soybean (cv. Essex) to elevated SO_2 under two atmospheric CO_2 concentrations, *Environ. Exp. Bot.*, 37, 85, 1997.
65. Heagle, A. S., Spencer, S., and Letchworth, M. B., Yield response of winter wheat to chronic doses of ozone, *Can. J. Bot.*, 57, 1999, 1979.
66. Mulchi, C. L. and Aycock, M. K., Jr., Response of Maryland tobacco to chronic ozone stress in the field, *Tob. Sci.*, 30, 30, 1986.
67. Agrawal, M., Nandi, P. K., and Rao, D. N., Effects of ozone and sulphur dioxide pollutants separately and in mixture on chlorophyll and carotenoid pigments of *Oryza sativa*, *Water Air Pollut.*, 18, 449, 1982.
68. Greenland, D. J., The Sustainability of Rice Farming, NRC Research Press, Canada, 1997, 320 pp.
69. Jeong, Y. H., Nakamura, H., and Ota, Y., Physiological studies on photochemical oxidant injury in rice plants (II), *Jpn J. Crop Sci.*, 50, 560, 1981.
70. Nakamura, H. and Ota, Y., An injury to rice plants caused by photochemical oxidants in Japan, *Jnp. Agric. Res. Q.*, 12, 69, 1978.
71. Heggestad, H. E., Origin of Bel-W3, Bel-C and Bel-B tobacco varieties and their use as indicators of ozone, *Environ. Pollut.*, 74, 264, 1991.
72. Temple, P. J., Surano, K. A., Mutters, R. G., Bingham, G. E. and Shinn, J. H., Air pollution causes moderate damage to tomatoes, *Calif. Agric.*, March-April, 20–22, 1985.
73. Heggestad, H. E., Howell, R. K., and Bennett, J. H., The effects of oxidant air pollutants on soybean, snap beans, and potatoes, in U.S. EPA Ecological Research Series, EPA-600/3-77-128, 1977, 38 pp.
74. Howitt, R. E., Gossard, T. W., and Adams, R. A., The economic effects of air pollution on annual crops, *Calif. Agric.*, March-April, 22–24, 1985.
75. Schmieden U. and Wild, A., The contribution of ozone to forest decline, *Physiol. Plant.*, 94, 371, 1995.
76. Skelly, J. M., Chappelka, A. M., Laurence, J. A., and Fredericksen, T. S., Ozone and its known and potential effects on forests in eastern United States, in *Forest Decline and Ozone*, Sandermann, H., Wellburn, A. R., and Heath, R. L., Eds., Springer-Verlag, Berlin, 1997, 69.
77. Heath, R. L., Ozone and the physiological basis for injury to green plants, *What's New Plant Physiol.*, 6, 1, 1974.
78. Heck, W. W., Dunning, J. A., and Hindawi, I. J., Ozone: nonlinear relation of dose and injury in plants, *Science*, 151, 577, 1966.
79. Smith, G., Neyra, C., and Brennan. E., The relationship between foliar injury, nitrogen metabolism, and growth parameters in ozonated soybeans, *Environ. Pollut.*, 63, 79, 1990.
80. Whitaker, B. D., Lee, E. H., and Rowland, R. A., EDU and ozone protection: foliar glycerolipids and steryl lipids in snapbean exposed to O_3, *Physiol. Plant.*, 80, 286, 1990.

81. Johnston, J. W., Jr., Shriner, D. S., and Kinerley, C. K., The combined effects of simulated acid rain and ozone on injury, chlorophyll, and growth of radish, *Environ. Exp. Bot.*, 26, 107, 1986.
82. Dass, H. C. and Weaver, G. M., Modification of ozone damage to *Phaseolus vulgaris* by antioxidants, thiols, and sulfhydryl reagents, *Can. J. Plant Sci.*, 48, 569, 1968.
83. Kangasiarvi, J., Talvinen, J., Utrianen, M., and Karjalainen, R., Plant defence systems induced by ozone, *Plant Cell Environ.*, 17, 783, 1994.
84. Heath, R. L. and Taylor, G. E., Jr., Physiological processes and plant responses to ozone exposure, in *Forest Decline and Ozone*, Sandermann, H., Wellburn, A. R. and Heath, R. L., Eds., Springer-Verlag, Berlin, 1997, 317.
85. Jager, H. J., Biochemical indication of an effect of air pollution on plants, in *Monitoring of Air Pollutants by Plants*, Steubing, L. and Jager, H. J., Eds., W. Junk Publishers, The Hague, 1982, 99.
86. Kainulainen, P. Holopainen, J.K., Hyttinen, H., and Oksanen, J., Effects of ozone on the biochemistry and aphid infestation of Scots pine, *Phytochemistry*, 35, 39, 1994.
87. Schraudner, M., Trost, M., Kerner, K., Heller, W., Leonardi, S., Langebartels, C., and Sandermann, H., Jr., Ozone induction and function of polyamines in ozone-tolerant and ozone-sensitive tobacco cultivars, *Curr. Top. Plant Physiol.*, 5, 394, 1990.
88. Langebartels, C., Kerner, K., Leonardi, S., Schraudner, M., Trost, M., Heller, W., and Sandermann, H., Jr., Biochemical plant responses to ozone: I. Differential induction of polyamine and ethylene biosynthesis in tobacco, *Plant Physiol.*, 95, 882, 1991.
89. Tuomainen, J., Pellinen, R., Roy, S., Kiiskinen, M., Eloranta, T., Karjalainen, R., and Kangasijarvi, J., Ozone affects birch (*Betula pendula* Roth) phenylpropanoid, polyamine and active oxygen detoxifying pathways at biochemical and gene expression level, *J. Plant Physiol.*, 148, 179, 1996.
90. Tingey, D. T., Wilhour, R. G., Taylor, O. C., Heck, W. W., Krupa, S. V., and Lizon, S. N., Eds., The measurement of plant responses, in *Methodology for the Assessment of Air Pollution Effects on Vegetation*, Proc. APCA Specialty Conf. April 19–21, 1978, APCA, Minneapolis, MN, 1978, 7.1.
91. Miller, J. E., Booker, F. L., Fiscus, E. L., Heagle, A. S., Pursley, W. A., Vozzo, S. F. and Heck, W. W., Ultraviolet-B radiation and ozone effects on growth, yield, and photosynthesis of soybean, *J. Environ. Qual.*, 23, 83, 1994.
92. Scandalios, J. G., Molecular genetics of superoxide dismutases in plants, in *Oxidative Stress and the Molecular Biology of Antioxidant Defenses*, Scandalios, J. G., Ed., Cold Spring Harbor Laboratory Press, Cold Spring Harbor, 1997, 527.
93. Sharma, Y. K. and Davis, K. R., The effects of ozone on antioxidant responses in plants, *Free Radical Bio. Med.*, 23, 480, 1997.
94. Sharma, Y. K., Leon, J., Raskin, I., and Davis, K. R., Ozone-induced responses in *Arabidopsis thaliana:* the role of salicylic acid in the accumulation of defense-related transcripts and induced resistance, *Proc. Natl. Acad. Sci. U.S.A.*, 93, 5099, 1996.
95. Wellburn, A. R. and Wellburn, F. A. M., Air pollution and free radical protection responses of plants, in *Oxidative Stress and Molecular Biology of Antioxidant Defenses*, Scandalios, J. G., Ed., Cold Spring Harbor Laboratory Press, Cold Spring Harbor, 1997, 861.
96. Tingey, D. T. and Taylor, G. E., Jr., Variation in plant response to ozone: A conceptual model of physiological events, in *Effects of Gaseous Air Pollution in Agriculture and Horticulture*, Unsworth, M. H. and Ormrod, D. P., Eds., Butterworth Scientific, London, 1982, 113.
97. Leshem, Y. Y., Plant senescence processes and free radicals, *Free Radical Biol. Med.*, 5, 39, 1988.
98. Meinander, O., Somersalo, S., Holopainen, T., and Strasser, R. J., Scots pines after exposure to elevated ozone and carbon dioxide probed by reflectance spectra and chlorophyll *a* fluorescence transients, *J. Plant. Physiol.*, 148, 229, 1996.
99. Kraft, M., Weigel, H.-J., Mejer, G.-J., and Brandes, F., Reflectance measurements of leaves for detecting visible and non-visible ozone damage to crops, *J. Plant Physiol.*, 148, 148, 1996.
100. Gausman, H. W., and Quisenberry, J. E., Spectrophotometric detection of plant leaf stress, in *Environmental Injury to Plants*, Katterman, F., Ed., Academic Press, Harcourt Brace Jovanovich, New York, 1990, 257.
101. Schreiber, U., Hormann, H., Neubauer, C. and Klughammer, C., Assessment of photosystem II photochemical quantum yield by chlorophyll fluorescence quenching analysis, *Aust. J. Plant Physiol.*, 22, 209, 1995.

102. Guidi, L., Nali, C., Ciompi, S., Lorenzini, G., and Soldatini, G. F., The use of chlorophyll fluorescence and leaf gas exchange as methods for studying the different responses to ozone of two bean cultivars, *J. Exp. Bot.*, 48, 173, 1997.
103 Lichtenthaler, H. K., Fluorescence imaging as a diagnostic tool for plant stress, *Trends Plant Sci.*, 2, 316, 1997.
104. Barnes, J. D., Velissariou, D., Davison, A. W., and Holevas, C. D., Comparative ozone sensitivity of old and modern Greek cultivars of spring wheat, *New Phytol.*, 116, 707, 1990.
105. Grandjean, G. A. and Fuhrer, J., The response of spring wheat (*Triticum aestivum* L.) to O_3 at higher elevations, III. Responses of leaf and canopy gas exchange, and chlorophyll fluorescence to O_3 flux, *New Phytol.*, 122, 321, 1992.
106. Buschmann, C., Schweiger, J., Lichtenthaler, H. K., and Richter, P., Application of the Karlsruhe CCD-OMA LIDAR fluorosensor in stress detection of plants, *J. Plant Physiol.*, 148, 548, 1996.
107. Lang, M., Lichtenthaler, H., Sowinska, M., Heisel, F., and Miehe, J. A., Fluorescence imaging of water and temperature stress in plant leaves, *Plant Physiol.*, 148, 613, 1996.
108. Guri, A., Variation in glutathione and ascorbic acid content among selected cultivars of *Phaseolus vulgaris* prior to and after exposure to ozone, *Can. J. Plant Sci.*, 63, 733, 1983.
109. Beckerson, D. W. and Hofstra, G., Effects of sulphur dioxide and ozone, singly or in combination, on membrane permeability, *Can. J. Bot.*, 58, 451, 1980.
110. Tiedemann, A. V. and Pfahler, B., Growth stage-dependent effects of ozone on the permeability for ions and non-electrolytes of wheat leaves in relation to the susceptibility to *Septoria nodorum* Berk, *Physiol. Mol. Plant Pathol.*, 45, 153, 1994.
111. Horton, A. A. and Fairhurst, S., Lipid peroxidation and mechanisms of toxicity, in *Critical Reviews in Toxicology*, Vol. 18, in Golberg, L., Ed., CRC Press, Boca Raton, FL, 1987, 27.
112. Floyed, R. A., West, M. S., Hogsett, W. E., and Tingey, D. T., Increased 8-hydroxyguanine content of chloroplast DNA from ozone-treated plants, *Plant Physiol.*, 91, 644, 1989.
113. Schulte-Hostede, S., Darrall, N. M., Blank, L. W., and Wellburn, A. R., Eds., *Air Pollution and Plant Metabolism*, Elsevier Applied Science, London and NY., 1988, 381 pp.
114. Tingey, D. T., Standley, C., and Field, R. W., Stress ethylene evolution: a measure of ozone effects on plants, *Atmos. Environ.*, 10, 969, 1976.
115. Bors, W., Langebartels, C., Michel, C., and Sandermann, H. Jr., Polyamines as radical scavengers and protectants against ozone damage, *Phytochemistry*, 28, 1589, 1989.
116. Kramer, G. F., Lee, E. H., Rowland, R. A., and Mulchi, C. L., Effects of elevated CO_2 concentration on the polyamine levels of field-grown soybean at three O_3 regimes, *Environ. Pollut.*, 73, 137, 1991.
117. Yalpani, N., Enyedi, A. J., Leon, J., and Raskin, I., Ultraviolet light and ozone stimulate accumulation of salicylic acid, pathogenesis-related proteins and virus resistance in tobacco, *Planta*, 193, 372, 1994.
118. Lee, E. H. and Bennett, J. H., Superoxide dismutase: a possible protective enzyme against ozone injury in snapbeans (*Phaseolus vulgaris* L.), *Plant Physiol.*, 69, 1444, 1982.
119. Evans-Rice, C. A., Diplock, A. T., and Symons, M. C. R., Techniques in free radical research, in *Laboratory Techniques in Biochemistry and Molecular Biology*, Vol. 22, Burdon, R. H. and van Knippenberg, P. H., Eds., Amsterdam, 1991, 290 pp.
120. Pitcher, L. H., Brennan, E., Hurley, A., Dunsmuir, P., Tepperman, J. M., and Zilinskas, B. A., Overproduction of *Petunia* chloroplastic copper/zinc superoxide dismutase does not confer ozone tolerance in transgenic tobacco, *Plant Physiol.*, 97, 452, 1991.
121. Pitcher, L. H. and Zilinskas, B. A., Over expression of copper/zinc superoxide dismutase in the cytosol of transgenic tobacco confers partial resistance to ozone-induced foliar necrosis, *Plant Physiol.*, 110, 583, 1996.
122. Bowler, C., Slooten, L., Vandenbranden, S., De Rycke, R., Botterman, J., Sybesma, C., Van Montagu, M., and Inze, D., Manganese superoxide dismutase can reduce cellular damage mediated by oxygen radicals in transgenic plants, *EMBO J.*, 10, 1723, 1991.
123. Bowler, C., Montagu, M. van, and Inze, D., Superoxide dismutase and stress tolerance, *Annu. Rev. Plant Physiol. Plant Mol. Biol.*, 43, 83, 1992.
124. Alscher, R. G. and Hess, J. L., Eds., *Antioxidants in Higher Plants*, CRC Press. 1993, 174 pp.
125. Sandermann, H., Wellburn, A. R. and Heath, R. L., Eds., *Forest Decline and Ozone*, Springer-Verlag, Berlin, 1996, 400 pp.

126. Heggestad, H. E., Gish, T. J., Lee, E. H., Bennett, J. H., and Douglass, D. W., Interaction of soil moisture stress and ambient ozone on growth and yields of soybeans, *Phytopathology*, 75, 472, 1985.
127. Heggestad, H. E., Anderson, E. L., Gish, T. J., and Lee, E. H., Effects of ozone and soil water deficit on roots and shoots of field-grown soybeans, *Environ. Pollut.*, 50, 259, 1988.
128. Pryor, W. A., Ed., *Free Radicals in Biology*, Vol. 5. Academic Press, New York, 1982.
129. Packer, L., Oxidative stress and the antioxidant vitamins C and E in functional medicine, in *Oxidative Processes and Antioxidants*, Paoletti, R. et al., Eds., Raven Press, New York, 1994, 153.
130. Bingle, C. D., Measurement of antioxidant gene expression, in *Free Radicals, A Practical Approach*, Punchard, N. A. and Kelly, F. J., Eds., The Practical Approach Series, Series Editors: Rickwood, D. and Hames, B. D., Oxford University Press, Oxford, U.K., 1996, 287.
131. Butler, L. K., and Tibbitts, T. W., Stomatal mechanisms determine genetic resistance to ozone in *Phaseolus vulgaris* L., *J. Am. Soc. Hortic. Sci.*, 104, 213, 1979.
132. Adedipe, N. O. and Ormrod, O. P., Hormonal control of ozone phytotoxicity in *Raphanus sativus*, *Z. Pflanzenphysiol.*, 68, 254, 1972.
133. Downton, W. J. S., Loveys, B. R., and Grant, W. J. R., Stomatal closure fully accounts for the inhibition of photosynthesis by abscisic acid, *New Phytol.*, 108, 263, 1988.
134. Engle, R. R. and Gabelman, W. H., Inheritance and mechanism for resistance to ozone damage in onion, *Allium cepa* L., *Proc. Am. Soc. Hortic. Sci.*, 89, 423, 1966.
135. Dean, C. E., Stomate density and size as related to ozone-induced weather fleck in tobacco, *Crop Sci.*, 12, 547, 1972.
136. Guzy, M. R. and Heath, R. H., Response to ozone of varieties of common bean (*Phaseolus vulgaris* L.), *New Phytol.*, 124, 617, 1983.
137. Lee, E. H., Jersey, J. A., Gifford, C., and Bennett, J. H., Differential ozone tolerance in soybeans and snapbeans: analysis of ascorbic acid in O_3-susceptible and O_3-resistant cultivars by high-performance liquid chromatography, *Environ. Exp. Bot.*, 24, 331, 1984.
138. Smirnoff, N., Antioxidant systems and plant response to the environment, in *Environmental and Plant Metabolism*, Smirnoff, N., Ed., Bios Scientific Publ., U.K., 1995. 217.
139. Fridovich, I., Superoxide dismutases, *Annu. Rev. Biochem.,* 147, 1975.
140. Fridovich, I., Superoxide dismutases, *Adv. Enzymol.*, 41, 35, 1986.
141. Halliwell, B., The toxic effects of oxygen on plant tissues, in *Superoxide Dismutase*, Vol. I, Oberley, L. W., Ed., CRC Press, Boca Raton, FL, 1982, 89.
142. Halliwell, B. and Gutteridge, J. M. C., *Free Radicals in Biology and Medicine*, Clarendon Press, Oxford, 1989.
143. Hassan, H. M. and Schellhorn, H. E., Superoxide dismutase: an antioxidant defense enzyme, in Oxy-Radicals in *Molecular Biology and Pathology*, Cerutii, P. A., Fridovich, I., and McCord, J. M., Eds., Proc. of UpJohn-UCLA Symposium, Jan. 24–30, 1988, Alan R. Liss, Park City, UT, 1988, 183.
144. Monk, L. S., Fagerstedt, K. V., and Crawford, R. M. M., Oxygen toxicity and superoxide dismutase as an antioxidant in physiological stress, *Physiol. Plant.*, 76, 456, 1989.
145. Pauls, K. P. and Thompson, J. E., Effects of cytokinins and antioxidants on the susceptibility of membranes to ozone damage, *Plant Cell Physiol.*, 23, 821, 1982.
146. Kendall, E. J. and McKersie, B. D., Free radical and freezing injury to cell membranes of winter wheat, *Physiol. Plant.*, 76, 86, 1989.
147. Lee, E. H., Upadhyaya, A., Agrawal, M., and Rowland, R. A., Mechanisms of ethylenediurea (EDU) induced ozone protection: reexamination of free radical scavenger systems in snap bean exposed to O_3, *Environ. Exp. Bot.*, 38, 199, 1997.
148. Caldwell, J. and Jakoby, W. B., *Biological Basis of Detoxication*, Academic Press, New York, 1983.
149. Bennett, J. H., Lee, E. H., and Heggestad, H. E., Biochemical aspects of ozone and oxyradicals: superoxide dismutase, in *Gaseous Air Pollutants and Plant Metabolism*, Koziol, M. J. and Whatley, F. R., Eds., Butterworths, London, 1984, 413.
150. Larson, R. A., The antioxidants of higher plants, *Phytochemistry*, 27, 969, 1988.
151. Chameides, W. L., The chemistry of ozone deposition to plant leaves: Role of ascorbic acid, *Environ. Sci. Technol.*, 23, 595, 1989.
152. Bannister, J. V., Bannister, W. H., and Rotilio, G., Aspects of the structure, function and application of superoxide dismutase, *CRC Crit. Rev. Biochem.*, 22, 111, 1987.

153. Yokota, A., Shigeoka, S., Onishi, T., and Kitaoka, S., Selenium as inducer of glutathione peroxidase in low CO_2 grown *Chlamydomonas reinhardtii*, *Plant Physiol.*, 86, 649, 1988.
154. Wingate, V. P., Lawton, M. A., and Lamb, C. J., Glutathione causes a massive and selective induction of plant defense genes, *Plant Physiol.*, 87, 206, 1988.
155. Tanaka, K., Machida, T., and Sugimoto, T., Ozone tolerance of spinach glutathione reductase in tobacco cultivars, *Agric. Biol. Chem.*, 54, 1061, 1990.
156. Tanaka, K., Mitsuko, A., Hikaru, S., and Kubo, A., Stress tolerance of transgenic *Nicotiana tabacum* with enhanced activities of glutathione reductase and superoxide dismutase, *Biochem. Soc. Trans.*, 24, 200S, 1996.
157. Foyer, C. H., Descourvieres, P., and Kunert, K. J., Protection against oxygen radicals: an important defence mechanism studied in transgenic plants, *Plant Cell Environ.*, 17, 507, 1994.
158. Rao, M. V., Paliyath, G., and Ormrod, D. P., Ultraviolet-B and ozone-induced biochemical changes in antioxidant enzymes of *Arabidopsis thaliana*, *Plant Physiol.*, 110, 125, 1996.
159. Rao, M. V. and Dubey, P. S., Biochemical aspects (antioxidants) for development of tolerance in plants growing at different low levels of ambient air pollutants, *Environ. Pollut.*, 64, 55, 1990.
160. Grill, D., Esterbauer, H., and Hellig, K., Further studies on effect of SO_2-pollution on the sulfhydryl-system of plants, *Phytopathol. Z.*, 104, 264, 1982.
161. Foyer, C. H. and Halliwell, B., The presence of glutathione and glutathione reductase in chloroplasts: a proposed role in ascorbic acid metabolism, *Planta*, 133, 21, 1976.
162. Mehlhorn, H., Cottam, D. A., Lucas, W., and Wellburn, A. R., Induction of ascorbate peroxidase and glutathione reductase activities by interactions of mixtures of air pollutants, *Free Radical Res. Commun.*, 3, 193, 1987.
163. Madamanchi, N. R. and Alscher, R. G., Metabolic bases for differences in sensitivity of two pea cultivars to sulfur dioxide, *Plant Physiol.*, 97, 88, 1992.
164. Pastori, G. M. and Trippi, V. S., Oxidative stress induces high rate of glutathione reductase synthesis in drought-resistant maize strain, *Plant Cell Physiol.*, 33, 957, 1992.
165. Paula, M. De, Perez-Otaola, M., Darder, M., Torres, G., and Martinez-Honduvilla, C. J., Function of the ascorbate-glutathione cycle in aged sunflower seeds, *Physiol. Plant.*, 96, 543, 1996.
166. Willekens, H., van Camp, W., van Montagu, M., Inze, D., Langebartels, C., and Sandermann, H., Jr., Ozone, sulfur dioxide, and ultraviolet B have similar effects on mRNA accumulation of antioxidant genes in *Nicotiana plumbaginifolia* L., *Plant Physiol.*, 106, 1007, 1994.
167. Torsethaugen, G., Pitcher, L. H., Zilinskas, B. A., and Pell, E. J., Overproduction of ascorbate peroxidase in the tobacco chloroplast does not provide protection against ozone, *Plant Physiol.*, 114, 529, 1997.
168. Mason, R. P., *In vitro* and *in vivo* detection of free radical metabolites with electron spin resonance, in *Free Radicals, A Practical Approach*, Punchard, N. A. and Kelly, F. J., Eds., The Practical Approach Series, Series Editors: D. Rickwood and B. D. Hames, Oxford University Press, Oxford, U.K., 1996, 11.
169. Mehlhorn, H., Tabner, B. J., and Wellburn, A. R., Electron spin resonance evidence for the formation of free radicals in plants exposed to ozone, *Physiol. Plant.*, 79, 377, 1990.
170. Naughton, D. P., Lynch, E., Hawkes, G. E., Hawkes, J., Blake, D. R., and Grootveld, M., Detection of free radical reaction products by high-field nuclear magnetic resonance spectroscopy, in *Free Radicals, A Practical Approach*, Punchard, N. A. and Kelly, F. J., Eds., The Practical Approach Series, Series Editors: Rickwood, D. and Hames, B. D., Oxford University Press, Oxford, U.K., 1996, 25.
171. Anderson, M. E., Glutathione, in *Free Radicals, A Practical Approach*, Punchard, N. A. and Kelly, F. J., Eds., The Practical Approach Series, Series Editors: Rickwood, D. and Hames, B. D. Oxford University Press, Oxford, U.K., 1996, 213.
172. Goldstein, S. and Czapski, G., Superoxide dismutase, in *Free Radicals, A Practical Approach*, Punchard, N. A. and Kelly, F. J., Eds., The Practical Approach Series, Series Editors: Rickwood, D. and Hames, B. D., Oxford Univ. Press, UK., 1996, 242.
173. St-Clair, D. K. and Chow, C. K., Glutathione peroxidase: activity and steady-state level of mRNA, in *Free Radicals, A Practical Approach*, Punchard, N. A. and Kelly, F. J., Eds., The Practical Approach Series, Series Editors: Rickwood, D. and Hames, B.D., Oxford Univ. Press, UK., 1996, 227.
174. Kaur, H. and Halliwell, B., Detection of hydroxyl radicals by aromatic hydroxylation, *Methods Enzymol.*, 233C, 67, 1994.

175. Karnosky, D. F., Consistency from year to year in the response of *Fraxinus pennsylvanica*, provence to ozone, INFRO Air Pollut. Mtg., Zabre, Poland, Aug. 27–29, 1979.
176. Robinson, J. M. and Rowland, R. A., Carbohydrate and carbon metabolite accumulation responses in leaves of ozone tolerant and ozone susceptible spinach plants after acute ozone exposure, *Photosyn. Res.*, 50, 103, 1996.
177. Tingey, D. T., Fites, R. C., and Wickliff, C., Differential foliar sensitivity of soybean cultivars to ozone associated with differential enzyme activities, *Physiol. Plant.*, 37, 69, 1976.
178. Asada, K., Formation and scavenging of superoxide in chloroplasts, with relation to injury by sulphur dioxide, in *Studies on the Effects of Air Pollutants on Plants and Mechanism of Phytotoxicity*, Res. Rep. Natl. Inst. Environ. Stud., Japan, 11, 165, 1980.
179. Chinoy, J. N., Ed., *The Role of Ascorbic Acid in Growth, Differentiation and Metabolism of Plants*, Kluwer Academic, Boston, 1984, 322 pp.
180. Lee, E. H., Saftner, R. A., Wilding, S. J., Clark, H. D., and Rowland, R. A., Effects of paclobutrazol on GA biosynthesis and fatty acid composition — A case study on the differential sensitivity to SO_2 stress in snap bean (*Phaseolus vulgaris* L.) plants, *Proc. Plant Growth Regul. Soc. Amer.*, 14, 295, 1987.
181. Miller, J. E., Pursley, W. A., and Heagle, A. S., Effects of ethylenediura on snap bean at a range of ozone concentrations, *J. Environ. Qual.*, 23, 1082, 1994.
182. Lee, E. H., Rowland, R. A., and Mulchi, C. L., Growth regulators serve as a research tool to study the mechanism of plant response to air pollution stimuli, *Br. Soc. Plant Regul.*, 20, 127, 1990.
183. Brennan, E., Leone, I., and Clarke, B., EDU: a chemical for evaluating ozone foliar injury and yield reduction in field-grown crops, *Int. J. Trop. Dis.*, 5, 35, 1987.
184. Kostka-Rick, R. and Manning, W. J., Dose-response to studies with ethylenediurea (EDU) and radish, *Environ. Pollut.*, 79, 249, 1993.
185. Eckardt, N. A. and Pell, E. J., Effects of ethylenediurea (EDU) on ozone-induced acceleration of foliar senescence in potato (*Solanum tuberosum* L.), *Environ. Pollut.*, 92, 299, 1996.
186. Gilbert, M. D., Elfving, D. C., and Lisk, D. J., Protection of plants against ozone injury using the antiozonant *N*-(1,3-dimethylbutyl)-*N*-phenyl-*p*-phenylenediamine, *Bull. Environ. Contamin. Toxicol.*, 18, 783, 1977.
187. Koiwai, A. and Kisaki, T., Effect of ozone on photosystem II of tobacco chloroplasts in the presence of piperonyl butoxide, *Plant Cell Physiol.*, 17, 1199, 1976.
188. Tonneijck, A. E. G. and van Dijk, C. J., Effects of ambient ozone on injury of *Phaseolus vulgaris* at four rural sites in the Netherlands as assessed by using ethylendiurea (EDU), *New Phyto.*, 135, 93, 1997.
189. Krizek, D. T., Semeniuk, P., and Wergin, W. P., Role of water stress and abscisic acid in modifying SO_2 sensitivity in coleus, *HortScience*, 18, 604, 1983.
190. Cathey, H. M. and Heggestad, H. E., Effects of growth retardants and fumigations with ozone and sulfur dioxide on growth and flowering of *Euphorbia pulcherrima* Willd, *J. Am. Soc. Hortic. Sci.*, 98, 3, 1973.
191. Mackay, C. E., Senaratna, T., McKersie, B. D., and Fletcher, R. A., Ozone induced injury to cellular membranes in *Triticum aestivum* L., and protection by the triazole S-3307, *Plant Cell Physiol.*, 28, 1271, 1987.
192. Rodgers, M. A. J. and Powers, E. L., Eds., Oxygen and Oxy-radicals in *Chemistry and Biology*, Academic Press, New York, 1981.
193. Rennenberg, H., Aspects of glutathione function and metabolism in plants, in *Plant Molecular Biology*, Wettztein, D. V. and Chua, N. H., Eds., Plenum Press, New York, 1987, 279.
194. Fiscus, E. L., Reid, C. D., Miller, J. E., and Heagle, A. S., Elevated CO_2 reduces O_3 flux and O_3-induced yield losses in soybeans: possible implications for elevated CO_2 studies, *J. Exp. Bot.*, 48, 313, 1997.
195. Barnes, J. D., Ollerenshaw, J. H., and Whitfield, C. P., Effects of elevated CO_2 and/or O_3 on growth, development and physiology of wheat, *Global Change Biol.*, 1, 129, 1995.
196. Balaguer, L., Barnes, J. D., Panicucci, A., and Borland, A. M., Production and utilization of assimilates in wheat (*Triticum aestivum* L.) leaves exposed to elevated O_3 and/or CO_2, *New Phytol.*, 129, 557, 1995.

CHAPTER 12

Defense Strategies against Ozone in Trees: The Role of Nutrition

Andrea Polle, Rainer Matyssek, Madeleine S. Günthardt-Goerg, and Stefan Maurer

CONTENTS

12.1 INTRODUCTION

Tropospheric ozone is a major air pollutant in industrialized countries. It is formed by photochemical oxidation of primary pollutants released into the air by burning of fossil fuels. In the presence of high irradiance the generation of ozone (O_3) is initiated by nitrogen dioxide (NO_2) and driven by volatile hydrocarbons and other components present in exhaust from traffic, power plants, or industrial productions.[1] Ozone is also a natural component in air at low concentrations of 5 to 15 ppb.[2] However, during the last 100 years these background levels have approximately doubled. Under clear and sunny weather conditions, O_3 rises to peak levels and occasionally reaches concentrations above 120 ppb.[1]

Ozone is highly phytotoxic. In aqueous phases it degrades into reactive oxygen species such as $O_2^{\cdot -}$ or H_2O_2.[3,4] Ozone itself or its oxidative degradation products easily oxidize cellular targets such as unsaturated fatty acids in membranes, thiol groups in enzymes, etc.[5] Antioxidative defense systems which are found in all aerobic organisms may prevent damage, if present at sufficient activities to counterbalance oxidative injury.[6] Sudden peak values of O_3 may, however, overwhelm protective systems and cause acute injury in sensitive plant species. Such increases in O_3 to high, immediately toxic peak concentrations occur occasionally and are typically confined to a local scale, whereas chronic exposure to persistently enhanced mean O_3 concentrations during diurnal

1-56670-341-7/00/$0.00+$.50

and seasonal courses is a general, regionally widespread phenomenon. It is questionable whether plants, in particular trees with their long reproduction cycles, have already adapted to the high stress levels imposed by O_3 or have sufficient metabolic flexibility to acclimate to such conditions. Protective measures against O_3 may be related to structural features limiting the access of O_3 to sensitive targets or physiological factors such as allocation of cellular resources to repair and detoxification processes.

There is now a large body of data showing that exposure to chronically elevated O_3 may cause changes in carbohydrate allocation patterns, decreases in photosynthesis, reductions in biomass, changes in growth patterns, premature senescence, etc.[5,7-10] These reactions are also accompanied by alterations in biochemical defense systems.[5,6,11] However, consistent O_3 responses have not always been found. A reason may be that the O_3 sensitivity is affected by the interaction of internal plant-specific factors on the one hand and external, environmental factors on the other hand. To date, little attention has been paid to the question of how defense mechanisms against O_3 may be affected by interaction with nutrition. The present chapter focuses on O_3-induced stress responses in tree species as affected by nutrition and developmental stage and aims at providing an integrative view from the cellular to the whole-plant level. Since there are only few and scattered data on this particular subject, the major body of this chapter will exemplify results of an experiment conducted on young birch trees (*Betula pendula*, Roth.) grown at high or low nutrient supply and under chronic O_3 exposure.

12.2 OZONE UPTAKE AS MODIFIED BY DEVELOPMENT, STRUCTURAL CHANGES, AND INJURY

Ozone is transported to the surface of plants by turbulent transport. In fully differentiated leaves, cuticles represent a nearly impermeable barrier[12] and O_3 is almost exclusively taken up via the open stomata. Therefore, stomatal conductance is an important factor determining the internal O_3 dose, i.e., the O_3 influx during exposure time.[13] The stomatal conductance, a feature given by the number of stomata per leaf area and their aperture, varies, however, largely during diurnal and seasonal courses and depends on water availability, nutrient supply, light regime, and developmental stage of a plant. In trees stomatal conductance is generally higher in deciduous than in evergreen species.[13] Thus, a given external O_3 concentration may lead to large differences in the internal O_3 dose between the two types of foliage, implying distinct differences in the need for defense systems between species, habitats, and under fluctuating environmental conditions.

With respect to the O_3 dose, it is important whether species with indeterminate shoot growth, i.e., species forming new leaves throughout the season (e.g., birch, poplar) or species with determinate growth, i.e., those forming only one or two flushes per season (e.g., conifers, beech) are considered, and during which ontogenic stage O_3 is present. During the initial stages of leaf formation, before the cuticle and outer epidermal cell wall have thickened and guard mother cells divide to form the stomata, the influence of O_3 is independent of stomatal regulation. At this early stage, the barrier properties of the cuticles may not be as efficient as in mature leaves. An example shows that in this ontogenic stage, the presence of chronic O_3 levels (40 ppb during the night, 90 ppb during the day) caused a reduction in leaf expansion in birch (Figure 12.1A) and resulted in smaller leaves (Figure 12.1B). These leaves displayed relatively higher densities of stomata, scales, small hairs, and veins than those from trees grown in filtered air.[14] Higher stomatal densities have also been found in various other birch clones after O_3 exposure.[15] The O_3-mediated reduction in leaf size was highly significant in high- but not in low-fertilized cuttings, which per se formed small leaves (Figure 12.1B). Still, in O_3-exposed leaves from low-fertilized birch, decreased guard cell sizes and increased stomatal density demonstrated that the influence of O_3 on differentiation was not necessarily bound to individual leaf expansion (Figure 12.1C vs 12.1B). In young, expanding leaves injury to subcellular structures was not yet apparent.

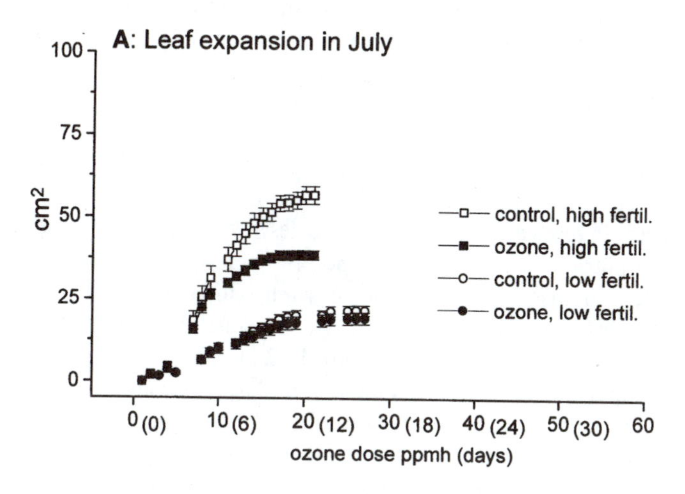

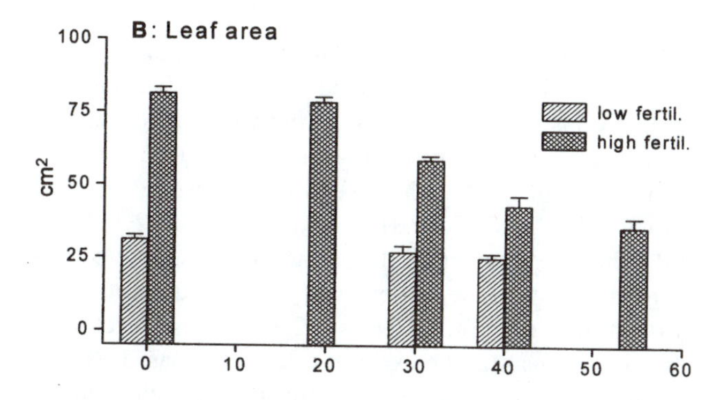

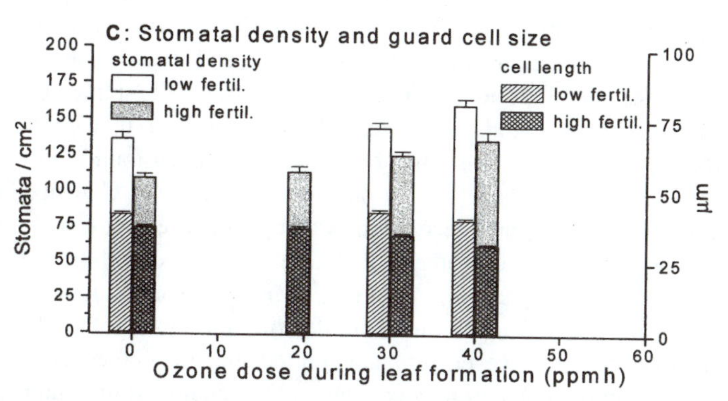

Figure 12.1 Leaf expansion (A), leaf area (B), stomatal density and size of guard cells (C) in high- and low-fertilized birch (*Betula pendula* Roth) trees as dependent on the external O_3 dose (concentration x time) during leaf formation. Young trees were grown from cuttings and fertilized regularly with a commercial fertilizer solution containing macro- and micronutrients either at a dilution of 0.005% (low fertilization) or 0.05% (high fertilization). The plants were exposed to filtered air (< 3 ppb, control) or to filtered air with added O_3 from before leaf flush to autumnal leaf loss in 20 field fumigation chambers in 1993. Ozone was generated from pure oxygen. Details of the experiment have been described elsewhere.[14,16,25] Leaves of the same age were investigated on individual trees. Data represent means ± standard error of $n = 10$ (A) or $n = 20$ trees (B and C). The number of determinations for stomatal density and guard cell size were 20 and 50, respectively, per leaf. The effects of O_3 and fertilization were significant ($P < 0.001$). Leaf expansion and leaf area were affected by significant interactions between O_3 and nutrition ($P < 0.005$).

In fully differentiated birch leaves, incipient subcellular structural changes appear in spongy parenchyma cells situated in intercostal fields adjacent to the free air space in the vicinity of stomatal openings.[16] Because of its high reactivity, O_3 is probably decomposed in the cell wall matrix and does not enter the intracellular space.[17] The diffusive pathway of O_3 within the leaf is likely to be small. It has been reported that in spruce the lignin content of guard cells is lower in O_3-exposed needles, whereas the mesophyll, which also contains lignin, did not show this reduction, pointing to O_3-induced disturbances at distinct locations.[18] Ozone-induced cell wall responses have been studied by cytochemistry in birch and several other species.[16,19] Under chronic O_3 exposure of birch and irrespective of the nutritional state of the plants, the outermost pectinaceous layer (calcium pectate) of spongy parenchyma cells was found to swell and protrude into the free air spaces, sometimes forming bubbles and at later stages wartlike structures (Figure 12.2B as compared with control in filtered air Figure 12.2A and D, vs. control 12.2C). The cytosol, the nucleus, and the mitochondria became relatively dark, whereas the chloroplasts appeared more translucent (Figure 12.2B vs. 12.2A). It is not yet known whether such initial ultrastructural changes are reversible; but after continued exposure the delimitation of the membranes became less distinct and the ultrastructural changes also spread to the palisade parenchyma. Only at this point did initial O_3 symptoms become macroscopically visible as light green dots on the adaxial leaf side in transmitted light.

With the appearance of initial O_3 symptoms and their development into stipplings and collapse of individual cells (Figure 12.3A), it was observed histochemically that starch granules remained accumulated along small leaf veins; this may indicate inhibited phloem loading (Figure 12.2F vs. 12.2E)[14,20-23] simultaneously with a reduction in CO_2 assimilation (see below). Autoradiography after $^{14}CO_2$ feeding showed the irregular CO_2 uptake in leaves with visible O_3 symptoms as compared with leaves from trees exposed to filtered air (Figure 12.2H vs 12.2G). In contrast to starch accumulation along small veins, the O_3-induced decline in the mesophyll cells was characterized by lowered starch content. At this stage the nucleus was condensed and the cell walls thickened by cellulose and pectin deposition (Figure 12.2D vs. 12.2C). These reactions that precede cell collapse led to an increase in dry mass at the expense of intercellular air spaces (Figure 12.3B). As injury proceeded, the cell-to-cell contact was interrupted and finally phenolic polymers appeared as a result of oxidative processes in cytosol and the vacuoles. The latter processes were partly caused by membrane disintegration irrespective of the species and the preceding stress.[24]

The sensitivity and appearance of initial leaf symptoms and their further development to cell injury and stippling varies inter-[21,24] and intraspecifically.[19] In birch, the stomatal density is increased by both low nutrition and exposure to O_3.[14,15,25] However, the role of O_3 in leaf differentiation cannot be generalized since in poplar reduced stomatal densities have been found.[20,26] Conflicting results have also been reported on the impact of O_3 on gas exchange: net CO_2 uptake rate, stomatal conductance, and water-use efficiency decreased, increased, or even remained unaffected.[5,8] One may ask whether nutrition can explain some of the variability found in the gas exchange behavior under O_3 stress. High nutrition typically stimulates leaf metabolism including CO_2 uptake rates.[27] This may increase the capacity for repair and detoxification processes[28] and, thereby, enhance the O_3 tolerance of photosynthesis. However, stomatal conductance may be enhanced as well[29] so that an increase in O_3 uptake into the leaves may counteract the benefits of high nutrition. By contrast, low nutrition may reduce the influx of O_3 and, thereby, the risk of injury. Moreover, the sensitivity of stomatal regulation is known to be mediated by nutrition.[30]

In spite of increased stomatal densities in birch leaves grown with low nutrient supply, under the influence of O_3 stomatal conductance decreased to the level of trees at high nutrition.[31] Thus, under O_3 impact partial stomatal closure in low-fertilized plants compensated for the increase in stomatal density.[25,32] Apparently, this effect was less pronounced in leaves of birch grown with high nutrient supply than in those from plants with low nutrient supply, thereby resulting in similar O_3 uptake rates in both nutrient regimes (Figure 12.4A and B).

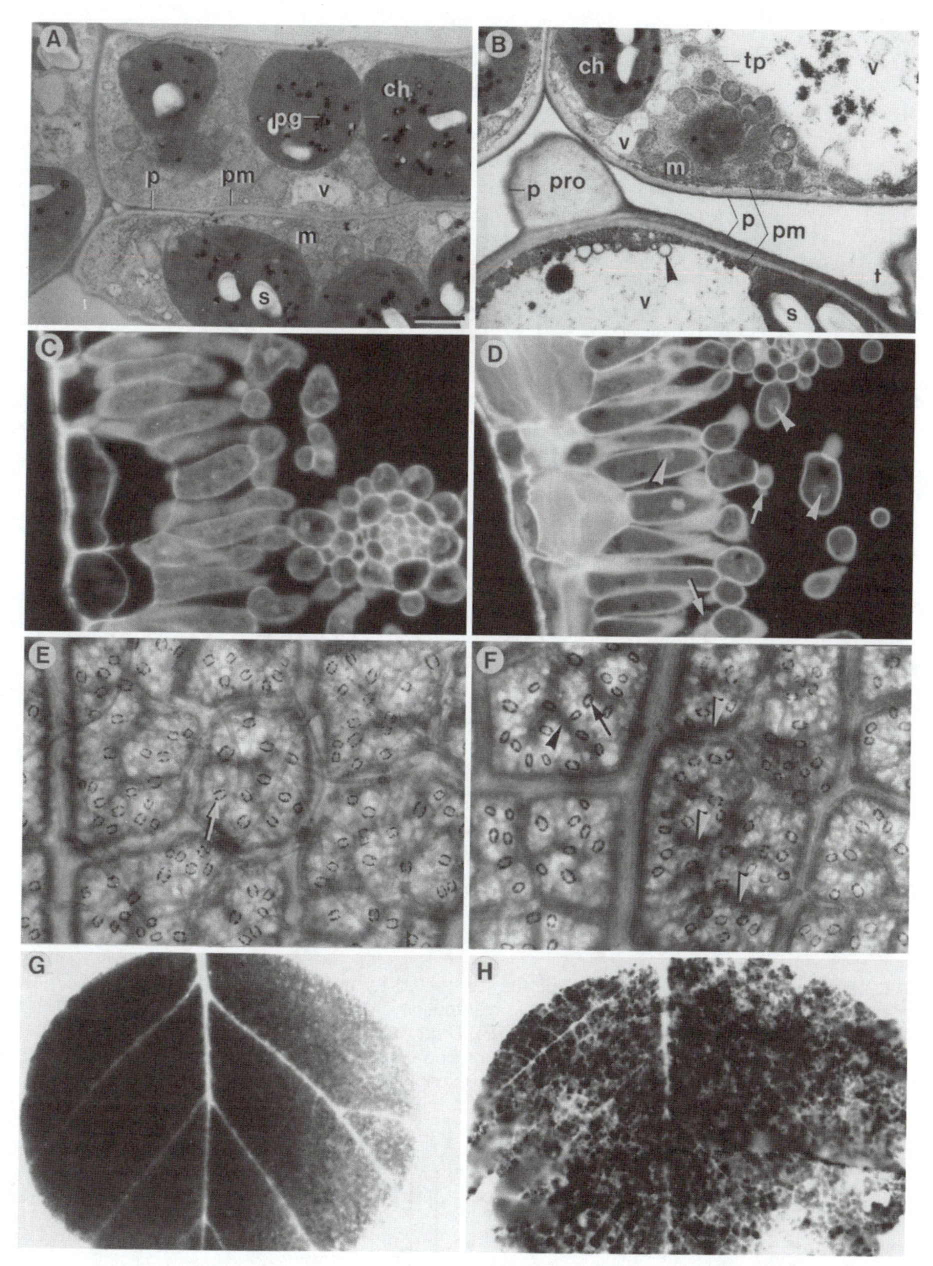

Figure 12.2 Birch (*Betula pendula*) leaf structures and ultrastructures after growth in filtered air (A, C, E, G) or in filtered air with 90 ppb O_3 added during daylight hours and 40 ppb during night (B, D, F, H). (A and B) TEM micrographs of palisade parenchyma cells in 26-day-old birch leaves from the high fertilization regime (bar = 1.5 μm); (A) control with distinct mitochondria (m) and chloroplasts (ch), small plastoglobuli (pg) and vacuole (v), and thin pectin layer (p). (B) With initial visible O_3 symptoms (O_3 dose = 43 ppmh): large vacuoles with tannin deposit (upper cell) and lipid droplet (lower cell). In the lower cell, plasmamembrane (pm) confining a darkened cytoplasm, tonoplast (tp) with proliferations (arrow head), thickened pectin layer with bubblelike projections (pro). (C and D) Cross sections stained with coriphosphine for pectinaceous substances (white), which in (D) are increased in the cell walls (sometimes forming bubbles or warts as indicated by arrows) and in the nuclei (arrow heads), bar = 22 μm. For technical details see Reference 16. (E and F) Surface view of lower leaf surface, stained for starch with I/KI, stomata with black amylopectin granules (arrows), and starch accumulated along small leaf veins (arrow heads), bar = 0.15 mm. (G and H) Autoradiographies 12 h after feeding with $^{14}CO_2$; (H) irregular CO_2-uptake.

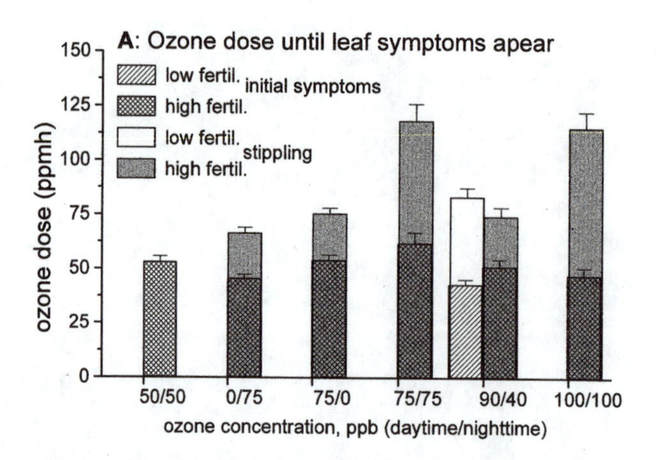

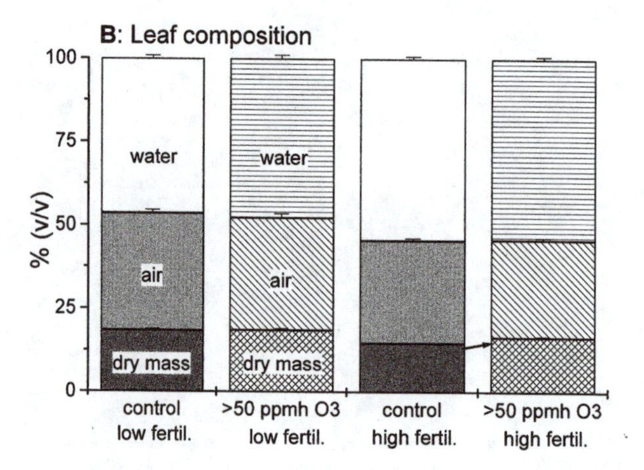

Figure 12.3 Development of visible leaf symptoms as related to the external O_3 dose and the O_3 exposure regimes (A) and effects on the relative distribution of water, air, and dry matter within foliar volume elements in leaves of *Betula pendula* (B). Data are means of $n = 10$ (A) and 40 (B). The effects of O_3 on the development of leaf symptoms and dry matter were significant ($P < 0.002$). The effects of fertilization were significant ($P < 0.03$), except on the appearance of stipplings.

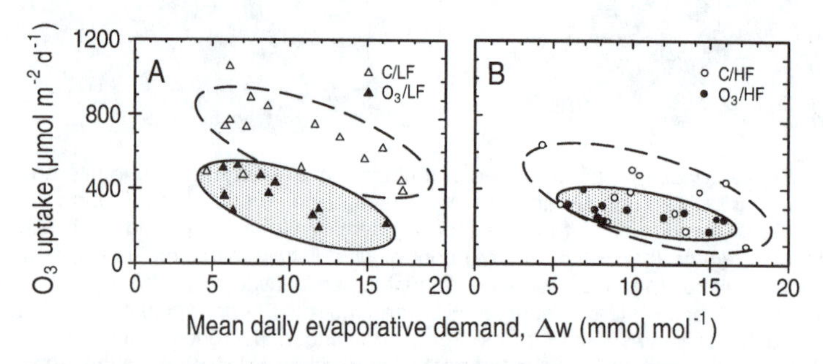

Figure 12.4 Daily O_3 uptake (i.e., O_3 influx density) into birch leaves as dependent on the mean daily evaporative demand of the ambient air (expressed as the difference in the leaf/air mole fraction of water vapor): (A) Low-fertilized plants; (B) high-fertilized plants. For the control in filtered air at each nutrient supply (open symbols), the potential O_3 uptake was calculated assuming the same O_3 regime as applied to the O_3-exposed plants (closed symbols): Abbreviations: C = control plants in filtered air; O_3 = plants exposed to O_3 (90 ppb during daylight hours, 40 ppb at night); LF = low-fertilized plants; HF = high-fertilized plants. (Modified from Maurer et al.[31])

Under O_3 stress, stomatal narrowing may be initiated by an increasing CO_2 concentration in the intercellular space of the mesophyll (*ci*) as frequently observed in O_3-exposed leaves.[33-35] In O_3-exposed birch, ci increased regardless of nutrition or the extent of stomatal closure.[31] Elevated ci at lowered stomatal conductance can only be explained by a decline in the enzyme-driven CO_2 consumption. In fact, O_3 made the rate of CO_2 uptake decrease relative to the stomatal conductance.[32] Thus, CO_2 uptake was limited by a diminished CO_2 demand in the mesophyll rather than by lowered stomatal conductance. Therefore, the increase in $\delta^{13}C$ in the plant biomass observed under O_3 stress[36] cannot be caused by stomatal limitation of photosynthesis[37] but was probably a result of stimulated phosphoenol pyruvate carboxylase (PEPC) activity (see below). The observation that stomatal narrowing is preceded by a decline of the adjacent mesophyll cells cells and parallelled by raised *ci* suggests that the decreased aperture is a consequence of injury. It is likely that the supply of guard cells with ions, hormones, etc., necessary for stomatal regulation is disturbed when neighboring cells have been damaged. But more research is needed on the osmotic and hormonal control of the stomatal width under O_3 stress.[38] Furthermore, the accumulation of immobile amylopectin in narrowed guard cells (Figure 12.2F vs. 12.2E) may also be interpreted as a defense strategy to reduce high apoplastic sucrose levels when mesophyll sucrose efflux exceeds translocation.[22,39] In this manner gas exchange for further assimilate production is restricted. It should be emphasized that the reduction in stomatal conductance found in low-fertilized birch resulted in a 50% decrease in the daily O_3 uptake relative to the potential O_3 influx in the absence of stomatal effects (see Figure 12.4).

The interaction between nutrition and O_3 effects on stomatal conductance is, however, species dependent. In contrast to birch, in alder O_3 caused increases in stomatal conductance and, thereby, in O_3 uptake, irrespective of the nitrogen supply.[40] If water supply is high, stomatal conductance may increase even in the absence of O_3 by low nutrition (see Figure 12.4).[41] It has typically been reported for conditions of low nutrient availability that CO_2 uptake rates remain low relative to stomatal conductance.[31,42,43]

12.3 PHOTOSYNTHESIS AND CARBON METABOLISM UNDER THE INFLUENCE OF O_3 AND NUTRITION

It has frequently been observed that O_3 leads to a decrease in CO_2 uptake rates[8,13,44] although the primary impact of O_3 is not in the chloroplasts but in cell walls and adjacent plasmalemma.[10,19] Functional and structural breakdown of the chloroplasts has been reported, e.g., photoinhibition, loss of ribulose 1,5 bisphosphate carboxylase/oxygenase (Rubisco) protein and activity.[5,45] In birch, the variable fluorescence (F_v/F_m) remained stable, until CO_2 uptake had almost reached compensation.[31] This observation indicates the dependence of CO_2 uptake rates on the collapse of entire mesophyll cells,[32] whereas photosystem II apparently stayed intact in the persisting clusters of green cells in the O_3-injured leaves. Inhibition of nitrate reductase by O_3 may be significant for plant nutrition,[46] as the balance between nitrate reductase and Rubisco activities may exert the ultimate control on the carbon and nutrient allocation of the whole plant.[47] Remarkably, O_3-exposed birch plants of low nutrition displayed increases in CO_2 uptake rates and Rubisco activity in young leaves relative to the corresponding control in filtered air.[22,31] Such effects were absent at high nutrition. Thus, nutrition determines the extent to which photosynthesis of newly formed leaves may compensate for the decline in older O_3-injured foliage.[8] By contrast, photosynthesis of unnodulated alder was more sensitive to O_3 at low nitrogen supply.[40] However, it is not known whether the limitation in a specific nutritional element such as nitrogen may have an effect similar to that caused by overall reduction in nutrient supply.

High nutrition could not prevent leaf injury in birch, but low nutrition delayed the impairment of photosynthesis and maintained the life span of O_3-exposed leaves almost throughout the entire growing season.[31] Leaves of high-fertilized trees displayed O_3 symptoms earlier (see Figure 12.3A), along with more pronounced declines in photosynthetic capacity, water-use efficiency, apparent CO_2 uptake efficiency, and quantum yield than did leaves from low-fertilized birch. Given the similar O_3 uptake under both nutrient regimes (see Figure 12.4B), leaf maintenance at low nutrition was apparently more efficient relative to leaves of high nutrition.

The influence of O_3–nutrient interactions on photosynthesis of birch was reflected in distinct changes in the carbohydrate metabolism of leaves. The levels of glucose, fructose, and sucrose were significantly enhanced by O_3 in young birch plants at low but not so at high nutrient supply.[22] Consistently, low nutrition was associated with an inhibited synthesis of sucrose as indicated by the reduced sucrose phosphate synthase activity and increased levels of fructose-2,6-bisphosphate[50] — a metabolite known to regulate sucrose synthesis.[47] In parallel to sucrose accumulation, enzymes of sucrose degradation like sucrose synthase and alkaline invertase were stimulated, probably contributing to the enhanced levels of glucose and fructose. The flux of assimilates was apparently redirected from sucrose synthesis to starch formation, as sustained by reduced starch phosphorylase and ADP-glucose-pyrophosphorylase activities, and to the glycolytic pathway. In relation to glycolysis, the induced high-PEPC activity was striking[22] as this enzyme initiates the anaplerotic synthesis of oxalacetate and malate through nonphotosynthetic CO_2 incorporation and, thereby, promotes the citrate cycle. The raised supply of C_4 acids may fuel the respiratory ATP production and provide substrate for repair and detoxification processes.[48,49] These interrelationships are consistent with the observed increase in $\delta^{13}C$ in the plant biomass, which can be explained by the stimulated PEPC activity and its low ^{13}C discrimination. Furthermore, increases in malate, respiration, and ATP-to-ADP ratio were found under O_3 stress.[22,36] Raised PEPC activity has been reported previously for O_3-exposed pine and poplar plants.[51-53]

Elevated carbohydrate levels as found in low-fertilized birch under O_3 stress may also contribute via end product inhibition to a decline in photosynthetic capacity.[54] However, in O_3-exposed birch of high nutrition sugar levels were not raised.[22] The decrease in photosynthetic capacity in these plants appeared to be determined by cell collapse rather than by diminished Rubisco activity.[22,31,32] Inositol levels were decreased by O_3 at both nutrient regimes, the cyclitols being regarded as scavengers of O_3-induced hydroxyl radicals,[55] although findings conflict about cyclitol responses to O_3.[52,56,57] Overall, the extent of nutrition did not prevent the decline in the leaf structure and photosynthesis of birch.

It is difficult to decide on the basis of carbohydrate responses alone whether high nutrition is advantageous for O_3 tolerance. One can state, however, that the carbohydrate metabolism responded to O_3 more sensitively in leaves of low nutrition, and that the life span of such leaves was longer than at high nutrition.[31] This latter finding is important with respect to the indeterminate shoot growth in birch: leaf formation and longitudinal growth of shoots continue throughout the growing season when nutrition is high, but cease early at low nutrition. Hence, it would appear advantageous to maintain the relatively fewer leaves formed at low nutrition even if they are injured. By contrast, at high nutrition the premature loss of the aging leaves on the lower shoot sections after the O_3 dose has become injurious might be balanced by new uninjured leaves formed at the elongating shoot tips.

Remarkably, protein levels, Rubisco activity, and photosynthetic capacity were higher in young O_3-exposed leaves on upper shoot parts relative to individuals in filtered air when nutrition was low, whereas such effects were absent at high nutrition.[31,50] At low nutrition, perhaps plant-internal nitrogen retranslocation to the young leaves compensated to some extent for the O_3-caused photosynthetic decline of the aging leaves. By contrast, as long as the nitrogen availability is high, the "opportunity costs"[47] seem to be lower in forming new leaves and abandoning the aging and injured ones instead of maintaining them. As a consequence of new leaf and side branch formation, the proportion of O_3-injured leaves in the whole-plant foliage area was decreased at high relative to low nutrition.[58,59] Regardless of nutrition, the proportion of nitrogen allocated to the whole-plant foliage was significantly increased under O_3 stress, in particular when taking into account the content and low retranslocation of nitrogen in the prematurely shed leaves. Contrasting with birch, the nitrogen retranslocated from O_3-exposed pine needles without visible injury reached an extent similar to that found during needle senescence.[60] In summary, nutrition is a driving factor in determining the response of photosynthesis and carbon metabolism to O_3.

12.4 ANTIOXIDANT DEFENSES

For the understanding of O_3 tolerance or protection it is not only important to determine the degree by which O_3 uptake can be diminished, but also whether photosynthetate is allocated to biochemical defenses contributing to O_3 detoxification. Although O_3 exposure may induce metabolic pathways involved in plant defenses from environmental stresses such as UV-B radiation, wounding, pathogens, etc, "specific O_3-tolerance traits" have not been identified.[61] Since O_3 may degrade into reactive oxygen species, there is increasing evidence that antioxidative systems play an important role in mediating protection from O_3 injury. Plant cells contain antioxidative enzymes and metabolites in all their subcellular compartments.[11] In chloroplasts and in the cytosol, these systems form the so-called ascorbate–glutathione cycle, which is composed of the antioxidants ascorbate and glutathione and enzymes scavenging reactive oxygen species — superoxide dismutases (SOD), peroxidases — as well as enzymes responsible for the regeneration of the antioxidants — dehydroascorbate reductase, monodehydroascorbate radical reductase, glutathione reductase.[62] The reductant necessary for the maintenance of this cycle is provided by NAD(P)H. Other subcellular compartments contain catalases (e.g., peroxisomes) and peroxidases (vacuole, apoplastic space) for the disposal of H_2O_2.[11] The observation that the apoplastic space contains ascorbate in the range of 0.05 to 5 m*M* is important with respect to protection from O_3 since ascorbate is highly reactive with several air pollutants ($K_{app(O_3)} = 6 \cdot 10^7\ M^{-1}s^{-1}$).[63]

More than 30 years ago, Freebairn[64] and Menser[65] demonstrated that tobacco leaves fed or sprayed with ascorbate solutions developed less foliar injury after O_3 exposure. More recently, Runeckles and Vaartnou[3] showed by means of electron spin resonance that O_3-induced formation of $O_2^{\cdot-}$ in plants was inhibited, if the leaves were infiltrated with ascorbate before O_3 exposure. Mutants of *Arabidopsis*, which contain diminished concentrations of ascorbate, were especially sensitive against oxidative stress including O_3.[66] In general, it seems that high antioxidative capacities present in plants before the onset of O_3-stress may protect from O_3 injury. This is an important qualitative difference as compared with the question whether plants can increase their antioxidative capacities in response to O_3, thereby acquiring a higher level of protection. In fact, quite variable changes in antioxidative capacities have been observed in response to O_3.[11] A lack in stress responsiveness may be related to developmental stage. Casano et al.[67] found that the inducibility of SOD was significantly lower in old than in young leaves. In most plant species analyzed, the capacity of antioxidative systems declines with leaf age.[11] In parallel, it has frequently been reported that younger leaves are more resistant against O_3-induced injury than older leaves. An important question is whether such a response is also modified by nutrition. It has repeatedly been reported that deficiencies in distinct nutrients cause increased capacities of antioxidants.[68] This has been observed for Mg,[69] Mn,[71] and N deficiencies in combination with varied Mg supply.[72] When Mg-deficient leaves were exposed to the herbicide paraquat, they showed increased resistance.[73] By contrast, field data suggest that Mg deficiency predisposes trees for O_3 injury.[74]

The role of SOD in protection from O_3-mediated injury has been addressed frequently.[6,11] Age-dependent reductions in SOD activities have been observed[75] and may render older foliage more sensitive to stresses. Little is known about how the combination of nutrient limitations and O_3 affects SOD and other components of antioxidative systems. We found in leaves from high-fertilized birch plants exposed to filtered air an age-dependent decline in SOD activities (Figure 12.5A) but not in leaves from the low-fertilization treatment (Figure 12.5B). The content of ascorbate was less dependent on leaf age, whereas glutathione also declined with increasing age (Figure 12.5E and F). Furthermore, in parallel with leaf age, decreases in pigmentation were observed in high-fertilized plants (Figure 12.6A and C). When such plants were grown in the presence of chronic O_3 levels, the pigment content decreased more rapidly than under filtered air (Figure 12.6 A and C). This is in accordance with the development of leaf injury at the macroscopic and microscopic scale (see above). The question was whether SOD activity or antioxidant levels were associated with mediating O_3 protection. Interestingly, in young leaves of O_3-exposed, high-fertilized plants the activity of SOD was diminished as compared with controls, even though these leaves did not show visible injury (Figure 12.5A). In high-fertilized plants, elevated SOD activities appeared with development of foliar symptoms of injury (Figure 12.5A). These data show that O_3 exposure may result in any kind of response: increases, decreases, and no effects. Such conflicting results have frequently been reported in the literature.[11] One of the regular functions of SOD is the detoxification of superoxide radicals formed during photosynthesis when electrons are transferred to oxygen instead of $NADP^+$. In O_3-exposed leaves, low SOD activity might indicate that initially when symptoms of injury are not yet apparent, the turnover of $NADPH/NADP^+$ is accelerated because more reductant is needed to compensate oxidative stress. Under such conditions, there may be a higher availability of $NADP^+$, thereby outcompeting O_2 as an alternative electron acceptor. At beginning of injury, the demand for NADPH by the Calvin cycle may decline, requiring an increased detoxification of $O_2^{\cdot-}$ radicals since there was no evidence for a general decrease in PSII-mediated electron flux.[31] The highest SOD activity was found in injured leaves from high-fertilized birch, perhaps in response to unspecific oxidative degradation of cellular components. Taken together, these data suggest that in foliage of high-fertilized plants, SOD activities do not respond actively to protect plants but follow cellular events occurring as a result of O_3 stress.

In contrast to high-fertilized plants, in leaves of low-fertilized plants the age-dependent decline in SOD activity was not found reflecting a trend similar to that found for photosynthesis in filtered air.[31] High SOD activities and elevated ascorbate contents were maintained under low nutrient supply for all the leaf age classes (Figure 12.5B and D). In low-fertilized leaves the glutathione content was lower than in leaves from high-fertilized plants but O_3 induced a partial increase in this component, especially in older leaves (Figure 12.5F). Obviously, the response of antioxidants to O_3 is strongly affected by the interaction between nutrition and leaf age. The data support the hypothesis that high antioxidative capacity protects from O_3 injury.[11] Such elevated capacities were present in low-fertilized and to a lesser extent in leaves from high-fertilized plants. The differences in antioxidant contents and their modification by O_3 are striking, since the analyzed leaves of high-fertilized birch with lower antioxidative protection were shed earlier than the leaves of low-fertilized plants with high antioxidant capacities. It will be necessary to address these O_3–nutrient interactions in other species, especially in such ones with determinate growth. This may be important because young expanding foliage from spruce or beech contains lower antioxidative capacities than the analyzed birch leaves.[76,77] It has been proposed that this ontogenic stage may be the “Achilles’ heel” with respect to oxidative injury.[77]

Since O_3 is probably degraded in the cell wall, apoplastic defense mechanisms have attracted considerable attention in the last few years. Ozone-acclimated spruce trees grown at high altitude contain up to 5 m*M* apoplastic ascorbate and show increases in response to O_3 exposure.[78] In beech, O_3 induced a delayed response in ascorbate.[79] In an herbaceous species (tobacco), the ascorbate pool was rapidly depleted in presence of O_3.[80] In contrast to relatively O_3-resistant species such as beech or spruce, the O_3-sensitive birch clone did not contain significant concentrations of apoplastic ascorbate (Figure 12.7A). The apoplast contained predominately dehydroascorbate, which is not active in O_3 detoxification (Figure 12.7A). Little, if any apoplastic ascorbate has been found in poplar, a species which is also O_3 sensitive (Polle, unpublished results). In O_3-exposed birch leaves with visible injury (stipplings) the apoplastic matrix contained significant approximately threefold higher total ascorbate (= ascorbate + dehydroascorbate) concentrations than healthy leaves, an important fraction of this pool being in the reduced stage (Figure 12.7A). In low-fertilized leaves the O_3-induced increases in apoplastic ascorbate and dehydroascorbate were less pronounced than in high-fertilized leaves (Figure 12.7A). The concentrations of apoplastic glutathione were similar

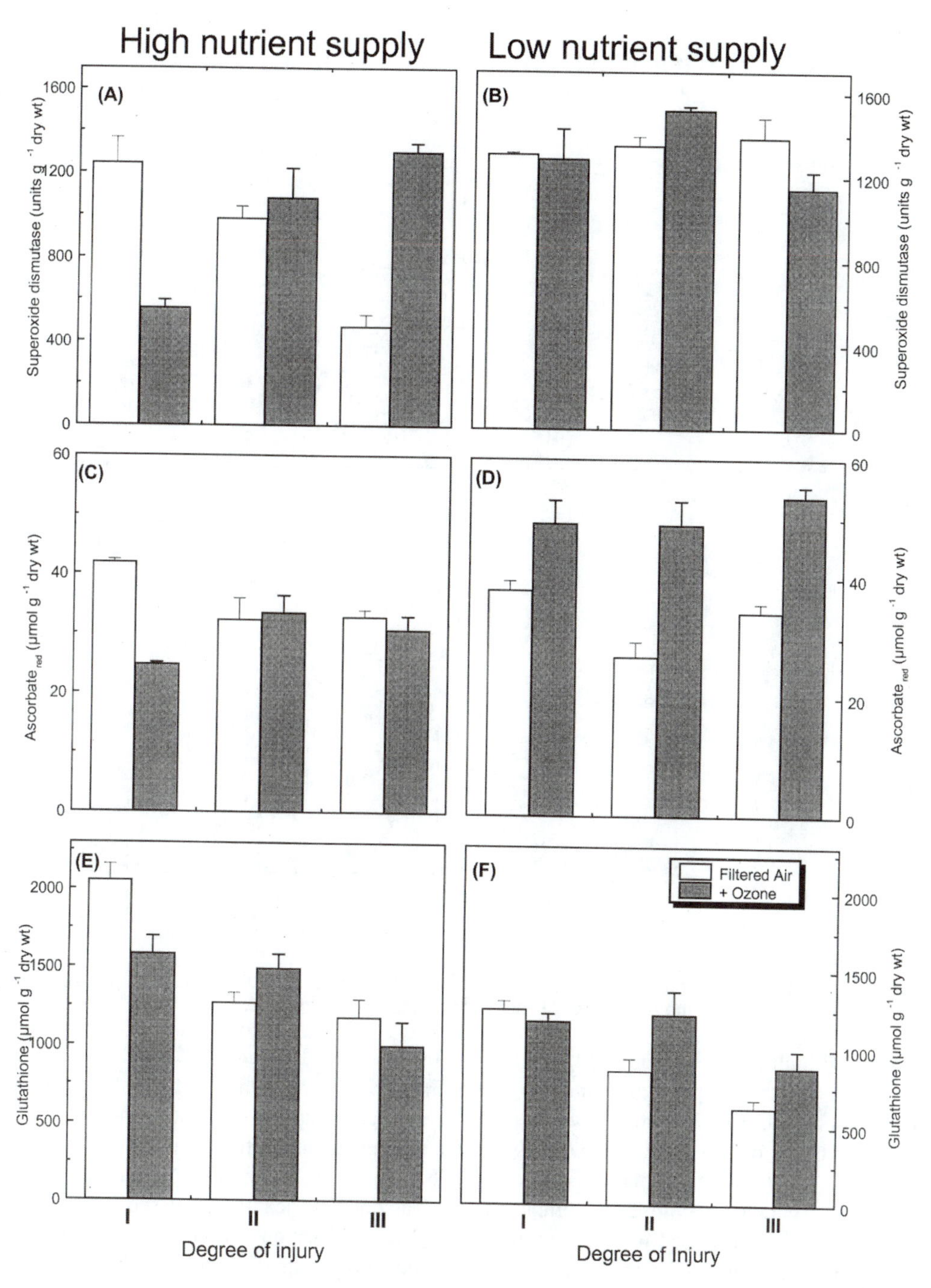

Figure 12.5 Response of SOD (A and B), ascorbate (C and D), glutathione (E and F) to O_3 in birch (*Betula pendula*) in dependence on leaf age and nutrient supply. The foliage was collected along the main axes of birch seedlings grown for about 4 months in the presence of O_3 (90/40 ppb, day/night) or filtered air (control). I, II, and III indicate leaf ages of about 4, 6, and 8 weeks, respectively, and correspond to increasing degree of injury for O_3-exposed plants. Limited nutrients were supplied by watering the plants with a 10-fold diluted fertilizer solution during the whole growth phase (for detailed growth conditions see Refernce 31). Data are means of six individual trees measured in three replicates (± SD).

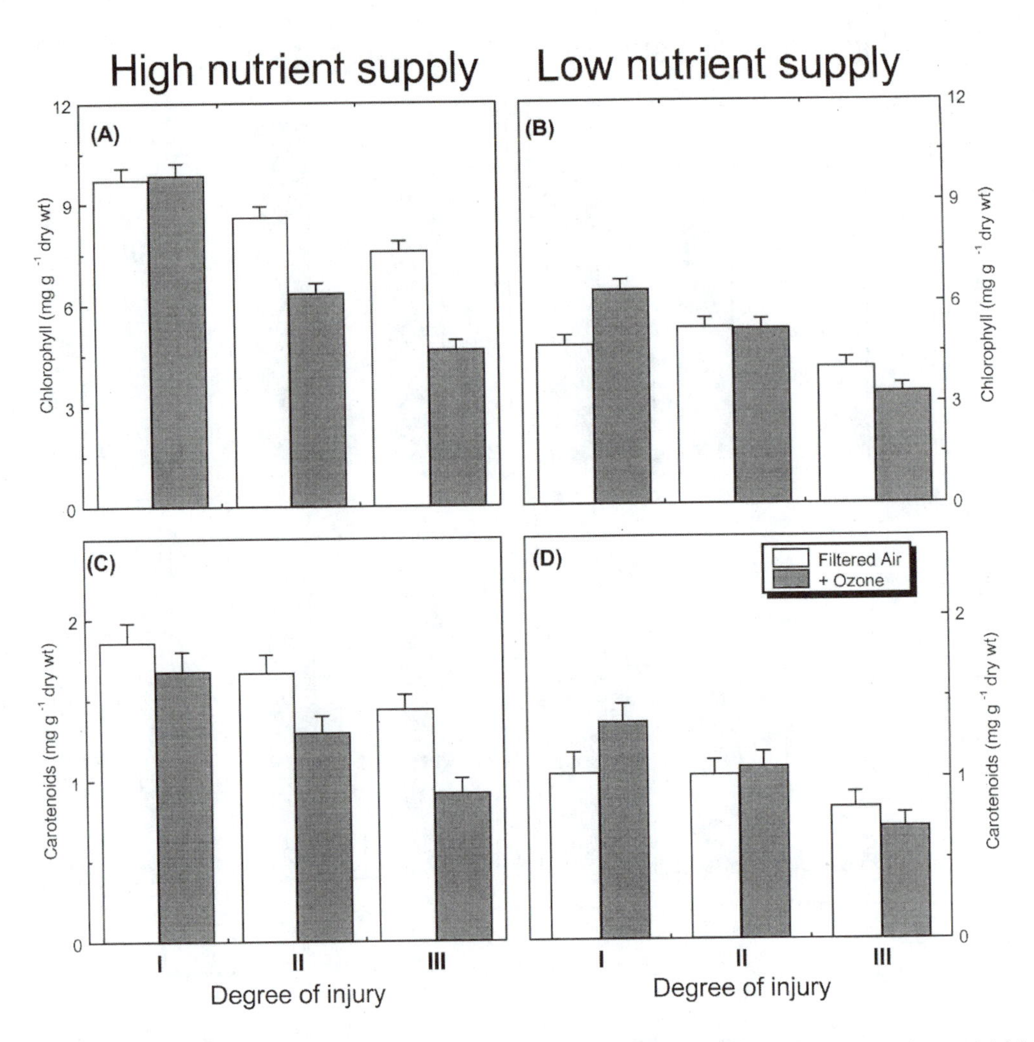

Figure 12.6 Response of chlorophyll (A and B) and carotenoids (C and D) to O_3 in birch (*Betula pendula*) in relation to leaf age and nutrient supply. For details see Figure 12.5.

in low- and high-fertilized plants (Figure 12.7C). In O_3-exposed foliage, the relative contents of antioxidants in the apoplast amounted to about 5 to 9% of the total foliar contents of these components (Figure 12.7B and D), which is relatively high compared with other studies.[80,81] The relative activities of apoplastic peroxidases — natural constituents in the apoplast — were in the same range as those of apoplastic antioxidants (Figure 12.7E), whereas glutathione reductase activity — a marker for symplastic components — was not significantly increased in O_3-exposed as compared with unstressed foliage (Figure 12.7F). This observation may indicate that O_3 caused a specific induction of apoplastic defenses. Unfortunately, it is not clear whether the relative increase in apoplastic antioxidants was caused by "easier" leakage of small solutes through slightly injured membranes than that of large proteins. However, regardless the causes of an increased presence of antioxidants in the apoplast, these will mediate some protection against unspecific oxidation.

The apoplastic phase also contains other solutes which may contribute to O_3 detoxification. For example, in spruce needles components like picein, p-hydroxyacetophenon, catechin, *p*-hydroxybenzoic acid glucoside, ferulic acid, kaempherol-3-glucoside, as well as unknown phenolic components have been found in significant concentrations.[11] Furthermore, the apoplastic defense systems can only operate if reductant is delivered at sufficiently high rates. The capacity and identity

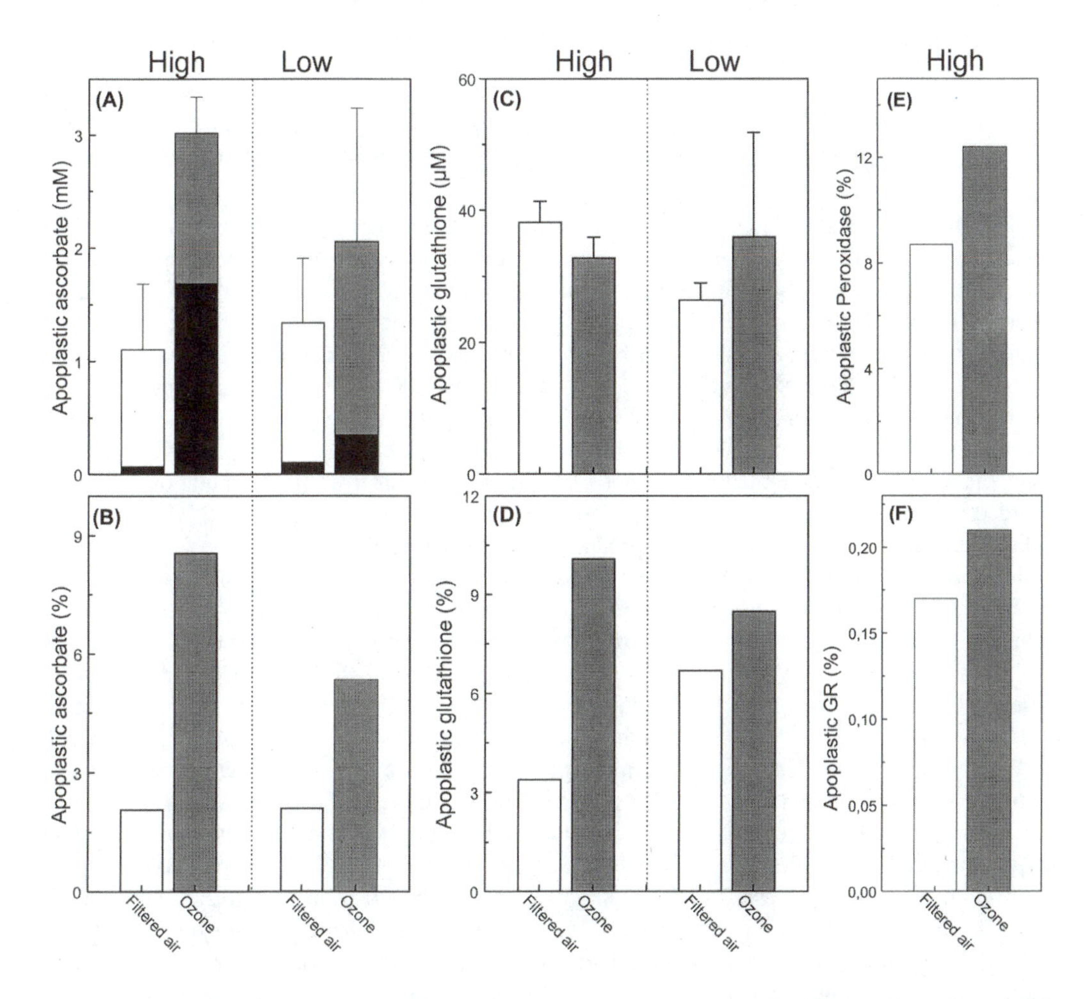

Figure 12.7 The effects of O_3 and nutrition on apoplastic ascorbate (A and B) and glutathione (C and D) in relation to apoplastic and symplastic enzymes in *Betula pendula*. The foliage was collected along the main axes of birch seedlings grown in the presence of O_3 (90/40 ppb, day/night) or filtered air (control). Leaves with visible injury symtoms were used for the analysis of apoplastic and symplastic components. The activities of peroxidase and glutathione reductase and the content of ascorbate and glutathione in foliar extracts were set as 100%. Ascorbate [(A) apoplastic concentration of ascorbate (black bars) and dehydroascorbate (white, gray bars), (B) relative occurrence in the apoplastic space], glutathione [(C) apoplastic concentration, (D) relative occurrence in the apoplastic space], peroxidase (E) and glutathione reductase activities (F), relative occurrence in the apoplastic space. Data indicate means of n = 6 (± SD).

of this system is still unclear. There is preliminary evidence that O_3 causes a rapid flux of electrons across the plasma membrane from the inner to the outer side (U. Heber, personal communication). Such a transmitting pathway might connect intracellular, symplastic with extracellular, apoplastic antioxidative systems. A better characterization of these systems is urgently needed in order to understand fully their role in O_3 detoxification. Furthermore, it will be necessary to resolve the reactions of the defense systems on a microspatial scale within the leaf because most O_3 that enters a leaf seems to react in the vicinity of the stomatal apertures. Since under both nutrient regimes the daily fluxes of O_3 into the leaves were similar to each other (see Figure 12.4), higher concentrations of apoplastic antioxidants can be expected to afford higher protection. The most important conclusion from the birch study is that low-fertilized plants allocate elevated amounts of substrate and energetic resources to detoxification systems probably in order to provide their leaves *a priori* with enhanced defense measures important for longer leaf persistence in a stressful environment.

12.5 OZONE–NUTRIENT INTERACTION IN WHOLE-PLANT CARBON ALLOCATION AND BIOMASS PRODUCTION

Both the short- and the long-term nutrient availability determines the fate of assimilated carbon for plant survival. During ontogenic development, carbon allocation is modified by chronic O_3 concentrations. Low nutrient availability and drought result in favored root growth,[82-84] whereas O_3 generally leads to an opposite effect resulting in decreased root–shoot ratios.[8,85] The O_3-induced changes in plant performance appear to be associated with an impeded assimilate translocation from the leaves to the roots.[9] The disturbance of the carbohydrate metabolism is also apparent from starch accumulation which occurs in O_3-injured leaves along the small veins (see Figure 12.2F).[22,23] Starch accumulation was observed under both nutrient regimes. Because of its specific localization along the veins, this phenomenon was not comparable with an overall accumulation of starch recorded in spruce mesophyll cells after an early necrosis of phloem cells by Mg or K deficiency.[86] A direct influence of O_3 and its reactive products on the small leaf veins cannot be ruled out since birch leaves have a high proportion of air space (see Figure 12.3B). Such an influence may also be relevant for stomatal regulation under O_3 stress as inhibited by abscisic (ABA) retranslocation from the leaves due to phloem dysfunction which may favor stomatal narrowing.[87] In leaves displaying visible O_3 injury, the cell walls in small leaf veins were similarly thickened as those in mesophyll cells, the cytoplasm was darkened, vacuoles were filled with dark or coarse tannin precipitation (Figure 12.8B vs. 12.8A, E) in contrast to the sandy tannin appearance in the control.

With respect to whole-plant carbon allocation, it has to be considered that birch grown at relatively low nutrition maintains a higher proportion of O_3-injured foliage area than that grown at high nutrition. Therefore, it would be expected that carbon supply to roots may be more limited at low as compared with high nutrient supply. In fact, the root–shoot biomass ratio (R/S), which was often reduced under O_3 stress,[8] was only slightly lowered by O_3 at high nutrition, whereas O_3 reversed R/S at low nutrition from values > 1[54] to < 1, i.e., to about the level of the high-nutritional plants.[58,88,89] Although R/S was altered dramatically by O_3 at low nutrition, the proportions of fine and coarse roots remained unchanged in the total root biomass as did the specific root length in each root class. However, the coarse roots displayed overall smaller cell sizes and fewer starch granules (Figure 12.8H vs. 12.8G). The fine roots appeared darker and contained phenolic substances (stained black in Figure 12.8F) in conducting tissues, endodermal cells, and the pericycle. Such symptoms were not found in roots from unstressed plants.

Furthermore, an important factor modifying carbon allocation is mycorrhizal infection. In birch, interactions among nutrition, O_3, and mycorrhizae have not been studied. In *Acer saccharum* seedlings, O_3 exposure can cause morphological changes of mycorrhizae and stimulate the fungal, presumably nonmycorrhizal biomass in the rhizosphere, perhaps reflecting increased risk of pathogenic infection.[90] In O_3-exposed *Picea abies*, mycorrhizae were reduced in calcareous rather than acidic soils[91]; moreover, decreases in root and soil respiration have been found.[92] Probably, low carbon allocation to the root of O_3-exposed trees inhibits mycorrhizal and rhizospheric activities.[93] It has been shown, however, recently that in O_3-exposed seedlings of *Pinus ponderosa* O_3 stimulated (as long as O_3 levels were not too high) the fungal and bacterial biomass in the rhizosphere and increased root respiration, the latter effect being driven by low rather than high nutrition.[94] In parallel, the respiratory CO_2/O_2 quotient was increased by O_3, indicating changing substrate in the belowground respiration. In mature trees rather than seedlings of *Quercus rubra*, O_3 increased the root respiration while decreased the production and turnover of the fine roots.[95] It was concluded that high root respiration favored nutrient uptake to meet the enhanced demand of the O_3-exposed foliage. It is unclear, however, whether mycorrhizae become a "burden" in trees under O_3 stress,

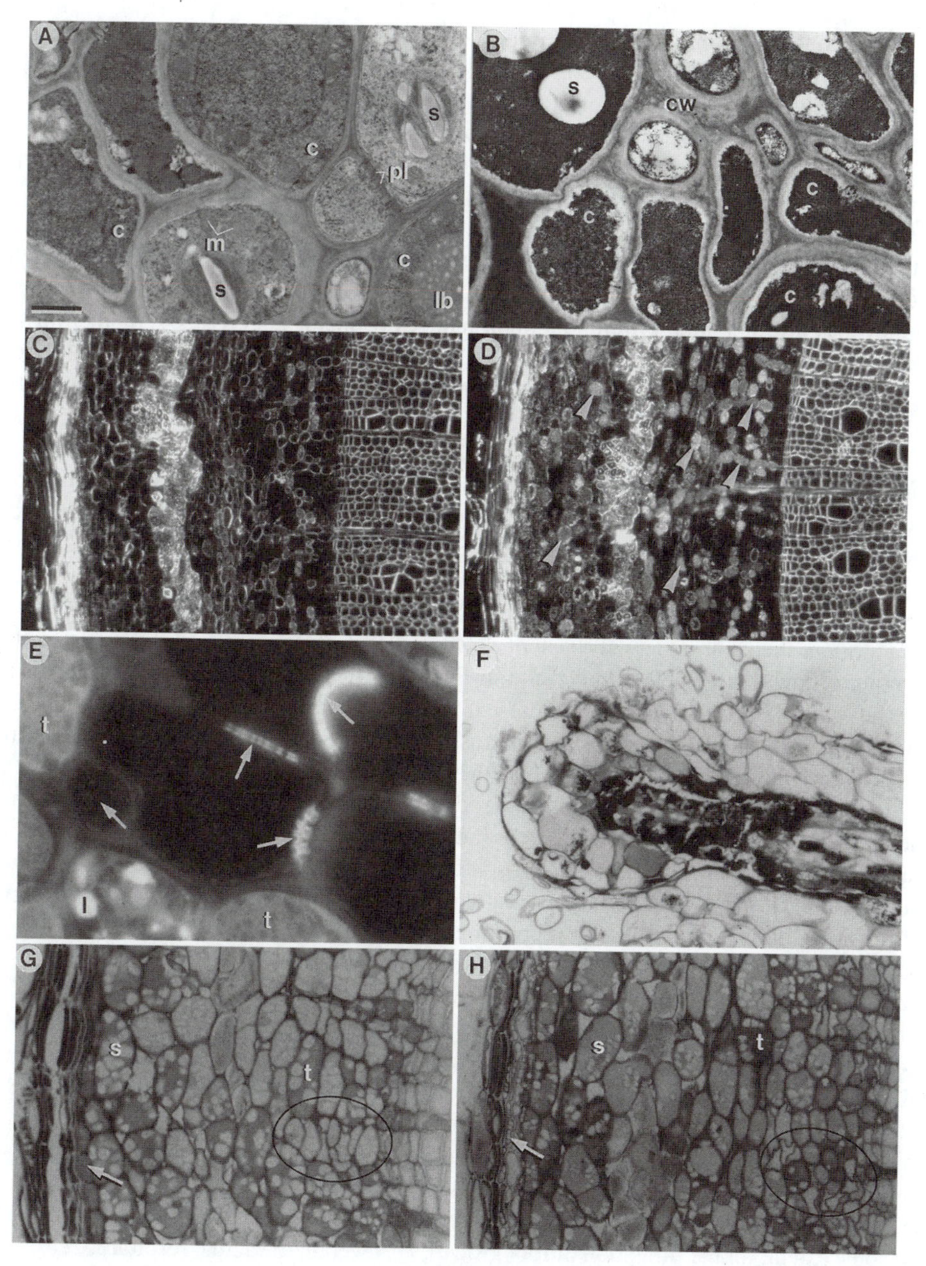

Figure 12.8 Transverse section of birch (*Betula pendula* Roth) leaves grown in filtered air (A, C, G) or filtered air with added O_3 (B, D, E, F, H) with low or high fertilization. Ozone was added during daylight hours (90 ppb) and night (40 ppb). The transverse section were prepared as described elsewhere.[16] (A and B) Bar = 1.3 μm; TEM micrographs of small leaf veins in 35-day-old birch leaves from the low-fertilization regime; (B) leaf with visible O_3 symptoms (stippling, O_3 dose at harvest = 58 ppm x h). In B (vs. control A) the cell walls (cw) are thickened, plasmodesmata (pl) and cytoplasmic lipid bodies (lb) are rare, starch grains (s) are large, the cytoplasm is electron dense and as dark as the mitochondria (m). (C and D) High fertilization, bar (see A) = 89 μm, stem section (age 150 days) viewed by light microscopy under dark field + phase contrast (unstained). Arrow heads in D (vs. control C) with an O_3 dose at harvest of 250 ppm x h denote phenolic (luminescent) cell content. (E) Detail from (D), stem phloem, bar = 7 μm. Arrows denote sieve pores narrowed by callose (white, stained by Aniline, viewed under UV fluorescent light) and an empty (black) companion cell; t = coarse tannin depositions, l = vacuolar lipid droplet. (F) Detail from fine root, lateral section, bar = 36 μm, phenolic compounds stained black by Os/KI. (G and H) Bar = 36 μm, coarse roots. H (vs. control G) shows declined phellogen cells (arrow), smaller storage starch grains (s), darkened vacuols (particularly tannin, t), and disordered phloem cell rows (circled).

taking into account the strong carbon sink represented by the fungus, the limited carbon allocation to the root, potential increases in root respiration, and, overall, the lowered photosynthetic capacity. Findings of unchanged R/S of mycorrhizal relative to nonmycorrhizal plants with reduced R/S under O_3 impact may support this view[96] and perhaps reflect the competition for assimilates required in the maintenance of the O_3-exposed foliage. As an issue also for herbaceous plants,[97] the nutritional impact on the cost–benefit relations of mycorrhizae under O_3 stress has not been clarified yet.

Stem growth can also be limited by O_3, although often radial rather than longitudinal increment appears to be affected.[8] One hypothesis is that O_3 and its reactive products may reach the phellogen from outside through the lenticells by diffusion through the intercellular space or the apoplast, where they may affect cellular differentiation or induce decline in stem tissues. Another possibility is that the supply of carbohydrates to the stem is limited via inhibited phloem loading. Microscopically, it can be observed that O_3 leads to a deterioration of the phellogen and phellem cells near the stem surface in young birch trees.[14] In the cortical parenchyma of the shoot more cells have their "disposal bags," the vacuoles, filled with phenolic substances (mostly tannin) at low rather than at high nutrient supply. Ozone-induced decline of cells becomes visible, when tannin precipitations become coarser and parenchyma cells become plasmolyzed and brownish,[14] which can be detected by their luminescence using the microscopical dark field technique (Figure 12.8D, arrow heads vs. control 12.8C). Similar cell decline occurs in the stem phloem parenchyma, and the rows of sieve elements (Figure 12.8, controls 12.8C and G) become disordered (12.8D and H). Irrespective of nutrition, the cambial activity is shut down earlier in the season and the xylem cells near the cambium show thicker cell walls (white in Figure 12.8D vs. 12.8C) than the younger xylem cells of the control. Before onset of autumnal leaf fall in early November, the stems from the control treatment still show open sieve pores. In contrast, the sieve pores in stems of the trees in the O_3 treatment, which have already partially lost their foliage, are narrowed by callose, have fewer protein bodies, and their companion cells appear dead without luminescent cytoplasm. The changed luminescence of the adjacent parenchyma cells indicates altered storage substances (Figure 12.8E).[19] Ozone-induced shutdown of the phloem transport forces the trees to suspend growth until the next growing season. This also inhibits retranslocation of assimilates. Cell decline was not observed at low fertilization, but xylem tissue was proportionally more decreased by low fertilization together with O_3 than the bark tissue.[19]

Ozone had minor impact on stem growth at low nutrition, and, therefore, the amount of respired CO_2 throughout the growing season per unit of volume increment was similar to the control in filtered air (Maurer and Matyssek, unpublished results). This ratio was enhanced by O_3; however, at high nutrition stem growth was significantly reduced and ceased earlier in the season than in the absence of O_3. Apparently, the proportion of growth respiration declined under O_3 stress relative to the respiratory maintenance costs which remained unchanged throughout the growing season (Maurer and Matyssek, unpublished results). The O_3-caused premature loss of the leaves at the lower stem probably was responsible, at high nutrition, for the lowered radial growth of the stem and, to some extent, for the limitation of root growth.[47,98] Overall, the actual respiratory costs related to wood formation did not seem to depend on nutrition or O_3 regime, which is consistent with findings about other factorial influences.[99,100]

Ozone has the potential of inhibiting the branching of the crown,[23] and so does low nutrition; consequently, suppressed branch formation by O_3 was most evident in birch of high nutrition.[58] Also, leaf expansion is inhibited by O_3 and low nutrition (Figure 12.1A and B), so again O_3 impact was significant only at high nutrition. In the latter nutrient regime, the lowered number of branches, apart from premature leaf loss and reduced leaf size, was mostly responsible for the smaller foliage area relative to plants in O_3-free air. At low nutrition, the foliage area was not reduced by O_3; however, the proportion of O_3-injured foliage was high. Thus, the carbon gain strongly depended on the photosynthetic performance of the injured leaves and on their maintenance.[58] The water loss was reduced per unit of whole-plant foliage area similar to the observation at the single leaf level, whereas no such effects were found at high nutrition. By modifying the crown architecture and the cost–benefit balance of water loss vs. carbon gain as related to the allometry, O_3 affects the mechanistic basis of plant competitiveness — mediated through nutrition.[101,102]

The concentrations of N, P, and K of whole birch plants were enhanced by O_3 at low nutrition, but differences diminished as high nutrition raised the whole-plant biomass and nutrient contents, with the greatest stimulations occurring in the absence of O_3.[58] Enhanced nutrient contents under O_3 have been reported previously,[103] although the overall findings conflict, due to changes in nutrient uptake vs. demand or plant-internal retranslocation.[8] Applying the concept of nutritional analysis by Timmer and Morrow[104] to birch with its indeterminate shoot growth, the response to nutrition would indicate nutrient demand at low nutrition. However, it should be noted that birch leaves did not develop visual symptoms of deficiency apart from decreased pigment contents (see Figure 12.6). The concentrations of Ca and Mg were not affected by O_3 at high and low nutrition. However, they declined as high nutrition increased the biomass and the nutrient contents of the plants, these changes being greatest in the absence of O_3. Thus, Ca and Mg seem to have been nonlimiting at low nutrient supply. Ozone did not fundamentally change the whole-plant interrelationships between levels and contents of nutrients and biomass production, but only decreased the extent of interaction between biomass production and nutrient relations.[58] Overall, the biomass production was driven by nutrition, whereas O_3 appeared to play a secondary role. Remarkably, the proportional limitation of annual growth by O_3 was similar in both nutrient regimes and tended to be smaller even at low nutrition. The whole-plant carbon balance of the second half of the growing season (after the O_3 dose had become a constraint on production) revealed that at low nutrition only a small proportion of the carbon gain was used for growth under O_3 stress, indicating high respiratory costs and explaining the low "water-use efficiency" of biomass production at the whole-plant level (Figure 12.9).

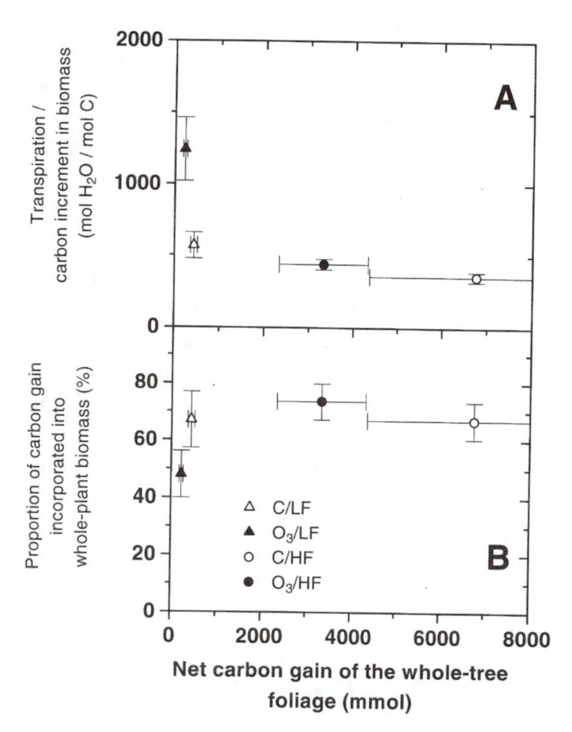

Figure 12.9 Transpiration as based on the carbon increment in the whole-plant biomass (A) and proportion of the carbon gain incorporated into the whole-plant biomass in relation to the net carbon gain of the foliage (B) in *Betula pendula* plants during the second half of the growing season (August 3 to October 3, 1993). (Calculated from data in Reference 58.) For abbreviations, see Figure 12.4.

It seems that birch can "choose" between two different "strategies" for assuring biomass production under O_3 stress, depending on the nutritional status and, as a consequence, on the rate of leaf formation. One has to be aware, however, that the extent of leaf longevity varies between years because of the additional variable impact of the accompanying environmental scenarios.[21] In indeterminate-growing trees, low nutrition can result in changes in whole-plant carbon allocation, and high respiratory costs and leaf maintenance; high nutrition allows for high leaf turnover[59] with minor impact on the whole-plant carbon allocation. Both nutritional strategies apparently allow for similar efficiencies in biomass production under O_3 stress,[105] and in these terms, high nutrition is not a prerequisite for high O_3 tolerance. Low nutrition did not override or balance the effects of O_3 stress on whole-plant carbon allocation, as suggested by Greitner and Winner,[40] Weinstein et al.,[106] or Mooney and Winner.[85] Considering the "bias" by nutrition on the carbon, water, and nutrient relations and on the allometric differentiation under O_3 impact, it is certainly imaginable that conflicting reports on responses to O_3 may relate to variable plant nutrition.[8] As well, conditions requiring the maintenance rather than the replacement of the O_3-injured foliage may render trees more susceptible to changes in the plant-internal resource allocation which, via altered root and crown architecture and related consequences for the "resource gathering capacity," can affect competitiveness.[107]

12.6 CONCLUSIONS

Since O_3 is taken up into the mesophyll via the open stomata, plants respond to O_3 fluxes rather than to ambient O_3 concentrations. The intrinsic O_3 exposure is determined by structural features of the leaf such as stomatal conductance, intracellular air spaces, thickness of cell walls, etc. A case study on birch shows that these features are strongly affected by nutrition on the one hand and are changed under the impact of O_3 on the other hand. Exposure to chronic O_3 levels frequently results in reduced O_3 fluxes into the leaf. Anatomical studies suggest that this influx is a consequence of injury rather than of regulated acclimation to O_3. Several lines of evidence suggest that O_3 injures pathways for assimilate transport, thereby affecting cellular carbohydrate metabolism. Low nutrient supply shifted assimilate resources to antioxidative defenses, thereby enabling a maintained life period of leaves. However, there was no evidence that O_3 as a single factor caused an induction of antioxidative defenses. In contrast, leaves from high-fertilized trees were shed earlier under O_3 stress so that the carbon gain overall rather depended on the uninjured foliage as compared with the low-fertilized plants, in which the carbon gain relied on the O_3-injured foliage. Ozone exposure also affected stem and roots growth. One reason was probably the diminished supply of carbohydrates. In the stem, effects on anatomical structures were also apparent. These O_3 effects were modified by the extent of nutrient supply. Since the case study was performend with birch, a species with indeterminate shoot growth and with young plants, further studies with older trees, preferably grown and exposed to experimentally elevated O_3 regimes under stand conditions and especially characterized by determinate growth patterns, are required. In addition to the variation in nutrients, other relevant factors such as light or water supply will have to be varied. Such factorial studies are important for identification of the most sensitive response mechanisms and for a scaling of O_3 effects from the cellular to the whole plant and up to the stand level.

ACKNOWLEDGMENTS

The work in the authors' laboratories has been funded by the DFG and the Commission of the European Communities (EUROSILVA). The financial support by the Swiss "Bundesamt für Bildung und Wissenschaft" through the EUREKA 447 EUROSILVA program is highly appreciated. We are grateful to Dr. W. Landolt for the fumigation management and to M. Eiblmeier, C. Rhiner, I. Kälin,

R. Gall, P. Bleuler, T. Koller, J. Bolliger, U. Bühlmann, and A. Kölliker for excellent technical assistance. The TEM micrographs (Figure 12.2A and B and 12.8A and B) have been kindly supplied by Dr. C. J. McQuattie, USDA Forest Service, Delaware, OH.

REFERENCES

1. Stockwell, W. R., Kramm, G., Scheel, H.-E., Mohnen, V. A., and Seiler, W., Ozone formation, destruction and exposure in Europe and the United States, in *Forest Decline and Ozone: A Comparison of Controlled Chamber and Field Experiments*, Sandermann, H., Wellburn, A., and Heath, D., Eds., Ecological Studies, Vol 127, Springer-Verlag, Heidelberg, 1997, 1.
2. Marenco, A., Gouget, H., Nedelec, P., and Pages, J-P., Evidence of a long term increase in tropospheric ozone from Pic Du Midi data series, consequences, positive radiative forcing, *J. Geophys. Res.*, 99, 16617, 1994.
3. Runeckles, V. C. and Vaartnou, M., EPR evidence for superoxide anion formation in leaves during exposure to low levels of ozone, *Plant Cell Environ.*, 20, 306, 1997.
4. Hoigné, J. and Bader, H., Ozonation of water: role of hydroxyl radicals as oxidizing intermediates, *Science*, 190, 782, 1975.
5. Heath, R. L. and Taylor, G. E., Physiological processes and plant responses to ozone exposure, in *Forest Decline and Ozone, A Comparison of Controlled Chamber and Field Experiments*, Sandermann H., Wellburn, A. R., and Heath, R. L., Eds., Ecological Studies, 127, Springer-Verla,g Berlin, 1997, 400.
6. Foyer, C., Descourvières, P., and Kunert, K. J., Protection against oxygen radicals: an important defence mechanism studied in transgenic plants, *Plant Cell Environ.*, 17, 507, 1994.
7. Chameides, W. L., Kasibhlata, P. S., Yienger, J., and Levy II, H., Growth of continental-scale metro-agro-plexes, regional ozone pollution, and world food production, *Science*, 264, 74, 1994.
8. Matyssek, R., Reich, P., Oren, R., and Winner, W. E., Response mechanisms of conifers to air pollutants, in *Ecophysiology of Coniferous Forests*, Smith, W. K. and Hinckley, T. M., Eds., Academic Press, New York, 1995, 255.
9. Rennenberg, H., Herschbach, C., and Polle, A., Consequences of air pollution on shoot–root interaction, *J. Plant Physiol.*, 148, 296, 1996.
10. Sandermann, H., Jr., Wellburn, A. R., and Heath, R. L., Forest decline and ozone: a synopsis, in *Forest Decline and Ozone, A Comparison of Controlled Chamber and Field Experiments*, Sandermann, H., Wellburn, A. R., and Heath, R. L., Eds., Ecological Studies 127, Springer-Verlag, Berlin, 1997, 367.
11. Polle, A., Photochemical oxidants:uptake and detoxification mechanisms, in *Responses of Plant Metabolism to Air Pollution*, DeKok, L. J. and Stulen, I., Eds. Backhuys Publishers, Leiden, 1998, 95.
12. Kerstiens, G. and Lendzian, K., Interactions between ozone and plant cuticles. I. Ozone deposition and permeability, *New Phytol.*, 112, 13, 1989.
13. Reich, P.B., Quantifying plant response to ozone: a unifying theory, *Tree Physiol.*, 3, 63, 1987.
14. Günthardt-Goerg, M. S., Matyssek, R., Scheidegger, C., and Keller, T., Differentiation and structural decline in the leaves and bark of birch (*Betula pendula*) under low ozone concentrations, *Trees*, 7, 104, 1993.
15. Pääkkönen, E., Paasisalo, S., Holopainen, T., and Kärenlampi, L., Growth and stomatal responses of birch (*Betula pendula* Roth.) clones to ozone in open-air and chamber fumigations, *New Phytol.*, 125, 615, 1993.
16. Günthardt-Goerg, M. S., McQuattie, C. J., Scheidegger, C., Rhiner, C., and Matyssek, R., Ozone induced cytochemical and ultrastructural changes, in leaf mesophyll cell walls, *Can. J. For. Res.*, 27, 453, 1997.
17. Urbach, W., Schmidt, W., Kolbowski, J., Rümmele, S., Reisberg, E., Steigner, W., and Schreiber, U., Wirkungen von Umweltschadstoffen auf Photosynthese und Zellmembranen von Pflanzen, in 1. Statusseminar der PBWU zum Forschungsschwerpunkt Waldschäden., Reuther, M. and Kirchner, M., Eds., GSF, München, 1989, 195.
18. Maier-Maercker, U., Image analysis of the stomatal cell walls of *Picea abies* (L.) Karst. in pure and ozone-enriched air, *Trees*, 12, 181, 1998.
19. Günthardt-Goerg, M. S., Different responses to ozone of tobacco, poplar, birch and alder, *J. Plant Physiol.*, 148, 207, 1996.

20. Günthardt-Goerg, M. S., Schmutz, P., Matyssek, R., and Bucher, J. B., Leaf and stem structure of poplar (*Populus* x *euramericana*) as influenced by O_3, NO_2, their combination, and different soil N supplies, *Can. J. For. Res.*, 26, 649, 1996.
21. Günthardt-Goerg, M. S., Maurer, S., Frey, B., and Matyssek, R., Birch leaves from trees grown in two fertilization regimes: Diurnal and seasonal responses to ozone, in *Responses of Plant Metabolism to Air Pollution*, De Kock, L.J., and Stulen, I., Eds., Backhuys, Leiden, the Netherlands, 1998, 315.
22. Landolt, W., Günthardt-Goerg, M. S., Pfenninger, I., Einig, W., Hampp, R., Maurer, S. and Matyssek, R., Effect of fertilization on ozone-induced changes in the metabolism of birch leaves (*Betula pendula*), *New Phytol.*, 137, 389, 1997.
23. Matyssek, R., Günthardt-Goerg, M.S., Saurer, M., and Keller,T., Seasonal growth, $\delta^{13}C$ of leaves and stem, and phloem structure in birch (*Betula pendula*) under low ozone concentrations, *Trees*, 6, 69, 1992.
24. Holopainen, T., Anttonen, S., Wulff, A., Palomäki, V., and Kärenlampi, L., Comparative evaluation of the effects of gaseous pollutiants, acidic deposition and mineral deficiencies: structural changes in the cells of forest plants, *Agric. Ecosyst. Environ.*, 42, 365, 1992.
25. Frey, B., Scheidegger, C., Günthardt-Goerg, M. S., and Matyssek, R., The effects of ozone and nutrient supply on stomatal response in birch (*Betula pendula*) leaves as determined by digital image-analysis and X-ray microanalysis, *New Phytol.*, 132, 135, 1996.
26. Matyssek, R., Günthardt-Goerg, M. S., Schmutz, P., Saurer, M., Landolt, W., and Bucher, J. B., Response mechanisms of birch and poplar to air pollutants, *J. Sustainable For.*, 6, 3, 1998.
27. Reich, P. B., Reconciling apparent discrepancies among studies relating life span, structure and function of leaves in contrasting plant life forms and climates, "The blind men and the elephant retold", *Functional Ecol.*, 7, 1, 1993.
28. Wolfenden, J. and Mansfield, T. A., Physiological disturbances in plants caused by air pollutants, *Proc. R. Soc. Edinburgh*, 97B, 1991, 117.
29. Schulze, E.-D. and Hall, A. E., Stomatal responses, water loss, and nutrient relations in contrasting environments, in *Encyclopedia of Plant Ecology, 12B, Physiological Plant Ecology II*, Lange, O. L., Nobel, P. S., Osmond, C. B., and Ziegler, H., Eds. Springer-Verlag, Berlin, 1982, 182,.
30. Schulze, E.-D., The regulation of plant transpiration: interactions of feedforward, feedback, and futile cycles, in *Flux Control in Biological Systems*, Schulze, E.-D., Ed., Academic Press, New York, 1994, 203.
31. Maurer, S., Matyssek R., Günthardt-Goerg, M. S., Landolt, W., and Einig, W., Nutrition and the ozone sensitivity of birch (*Betula pendula*), I. Responses at the leaf level, *Trees*, 12, 1, 1997.
32. Matyssek, R., Günthardt-Goerg , M. S., Keller, T., and Scheidegger, C., Impairment of the gas exchange and structure in birch leaves (*Betula pendula*) under low ozone concentrations, *Trees*, 5, 5, 1991.
33. Sasek, T. W. and Richardson, C. J., Effects of chronic doses of ozone on loblolly pine, photosynthetic characteristics in the third growing season, *For. Sci.*, 35, 745, 1989.
34. Schweizer, B. and Arndt, U., CO_2/H_2O gas exchange parameters of one- and two-year-old needles of spruce and fir, *Environ. Pollut.*, 68, 275, 1990.
35. Lippert, M., Steiner, K., Payer, H.-D., Simons, S., Langebartels, C., and Sandermann, H., Jr., Assessing the impact of ozone on photosynthesis of European beech (*Fagus sylvatica* L.) in environmental chambers, *Trees*, 10, 268, 1996.
36. Saurer, M., Maurer, S., Matyssek, R., Landolt, W., Günthardt-Goerg, M. S., and Siegenthaler, U., The influence of ozone and nutrition on $\delta^{13}C$ in *Betula pendula*, *Oecologia*, 103, 397, 1995.
37. Farquhar, G. D., Ehleringer, R.J., and Hubick, K. T., Carbon isotope discrimination and photosynthesis, *Annu. Rev. Plant Physiol. Plant Mol. Biol.*, 40, 503, 1989.
38. Lucas, P. W. and Wolfenden, J., The role of plant hormones as modifiers of sensitivity to air pollutants, *Phyton*, 36, 51, 1996.
39. Lu, P., Outlaw, W. H., Smith, B. G. and Freed, G. A., A new mechanism for the regulation of stomatal aperture size in intact leaves, accumulation of mesophyll derived sucrose in the guard cell wall of *Vicia faba*, *Plant. Physiol.,* 114, 109, 1997.
40. Greitner, C. S. and Winner, W. E., Nutrient effects on responses of willow and alder to ozone, in *Transaction, Effects of Air Pollution on Western Forests*, Olson, R. K., and Lefohn, A. S., Eds., Air & Waste Management Association, Anaheim, CA, 1989, 493.

41. Glatzel, G., Zur Frage des Mineralstoff- und Wasserhaushalts frischverpflanzter Fichten, *Central bl. Gesamte Forstwes.*, 90, 65, 1973.
42. Küppers, M., Zech, W., Schulze, E.-D. and Beck, E., CO_2-Assimilation, Transpiration und Wachstum von *Pinus sylvestris* L. bei unterschiedlicher Magnesiumversorgung, *Forstwiss. Central bl.*, 104, 23, 1985.
43. Beyschlag, W., Wedler, M., Lange, O. L., and Heber, U., Einfluß einer Magnesiumdüngung auf Photosynthese und Transpiration von Fichten an einem Magnesium-Mangelstandort im Fichtelgebirge, *Allg. Forstz*, 42, 738, 1987.
44. Pye, J. M., Impact of ozone on the growth and yield of trees, a review, *J. Environ. Qual.*, 17, 347, 1988.
45. Dann, M. S. and Pell, E. J., Decline of activity and quantity of ribulose bisphosphate carboxylase/oxygenase and net photosynthesis in ozone-treated potato foliage, *Plant Physiol.*, 91, 427, 1989.
46. Wellburn, F. A. M., Lau, K. K., Milling, P. M. K., and Wellburn, A. R., Drought and air pollution affect nitrogen cycling and free radical scavenging in *Pinus halepensis*, *J. Bot.*, 47, 1361, 1996.
47. Stitt, M., Rising CO_2 levels and their potential significance for carbon flow in photosynthetic cells, *Plant Cell Environ.*, 14, 741, 1991.
48. Wiskich, J. T. and Dry, I. B., The tricarboxylic acid cycle in plant mitochondria: its operation and regulation, in *Higher Plant Cell Respiration, Encyclopaedia of Plant Physiology*, Douce, R. and Day, D. A., Eds., New Series, Vol. 18, Springer-Verlag, Berlin, 1985, 281.
49. Martinoia, E. and Rentsch, D., Malate, compartimentation-responses to a complex metabolism, *Annu. Rev. Plant Physiol. Plant Mol. Biol.*, 45, 447, 1994.
50. Einig, W., Lauxmann, U., Hauch, B., Hampp, R., Landolt, W., Maurer, S., and Matyssek, R., Ozone-induced accumulation of carbohydrates changes enzyme activities of carbohydrate metabolism in birch leaves, *New Phytol.*, 137, 673, 1997.
51. Luethy-Krause, B., Pfenninger, I., and Landolt, W., Effects of ozone on organic acids in needles of Norway spruce and Scots pine, *Trees*, 4, 198, 1990.
52. Landolt, W., Günthardt-Goerg, M. S., Pfenninger, I., and Scheidegger, C., Ozone-induced microscopical changes and quantitative carbohydrate contents of hybrid poplar (*Populus* x *euramericana*), *Trees*, 8, 183, 1994.
53. Gerant, D., Podor, M., Grieu, P., Afif, D., Cornu, S., Morabito, D., Banvoy, J., Robin, C., and Dizengremel, P., Carbon metabolism, enzyme activities and carbon partitioning in *Pinus halepensis* Mill. to mild drought and ozone, *J. Plant Physiol.*, 148, 142, 1996.
54. Stitt, M. and Schulze, E.-D., Plant growth, storage, and resource allocation: from flux control in a metabolic chain to the whole-plant level, in *Flux Control in Biological Systems*, Schulze, E.-D., Ed. Academic Press, San Diego, 1994, 57.
55. Smirnoff, N. and Cumbes, Q. J., Hydroxyl radical scavenging activity of compatible solutes, *Phytochemistry*, 28, 1057, 1989.
56. Bücker, J. and Guderian, R., Accumulation of myo-inositol in *Populus* as a possible indication of membrane disintegration due to air pollution, *J. Plant Physiol.*, 144, 121, 1994.
57. Landolt, W., Pfenninger, I., and Lüthy-Krause, B., The effect of ozone and season on the pool sizes of cyclitols in Scots pine (*Pinus sylvestris*), *Trees*, 3, 85, 1989.
58. Maurer, S. and Matyssek, R., Nutrition and the ozone sensitivity of birch (*Betula pendula*), II. Carbon balance, water-use efficiency and nutritional status of the whole plant, *Trees*, 12, 11, 1997.
59. Tjoelker, M. G. and Luxmoore, R. J., Soil nitrogen and chronic ozone stress influence physiology, growth and nutrient status of *Pinus taeda* L. and *Liriodendron tulipifera* L. seedlings, *New Phytol.*, 119 69, 1991.
60. Baker, T. R. and Allen, H. L., Ozone effects on nutrient resorption in loblolly pine, *Can. J. For. Res.*, 26, 1634, 1996.
61. Sandermann, H., Ozone and plant health, *Annu. Rev. Phytopathol.*, 34, 347, 1996.
62. Asada K., Production and action of reactive oxygen species in photosynthetic tissues, in *Causes of Photo-Oxidative Stress and Amelioration of Defense Systems in Plants*, Foyer, C. and Mullineaux, P., Eds., CRC Press, Boca Raton, FL, 1994, 77.
63. Giamalva, D., Church, D., and Pryor, W. A., Comparison of the rates of ozonisation of biological antioxidants and loleate and linoleate esters, *Biochem. Biophys. Res. Commun.*, 133, 773, 1985.
64. Freebairn, H. T., The prevention of air pollution damage to plants by use of vitamin C sprays, *J. Air Pollut. Control Assoc.*, 10, 314, 1960.

65. Menser, A., Response of plants to air pollutants: III. A relation between acorbic acid levels and ozone susceptibility of light pre-conditioned tobacco plants, *Plant Physiol.*, 39, 564, 1964.
66. Conklin, P. L., Williams, E. H., and Last, R. L., Environmental stress sensitivity of an ascorbic acid-deficient *Arabidopsis* mutant, *Proc. Natl. Acad. Sci. U.S.A.*, 93, 9970, 1996.
67. Casano, L. M., Matrin, M., and Sabater, B., Sensitivity of superoxide dismutase transcrpit levels and activities to oxidative stress is lower in mature-senescent than in young barley leaves, *Plant Physiol.*, 106, 1033, 1994.
68. Polle, A. and Rennenberg, H., Photooxidative stress in trees, in *Causes of Photo-oxidative Stress and Amelioration of Defense Systems in Plants Foyer*, C. and Mullineaux, P., Eds., CRC Press, Boca Raton, FL, 1994, 199.
69. Osswald, W., Senger H., and Elstner, E. F., Ascorbic acid and glutathione content of spruce needles from different locations in Bavaria, *Z. Naturforsch.*, 42, 879, 1987.
70. Cakmak, I. and Marschner, H., Magnesium deficiency and high light intensity enhance activities of superoxide dismutase, ascorbate peroxidase and glutathione reductase in bean leaves, *Plant Physiol.*, 98, 1222, 1992.
71. Polle, A., Chakrabarti, K., Chakrabarti, S., Seifert, F., Schramel, P. and Rennenberg, H., Antioxidants and manganese deficiency in needles of Norway spruce (*Picea abies* L.) trees, *Plant Physiol.*, 99, 1084, 1992.
72. Polle, A., Otter, T., and Mehne-Jakobs, B., Effect of magnesium-deficiency on antioxidative systems in needles of Norway spruce (*Picea abies* L. Karst) grown with different ratios of nitrate and ammonium as nitrogen sources, *New Phytol.*, 128, 621, 1994.
73. Cakmak, I. and Marschner, H., Magnesium deficiency enhances resistance to paraquat toxicity in bean leaves, *Plant Cell Environ.*, 15, 955, 1992.
74. Schmieden, U. and Wild, A., The contribution of ozone to forest decline, *Physiol. Plant.*, 94 371,1995.
75. Polle, A., Defense against photooxidative damage in plants in *Oxidative stress and the Molecular Biology of Antioxidant Defenses,* Scandalios, J., Ed., Cold Spring Harbor Laboratory Press, Cold Spring Harbor, 1997, 623,.
76. Polle, A. and Morawe, B., Seasonal changes of antioxidative systems in foliar buds and leaves of field grown beech trees (*Fagus sylvatica* L.) in a stressful climate, *Bot. Acta*, 108, 314, 1995.
77. Polle, A., Kröniger, W., and Rennenberg, H., Seasonal fluctuations of ascorbate-related enzymes: acute and delayed effects of late frost in spring on antioxidative systems in needles of Norway spruce *(Picea abies* L.), *Plant Cell Physiol.*, 37, 717, 1996.
78. Polle, A., Wieser, G., and Havranek, W. M., Quantification of ozone influx and apoplastic ascorbate content in needles of Norway spruce trees (*Picea abies* L., Karst.) at high altitude, *Plant Cell Environ.*, 18, 681, 1995.
79. Luwe, M. and Heber, U., Ozone detoxification in the apoplast and symplast of spinach, broad bean and beech leaves at ambient and elevated concentrations of ozone in air, *Planta*, 197, 448, 1995.
80. Luwe, M., Takahama, U., and Heber, U., Role of ascorbate in detoxifying ozone in the apoplast of spinach (*Spinacia oleracea*) leaves, *Plant Physiol.*, 101, 969, 1993.
81. Polle, A., Chakrabarti, K., Schürmann, W., and Rennenberg, H., Composition and properties of hydrogen peroxide decomposing systems in extracellular and total extracts from needles of Norway spruce (*Picea abies* L., Karst.), *Plant Physiol.*, 94, 312, 1990.
82. Gedroc, J. J., McConnaughay, K. D. M., and Coleman, J. S., Plasticity in root shoot partitioning: optimal, ontogenetic, or both? *Functional Ecol.*, 10, 44, 1996.
83. Canham, C. D., Berkowitz, A. R., Kelly, V. R., Lovett, G. M., Ollinger, S. V., and Schnurr, J., Biomass allocation and multiple resource limitation in tree seedlings, *Can. J. For. Res.*, 26, 1521, 1996.
84. Waring, R. H., Characteristics of trees predisposed to die, *BioScience*, 37, 569, 1987.
85. Mooney, H. A. and Winner, W. E., Partitioning response of plants to stress, in *Response of Plants to Multiple Stresses*, Mooney, H. A, Winner, W. E, and Pell, E. J., Eds, Academic Press, San Diego, 1991, 129.
86. Fink, S., Pathological anatomy of conifer needles subjected to gaseous pollutants or mineral deficiencies, *Aquilo Ser Bot.*, 27, 1, 1989.
87. Neals, T.F. and McLeod, A. L., Do leaves contribute to the abscisic acid present in the xylem of "droughted"sunflower plants? *Plant Cell Environ.*, 14, 979, 1992.

88. Horton, S. J., Reinert, R. A., and Heck, W. W., Effects of ozone on three open-pollinated families of *Pinus taeda* L. grown in two substrates, *Environ. Pollut.*, 65, 279, 1990.
89. Chevone, B. I., Young, Y. S., and Reddick, G. S., Acidic precipitation and ozone effects on growth of loblolly and shortleaf pine seedlings, *Phytopathology*, 74, 756, 1984.
90. Duckmanton, L. and Widden, P., Effect of ozone on the development of vesicular-arbuscular mycorrhizae in sugar maple saplings, *Mycologia*, 86, 181, 1994.
91. Blaschke, H. and Weiss, M., Impact of ozone, acid mist and soil characteristics on growth and development of fine roots and ectomycorrhizae of young clonal Norway spruce, *Environ. Pollut.*, 64, 225, 1990.
92. Edwards, N. T., Root and soil respiration reponses to ozone in *Pinus taeda* L. seedlings, *New Phytol.*, 118, 315, 1991.
93. Andersen, C. P. and Rygiewicz, P. T., Stress interactions and mycorrhizal plant response: understanding carbon allocation priorities, *Environ. Pollut.*, 73, 217, 1991.
94. Andersen, C. P. and Scagel, C. F., Nutrient availability alters belowground respiration of ozone-exposed ponderosa pine, *Tree Physiol.*, 17, 377, 1997.
95. Kelting, D. L., Burger, J. A., and Edwards, G. S., The effects of ozone on the root dynamics of seedlings and mature red oak (*Quercus rubra* L.), *For. Ecol. Manage.*,79, 197, 1995.
96. Mahoney, M. J., Chevone, B. I., Skelly, J. M., and Moore, L. D., Influence of mycorrhizae on the growth of loblolly pine *Pinus taeda* seedlings exposed to ozone and sulphur dioxide, *Phytopathology*, 75, 679, 1985.
97. Miller, J. E., Shafer, S. R, Schoeneberger, M. M., Pursley, W. A., Horton, S. J., and Davey, C. B., Influence of a mycorrhizal fungus and/or *Rhizobium* on growth and biomass partitioning of subterranean clover exposed to ozone, *Water Air Soil Pollut.*, 96, 233, 1997.
98. Dickson, R. E. and Isebrands, J. G., Leaves as regulators of stress response, in *Response of Plants to Multiple Stresses*, Mooney, H. A., Winner, W. E., and Pell, E. J., Eds., Academic Press, San Diego, 1991, 4.
99. Carey, E. V., Callaway, R. M., and Delucia, E. H., Stem respiration of ponderosa pine grown in contrasting climates: implications for global climate change, *Oecologia*, 111, 19, 1997.
100. Matyssek, R. and Schulze, E.-D., Carbon uptake and respiration in above-ground parts of a *Larix decidua* x *leptolepis* tree, *Trees*, 2, 233, 1988.
101. Küppers, M., Canopy gaps, competitive light interception and economic space filling — a matter of whole-plant allocation, in *Exploitation of Environmental Heterogenity by Plants — Ecophysiological Processes Above And Below Ground*, Caldwell, M. M. and Pearcy, R. W., Eds., Academic Press, San Diego, 1994, 111.
102. Tremmel, D. C. and Bazzaz, F. A., Plant architecture and allocation in different neighborhoods — implications for competitive success, *Ecology*, 76, 262, 1995.
103. Keller, T. and Matyssek, R., Limited compensation of ozone stress by potassium in Norway spruce, *Environ. Pollut.*, 67, 1, 1990.
104. Timmer, V. R and Morrow, L. D., Predicting fertilizer growth response and nutrient status of jack pine by foliar diagnosis, in *Forest Soils andTtreatment Impacts*, Stone, E. L., Ed., Proceedings of the Sixth North Amercican Forest Soils Conference, Knoxville, TN, 1983, 335.
105. Matyssek, R., Maurer, S., Günthardt-Goerg, M. S., Landolt, W., Saurer, M., and Polle, A., Nutrition determines the "strategy"of *Betula pendula* for coping with ozone stress, *Phyton*, 37, 157, 1997.
106. Weinstein, D. A., Beloin, R. M., and Yanai, R. D., Modeling changes in red spruce carbon balance and allocation in response to interacting ozone and nutrient stresses, *Tree Physiol.*, 9, 127, 1991.
107. Nebel, B. and Fuhrer, J., Inter- and intraspecific differences in ozone sensitivity in semi-natural plant communities, *Angew. Bot.*, 68, 116, 1995.

CHAPTER 13

Use of Protective Chemicals to Assess the Effects of Ambient Ozone on Plants

William J. Manning

CONTENTS

13.1 INTRODUCTION

Concentrations of tropospheric ozone (O_3) that exceed normal background (25 to 40 ppbv) frequently occur during plant growing seasons in developed and industrializing regions of the world, allowing O_3 to become the world's most important and all-pervasive air pollutant.[1-6] Tropospheric O_3 concentrations are expected to increase by 0.3 to 1.0% per year for the next 50 years.[1] In order to understand the magnitude of the effects of ambient O_3 on plants, and to provide the data needed to devise adequate air quality standards, it is essential that we have accurate and reliable methods to assess the effects of ambient O_3 on vegetation.[7]

A great many experiments have been conducted and much has been published about the effects of O_3 on crops and forest trees.[8-10] Reports of results either come from studies done in controlled

1-56670-341-7/00/$0.00+$.50

exposure chambers, emphasizing short-term dose–response experiments, or from field chambers designed to assess long-term effects on growth and reproduction. Results from controlled exposure chambers lack relevance to ambient conditions, and field chambers may have effects of their own ("chamber effects") that can result in under- or overassessment of the effects of ambient O_3.

Open-top field exposure chambers (OTCs) were developed to assess the effects of either ambient O_3 alone or ambient O_3 plus increments of additional O_3 on plants.[11] Activated charcoal filters are used to remove most of the O_3 in some OTCs, but not in others. Comparisons are made between results from charcoal-filtered OTCs (CF) and non-filtered OTCs (NF). If results from NF OTCs are lower than from CF OTCs, then the differences are attributed to the effects of exposure to ambient or ambient plus additional O_3. Problems with this method may arise when a third treatment: the ambient air (AA) nonchamber plot is also used. In theory, results from NF OTCs and AA plots should be the same. When this is not the case, it casts doubt on the reliability of the OTC method.[12]

OTCs have certain design problems that can affect results. Air moves horizontally across and through the plant canopy and then upward and out through the open top of the chamber. In nature, air moves over the plant canopy and then down into it. The air movement system in the chambers runs continuously, breaking plant leaf boundary layers at all times, and resulting in increased O_3 uptake. Charcoal filters may alter dry deposition of sulfur and nitrogen compounds and/or increase nitrogen dioxide in the chambers. These chamber effects affect plant responses to O_3, thus decreasing the relevancy of the results achieved in them to ambient conditions.[11-16] Despite these problems, OTCs have been used all over the world, and results from them form the basis for much of the literature on the effects of O_3 on vegetation.

Due to the technical problems and limitations of controlled exposure and OTC methods, the key question — "Does ambient O_3 affect plant growth and yield?" — remains unanswered. To answer this question in the most relevant way possible will require the use of other methods of assessment — ones that do not have appreciable effects of their own on plant growth, and also allow integration of the effects of fluctuating concentrations of ambient O_3 with natural conditions of light, temperature, air movement, moisture, relative humidity, nutrients, etc.[12]

An alternative method would involve the application of protective chemical compounds to plants in the field to prevent injury from exposure to ambient O_3. This would allow the plants to grow under completely natural conditions, without any confounding chamber effects.

13.2 PROTECTIVE CHEMICALS

13.2.1 Historical Perspective

It has been known for many years that a diverse group of chemical compounds, such as antioxidants, antisenescence agents, antitranspirants, dusts, growth regulators, growth retardants, and pesticides, will provide varying degrees of short-term protection for plants from O_3 injury (Table 13.1). Most of these protectant chemicals were used in short-term single application studies designed to identify those that were effective in preventing acute O_3 injury. Many that were effective also had other effects on plants that made them unsuitable for research on the long-term effects of chronic exposure to ozone under ambient conditions.[17-20] The development of the OTC method also led to a decrease in interest in the use of protective chemicals. As problems with the OTC method became apparent, there has been a resurgence of interest in protective chemicals, especially in areas where technology is a limiting factor.

13.2.2 Considerations for Use

In order to use a protective chemical successfully to assess the long-term effects of O_3 on plants, there has to be a clear and complete determination of the effects of repeated applications of the

Table 13.1 Examples of Chemicals Used to Protect Plants from Ozone Injury

Antioxidants	
Ascorbic acid	K-, N-ascorbate
Butox	Piperonyl butoxide
DPA	Diphenylamine
EDU	Ethylenediurea *N*-[2-(2-oxo-1-imididazolidinyl) ethyl]-*N*-phenylurea
NBC	Nickel-*N*-dibutyldithiocarbomate
Antisenescence Agents	
Polyamines	Putrescine, spermidine, spermine
Antitranspirants	
Folicote	Paraffinic hydrocarbon waxes
Wilt-Pruf	
Dusts	Charcoal, diatomaceous earth, ferric oxide, kaolin
Growth Regulators	
Cytokinins	
BA	6-Benzylamine purine
Kinetin	*N*-6-Benzyladinine
Growth Retardants	
CBCP	2,4-Dichloro-benzyl tributyl phosphomium chloride
SADH	Succinic acid, 2,2-dimethyl hydrazide
Pesticides	
Fungicides	
Benomyl	Methyl-1-butyl-carbamyl-2-benzimidazole
Carboxin	5,6-Dihydro-2-methyl-1,4-oxathin-3-carboxanilide
Dithiocarbamates	Ethylene bis dithiocarbamates
Maneb (manganese)	
Zineb (zinc)	
Herbicides	
Diphenamid	*N,N*-Dimethyl-2,2-diphenylacetamide
Isopropalin	2,6-Dinitro-*N,N*-dipropyl cumidine
Insecticides	
Spectracide 25	Diazinon

chemical on the plant, both in the presence and absence of O_3. Classic dose–response studies must be done to determine when the chemical itself affects the plant and when O_3-injury suppression occurs without chemical effects on the plant. How many applications to make, at what intervals, how much to apply, and what application route to use must all be accurately determined. If this is not done, results are often ineffectual or are confounded by phytotoxicity from excess applications of the chemical.

The antiozonant compound EDU (ethylenediurea) can be used to illustrate the need for basic toxicology studies before use in the field to assess ambient O_3 injury. Carnahan et al.[21] did a dose–response study with EDU to determine the optimal concentration for prevention of acute O_3 injury in pinto bean seedlings. They determined this to be 500 $\mu g\ ml^{-1}$ (500 ppm) applied once as a foliar spray 24 h before exposure to O_3. Other investigators then designed field studies that involved repeated applications of EDU as a foliar spray, at 500 ppm, to suppress ambient O_3 injury and allow determination of O_3 effects on plant yields. They did not do new dose–response studies with EDU to determine if this was appropriate and if phytotoxicity would result. Heggestad[22] did this with his field studies with EDU and cotton, potatoes, and sweet corn. He used the 500 ppm EDU rate every week and effectively overdosed the plants, resulting in phytotoxicity. This phytotoxicity was logically more evident when ambient O_3 concentrations were quite low. He and others have used this type of results to disparage and suppress all results achieved with protective chemicals.[22-24]

Consideration should also be given to the method used to apply a protective chemical to prevent O_3 injury. Many of them become systemic within the plant as a result of being applied as a drench to soil or potting medium. When applied to leaves in spray applications, they become systemic within sprayed leaves, but are not translocated to newly emerging nonsprayed leaves. Care must be taken in applying the materials too often, or phytotoxicity may occur. Long-term use of soil drench applications may result in phytotoxicity due to uneven release of the material from organic matter. Low soil moisture levels could allow accumulation of the material. Stem injection may be the most effective method of introducing protective chemicals into tree seedlings.[25]

Using EDU as an example, a comparison of application methods and plant cultivar responses can be quite useful in the preliminary stages of research. Single foliar spray or pot drench applications of EDU were made to Bel-W3 (O_3-sensitive) and Bel-B (ozone-tolerant) tobacco (*Nicotiana tabacum* L.) seedlings (four-leaf stage) at 0, 125, 250, 500, 1000, and 2000 ppm. Foliar sprays did not cause any injury, but soil drench applications at 500, 1000, and 2000 ppm caused phytoxicity on Bel-W3 seedlings (Manning, unpublished). Weidensaul[26] has reported safe use of foliar sprays on pinto bean leaves up to 5000 ppm.

Plant species, cultivars, lines, selections, clones, etc. differ in their responses to O_3 and in the degree of protection they receive from applications of protective chemicals.[2,11] Ozone-tolerant species of several plants, most notably potato (*Solanum tuberosum* L.), are not affected by either O_3 or EDU.[27-29] Kostka-Rick and Manning[30] found variations in radish (*Raphanus sativus* L.) cultivar responses to O_3.

13.2.3 Examples

The antioxidant ascorbic acid, the systemic fungicide benomyl, and the antiozonant EDU have been the most widely used protective chemicals in research on ambient O_3 effects on plant growth and yield.

13.2.3.1 Ascorbic Acid

13.2.3.1.1 Characteristics — Ascorbic acid (vitamin C) is a naturally occurring antioxidant in plants.[31] Ascorbic acid in plant cell walls may limit the amount of O_3 that penetrates the wall to reach more vulnerable cell contents. Absorption of exogenously applied ascorbic acid or K- or Ca-ascorbate by leaves may result in increased concentrations of ascorbic acid in cell walls and may result in partial protection from O_3 injury.[32, 33]

13.2.3.1.2 Evaluation and Use — Potassium ascorbate foliar sprays were used in the field in Southern California to protect pinto beans, petunias, celery, romaine lettuce, and three types of citrus from ambient O_3.[32,33] Spinach was not protected. When K-ascorbate was applied twice a week to lettuce, 66% protection was obtained, increasing to 82% with five applications per week.[32] Applications of K-ascorbate to petunia leaves resulted in an increase in leaf number and weight.[33]

The Pfizer Chemical Division developed a commercial product called Ozoban for suppression of O_3 injury in plants. It was originally marketed for use as an "antioxidant spray to reduce yield loss caused by O_3 damage to Thompson seedless grapes" in California. Pfizer claimed that it acts "systematically through its absorption by the leaves and provides hugh levels of long-lasting oxidant protection." Pfizer described Ozoban as an isomer of ascorbic acid, but its chemical formula and structure are proprietary. Sodium erythorbate, however, is listed as the active ingredient of Ozoban. Ozoban was used by Brewer in extensive field trials with Thompson seedless grapes in Southern California and was found to reduce O_3 stipple and increase yields. One spray every 4 weeks at

1.25 lbs $acre^{-1}$ was determined to be effective. Not only were yields increased, but grape berry size, sugar content, pH, and titratable acidity were not affected.[34]

Sodium erythorbate has been used to protect young short-leaf pines from O_3 injury in east Texas. In a 1.5-year study, tree growth improved when sodium erythorbate was applied at rates up to 1030 ppm.[35] After 31 months, protection from ozone injury was provided by rates of sodium erythorbate up to 1545 ppm.[36]

13.2.3.2. Benomyl

13.2.3.2.1. Characteristics — Benomyl (methyl-1-butylcarbamoyl-2-benzimidazole carbamate), and several related compounds that contain the benzimidazole moiety, is primarily a systemic fungicide that also has antisenescence and antiozonant properties.[37,38] This is due to the presence of the benzimidazole moiety in the compound. Benomyl is systemic in plants when applied as a soil or potting medium drench and within leaves when applied as a foliar spray. The systemic presence of the benzimidazole moiety in various plant tissues can be determined by a bioassay that is based on its activity as a fungicide. An example involves placement of disks from treated leaves onto the surface of agar in petri dishes previously seeded with dormant conidia of fungi, such as species of *Penicillium* or *Botrytis*. Benzimidazole in the leaf disks will inhibit spore germination. By comparing the width of the zones of inhibition to results from a standard curve, a quantitative determination of persistence and distribution in relation to O_3 injury suppression can be determined.

13.2.3.2.2 Evaluation and Use — Benomyl has been used to protect a wide variety of plants from O_3 injury (Table 13.2). Many of these were short-term studies with potted plants in greenhouses. Benomyl did not protect plants from peroxyacetyl nitrate (PAN), a component of photochemical oxidant air pollution[39] and increased sensitivity to PAN in petunia, when incorporated into the potting medium at a high concentration.[40]

Table 13.2 Examples of the Use of Benomyl to Determine the Effects of Ozone on Plants

Plant	Ref.
Azalea (*Rhododendron* sp.)	66
Bean (*Phaseolus vulgaris*)	37, 45, 67
Grape (*Vitis labrusca*)	42, 43
(*V. vinifera*)	44
Poinsettia (*Euphorbia pulcherrima*)	68
Tobacco (*Nicotiana tabacum*)	69, 70
Turf grasses (*Poa* spp.)	71, 72

Oxidant stipple of grape was one of the first reported plant diseases caused by O_3.[41] Benomyl has been used extensively to suppress O_3 injury in grape leaves.[42-44] Benomyl was found to be more effective than EDU in field trials in Germany.[44] Benomyl also decreases the incidence of powdery mildew disease on grape leaves. Where powdery mildew is a problem, it may not be possible to separate the fungicidal effects of benomyl from its antisenescence and antiozonant effects. This could be a confounding factor in studies on O_3 effects on grape growth and yield.

Snap beans (*Phaseolus vulgaris*) have few plant diseases that would be affected by foliar sprays of benomyl. Foliar sprays of benomyl were applied to an O_3-tolerant and an O_3-sensitive cultivar in the field for the length of the plant life cycles. Benomyl-treated, O_3-sensitive bean plants had improved pod yields and increased fresh and dry total plant weights, compared with nontreated plants. Benomyl had no appreciable effects on the ozone-tolerant cultivar.[45] These results suggest that where crop biology and disease incidence are well known, benomyl could be used successfully to protect plants from O_3 injury and to determine yield losses.

13.2.3.3 EDU

13.2.3.3.1 Characteristics — In 1978, Carnahan et al.[21] reported a new chemical compound *N*-[-2-(2-oxo-1-imidazolidinyl)ethyl-]-*N*-phenylurea (abbreviated EDU for ethylendiurea) that prevented acute O_3 injury in pinto bean leaves when applied as a spray or potting medium drench. In dose–response studies, they found that increasing EDU concentrations resulted in increasing protection from O_3. The optimal concentration for a single application to prevent acute ozone injury in pinto bean was determined to be 500 ppm.

EDU has been widely used and appears to suppress acute and chronic O_3 injury in a wide range of plants, without appreciable effects of its own, other than delaying senescence, even in the absence of O_3-induced stress.[46-49] EDU is not a pesticide, growth regulator, or a plant nutrient. It is systemic in plants from root or stem uptake or translaminar from foliar sprays.[25,26] Repeated applications are required to protect newly emerging leaves. EDU appears to be specific for the suppression of O_3 injury, having no effects on PAN or sulfur dioxide injury.[50,51]

While it is known that EDU moves systemically in plants, there was no chemical or bioassay for EDU or its possible breakdown products in plant tissues, until recently. By using hydroponically grown bean plants, and HPLC, it has been reported that EDU concentrates in plant leaves and persists for 10 days or more.[52,53] By using protoplasts and cell cultures, it was shown that EDU did not enter cells, but remained in the apoplast, suggesting a direct role of EDU itself in protection from O_3 injury.[53]

There have been many investigations of the mode of action of EDU, but there is no clear picture about how EDU protects plants from O_3. Results from studies on EDU-induced changes in antioxidants and enzymes are inconsistent and have recently been reviewed.[49,53]

EDU affects different plants differently, providing varying degrees of protection from O_3 injury, depending on plant age, soil moisture levels, plant nutrition, etc. This means that appropriate dose–response studies must be done in advance to find the application rates and intervals between applications that are effective in protecting plants from O_3 injury, without interfering with plant growth itself.

13.2.3.3.2 Evaluation and Use — Since 1978, EDU has been used widely and extensively in studies designed to protect plants, mostly annual crops, from O_3 (Table 13.3). Results from field experiments have sometimes been contradictory or negative, when prior dose–response studies were not done.

Kostka-Rick and Manning[54,55] conducted extensive greenhouse tests with EDU on radish and snap bean to determine optimal rates for application of EDU as a soil drench for protection from O_3 injury, without affecting plant growth. Then they conducted field studies on the effects of ambient O_3 on the dynamics of biomass partitioning of radish and bean, using EDU to reveal the effects.[56,57] They were able to demonstrate that EDU can protect plants from ozone injury, without interfering with plant growth itself.

Damicone[58] used EDU soil drenches to detect ambient ozone effects on yield of two ozone-sensitive and two O_3-tolerant soybean plant introduction lines (Table 13.4). EDU increased the number of pods per plant, the number of seeds per plant, and the total seed weight per plant for the two O_3-sensitive lines, but did not significantly affect these parameters in the two O_3-tolerant lines. A combination of EDU with a pair of closely-related plants, that differ in O_3 sensitivity provides a good field method for assessing ambient O_3 injury effects on plants.

EDU has also been used in conjunction with OTCs. Hassan et al.[59] used EDU to protect radish and turnip plants from O_3 in the field in Egypt. They also found that EDU had no effects on plant growth in the absence of ozone in a charcoal-filtered air OTC. EDU is a low-cost, low-technology

Table 13.3 Examples of the Use of EDU to Determine the Effects of Ozone on Plants

Plant	Yield Reductions	Ref.
Onion (*Allium cepa*)	Up to 37%	73
Potato (*Solanum tuberosum*)	Up to 19%	28
	Up to 35%	27
Soybean (*Glycine max*)	Up to 20 %	58
	None detected	74
Tomato (*Lycopersicon esculentum*)	Up to 36%	75
White (navy) bean	Up to 36%	76
	Up to 24%	77
	Up to 35%	78
Watermelon (*Citrullus vulgaris*)	Up to 30%	79

Table 13.4 Influence of EDU on Yield of Two Ozone-Sensitive (OS) and Two Ozone-Tolerant (OT) Soybean Genotypes in the Field.

Genotype and Treatments	No. Pods per Plan*	No. Seeds per Plant	Seed Weight per Plant
Maturity Group O			
(153.317) OT			
No EDU	30.0	67.7	14.4
EDU	30.6	67.8	14.7
(153.283) OS			
No EDU	55.9a	128.5a	20.2a
EDU	69.2b	159.9b	27.2b
Maturity Group I			
(189.907) OT			
No EDU	45.2	103.1	20.8
EDU	39.8	107.1	21.7
(153.284) OS			
No EDU	47.2a	108.5a	17.4a
EDU	62.6a	142.5b	23.2b

* Mean values of 60 plants.
Values followed by different letters are significantly different from each other, P = 0.05.
Source: Damicone, J. P., Ph.D. dissertation, University of Massachusetts, Amherst, 1985.

tool for assessing O_3 injury in the field. It has potential for use in remote or developing regions where electricity is not available or reliable and where funding is limited. EDU-treated and non-treated Bel-W3 (O_3-sensitive) and Bel-B (O_3-tolerant) tobacco seedlings were used in Warsaw and Kiev to detect and verify phytotoxic concentrations of O_3.[60, 61]

EDU has also been used as part of a regional O_3 bioindicator program in most of Europe.[62-64] EDU is applied as soil or pot drenches at 150 ppm at 14-day intervals. Plants used include radish and tobacco[62] and subterranean and white clover, bean, soybean, tomato, and watermelon.[63, 64]

13.3 DISCUSSION AND CONCLUSIONS

It is apparent from many reports in the literature that a variety of protective chemicals can be used to protect plants from O_3 injury under field conditions, in commercial agriculture, forests, and wilderness areas. The use of protective chemicals allows the treated plants and controls to grow under natural conditions and to be exposed to fluctuating natural concentrations of ambient O_3, without the confounding effects of devices, such as OTCs. Successful use of the protective chemical method requires extensive plant toxicology studies before field work is done to establish dose–response rates in the presence and absence of O_3. This allows establishment of appropriate treatment rates and intervals between treatments. A summary of the comparative advantages and disadvantages of the use of protective chemicals to assess O_3 injury on vegetation is given in Table 13.5

Table 13.5 Comparative Advantages and Disadvantages of the Use of Protective Chemicals to Assess Ozone Injury on Plants

Advantages
No chambers or apparatus required, ambient conditions prevail
No microclimate or chamber effects
Plant numbers and plot sizes can be varied according to the requirements of the experiment
High degree of replication possible
Low costs for materials and labor, low technological input
Protocol is simple and uncomplicated, requiring little equipment
Disadvantages
Ozone dose–response studies cannot be done, unless coupled with OTCs
Ambient ozone concentrations and environmental variables need careful monitoring
Repeated applications of chemicals could cause phytotoxicity, especially under dry soil conditions
Extensive plant toxicology studies are needed before the start of field experiments
Results vary from year to year; this is also an advantage, as this reflects annual differences in ozone and environmental conditions

Ascorbic acid and sodium erythorbate as antioxidants should have few if any confounding effects on plant growth. Ascorbic acid and ascorbate have been used in the past to reduce O_3 injury on a variety of plants. Sodium erythorbate shows promise in reducing O_3 injury on several species of pine. Sodium erythorbate appears to be more effective than ascorbic acid or ascorbate and appears to merit more use and evaluation.

Benomyl prevents O_3-caused senescence in plants. Its breakdown product is known and can be detected chemically or by bioassay, allowing monitoring of uptake and persistence of the benzimidazole moiety in relation to degree of O_3 injury suppression. As a systemic fungicide, however, it may also reduce fungal diseases in treated plants and confound efforts to separate factors affecting plant growth. Where crop biology and disease incidence are well known and characterized, benomyl could be used successfully to protect plants from O_3 and to determine yield or growth losses.

The antiozonant EDU has been the most extensively used of the three protective chemicals considered.[65] Methods now allow tracking of EDU in the plant and will lead to determination of its mode of action. It is evident from the substantial literature that, when EDU is used properly, it is a very powerful tool for verifying incidence of phytotoxic concentrations of ambient O_3 and detecting losses in yield and growth under field conditions. It has great potential for identifying O_3-sensitive vegetation in remote areas and assessing O_3 effects on plant ecosystems.

Protective chemicals can and should be used to assess the effects of O_3 on vegetation under ambient conditions, without the confounding effects of the other factors involved in the use of other methods.

REFERENCES

1. Chameides, W. L., Kasibhata, P. S., Yienger, J., and Levy, H., Growth of continental-scale metro-agroplexes, regional pollution, and world food production, *Science*, 264, 74, 1994.
2. Krupa, S. V. and Manning, W. J., Atmospheric ozone formation and effects on vegetation, *Environ. Pollut.*, 50, 101, 1988.
3. Runeckles, V. C. and Krupa, S. V., The impact of UV-B radiation and ozone on terrestrial vegetation, *Environ. Pollut.*, 83, 191, 1994.
4. Stockwell, W. R., Kram, G., Schel, H. E., Mohnen, V. A., and Seilev, W., Ozone formation, destruction, and exposure in Europe and the United States, in *Ozone and Forest Decline: A comparison of Controlled Chambers and Field Experiments*, Sandermann, H., Wellburn, A. R., and Heath, R. L., Eds., Ecol. Studies, Vol. 127, Springer-Verlag, Berlin, 1997.
5. Taylor, G. E., Jr., Johnson, D. J., and Andersen, C. P., A commissioned review — Air pollution and forest ecosystems: a regional to global perspective, *Ecol. Appl.*, 4, 662, 1994.
6. Tingey, D. T., Olszyk, D. M., Herstrom, A. E., and Lee, E. H., Effects of ozone on crops, in *Tropospheric Ozone,* McKee D. J., Ed., Lewis Publishers, Boca Raton, FL, 1994.
7. Krupa, S. V., Grunhage, L., Jager, H.-J, Nosal, M., Manning, W. J., Legge, A. H., and Hanewald, K., Ambient ozone and adverse crop response: a unified view of cause and effect, *Environ. Pollut.*, 87, 119, 1995.
8. Heck, W. W., Taylor, O. C., and Tingey, D. T., Eds., *Assessment of Crop Loss from Air Pollutants*, Elsevier Applied Science, London, 1988.
9. Flagler, R. B., Ed., *The Response of Southern Commercial Forests to Air Pollution*, Air and Waste Management Association, Pittsburgh, PA, 1992.
10. Chappelka, A. H. and Chevone, B. I., Tree response to ozone, in *Surface Level Ozone Exposures and Their Effects on Vegetation*, Lefohn, A. S., Ed., Lewis Publishers, Chelsea, MI, 1992.
11. Heagle, A. S., Body, D. E., and Heck, W. W., An open-top field chamber to assess the impact of air pollution on plants, *J. Environ. Qual.*, 2, 365, 1973.
12. Manning, W. J. and Krupa, S. V., Experimental methodology for studying the effects of ozone on crops and trees, in *Surface Level Ozone Exposures and Their Effects on Vegetation*, Lefohn, A. S., Ed., Lewis Publishers, Chelsea, MI, 1992.
13. Bytnerowicz, A., Olszyk, D. M., Dawson, P. J., and Morrison, C. L., Effects of air filtration on concentration and deposition of gaseous and particulate air pollutants in open-top chambers, *J. Environ. Qual.*, 18, 268, 1989.
14. Olszyk, D. M., Dawson, P. J., Morrison, C. L., and Takemoto, B. K., Plant response to non-filtered air vs. added ozone generated from dry air or oxygen, *J. Air Waste Manage. Assoc.*, 40, 77, 1990.
15. Fuhrer, J., Effects of ozone on managed pasture. I. Effects of open-top chambers on microclimate, ozone flux, and plant growth, *Environ. Pollut.*, 86, 297, 1994.
16. Grantz, D., Ozone uptake by San Joaquin valley crops, *Proc. Air Waste Manaeg. Assoc. Ann. Mtg.*, Paper 94-TA36.01, 1994.
17. Bialobok, S., Controlling atmospheric pollution, in *Air Pollution and Plant Life*, Treshow, M., Ed., John Wiley & Sons, New York, 1984.
18. Guderian, R., Tingey, D. T., and Rabe, R., Effects of photochemical oxidants on plants, in *Photochemical Oxidants*, Guderian, R., Ed., Springer-Verlag, Berlin, 1985.
19. Kender, W. J. and Forsline, P. J., Remedial measures to reduce air pollution losses in horticulture, *Hortic. Sci.*, 18, 680, 1983.
20. Ormrod, D. P. and Beckerson, D. W., Polyamines, *Hortic. Sci.*, 21, 1070, 1986.
21. Carnahan, J. E., Jenner, E. L., and Wat, E. K. W., Prevention of ozone injury in plants by a new protective chemical, *Phytopathology*, 68, 1225, 1978.

22. Heggestad, H. E., Reduction in soybean seed yields by ozone air pollution? *J. Air Pollut. Control Assoc.*, 38, 1040, 1988.
23. Heagle, A. S., Ozone and crop yield, *Annu. Rev. Phytopath.*, 27, 397, 1989.
24. Ormrod, D. P., Marie, B. A., and Allen, O. B., Research approaches to pollutant crop loss functions, in *Assessment of Crop Loss from Air Pollutants*, Heck, W. W., Taylor, O.C., and Tingey, D.T., Eds., Elsevier Applied Science, London, 1988, chap. 2.
25. Roberts, B. R., Wilson, L. R., Cascino, J. J., and Smith, G. P., Autoradiographic studies of ethylenediurea distribution in woody plants, *Environ. Pollut.*, 46, 81, 1987.
26. Weidensaul, T. C., *N*-[2-(2-oxo-1-imidazolidinyl) ethyl]-*N*-phenylurea as a protectant against ozone injury to laboratory-fumigated pinto bean plants, *Phytopathology*, 70, 42, 1980.
27. Clarke, B. B., Greenhalgh-Weidman, B., and Brennan, E. G., An assessment of the impact of ambient ozone on field-grown crops in New Jersey using the EDU method: Part 1 — white potato (*Solanum tuberosum*), *Environ. Pollut.*, 66, 351, 1990.
28. Foster, K. W., Guerard, R. J., Oshima, R. J., Bishop, J. C., and Timm, H., Differential ozone susceptibility of Centennial Russet and White Rose potato as demonstrated by fumigation and antioxidant treatment, *Am. Potato J.*, 60, 127, 1983.
29. Brennan, E. G., Clarke, B. B., Greenhalgh-Weidman, B., and Smith, G., An assessment of the impact of ambient ozone on field-grown crops in New Jersey using the EDU method: Part 2 — Soybean (*Glycine max*), *Environ. Pollut.*, 66, 361, 1990.
30. Kostka-Rick, R. and Manning, W. J., Dose-response studies with ethylenediurea (EDU) and radish, *Environ. Pollut.*, 79, 249, 1993.
31. Chameides, W. L., The chemistry of ozone deposition to plant leaves: role of ascorbic acid, *Environ. Sci. Technol.*, 23, 595, 1989.
32. Freebairn, H. T., Uptake and movement of 1-C 14 ascorbic acid in bean plants, *Plant Physiol.*, 16, 517, 1963.
33. Freebairn, H. T. and Taylor, O. C., Prevention of plant damage from airborne oxidizing agents, *Proc. Am. Soc. Hortic. Sci.*, 76, 693, 1960.
34. Hall, K., New compound counteracts smog, *Calif. Grape Grower*, December, 26, 1987.
35. Flagler, R. B. and Toups, B.G., Use of sodium erythorbate to determine the effects of ambient ozone on shortleaf pine, in *The Response of Southern Commercial Forests to Air Pollution*, Flagler, R.B., Ed., Air & Waste Management Association, Pittsburgh, PA, 1992.
36. Flagler, R. B., Lock, J. E., and Elsik, C. G., Assessing the response of short leaf pine to ozone using sodium erythorbate and ethylenediurea, *Proc. Air Waste Manage. Assoc. Annual Mtg.*, paper 94-TA36.05, 1994.
37. Pellissier, M., Lacasse, N. L., and Cole, H. Jr., Effectiveness of benzimidazole, benomyl, and thiabendizole in reducing ozone injury to pinto beans, *Phytopathology*, 580, 1972.
38. Tomlinson, H. and Rich, S., Relating ozone resistance and antisensescence in beans treated with benzimidazole, *Phytopathology*, 63, 208, 1973.
39. Pell, E. J., Influence of benomyl soil treatment on pinto bean plants exposed to peroxacetyl nitrate and ozone, *Phytopathology*, 66, 731, 1976.
40. Pell, E. J. and Gardner, W., Enhancement of peroxyacetyl nitrate injury to petunia foliage by benomyl, *Hortic. Sci.*, 14, 61, 1979.
41. Richards, B. L., Middleton, J. T., and Hewitt, W. B., Air pollution with relation to agronomic crops. V. Oxidant stipple of grape, *Agron. J.*, 50, 559, 1958.
42. Kender, W. J., Taschenbert, E. F., and Shaulis, N. J., Benomyl protection of grapevines from air pollution injury, *Hortic. Sci.*, 8, 396, 1973.
43. Musselman, R. C., Protecting grapevines from ozone injury with ethylenediurea and benomyl, *Am. J. Enol. Viticult.*, 36, 38, 1985.
44. Tiedemann, A. V. and Herrman, J. V., First record of grapevine oxidant stipple in Germany and effects of field treatments with ethylenediurea (EDU) and benomyl on disease, *Z. Pflanzenbau Pflanzenschutz*, 99, 533, 1992.
45. Manning, W. J., Feder, A. A., and Vardaro, P. M., Suppression of oxidant injury by benomyl: effects on yields of bean cultivars in the field, *J. Environ. Qual.*, 3, 1, 1974.
46. Miller, J. E., Pursley, W. A., and Heagle, A. S., Effects of ethylenediurea on growth, biomass partitioning and yield of snapbean at a range of ozone concentrations, *J. Environ. Qual.*, 23, 1082, 1994.

47. Wang, C.Y. and Baker, J. E., Extending vase life of carnations with aminooxyacetic acid, polyamines, EDU and CCCP, *Hortic. Sci.*, 15, 805, 1980.
48. Brunschon-Harti, S., Fangmeir, A., and Jager, H.-J., Influence of ozone and ethylene diurea (EDU) on growth and yield of bean (*Phaseolus vulgaris*) in open-top chambers, *Environ. Pollut.*, 90, 89, 1995.
49. Brunschon-Harti, S., Fangmeier, A., and Jager, H.-J., Effects of ethylenediurea and ozone on the antioxidant systems in beans (*Phaseolus vulgaris*), *Environ. Pollut.*, 95, 1995.
50. Cathey, H. M. and Heggestad, H. E., Ozone and sulfur dioxide sensitivity of petunia: Modification by ethylenediurea, *J. Am. Soc. Hortic. Sci.*, 107, 1028, 1982.
51. Cathey, H. M. and Heggestad, H. E., Ozone sensitivity of herbaceous plants: Modification by ethylenediurea , *J. Am. Soc. Hortic. Sci.*, 107, 1035, 1982.
52. Regner-Joosten, K., Manderscheid, R., Bergmann, R., Bahadir, M., and Weigel, H. J., An HPLC method to study the uptake and partitioning of the antiozonant EDU in bean plants, *Angew. Bot.*, 68, 151, 1994.
53. Gatta, L., Mancino, L., and Federico, R., Translocation and persistence of EDU (ethylenediurea) in plants: the relationship of its role in ozone damage, *Environ. Pollut.*, 96, 445, 1997.
54. Kostka-Rick, R. and Manning, W. J., Radish (*Raphanus sativus*): a model for studying plant responses to air pollutants and other environmental stresses, *Environ. Pollut.*, 82, 107, 1993.
55. Kostka-Rick, R. and Manning. W. J., Dose-response studies with the antiozonant ethylenediurea (EDU) applied as a soil drench to two growth substrates on greenhouse grown varieties of *Phaseolus vulgaris*, *Environ. Pollut.*, 82, 63, 1993.
56. Kostka-Rick, R., Manning, W. J., and Buonaccorsi, J. P., Dynamics of biomass partitioning in field-grown radish varieties, treated with ethylenediurea, *Environ. Pollut.*, 80, 133, 1993.
57. Kostka-Rick, R. and Manning, W. J., Dynamics of growth and biomass partitioning in field-grown bush bean (*Phaseolus vulgaris*) treated with the antiozonant ethylenediurea (EDU), *Agric. Ecosyst. Environ.*, 47, 195, 1993.
58. Damicone, J. P., Growth, Yield and Foliar Injury Response of Early-Maturing Soybean Genotypes to Ozone and *Fusarium oxysporum*, Ph.D. dissertation, University of Massachusetts, Amherst, 1985.
59. Hassan, I. A., Ashmore, M. R., and Bell, J. N. B., Effect of ozone on radish and turnip under Egyptian field conditions, *Environ. Pollut.*, 89, 107, 1995.
60. Bytnerowicz, A., Manning, W. J., Grosjean, D., Chmielewski, W., Dmuchowski, W., Grodzinska, K., and Godzik, B., Detecting ozone and demonstrating its phytotoxicity in forested areas of Poland: a pilot study, *Environ. Pollut.*, 80, 301, 1993.
61. Blum, O., Bytnerowicz, A., Manning, W. J., and Popovicheva, L., Ambient tropospheric ozone in the Ukranian Carpathian Mountains and Kiev Region: Detection with passive samplers and bioindicator plants, *Environ. Pollut.*, 98, 299, 1998.
62. Manes, F., Altieri, A., Tripodo, P., Booth, C. E., and Unsworth, M. H., Bioindication study of effects of ambient ozone on tobacco and radish plants, using a protective chemical (EDU), *Ann. Bot.*, 48, 133, 1990.
63. Sanders, G. E., Booth, C. E., and Weigel, H. J., The use of EDU as a protectant against ozone, in *Effects of Air Pollution on Agricultural Crops in Europe*, Report no. 46, UN/ECE ICP Crops Program, Proc. Final Symposium of the European Open-Top Chamber Project, Tervuren, Belgium, 1992.
64. UN/ECE , *ICP-Crops Experimental Protocol*, ICP-Crops Coordination Centre, the Nottingham Trent University, Nottingham, U. K., 1996.
65. Clark, B. B., Brennan, E., and Rebbeck, J., EDU: A tool for assessing crop loss due to ambient oxidants, *Phytopathology*, 74, 843, 1984.
66. Moyer, J. W., Cole, H. Jr., and Lacsasse, N. L., Suppression of naturally occurring oxidant injury on azalea plants by drench or foliar spray treatments with benzimidazole or oxathiin compounds, *Plant Dis. Rept.*, 58, 136, 1974.
67. Manning, W. J., Feder, W. A., and Vardaro, P. M., Benomyl in soil and response of pinto bean plants to repeated exposures to a low level of ozone, *Phytopathology*,1539, 1973.
68. Manning, W. J., Feder, W. A., and Vardaro, P. M., Reduction of chronic ozone injury on poinsettia by benomyl, *Can. J. Plant Sci.*, 53, 833, 1973.
69. Taylor, G. S., Tobacco protected against fleck by benomyl and other fungicides, *Phytopathology*, 60, 578, 1970.

70. Reinert, R. A. and Spurr, H. W., Jr., Differential effects of fungicides on ozone injury and brown spot disease of tobacco, *J. Environ. Qual.*, 1, 450, 1972.
71. Moyer, J. W., Cole, H., Jr., and Lacasse, N. L., Reduction of ozone injury on *Poa annua* by benomyl and thioallophonate-ethyl, *Phytopathology*, 64, 584, 1974.
72. Papple, D. J. and Ormrod, D. P., Comparative efficacy of ozone-injury suppression by benomyl and carboxin on turf grasses, *J. Am. Soc. Hortic. Sci.*, 102, 792, 1977.
73. Wukasch, R. T. and Hofstra, G., Ozone and *Botrytis* spp. interaction in onion leaf die-back: field studies, *J. Am. Soc. Hortic. Sci.*, 102, 543, 1977.
74. Smith, G. S., Greenhalgh, B., Brennan, E., and Justin, J., Soybean yield in New Jersey relative to ozone pollution and antioxidant application, *Plant Dis.*, 71, 121, 1987.
75. Legassicke, B. C. and Ormrod, D. P., Suppression of ozone-injury on tomatoes by ethylenediurea in controlled environments and the field, *Hortic. Sci.*, 16, 183, 1981.
76. Hofstra, G., Littlejohns, D. A., and Wukasch, R. T., The efficacy of the antioxidant ethylenediurea (EDU) compared to carboxin and benomyl in reducing yield losses from ozone in navy bean, *Plant Dis. Rept.*, 62, 350, 1978.
77. Temple, P. J. and Bisessar, S., Response of white bean to bacterial blight, ozone, and antioxidant protection in the field, *Phytopathology*, 69, 101, 1979.
78. Toivonen, P. M. A., Hofstra, G., and Wukasch, R. T., Assessment of yield losses in white beans due to ozone using the antioxidant EDU, *Can. J. Plant Pathol.*, 4, 381, 1982.
79. Fieldhouse, D. J., Chemical control of ozone damage on watermelon, *Hortic. Sci.*, 13, 343, 1978.

CHAPTER 14

Sources, Atmospheric Transport, and Sinks of Tropospheric Nitrous and Nitric Acids

Ralf Zimmerling and Ulrich Dämmgen

CONTENTS

14.1 INTRODUCTION

Atmospheric deposition of reactive nitrogen species has resulted in major problems such as acidification and eutrophication of near-natural terrestrial and aquatic ecosystems. Nitrogen depo-

1-56670-341-7/00/$0.00+$.50

sition mainly occurs by sedimentation of nitrate and ammonium N with coarse particles both wet and dry, by dry deposition of nitrate and ammonium in fine particles, as well as by dry deposition of gaseous ammonia (NH_3), nitrogen dioxide (NO_2), nitrous acid (HONO), and nitric acid ($HONO_2$). The role of organic species such as organic nitrates and amines is uncertain and is still being discussed.[1] Much is known about PAN. At least in Europe, the contribution of these species to atmospheric deposition seems to be of minor importance.

Comparatively much is known about the sources and fate of atmospheric NH_3 and nitrogen oxide (NO). Whereas NH_3 is deposited dry or wet and dry as ammonium, most of the NO undergoes chemical reaction to NO_2, HONO, and $HONO_2$ before it is deposited. Both their atmospheric chemistry and their transport mechanisms seem to be known in principle. However, details of the heterogeneous formation HONO are still missing. Few experiments deal with vertical fluxes of gaseous HONO and of nitrite in wet and dry particles.[2] The dry deposition of $HONO_2$ has been investigated for more than 10 years. Its in-cloud and subcloud chemistries seem to be known.

Since the detailed descriptions of Finlayson-Pitts and Pitts[3] and Warneck,[4] the development and application of new measurement techniques (see Section 14.4.1) led to a whole host of new results concerning sources and atmospheric chemistry of HONO and $HONO_2$. This chapter compiles the results of recent measurements and models of concentrations and fluxes of nitrous and nitric acids. It focusses on those processes which contribute significantly to the concentrations of HONO and $HONO_2$ in the lower troposphere and to the deposition of these acids to terrestrial ecosystems.

14.2 SOURCES OF NITROUS AND NITRIC ACIDS

In the polluted atmospheres of Europe and North America, oxidized N is mainly emitted as NO from internal combustion engines and from microbial denitrification in agricultural soils. In these regions, natural sources of NO, such as lightnings, oxidation of NH_3, or microbial activities in natural soils, are of minor importance. Nearly all NO reacts according to Figure 14.1 to form NO_2, and consequently the reactive acids, their salts, and solutions. The principle reaction patterns already reviewed in papers[3,4] have not changed.

HONO, $HONO_2$, and peroxonitric acid (HO_2NO_2) are the three acids of oxidized N present in the troposphere. HO_2NO_2 is mainly found in the upper troposphere, it is thermally unstable.[4] The main formation reaction leads to an equilibrium:

$$NO_2 + HO_2 \rightleftharpoons HO_2NO_2 \tag{14.1}$$

The contribution of HO_2NO_2 to N deposition is irrelevant.

14.2.1 Nitrous Acid

14.2.1.1 Direct Emissions

It has been know for some time that HONO is emitted directly from internal combustion engines in addition to NO_x.*[5,6] In the exhausts of both spark-ignition and diesel vehicles, HONO was measured at a rate of $\chi_{HONO}/\chi_{NOX} \approx 1$ to $8 \cdot 10^{-8}$ depending on the efficiency of the engine.[7,8] Winer and Biermann[9] and Li et al.,[10] concluded from these data that in urban atmospheres one third to more than one half of the HONO concentrations observed may originate from automobiles. In contrast to this, Andrés-Hernández et al.[11] and Hoek et al.[12] concluded from concentration measurement in Italy and in the Netherlands that direct emissions of HONO are of minor importance. Another direct source of HONO is the biomass burning in tropical regions.[13]

* NO_x is the sum of NO and NO_2.

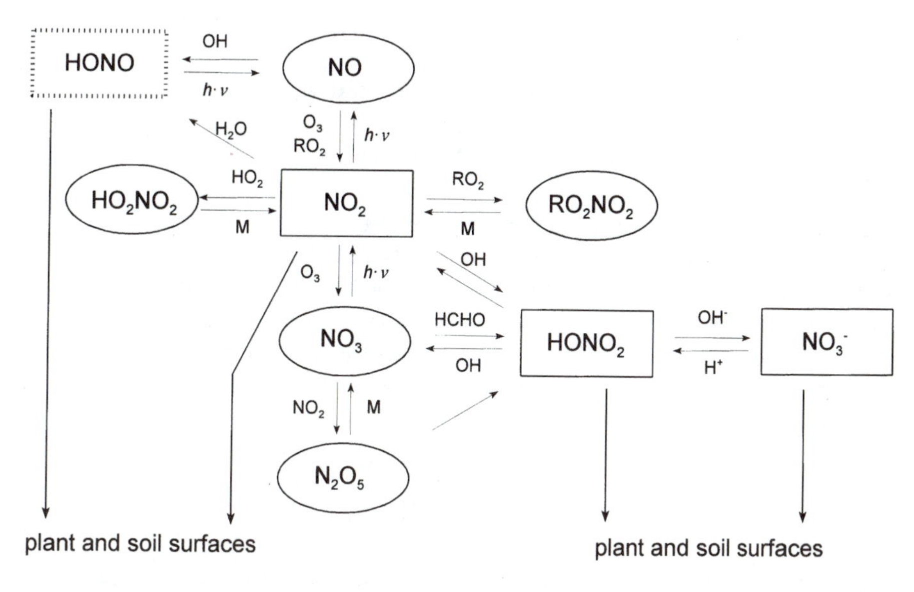

Figure 14.1 Major reaction pathways for oxidized nitrogen in the troposphere. (Rectangles: species relevant to deposition; Dotted lines: thermally or photochemically unstable species. (Adapted from Warneck, P., 1988.[4])

14.2.1.2 Atmospheric Formation

Two gas-phase reactions are frequently cited for the formation of HONO from nitrogen oxides in the troposphere:

$$NO + NO_2 + H_2O \rightarrow HONO \tag{14.2}$$

$$NO_2 + H_2O \rightarrow HONO + HONO_2 \tag{14.3}$$

Reaction 14.2 is third order and should be extremely slow at ambient concentrations. Although Raschig and Prahl[14] had shown that N_2O_3 is nonexistent, a number of authors still use it as an intermediate. In any case, this reaction is too slow ($k \approx 2 \cdot 10^{-9}$ ppm^{-2} s^{-1}) to contribute significantly to the HONO production rates currently observed.[15-17] Reaction 14.3 should also be of third order. Its rate constants are so small that there still is some controversy if it plays any role at all in the formation of HONO.[18,19]

However, all (nighttime) kinetic field measurements and chamber experiments show that the formation of HONO is first order with respect to both NO_2 and H_2O[17] $k = 1.0 \cdot 10^{-8}$ ppm s^{-1}, [6,19-25] and is in most of the cases a heterogeneous reaction taking place in droplets or at the surfaces of aerosols and soil particles. Indeed, "every moist surface exposed to an atmosphere containing NO_2 constitutes a source of HONO."[26] Linear correlations between the concentrations of NO_2 and HONO have been observed frequently.[11,12,22,27,28] However, it is still not clear to which extent Reactions 14.2 and 14.3 contribute to the heterogeneous formation of HONO in droplets or on wet surfaces.[11,29] Sakamaki et al.[17] determined an overall reaction rate $k = 4 \cdot 10^{-5}$ min^{-1} for

$$NO_2 \xrightarrow{\text{wall}} \rightarrow \rightarrow HONO \tag{14.4}$$

In coastal areas reactions with N_2O_5 on wet surfaces (a monolayer of water is sufficient) of NaC1 aerosols may contribute significantly to the nighttime formation of HONO[30]:

$$N_2O_5(g) + NaC1(s) + H_2O \rightarrow HONO + NaOH(s) + NOC1 + O_2$$
$$NOC1 + H_2O \rightarrow HONO + HC1 \qquad (14.5)$$

This reaction is somewhat obscure and cited nowhere else.

HONO formed in aerosol or cloud droplets is readily volatilized in the resulting acidic solution (pK~ 4.2), so that in equilibrium less than 1% of the total HONO is in solution.[31]

Measurements at an urban station (main source of nitrogen oxides NO_x: local automotive traffic; no significant SO_2 concentration) have shown that both HONO and particulate $NO_2^{\bullet}$ may be formed. During wintertime the concentrations of both species exhibit a similar concentration variation.[32] NO is not directly involved in the formation of HONO.[24] Therefore, the recombination reaction of the radicals NO and OH

$$NO + OH \rightarrow HONO \qquad (14.6)$$

is unlikely to be a major formation reaction of HONO in unpolluted areas. However, for central urban sites, no clear relationship between the concentrations of HONO and NO_2 was observed. Here, daytime HONO concentrations are attributed to significant production rates by direct combination of OH and NO as in Equation 14.6.[2]

Simon and Dasgupta[32] developed an automated measuring technique which allows a sensitive and precise determination of gaseous HONO and of $NO_2^{\bullet}$ in aerosols. From indoor measurements they concluded that the formation of HONO is influenced not only by the reaction of NO_2 with water vapor, but also and especially by the reaction of S(IV) species (SO_2) with NO_2. It is doubtful that these results can be transferred to the free atmosphere.

14.2.2 Nitric Acid

The oxidation of most nitrogen species emitted by natural or anthropogenic sources finally leads to nitric acid or nitrates involving numerous reactions (Figure 14.1). Although direct emissions of $HONO_2$, e.g., by exhaust gases of internal combustion engines, cannot be excluded,[10,33-35] their contribution to the overall formation of $HONO_2$ is likely to be small. The prevailing formation reaction at daytime is the reaction of NO_2 with OH radicals:

$$NO_2 + OH \rightarrow HONO_2 \qquad (14.7)$$

During nighttime, reactions including the formation of nitrate radicals (NO_3) are important and contribute a reasonable share to the overall HNO_3 formation.[3]

$$NO_2 + O_3 \rightarrow NO_3 + O_2 \qquad (14.8)$$

The NO_3 radical is only stable at night because it is rapidly photolyzed ($\lambda < 630$ nm); its lifetime during day is limited to approximately 5 s.[36] However, it can also contribute to other atmospheric reactions during overcast conditions and inside canopies when photolysis is reduced. As a highly reactive species, NO_3 may react with NO_2 (Equation 14.9) and with organic gases like dimethylsulfide, aldehydes, or alkenes (e.g., terpenes and isoprenes from forests) Equation 14.10).

$$NO_2 + NO_3 \rightleftharpoons N_2O_5 \qquad (14.9)$$

$$RH + \cdot NO_3 \rightarrow HONO_2 + \cdot R \tag{14.10}$$

The equilibrium in Reaction 14.9 is temperature dependent; increasing air temperatures result in decreasing concentrations of N_2O_5 as a result of thermolysis. As it is photolytically more stable than NO_3, N_2O_5 formed under cooler conditions may serve as a source of NO_3 and hence of $HONO_2$ under subsequent warmer conditions. The hydrolysis of N_2O_5 yields nitric acid:

$$N_2O_5 + H_2O \rightarrow NH_3 \tag{14.11}$$

This reaction occurs in the gas phase as well as on the wet surfaces of particles. The homogeneous oxidation pathway alone is too slow to explain the $HONO_2$ formation at night. Therefore, it is assumed that both homogeneous and heterogeneous reactions are of importance.[37] NO_3 may further react with organic air constituents like aldehydes, alkanes, and alkenes contributing significantly to the formation of $HONO_2$, especially in polluted regions[3] (Equation 14.10).

NO_3 may also be scavenged by hydrometeors yielding aqueous nitrate.[3] At rural sites, nighttime measurements have shown that the heterogeneous reaction of NO_3 and of N_2O_5 with particles is responsible for most of the formation of particulate nitrate. The contributions of NO_3 and of N_2O_5 to the formation of particulate nitrate were 80 and 10%, whereas the neutralization of $HONO_2$ contributed less than 5%.[38] As it is well known that particulate nitrate may serve as source of $HONO_2$ due to the thermolysis of NH_4NO_3 and to the reaction of nitrate with strong acids like sulfuric acid, the formation of particulate nitrate can be regarded as an indirect source of $HONO_2$:

$$NH_4NO_{3(s,1)} \rightleftharpoons HONO_{2(g)} + NH_{3\,(g)} \tag{14.12}$$

$$\begin{aligned} NaNO_{3(s,1)} + H_2SO_{4(1)} &\rightarrow HONO_{2(g)} + NaHSO_{4(s,1)} \\ NaNO_{3(s,1)} + H_2SO_{4(1)} &\rightarrow HONO_{2(g)} + NaSO_{4(s,1)} \end{aligned} \tag{14.13}$$

14.3 PHYSICAL AND CHEMICAL PROPERTIES OF GASEOUS NITROUS AND NITRIC ACIDS

14.3.1 Nitrous Acid

In principle, HNO_2 may exist in two tautomers, of which HONO prevails by far.

$$H - ONO \rightleftharpoons \begin{array}{c} H \\ | \\ ONO \end{array} \tag{14.14}$$

Of the first tautomer, *cis*- and *trans*-HONO can be identified spectroscopically. At 23°C, the ratio of concentrations in equilibrium is $\chi_{cis}/\chi_{trans} \approx 0.44$.[15] The isomers equilibrate rapidly so that reaction kinetics for one of them cannot be determined.[15]

HONO dissolves in water.[39,40]

$$\begin{aligned} HONO_{(g)} &\rightleftharpoons HONO_{(aq)} \\ \Delta H = -39.7\ \text{kJ mol}^{-1},\ K &= 5.1 \cdot 10^{-4}\ \text{mol l}^{-1} \end{aligned} \tag{14.15}$$

Its equilibrium concentration and thus its vapor pressure above bulk water is pH-dependent:[31]

$$HONO + H_2O \rightleftharpoons ONO^- + H_3O^+ \tag{14.16}$$

$$p_{HONO} = \frac{([HONO]+[ONO^-]) \cdot [H^+]}{H_{HONO} \cdot K + H_{HONO} \cdot [H^+]} \tag{14.17}$$

with Henry's constant $H_{HONO} = 49$ mol l^{-1} bar^{-1} and $K = 5.1 \cdot 10^{-4}$ mol l^{-1}. In clouds, with nitrite concentrations of 10^{-6} mol l^{-1} and pH = 4, equilibrium gas-phase concentrations of about 3 μg m^{-3} result. However, Henry's law is not applicable to clouds (droplet size, nonequilibrium conditions).[41,42]

Measurements with a resolution in time of a few minutes have shown that the concentration of nitrite in aerosols may reach 1 μg m^{-3} NO_2^-.[32]

In solution, HONO is a very weak acid and may be protonized at "normal" rain or cloud pH:

$$HONO + H_3O) \rightleftharpoons H_2ONO^+ + H_2O \tag{14.18}$$

HONO is scavenged or formed in clouds and may lead to nitrite concentrations which are in the same order of magnitude as nitrate concentrations.[31] In contrast to this, Fuzzi et al.[41] report that the contributions to the overall N content of clouds by HONO were negligible at Kleiner Feldberg (Germany). In precipitation, concentrations of NO_2^- are only a small fraction of those of NO_3^-.[43]

Little is known about further sinks for HONO in solution: nitrite is oxidized in solution according to Martin et al.:[44]

$$NO_2^- + O_3 \rightarrow NO_3^- + O_2 \tag{14.19}$$

$$NO_2^- + H_2O_2 \rightarrow NO_3^- + H_2O \tag{14.20}$$

There is an obvious lack of reliable data for the fate of nitrite in the condensed phase, because its oxidation to nitrate occurs very fast. If the measurements of Cape et al.[31] can be transferred to other regions, a considerable share of bulk nitrate deposition results from oxidized HONO. However, it should be kept in mind that clouds in background air show a different behavior from more-polluted clouds or fogs where metal ions and soot may serve as catalysts. Gaseous nitrous acid photolyzes to give OH and NO ($\lambda < 400$ nm). The quantum yield is nearly unity.[45] In solution, the solvent cage effect causes a significant recombination of the fragments. Therefore, the overall efficiency of the HONO photolysis including the escape from the solvent cage is near 10% (effective quantum yield ~ 0.1).[3] The reaction

$$HONO \rightarrow NO + NO_2 + H_2O \tag{14.21}$$

is slow and seems to be of minor importance in ambient air.[15] However, it serves to explain concentration inside automobiles.[46] In ambient air, the formation and destruction reactions lead to the typical diurnal pattern of concentrations, with steadily increasing concentrations after sunset and a sharp decline to almost zero after sunrise. Daylight equilibrium concentrations without consideration of direct emission or heterogeneous reactions range between 10 and 100 ppt (approximate model calculations by Neftel et al.[28]). These results have been verified by other measurements.[8,27,28,47,48] The photodissociation of HONO is a major source of OH radicals in polluted urban atmospheres after sunrise.

14.3.2 Nitric Acid

The atmospheric chemistry of $HONO_2$ is coined by its solubility in water, its acid properties, and its vapor pressure. $HONO_2$ dissolves in water ($H = 2.1 \cdot 10^5$ mol l^{-1} bar^{-1}).[39] It is scavenged efficiently by water droplets both in clouds and below clouds. Gaseous $HONO_2$ may be produced as acidic cloud droplets containing nitrate evaporate.[49] $HONO_2$ forms a strong acid. In concentrations relevant to atmospheric chemistry it is totally dissociated. Its solubility and reactivity lead to an efficient removal by dry deposition on vegetation and on solid and moist surfaces (see Section 14.5.2).

As strong acid, it reacts with atmospheric bases like NH_3 or alkaline particles forming nitrates: in the gas phase, neutralization of $HONO_2$ with NH_3 yields particulate NH_4NO_3 resulting in the formation of aerosols or of layers of NH_4NO_3 on existing surfaces.[50,51]

$$HNO_{3(g)} + NH_{3(g)} \rightleftharpoons NH_4NO_{3(s)} \tag{14.22}$$

This reaction is reversible. The concentrations of the species involved may vary depending on air temperature and on humidity. Reactions of $HONO_2$ in solution are normally irreversible due to its comparatively low vapor pressure. The reaction of $HONO_2$ with chloride in particles or in solution can lead to the (pH-dependent) liberation of hydrochloric acid and the formation of the respective nitrates:

$$HONO_{2(g)} + NaCl_{(s,l)} \rightarrow HCl_{(g)} + NaNO_{3(s,l)} \tag{14.23}$$

The photolysis of $HONO_2$ is restricted to wavelengths $\lambda < 320$ nm and therefore irrelevant for tropospheric chemistry.[3]

14.4 AMBIENT CONCENTRATIONS AND LATERAL TRANSPORT

14.4.1 Measurement Techniques

In contrast to the monitoring of concentrations of the so-called criteria pollutants, no measuring techniques with reasonable costs and labor effort and adequate temporal and chemical resolution are commercially available for HONO and $HONO_2$. Concentrations of HONO and $HONO_2$ are determined using accumulating sampling techniques like impregnated filters or denuders as well as direct spectroscopic techniques like Fourier transform infrared spectroscopy FTIR ($HONO_2$), tunable diode laser spectrometry TDLAS ($HONO_2$), differential optical absorption spectrometry DOAS (HONO), or laser-photolysis fragment fluorescence ($HONO_2$).[52,53] Each method exhibits different resolutions in time and space, different limits of detection (LOD), and requires specific efforts. Intercomparisons for HONO and $HONO_2$ detection methods have shown the fields of applicability of each system.[54-57]

Filters and filter packs are the simplest technique to accumulate air constituents for a subsequent analysis. Normally, a pack of at least two filters is exposed: the first filter is chemically inert and collects particles. The acidic gases (e.g., $HONO_2$) are collected on a second filter with basic properties (e.g., Na_2CO_3 coating, nylon). The applicability of filters depends on the goal of the measurements (precision, resolution in time, duration of the measuring campaign) and on the sorption properties and the efficiencies of the filters. Insufficient selectivity and reproducibility as well as the formation of artifacts restrict their applicability. The major advantages of filters are their high accumulation capacity, their low price, and their simple handling. Filter packs have not been used for the determination of HONO concentrations. However, various filter techniques have been applied to measure concentration of $HONO_2$. Here, filter pack measurements under cool and humid conditions at low concentration levels of NH_4NO_3 normally yield correct concentrations. Nevertheless, comparatively high concentrations of NH_4NO_3 and frequently changing meteorological conditions may lead to an overestimation of $HONO_2$ concentrations due to the thermolysis of NH_4NO_3.[54]

Denuder techniques separate gases from particles by diffusion from a laminar airflow. Normally, a series of denuders with subsequent filter pack is used to measure the concentration of HONO, $HONO_2$, and particulate nitrate simultaneously and discontinuously.[52,53,58] Annular denuders are usually preferred to simple denuders. Denuders have a typical resolution in time of several hours to a few days. Basic coatings for the sampling of HONO and $HONO_2$ are Na_2CO_3, K_2CO_3, or Na_2CO_3/glycerol; more selective agents like NaF and NaCl are sometimes used to avoid sampling artifacts. Artifacts from NO_2 in the presence of SO_2 can be minimized using tetrachloromercurate-coated predenuders.[59] Continuous measurements of $HONO_2$ concentrations are achieved using wet denuder or related techniques.[28,32,60]

The application of spectroscopic techniques is normally restricted to special investigations.[36] However, DOAS is increasingly used for routine observations.

14.4.2 Concentrations of Nitrous Acid in Ambient Air

Results of concentration measurements in ambient air which have been performed and published in the last 15 years are compiled in Table 14.1.

A diurnal cycle which exemplary reflects the source and sink reactions during summer is presented in Figure 14.2. The annual cycle (Figure 14.3) exhibits maximum concentrations during the winter months. During the summer months, daily means cannot only be understood from nighttime formation only. This is underlined by the measurements at Britz, Germany[61] and Speulderbos, the Netherlands.[62] For northeastern Brandenburg, Germany, it should be kept in mind that the extensive use of lignite as fuel is to be considered as a direct source of HONO. Due to the distinct annual and diurnal cycles of HONO concentrations, a comparison between different stations and a classification is difficult. Table 14.2 is an attempt to summarize the results listed in Table 14.1. The somewhat high concentrations measured by Nash[63] were the first to be published. Because of the technique applied, they are likely to contain artifacts from the adsorption of other nitrite-forming species.

14.4.3 Concentrations of Nitrous Acid in Confined Environments (Rooms, Cars)

The reaction of NO_2 on wet surfaces resulting in the formation of HONO also takes place in rooms and automobiles. Concentrations up to 100 ppb (peak) and 40 ppb (24-h mean) were observed during the use of unvented gas space heaters inside homes. Measurements inside automobiles show similar results. In comparison with outdoor situations, it is obvious that the HONO exposure of humans is dominated by indoor situations.[26,46,64,65]

14.4.4 Lateral Transport of Nitrous Acid

Lateral transport is probable if HONO concentrations are depending on wind directions. Figure 14.4 illustrates the HONO concentrations in southeast Lower Saxony, Germany. A similar pattern was observed for SO_2 which is advected from distant sources. Thus, the same transport mechanism is likely for HONO or its precursors. Lammel and Perner[20] proved that nitrite is present in aerosols which are liable to lateral transport, and that aerosol nitrite is a source of gaseous HONO. Simultaneous measurements of concentrations of HONO were carried out by Hoek et al.[12] at several locations in the Netherlands. Significant differences of the concentration levels at different sizes were explained with different concentration levels of precursors (NO_x).

Table 14.1 Concentrations HONO in Ambient Air

Location	Time and Details of Methods	Urban/Rural/ Remote	Mean Concentrations (µg m^{-3})			Ref.
South-West Desert, UT, USA	I to II/86, s, AD	Remote	n+d:		0.14 (0.0 – 0.5)	126
Whitaker Forest, CA, USA	1988 to 1990 c, ac	Remote	n+d:	89:	0.08	127
				90:	0.19	
				89:	0.08	
				90:	0.14	
Alert, Northwest Territories, Canada	I-IV/92, AD	Remote	n+d:		0.2 max. 0.8	128
			n*:		0.7 max. 3.2	
Crossfield, Alberta, Canada	1985 to 1987, 12-h, c, AD	Remote	n+d:		0.38 (0.003–2.1	74
Shenandoah National Park, VA, USA	IV to IX/87, c, AD	Forest	n+d:		0.2 (0–2.0)	129
Mt. Michell State Park, NC, USA	V/88 to IX/89, c, AD	Forest	n+d:	88:	0.3 (0.09–1.53)	78
				89:	0.3 (0.02–0.70	
Wast Hills, UK	VIII/93, IV to V/94, s, 4-h, AD	Semiurban	d:	93:	0.12–0.59	2
				94:	0.15–0.86	
Nottingham, UK	VIII/94, s, 4-h, AD	Semirural	d:		0.17–2.0	2
Birmingham, UK	III/94, c, dc, 2 days, WD	Suburban	n+d:		0.1–0.7	2
Britz, Brandenburg, Germany	VI/96 to V/97, c, weekly day and night, AD	Forest	n+d:		0.8	61
			n:		1.0	
			d:		0.7	
			Summer	n:	0.31	
				d:	0.10	
			Autumn	n:	0.66	
				d:	0.46	
			Winter	n:	2.41	
				d:	1.91	
			Spring	n:	0.45	
				d:	0.14	
San Dimas Experimental Forest, CA, USA	VII to IX/89, VI to X/90, VII to IX/91, s, AD	Forest, urban	n:	89:	0.9 (0.3–1.4)	70
			d:	90:	0.8 (0.2–2.0)	
				91:	0.8 (0.2–2.0)	
				89:	0.2 (0.0–0.5)	
				90:	0.3 (0.0–0.6)	
				91:	0.5 (0.0–1.5)	
Speulderbos, the Netherlands	1987 to 1989, c, AD	Forest	n+d:	0.6		130
Speulderbos, the Netherland	sXI/92 to IX/93, WD	Forest	n+d:	0.86		62
Petten, the Netherlands	1987 to 1989, c, AD	Rural, marine	n+d:	1.0		130
Mainz, Germany	IX/87 to V/88, s, DOAS	Suburban	n:	1.1 (0.1–5.0)		20
Deurne, the Netherlands	X/87 to IV/90, s, AD	Rural	d:	1.3 (0.0–10.6)		12
Rotenkamp near Braunschweig, Lower Saxony, Germany	IV/91 to V/93	Rural	n+d:	1.4 (max. 17.6)		47
Braunschweig, Germany	VI to VIII/92, c, AD	Suburban	n+d:	0.6		47
Chicago, IL, USA	IV/90 to III/91, s, AD	Urban	n+d:	0.99 1.		131
			n:	01 max. 2.39		
			d:	0.97 max. 2.57		
Bilthoven, the Netherlands	X/87 to IV/90, s, AD	Semirural	n+d:	2.0 (0.0–21.2)		12
Warren, MI, USA	XII/87 to IV/88, c, AD	Suburban	n+d:	2.0		133
Rome, Italy	II/85, xx h, s, AD	Rural	n+d:	2.0		133
Rome, Italy	I to II/86, xx h, s, AD	Rural	n+d:	1.8		134
			n:	2.5		
			d:	1.1		

Table 14.1 continued

Location	Time and Details of Methods	Urban/Rural/ Remote	Mean Concentrations (μg m⁻³)			Ref.
Rome, Italy	III to IV/86, AD	Rural	n:		1.1	135
			d:		0.8	
Rome, Italy	IX/88, s, AD	Urban	n:		9.2 (6.3–12.6)	26
Claremont, CA, USA	IX/85, s dc, DOAS	Urban	n:		1,7(1.0–4.8)	136
Göteborg, Sweden	I to IV/84, s, AD	Urban	n:		0.5 (0.02–2.4)	18
			d:		0.5 (0.02–1.9)	
Müncheberg, Brandenburg, Germany	IV/95 to V/97, c, AD	Rural	n+d:		0.9	61
			Summer n+d:		0.3	
			Autumn n+d:		0.8	
			Winter n+d:		2.1	
			Spring n+d:		0.5	
Southeast England	s, AD	Rural	n:		1.2 (0.4–2.5)	27
			d:		0.2 (0.0–0.7)	
Petten, the Netherlands	2 days in VI to VII/92, c, dc, WD	Rural	n+d:		0.5–2.5	137
Speulderbos, the Netherlands	I to XII/89, c, dc, ac, WD	Forest				84
Boston, MA, USA	Summer 1994, s, HD, AD	Urban	n+d:		1.1	138
Research Triangle Park, NC, USA	X to XI/861; I/87, s, AD	Urban	n+d:	88:	1.7 (0.7–3.7)	139
				89:	1.0 (0.6–1.6)	
Claremont, CA, USA	IX 1985, s		n+d:		3.4 (1.0–9)	68
Los Angeles, CA, USA	VII to VIII/80, s, DOAS	Urban	n:		6.4 (0.8–15)	33
Porton, southern England	X/72 to I/73, s, DOAS	Rural	d:		7.2 (0.8–21)	63
Long Beach, CA, USA	XI to XII/87, s, dc, AD, DOAS	Urban	n+d:		9.0	
			n:		13.1 (4–21)	
			d:		< 2	

Abbreviations: continuously: c, sporadically: s, diurnal cycle: dc, annual cycle: ac, annular denuder: AD, denuder: D, honeycomb denuder: HD, wet denuder: WD, differential optical absorbtion spectroscopy: DOAS, nights only: n, days only: d, all day: n+d, range: ( ), max: maximum, * polar night.

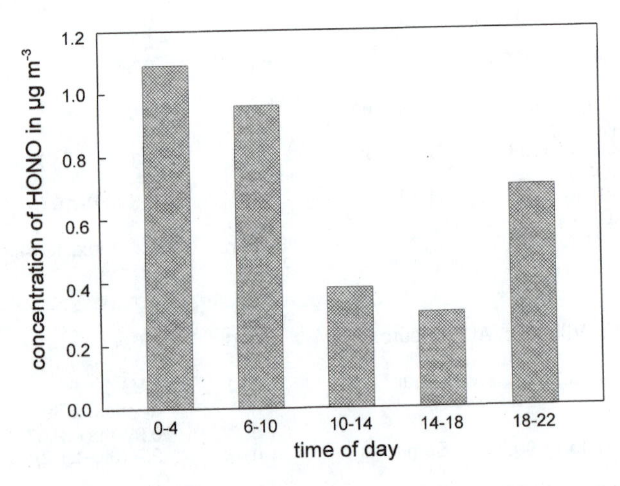

Figure 14.2 Diurnal cycle of HONO concentrations in rural southeastern Lower Saxony, Germany. (After Zimmerling et al., 1996.[48])

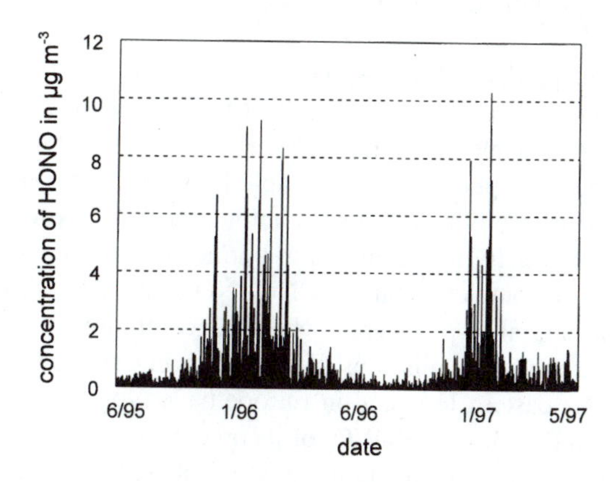

Figure 14.3 Annual cycle of the daily means of HONO concentrations in rural East Brandenburg, Germany. (After Zimmerling et al. 2000.[61])

Table 14.2 Typical Ranges of Concentrations of HONO Depending on Air Quality

Pollution Climate	Concentration ($\mu g\ m^{-3}$ HONO)
Remote	<0.1
Rural	0.1–1.5
Moderately polluted	1.5–4
Heavily polluted	4–16

After Legge, A.H., 1988.[74]

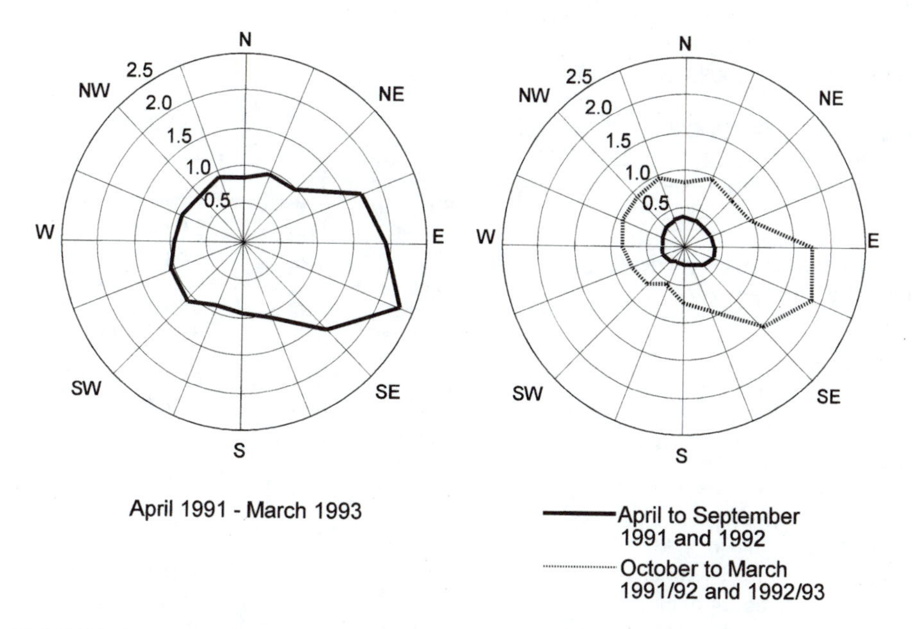

Figure 14.4 HONO concentrations (in $\mu g\ m^{-3}$) in rural southeastern Lower Saxony, Germany, as a function of wind direction. (After Zimmerling et al., 1996.[47])

14.4.5 Concentrations of Nitric Acid in Ambient Air

As $HONO_2$ is the final product of atmospheric NO_x oxidation, its concentrations are highly dependent on anthropogenic NO emissions and reflect the degree of pollution by automobiles in particular. Thus, it is understandable that atmospheric $HONO_2$ was first dealt with at locations with high NO_x emissions and intensive photochemistry. Table 14.3 compiles concentration measurements since 1984 for $HONO_2$. The results reported are confined to North America and Europe and the adjacent remote regions potentially reflecting natural atmospheres. Few measurements have been carried out during winter or complete years. As NO, NO_2, and OH are ubiquitous in principle, $HONO_2$ is found even in natural tropospheres, where concentrations range below 0.1 µg m^{-3}. In regions influenced by anthropogenic emissions, $HONO_2$ concentrations are one order of magnitude higher. Rural regions in the eastern U.S. and in Europe have typical annual mean concentrations between 1 and 3 µg m^{-3}. Lateral flow of $HONO_2$ or its precursors may coin $HONO_2$ concentrations even in rural areas.[66,67] Few measurements in urban atmospheres are available.

Table 14.3 Concentrations of $HONO_2$ in Ambient Air

Location	Time and Details of Measurement	Urban/ Rural/ Remote	Mean Concentrations (µg m^{-3})			Ref.
Baring Head, New Zealand	VII to VIII/91, 24-h; IX to XI/91, 24-h; I to II/92, 8-h; c, FI	Remote, marine	Total	n+d:	0.05	140
			Marine air:			
			Spring	n+d:	0.03	
			Summer	n+d:	0.03	
			Winter	n+d:	0.02	
Atlantic Ocean	IX to X/88, (several days), c, dc, FI, LPFF	Remote, marine	n+d:	n–0.75		35
Mauna Loa, HA, USA	V to X/88, 24-h, FI	Remote, marine	n+d		0.33	141
San Nicholas Island, CA, USA	1986, DDM	Remote, marine	n+d:		0.3 max. 4.7	67
Rörvik, Sweden	XI/81 to X/82, 24-h, AD	Remote, marine	n+d:	1.6		76
			Summer	n+d:	1.1 max. 5.4	
			Winter	n+d:	2.0 max. 8.8	
Areâo near Aveiro, Portugal, Atlantic coast	XI/93 to VIII/94, s, 24-h, AD	Remote, marine	Autumn	n+d:	0.06 ± 0.05	142
			Winter	n+d:	0.06 ± 0.06	
			Spring	n+d:	0.06 ± 0.08	
			Summer	n+d:	0.14 ± 0.11	
			Marine	n+d:	0.023 ± 0.036	
			Cont.	n+d	0.17 ± 0.12	
Arctic troposphere, Canada	III to IV/88, 24-h, FI	Remote	Winter	n:	0.04	143
			Spring	n+d:	0.12	
Schefferville, Canadian Taiga	13 days in VII to VIII/90, s, MC	Remote	d:		0.25 max. 1.13	144
Cree Lake, Canada	1983 to 1985, c, ac, 24-h, FI	Remote	n+d:		0.10 max. 1.46	82
Bay d'Espoir Canada	1983 to 1985, c, ac, 24-h, FI	Remote	n+d:		0.11 max. 1.45	82
Experimental Lakes Area, Canada	1983 to 1987, c, ac, 24-h, FI	Remote	n+d:	0.30 max. 4.69		82

Table 14.3 continued

Location	Time and Details of Measurement	Urban/ Rural/ Remote	Mean Concentrations ($\mu g\ m^{-3}$)			Ref.
Forêt Montmorency, Canada	1983 to 1987, c, 24-h, FI	Remote	n+d:		0.35 max. 4.88	82
Kejimkujik, Canada	1983 to 1987, c, 24-h, FI	Remote	n+d:		0.33 max. 4.47	82
Fortress Mountain, Canada	1983 to 1987, c, 24-h, AD	Remote	n+d:		0.2	74
			d:		0.22	
			n:		0.18	
Bilthoven, the Netherlands	X/87 to IV/90, c, AD	Semirural	n+d:		0.3 (0.0–13.9)	12
Kejimkujik, Canada	V/92 to XII/94, s, AD	Rural	n+d:		0.32 (4/16)	145
Sutton, Canada	V/92 to XII/94, s, AD	Rural	n+d:		0.50 (4.03)	145
Säby, Sweden	4 months in XI/93 to VIII/94, s, AD	Rural, marine	n+d:	0.6		86
			Spring	n+d:	0.22	
			Summer	n+d:	1.44	
			Autumn	n+d:	0.45	
			Winter	n+d:	0.41	
Algoma, Canada	983 to 1987, c, ac, 24-h, FI	Remote	'n+d:		0.62 max. 1033	82
Chalk River, Canada	1983 to 1987, c, ac, 24-h, FI	Remote	n+d:		0.59 max. 8.00	82
Lista, Norway	4 months in XI/93 to VIII/94, s, AD	Rural, marine	n+d:		0.5	86
			Spring	n+d:	0.36	
			Summer	n+d:	0.77	
			Autumn	n+d:	0.50	
			Winter	n+d:	0.18	
Deurne, the Netherlands	X/87 to IV/90, s, AD	Rural	d:		0;7 (0.0–9.2)	12
Hubbard Brook, NH, USA	7/89 to 12/91, c, weekly, FI	Rural	90			146
			n+d:		0.90	
			Winter	n+d:	1.09	
			Spring	n+d:	1.07	
			Summer	n+d:	0.69	
			Autumn	n+d:	0.76	
			91			
			n+d:		0.74	
			Winter	n+d:	0.91	
			Spring	n+d:	0.68	
			Summer	n+d:	0.61	
			Autumn	n+d:	0.77	
Sutton, Canada	1986 to 1987, c, ac, 24-h, FI	Remote	n+d:		0.97 max. 6.48	82
Whiteface Mountain, NY, USA	VIII/84 to VIII/87, weekly, c, ac, FI	Remote	85			71
			n+d:		0.8	
			Spring	n+d:	0.9	
			Summer	n+d:	0.4	
			Autumn	n+d:	1.0	
			Winter	n+d:	0.9	
			86			
			n+d:	1.1		
			Spring	n+d:	1.3	
			Summer	n+d:	1.3	
			Autumn	n+d:	0.9	
			Winter	n+d:	0.8	

Table 14.3 continued

Location	Time and Details of Measurement	Urban/ Rural/ Remote	Mean Concentrations ($\mu g\ m^{-3}$)			Ref.
Mt. Mitchell, NC, USA	V to IX/88, V to IX/89, 24-h, c, AD	Remote	n+d:	88	1.14 max. 5.6	78
				89	1.40 max. 2.6	
North Fork Reservoir, NC, USA	VIII to IX/86, 12-h, AD	Forest	n+d:		0.85	147
			d:		1.4	
			n:		0.3	
Speulder, the Netherlands	XI/92 to IX/93, c, ac, dc, WD	Forest	n+d:		0.86	62
Black Forest, Germany	2 days, 24-h, s, dc, LPFF	Forest	n+d:		1.6	79
Whitaker Forest, CA, USA	Summer 1988 (3 days), 1989 (5 days), 1990 (12 days), 12-h, AD	Forest	n:	88:	0.28	127
				89:	0.65	
				90:	0.60	
			d:	88:	1.11	
				89:	1.76	
				90:	1.97 max. 4.15)	
Walker Branch Watershed, TN, USA	VII/81 to VII/83, weekly, c, FI	Forest	n+d:		0.87	148
Melpitz, Germany	VIII/91 to IX/91 (12 days), s, FI	Rural	n+d:		1.24±1.54	149
Egbert, Canada	V/92 to XI/94, s, AD	Rural	n+d:		1.24 ± 1.54	149
Bayerischer Wald, Germany	VI/90 (6 days), s, FI, several heights in and above the forest	Forest	n+d:		1.8 (50 m)	150
Harwell Laboratory, Great Britain	1984, s, a, 8 to 16-h, FI	Rural	n+d:		1.65 (0.2–12.5)	111
			Summer	n:	0.91	
				d:	3.20	
			Winter	n:	0.66	
				d:	1.20	
Oak Ridge, TN, USA	VIII/84 to VIII/87, weekly, c, a, FI	Forest	85			71
			n+d:		1.7	
			Spring	n+d:	2.1	
			Summer	n+d:	1.8	
			Autumn	n+d:	1.0	
			Winter	n+d:	1.8	
			86			
			n+d:		1.8	
			Spring	n+d:		
			Summer	n+d:	2.4	
			Autumn	n+d:	1.8	
			Winter	n+d:	1.6	
					1.2	
Ulborg, Denmark	10 days in V/91, s, D	Forest	n+d:		0.7 (0.1–1.9)	151
Britz, Brandenburg, Germany	VI/96 to V/97, weekly, c, q0-h, AD	Forest	n+d:		0.7 0.	61
			n: d:		6 0.9	
Hamilton, Canada	V/92 to XII/94	Rural	n+d:		2.01 (8.76)	145

Table 14.3 continued

Location	Time and Details of Measurement	Urban/ Rural/ Remote	Mean Concentrations ($\mu g\ m^{-3}$)			Ref.
Stafford Springs, CT, USA	VII/89 to XII/91, weekly, c	Rural	90			146
			n+d:		4.46	
			Winter	n+d:	616	
			Spring	n+d:	5.78	
			Summer	n+d:	3.98	
			Autumn	n+d:	1.93	
			91			
			n+d:		2.25	
			Winter	n+d:	2.53	
			Spring	n+d:	2.20	
			Summer	n+d:	1.83	
			Autumn	n+d:	2.44	
West Point, NY, USA	VII/89 to XII/91, weekly, c	Rural	90			146
			n+d:		2.46	
			Winter	n+d:	2.34	
			Spring	n+d:	2.99	
			Summer	n+d:	2.52	
			Autumn	n+d:	1.99	
			91			
			n+d:		2.36	
			Winter	n+d:	2.09	
			Spring	n+d:	2.55	
			Summer	n+d:	2.87	
			Autumn	n+d:	1.91	
State College, PA, USA	VII/89 to XII/91, weekly, c	Rural	90			146
			n+d:		2.71	
			Winter	n+d:	2.56	
			Spring	n+d:	2.87	
			Summer	n+d:	3.09	
			Autumn	n+d:	2.30	
			91			
			n+d:		2.71	
			Winter	n+d:	2.23	
			Spring	n+d:	3.03	
			Summer	n+d:	3.68	
			Autumn	n+d:	1.90	
Windsor, Canada	V/92 to XII/94	Rural	n+d:		2.71 (18.02)	145
Shenandoah National Park, VA, USA	IX to XI/87, weekly c, AD	Forest	n+d:		2.8 (0.0–8.9)	129
Oak Ridge, TN, USA	IV to X/87, FI	Forest	n+d:		3.0	77
Solling, Germany	IV to X/87, FI	Forest	n+d:		5.8	77
Birch Lake, MN, USA	1991 to 1993, c, weekly, FI	Forest	n+d:	91:	0.45	73
				92:	0.45	
				93:	0.43	
Cedar Creek, USA	1991 to 1993, c, weekly, FI	Forest	n+d:	91:	0.81	73
				92:	0.86	
				93:	0.63	
Ely, MN, USA	1991 to 1993, c, weekly, FI	Forest	n+d:	91:	0.49	73
				92:	0.50	
				93:	0.55	
Finland, MN, USA	1991 to 1993, c, weekly, FI	Forest	n+d:	91:	0.83	73
				92:	0.77	
				93:	0.80	

Table 14.3 continued

Location	Time and Details of Measurement	Urban/ Rural/ Remote	Mean Concentrations ($\mu g\ m^{-3}$)			Ref.
Koch, MN, USA	1991 to 1993, c, weekly, FI	Grassland, suburban	n+d:	91:	0.83	73
				92:	0.77	
				93:	0.80	
Marcell, MN, USA	1991 to 1993, c, weekly, FI	Forest	n+d:	91:	0.51	73
				92:	0.56	
				93:	0.52	
Sandstone, MN, USA	1991 to 1993, c, weekly, FI	Forest	n+d:	91:	0.70	73
				92:	0.75	
				93:	0.64	
Eastern USA, 50 NDDN sites	I to XII/89, weekly, FI			annual means (all stations):		72
				n+d:	(0.7–3.6)	
Crossfield, Canada	1985 to 1987, 12-h, c, AD	Rural	n+d:		0.28	74
			d:		0.32	
			n:		0.24	
Edmonton, Canada	XI/82 to X/83, 24-h, c, ac, FI	Suburban	n+d:		0.30	152
Great Domsey, UK	VIII/87 to I/88, 30 day, s, FI	Rural	n+d:		0.38 (0.05–3.2)	153
Southeast England, four sites	s, dc, AD	Rural to semirural	n+d:		0.56 (0.18–1.77)	27
			d:		2.32 (0.20–8.04)	
			n:		0.95 (0.09–2.17)	
Southeast England, five sites	VII/86 to I/88, 24-h, s, d, weekly, DI	Rural semirural, urban	VIII to XI/86:			76
			Overall	n+d:	0.67	
			Rural	n+d:	0.43	
			Semirural	n+d:	0.80	
			Urban	n+d:	0.80	
			Marine	n+d:	0.66	
			II to IV/87:			
			Overall	n+d:	0.43	
			Rural	n+d:	0.35	
			Semirural	n+d:	0.55	
			Urban	n+d:	0.46	
			II/87 to I/88 (147 days): Rural/semirural n+d:		1.01 (0.05–9.66)	
Rotenkamp near Braunschweig, Germany	IV/91 to V/93, c, 24-h, ad, AD	Rural	n+d:		0.7 (0–5.7)	48
			Spring	n+d:	1.0	
			Summer	n+d:	1.1	
			Autumn	n+d:	0.8	
			Winter	n+d:	0.4	
Near Roskilde, Denmark	III to VI/93, s, 24-h, D	Rural	n+d:		0.86	1
Müncheberg, Brandenburg, Germany	VI/95 to V/97, c, 24-h, ac, AD	Rural	n+d:		0.8	61
			Spring	n+d:	0.9	
			Summer	n+d:	1.1	
			Autumn	n+d:	0.7	
			Winter	n+d:	0.7	
Cabauw, the Netherlands	30 day, 12 h, s, concentration profiles, AD	Rural	n+d:		1.1 max.12.20	18
Longwoods, Canada	1983 to 1987, c, ac, 24-h, FI	Rural	n+d:		1.63 max. 11.42	82
Wolkersdorf, Austria	XI/90 to X/91, c, AD	Rural	n+d:		1.8	83

Table 14.3 continued

Location	Time and Details of Measurement	Urban/ Rural/ Remote	Mean Concentrations (μg m⁻³)			Ref.
Panola, GA, USA	VIII/85 to VIII/87, weekly c, ac, FI	Rural	86			71
			n+d:		1.8	
			Spring	n+d:	2.0	
			Summer	n+d:	2.1	
			Autumn	n+d:	1.8	
			Winter	n+d:	1.4	
Ontario, Canada	I to II/84, 6-h, s, FI	Rural	n+d:		2.2	87
West Point, NY, USA	VIII/84 to VIII/87, weekly c, ac, FI	Rural	85			71
			n+d:		2.5	
			Spring	n+d:	3.5	
			Summer	n+d:	2.4	
			Autumn	n+d:	1.6	
			Winter	n+d:	2.3	
			86			
			n+d:		2.4	
			Spring	n+d:	2.9	
			Summer	n+d:	2.6	
			Autumn	n+d:	2.2	
			Winter	n+d:	1.7	
State College, PA, USA	VIII/84 to VIII/87, weekly c, ac, FI	Rural	85			71
			n+d:		3.3	
			Spring	n+d:	3.7	
			Summer	n+d:	3.1	
			Autumn	n+d:	2.8	
			Winter	n+d:	3.7	
			86			
			n+d:		2.7	
			Spring	n+d:	2.7	
			Summer	n+d:	2.4	
			Autumn	n+d:	2.0	
			Winter	n+d:	3.5	
Argonne, IL, USA	VIII/84 to VIII/87, weekly c, ac, FI	Rural	85			71
			n+d:		2.5	
			Spring	n+d:	3.2	
			Summer	n+d:	2.9	
			Autumn	n+d:	2.9	
			Winter	n+d:	2.1	
			86			
			n+d:		2.4	
			Spring	n+d:	4.0	
			Summer	n+d:	3.6	
			Autumn	n+d:	2.9	
			Winter	n+d:	2.5	
Bondville, IL, USA	VIII/84 to VIII/87, weekly c, ac, FI	Rural	86			71
			n+d:		3.4	
			Spring	n+d:	3.5	
			Summer	n+d:	2.9	
			Autumn	n+d:	2.1	
			Winter	n+d:	4.0	
Puszta, Hungary	II/82 to V/83, c, ac, FI	Rural	n+d:		3.7	154
Petten, the Netherlands	2 days in VI to VII/92, c, d, WD	Rural	n+d:		1-10	137
Nottingham, UK	VIII/94, s, 4-h, AD	Semirural	d:		0.7-1.8	2

Table 14.3 continued

Location	Time and Details of Measurement	Urban/ Rural/ Remote	Mean Concentrations (µg m^{-3})			Ref.
Eastern USA, 50 NDDN sites	89, weekly, c, FI	Rural to urban	Ranges for all sites:			72
			n+d:		(0.7–3.6)	
			Spring	n+d:	(0.6–3.9)	
			Summer	n+d:	(0.6–4.2)	
			Autumn	n+d:	(0.6–3.2)	
			Winter	n+d:	(1.0–3.2)	
Warren, MI, USA	VI/81 to VI/82, c, 24-h/48-h/72-h, DDM	Suburban	n+d:		1.68	81
			Spring	n+d:	1.07	
			Summer	n+d:	2.69	
			Autumn	n+d:	1.16	
			Winter	n+d:	1.78	
Claremont,CA, USA	8 days in IX/85, c, dc, AD	Suburban	n+d:		4.8 (0.2–32.0)	66
Burbank, CA, USA	1986, c, FI	Rural to urban	n+d:		6.7 max. 17.4	66
Downtown Los Angeles, CA, USA	1986, c, FI	Rural to urban	n+d:		6.0 max. 16.6	66
Hawthorne, CA, USA	1986, c, FI	Rural to urban	n+d:		3.1 max. 13.0	66
Long Beach, CA, USA	1986, c, FI	Rural to urban	n+d:		3.4 max. 15.6	66
Anaheim, CA, USA	1986, c, FI	Rural to urban	n+d:		3.2 max. 13.5	66
Upland, CA, USA	1986, c, FI	Rural to urban	n+d:		6.0 max. 18.8	66
Rubidoux, CA, USA	1986, c, FI	Rural to urban	n+d:		1.7 max. 6.7	66
Tanbark Flat, CA, USA	1986, c, FI	Rural to urban	n+d:		6.9 max. 21.0	66
Tanbark Flat, CA, USA	Summer 1989, 1990, 1991, s, 12-h, AD	Urban to Rural	d:	89:	20.3 (8.6–27.3)	70
				90:	17.2(2.7–46.7)	
				91:	11.3(3.0–15.1)	
			n:	89:	3.1 (2.0–4.5)	
				90:	1.8 (0.4–4.1)	
				91:	2.1 (0.6–4.2)	
Wast Hills, UK	VIII/93, IV to V/94, s, 4-h, AD	Semiurban	d:	93:	0.2–2.2	2
				94:	0.3–1.4	
Braunschweig, Germany	IV to VIII/93, c, weekly, AD	Suburban	n+d:		1.6	48
Exelberg near Vienna, Austria	VII/86 to VIII/88, s, 4-h, AD	Suburban	n+d:	VII to VIII/86:	6.1	67
				VII to VIII/87:	4.3	
			n:	VII to VIII/86:	4.1	
				VII to VIII/87:	3.5	
			d:	VII to VIII/86:	8.0	
				VII to VIII/87:	5.0	
Chicago, IL, USA	IV/90 to III/91, s, 12 h, AD	Urban	n+d:		0.81	131
			n:		0.52 max. 4.85	
			d:		1.1 max 2.7	
Warren, MI, USA	XII/87 to IV/88, 24-h, s, AD	Urban	n+d:		1.14 ± 0.72	132

Abbreviations: continuously: s, sporadically: s, diurnal cycle: dc, annual cycle: ac, filter: FI, annular denuder: AD, denuder: d, denuder difference method: DDM, wet denuder: WD, laser-photolysis fragment-fluorescence: LPFF, nights only: n, days only: d, all day: n+d, range: ( ), maximum: max.

Highest pollution levels are reported for Southern California due to its frequent photosmog: concentration levels of $HONO_2$ near Los Angeles may exceed 40μg m^{-3} (24-h average) (see Table 14.3).[68-70] High emissions of NO_x and of volatile organic compounds (VOC) combined with intensive radiation, high air temperatures, and low rainfall frequency are the main prerequisites for these situations. Concentration levels (annual means) obtained for less-polluted sites in the U.S. are in the range of 0.5 to 4 μg m^{-3} $HONO_2$[31,72,73] and comparable to those found at most European stations (see Table 14.3).

Legge[74] gives typical ranges of the concentration of $HONO_2$ (Table 14.4). In contrast to HONO, remote regions can be identified for $HONO_2$, e.g., Arctic troposphere, Atlantic or Pacific Oceans. However, there are hardly differences between mean concentrations found for rural, suburban, and urban locations, if mesoscale transport of $HONO_2$ or of its precursors from heavily polluted regions can be excluded.[75-78] Thus, most of the concentration data published for North America and Europe can be regarded as rural to moderately polluted; concentrations (daily means) of more than 10 μg m^{-3} have rarely been observed.

$HONO_2$ concentrations exhibit a typical diurnal cycle (Figure 14.5) reflecting both the photochemical activity of the troposphere, which depends on the availability of the OH radical, and the efficiency of the depletion processes; the concentrations of OH increase after sunrise, reaching a maximum during afternoon, and decrease after sunset. Nitric acid concentrations during sunlight confirm the importance of the photochemical formation pathway.[62,76,79] The fact that the concentrations of $HONO_2$ do not become negligible during nighttime may be explained by limited turbulent exchange and by nighttime formation.[62,67,76,79] If photochemical production is the main formation pathway of $HONO_2$, one should expect a distinctive annual cycle with maximum concentrations during spring and summer and minimal concentrations during winter (Figure 14.6). Such annual cycles have been observed frequently.[47,61,71,72,80-84]

Table 14.4 Typical Ranges of Annual Mean Concentrations of $HONO_2$ Depending on Air Quality

Pollution Climate	Concentration in μg m^{-3} $HONO_2$
Remote	<0.08–0.25
Rural	0.25–10
Moderately polluted	2.5–25
Heavily polluted	25–130

Source: After Legge, A.H., 1988.[74]

However, contradictory results with higher concentrations of $HONO_2$ were found during winter and have not been explained satisfactorily so far.[71,72,75,82] Also, Dasch and Cadle[85] found 2 weeks with elevated nitric acid concentrations during a wintertime study in Michigan without a plausible explanation. Above snow-covered surfaces photolysis is enhanced by reflection; VOC/NO_x mixtures can then readily produce $HONO_2$. It is not known to what extent processes other than photochemistry (including direct emission and liberation from nitrate salts) contribute significantly to $HONO_2$ production.

14.4.6 Lateral Transport of Nitric Acid

$HONO_2$ or its precursors may be subject to mesoscale or long-range transport depending on wind direction. At a remote to rural station at southwestern Sweden, elevated concentrations of $HONO_2$ for southwesterly winds were observed,[75] which we attribute to long-range transport from Great Britain or continental northwestern Europe. Measurements in rural southeastern England yielded higher concentrations of $HONO_2$ for periods influenced by continental air masses.[76] Both

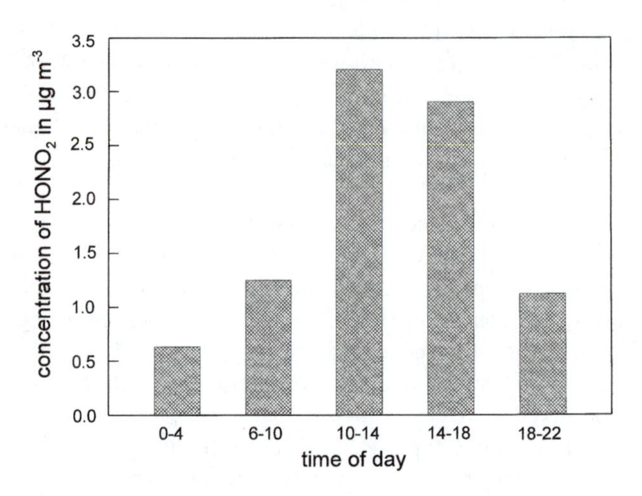

Figure 14.5 Diurnal cycle of $HONO_2$ concentrations in rural southeastern Lower Saxony, Germany. (After Zimmerling et al., 2000[61]).

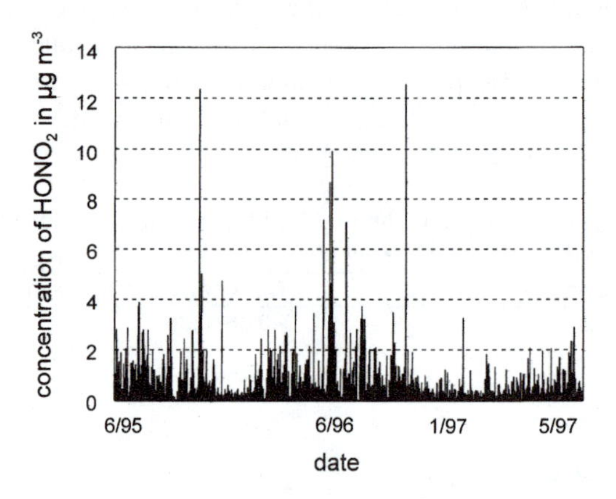

Figure 14.6 Annual Cycle of the daily means of $HONO_2$ concentrations in rural East Brandenburg, Germany. (After Zimmerling, et al., 2000[61]).

horizontal advection and thermolysis of NH_4NO_3 may contribute to this effect. Foltescu et al.[86] assigned concentrations measured at rural and marine sites in Scandinavia (North Sea) to certain air masses of different origin using trajectory analyses (Table 14.5). Here, concentrations for westerly winds and for easterly winds in wintertime can only be explained by lateral transport processes.[75] In contrast to the corresponding results for HONO, no dependency of $HONO_2$ concentrations on wind direction was found in rural southeastern Lower Saxony (Germany) (Figure 14.7).[48] Periods with easterly winds exhibited only slightly elevated concentrations. It is not clear to what extent this is caused by long-range transport of precursors from the industrial regions of Saxony-Anhalt and Saxony, by enhanced photochemical formation (high-pressure situations with sunny and warm weather are normally dominated by easterly winds), or by a high emission density of ammonia. During summer photosmog situations in the Vienna basin, a dependency of the concentrations of $HONO_2$ on wind velocities was observed for air masses with urban origin: higher wind velocities resulted in lower concentrations of $HONO_2$ due to incomplete oxidation of NO_2.[67] For a wintertime study a trajectory analysis for NO_x, HNO_3, NO_3^-, PAN, and SO_4^{2-} yielded a similar dependency of the concentrations on wind direction.[87]

Table 14.5 Mean Concentration of $HONO_2$ at Lista (South Norway) for Air Masses of Different Origin (concentrations in μg m⁻³ $HONO_2$)

Continental (Different Wind Direction)	Modified Marine	Marine	Marine Polar
North: 0.45	0.41	0.14	0.18
East: 0.72			
Summer: 0.68	Summer: 0.68	Summer: 0.23	
Winter: 0.81	Winter: 0.23	Winter: 0.09	
West: 1.98			

Source: After Foltescu et al. 1996.[86]

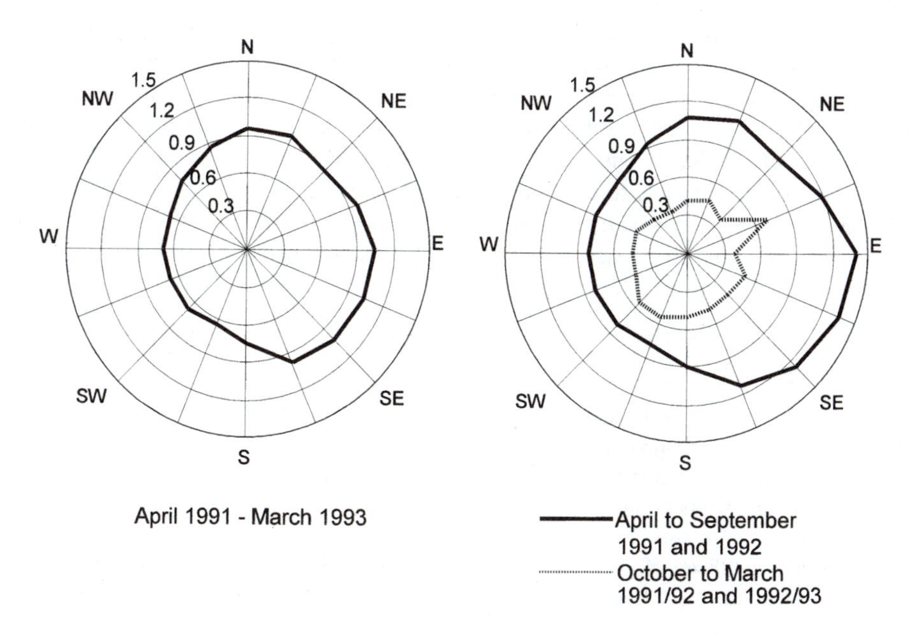

Figure 14.7 $HONO_2$ concentrations (in μg m⁻³) in rural southeaster Lower Saxony, Germany, as a function of wind direction. (After Zimmerling et al., 2000[61]).

Lateral transport of $HONO_2$ is limited by depletion processes (wet and dry deposition). The extent of depletion processes is mainly determined by meteorology, especially by rainfall and atmospheric stratification. Rain reduces the concentration of water-soluble species by washout processes. The formation of inversion layers aggravates the exchange between air and surfaces and favors the conservation of pools of air constituents above those inversion layers. During lateral flow, $HONO_2$ is in equilibrium with NH_4NO_3 (no deposition, no influence of direct emissions) the concentrations of which depend on air temperature and humidity.[67]

14.5 DRY DEPOSITION OF NITROUS AND NITRIC ACIDS

In plant–soil systems, dry deposition occurs on external surfaces (cuticula, soil surface) and on internal surfaces (mesophyll via stomata). In principle, these fluxes cannot be determined adequately with direct measurement at the surfaces themselves. Indirect experimental methods and models have to be applied.[88,89]

14.5.1 Micrometeorological Methods of Flux Determination

Micrometeorological methods make use of a concept which presupposes that, in a certain atmospheric layer and in steady-state net vertical fluxes of matter, heat and momentum are constant with height; this constant flux layer (CFL) is horizontally homogeneous and free of internal sources and sinks. Its extension depends on the fetch, which is a function of the turbulent properties of the atmosphere and the roughness of the canopy. Within this layer, flux densities can be obtained from vertical profiles of mean concentrations, wind velocities, and air temperatures (gradient method), or from fluxes of these entities in individual eddies (eddy correlation methods). The application of the eddy correlation method depends on the availability of fast sensors (resolution in time: 5 to 20 s^{-2}). For measurements of the concentrations of HONO and $HONO_2$ such sensors are rarely available. Therefore, most determinations have to rely on gradient techniques whose typical resolution in time is on the order of about 1 h. As both HONO and $HONO_2$ exhibit distinct diurnal and annual cycles, and their reaction rates are high compared with the resolution in time of accumulating procedures, sophisticated techniques are required. If the determination of profiles is impossible, the flux densities of air constituents can be modeled using meteorological standard measurements and adequate concentration measurements (inferential method).

Direct methods (eddy correlation methods) determine the fluctuations of both concentration and vertical wind velocity, from which the vertical flux is obtained according to

$$F_c = \overline{w' \cdot c'} \tag{14.24}$$

where F_c = vertical flux density of the trace gas
w' = fluctuation of the vertical wind velocity
c' = fluctuation of the concentration of the trace gas
overbar = temporal mean

In principle, vertical fluxes of trace gases F_c are determined indirectly using the following relation:

$$\frac{F_c}{\Delta c} = \frac{F_{H_2O}}{\Delta c_{H_2O}} = \frac{F_h}{\Delta \theta} = \frac{F_m}{\Delta u} \cdot \frac{\Phi_m}{\Phi_h} \tag{14.25}$$

where F_c = vertical flux density of the trace gas
F_{H_2O} = vertical flux density of water vapor
F_h = vertical flux density of sensible heat
F_m = vertical flux density of momentum
c = concentration of the trace gas
c_{H_2O} = concentration of water vapor
θ = potential air temperature
u = horizontal wind velocity
Φ_m, Φ_h = stability functions

c, c_{H_2O}, and u have to be determined for two given height. This chain of assumptions has been proved to be correct by experiments.[90-92]

In practice, one needs the determination of a concentration gradient, a temperature or a wind velocity gradient, and the vertical fluxes of heat or of momentum. This can be achieved by several methods which, in principle, lead to identical results. The major problem is to select sampling and analytical procedures that guarantee an adequate resolution in time in combination with an adequate analytical precision.

Single concentration measurements and standard meteorological and canopy parameters can be combined to model fluxes using the following general approach (inferential model, resistance model):

$$F_c = -\frac{c_z - c_b}{R_{c,\,atm}(z)} = -\frac{c_b - c_s}{R_{c,\,LBL}} = -\frac{c_s - c_p}{R_{c,\,C}} = \frac{c_z - c_p}{R_{c,\,atm}(z) + R_{c,\,LBL} + R_{c,\,C}} \quad (14.26)$$

where
c_z = concentration in ambient air at a reference height z
c_b = concentration immediately above the quasi-laminar boundary layer
c_s = concentration immediately above all surfaces of the system
c_p = concentration within the plant/soil system
$R_{c,\,atm}$ = atmospheric columnar resistance (transport by turbulent diffusion)
$R_{c,\,LBL}$ = columnar resistance of quasi-laminar boundary layer (transport by molecular diffusion)
$R_{c,\,C}$ = total surface columnar resistance of plant/soil system including external and internal surfaces

This approach is sensible if the internal concentration c_p (the so-called compensation point) of the plant/soil system is zero or known. The atmospheric columnar resistance has to be determined from the measurement of meteorological parameters. The determination of $R_{c,LBL}$ and $R_{c,C}$ presupposes detailed vegetation models, because they depend both on the biological species and on the physical and chemical surface properties involved. Table 14.6 gives the order of magnitude of the surface resistances of relevant chemical species and receptor systems.

Table 14.6 Examples of Stomatal and Surface Resistances (s m^{-1}) for SO_2, NO_2, HONO, and $HONO_2$ during Daytime under Different Pollution Climates

	Moist,[a] Moderate Temperature		Winter, Snow-Covered Surface		Dry, Warm Summer,[b] Well-Watered Soil	
	Coniferous Forest	Grassland	Coniferous Forest	Grassland	Coniferous Forest	Grassland
R_{stom}	200	60	400	—	150	50
SO_2	0	0	500	500	300	100
NO_2	320	100	640	1000	240	8
HONO	0	0500	500	300	100	
$HONO_2$	0	0	10	50	0	0

[a] Rain or relative humidity > 90%.
[b] Relative Humidity < 60%.
Source: After Erisman et al., 1994.[99]

14.5.2 Vertical Fluxes of Nitrous Acid

Harrison and Kitto[93] carried out flux measurements over grassland in southeastern England using annular denuders. Fluxes at Halvergate were calculated according to the Bowen ration method (concentrations at three measuring heights). Fluxes in Essex were calculated from measurements of concentrations, wind velocities, and air temperatures at for heights. Both upward and downward fluxes were observed as a result of efficient deposition of HONO on surfaces, and of fast production of HONO by a heterogenous reaction of NO_2 on moist surfaces including resuspended soil particle surfaces. During daytime concentrations of HONO frequently were below the limit of detection

(LOD)~ 0.4 μg m^{-3} HONO). Therefore, most fluxes were calculated from nighttime measurements. Correlation calculations between flux densities of HONO and concentrations of NO_2 exhibited a more or less pronounced relationship: during periods with elevated concentrations of NO_2 (> 20 μg m^{-3}) emission of HONO dominated. Bidirectional fluxes of HONO were measured above grassland; the deposition velocities determined for periods with downward fluxes were about two thirds of the deposition velocities of $HONO_2$ at the same time and locations.[2]

Above an intensively managed grassland in Switzerland, flux measurements were carried out by the gradient methods:[28,94] concentrations of HONO were measured at two heights using the wet effluent diffusion denuder.[95] A large scatter of results which were often in the range of the LOD during daytime was observed. Therefore, no fluxes were calculated. However, although the correlation between the concentration of HONO and the concentration of NO_2 established by Harrison and Kitto[93] was confirmed, no emission profiles of HONO were observed. Methodological problems concerning these results are discussed.

Flux measurements above a winter barley field over the summer months using a gradient method (ratiometric method)[88] and annular denuders (two measuring heights) for concentration measurements were carried out in southeastern Lower Saxony.[97] The quasi-continuous determination of the flux densities was achieved by classifying the atmospheric exchange properties using an online calculated atmospheric conductivity to switch pairs of denuders on and off. The measurements were performed to prove the applicability of the procedure in principle, but suffered from insufficient fetch. About one half of the concentration differences was resolved with sufficient analytical precision; the mean concentration difference was 0.08 μg m^{-3} m^{-1} HONO. Both deposition and emission profiles were observed. Distinctive steeper concentration gradients were found during periods with low atmospheric conductivity only (e.g., during nighttime).

Vertical fluxes of HONO can be obtained by inferential modeling typically taking into account the solubility of HONO in water, its reactivity, and its relative diffusivity compared with water vapor. For different land-use types and with consideration of the seasonal variation the surface resistances of HONO and SO_2 are thought to be similar, and considerably lower than that of NO_2. It is obvious that under dry conditions stomatal uptake dominates dry deposition, whereas wet surfaces absorb HONO efficiently.[98,99]

In Europe, an inferential model was first applied to HONO depositions in the Netherlands.[100] It combined measured concentrations (average concentration for the period 1987 to 1989: 1.0 μg m^{-3} HONO) and modeled deposition velocities (0.7 cm s^{-1}< v_D < 0.84 cm s^{-1}), which resulted in annual fluxes in the range of about 0.5 kg ha^{-1} a^{-1} N < F_c < 1.0 kg ha^{-1} a^{-1} N. For Speulder annual inputs of 0.7 kg ha^{-1} a^{-1} N HONO were modeled both for 1988 and 1989.[101] Measurements of concentrations and meteorological parameters in Speulderbos in the Netherlands from November 1992 to September 1993 resulted in an annual flux of 1.4 kg ha^{-1} a^{-1} HONO-N (average concentration: 1.1 μg m^{-3} HONO). At the same time, the annual fluxes of $HONO_2$ and NO_2 were 2.0 kg ha^{-1} a^{-1} N each (average concentration of $HONO_2$: 0.86 μg m^{-3}). A comparison of the depositions determined from inferential modeling and throughfall measurements yielded that about 11% of the total deposition was due to stomatal uptake ($HONO_2$: 0%; NO_2: 100%).[102]

In comparatively unpolluted areas, HONO contributes significantly to the dry deposition of total NO_y.* [30] Based on the hypotheses that for HONO deposition dominates if NO_2 concentrations fall below 20 μg m^{-3}, a dry deposition of HONO into forest is likely. Assuming a daytime concentration of 1 μg m^{-3} HONO, a deposition of about 3 kg ha^{-1} a^{-1} HONO-N was calculated for forest in northern England.[93]

The results of these measurements and calculations can be summarized as follows: radiation enhances both atmospheric turbulence and the photolysis of HONO. Therefore, during summer low concentrations of HONO generally lead to small vertical flux densities of HONO. During daytime, concentrations are generally in the range of the LODs of the respective methods. This

* NO_y denotes the sum of total nitrogen species ($NO_y = NO + NO_2 + NO_3 + NO_2O_5 + HONO + HONO_2 + PAN +$ particulate NO_3^- +...).

results in a comparatively large scatter of the results. The same applies to the diurnal cycle of the concentrations of HONO: during spring, summer, and autumn, higher concentrations of HONO during nighttime normally coincide with periods of inefficient atmospheric exchange; during periods with intensive atmospheric exchange the concentrations of HONO are generally low. The contribution of HONO to the deposition or emission of N species during these seasons is small and almost likely to be negligible. No flux measurements have been published for wintertime. Although the sink strength of the vegetation for gases during winter is small, elevated concentrations in the range of 1 to 10 $\mu g\ m^{-3}$ and a higher reactivity of HONO than that of NO and NO_2 may result in a significant contribution to the N deposition during winter.

Both net emission and net deposition of HONO were observed, but the cumulative flux densities over longer periods were close to zero. If HONO is produced by the reaction of NO_2 on surfaces of soil-derived particles, the concentration profiles measured may be the result of emission, deposition, as well as chemical reaction. A correction of the concentration profiles considering chemical reactions is necessary for realistic description of HONO exchange.

14.5.3 Vertical Fluxes of Nitric Acid

Measurements of vertical fluxes of $HONO_2$ were first carried out by Huebert and co-workers.[103,104] $HONO_2$ is mainly deposited on vegetation and soil surfaces; if chemical reactions are negligible all vertical concentration profiles indicate deposition.[105,106] The deposition of $HONO_2$ through the stomata is of minor importance.[102,107] Furthermore, it was shown that the deposition of $HONO_2$ is hardly influenced by surface processes: the high reactivity of $HONO_2$ (solubility in water, strong acid) leads to it negligible surface resistance.[104,108-112] The only exception is the deposition on snow-covered surfaces: in laboratory experiments, Janson and Franat[113] found a decrease of the surface resistance $R_{c,C}$ from 5000 s m^{-1} at -18°C to 175 s m^{-1} at -1°C. Even for dry surfaces the surface resistance is infinite.[114] However, for most cases the application of the resistance approach for the modeling of the dry deposition of $HONO_2$ requires "only" correct resistances $R_{c,atm}$ and $R_{c,LBL}$, which makes $HONO_2$ an ideal candidate for inferential models. The dry deposition of $HONO_2$ is mainly limited by the turbulent exchanges between atmosphere and canopy. This is reflected by the dependency of the dry deposition velocity on other parameters describing atmospheric exchange conditions (wind velocity: Table 14.7[115]; friction velocity[114]; atmospheric conductivity.[116]) Meixner[117] emphasized that for a successful application of an inferential model two problems still have to be considered:

- For periods with small $R_{c,atm}$ a correct modeling of $R_{c,LBL}$ is very important, but becomes a problem "since our understanding of the physical process underlaying $R_{c,LBL}$ is limited."
- The application of inferential methods presupposes a flux that is independent of height. For the dry deposition of $HONO_2$ this means that influence of the thermolysis of NH_4NO_3 and of the neutralization of $HONO_2$ by NH_3 on the vertical concentrations profile of $HONO_2$, as well as potential formation reactions, has to be negligible.

Measurements have shown that the latter is frequently not the case: disturbances caused by chemical reactions within the cycle of $HONO_2$ - NH_3 - NH_4NO_3 can change the vertical concentration profiles of all species involved. These reactions can occur very fast with a rate comparable with the rate of vertical atmospheric exchange. However, formation reactions do not interfere.[118]

Atmospheric resistances are not only controlled by meteorological parameters but also by the physical and chemical properties of the surface investigated: the rougher the surface the higher the deposition velocity (see Table 14.7). For a short grassland surface and a forest canopy with a height of about 10 m deposition velocities differ by a factor of 5 to 10. The results of flux measurement of $HONO_2$ are compiled chronologically in Table 14.8. The flux data published can be categorized into data derived from inferential methods and data from micrometeorological measurements.

Table 14.7 Deposition Velocities of $HONO_2$ onto Types of Vegetation Differing in Height and Roughness (deposition velocity in cm s^{-1})

	Short Grass	Moorland	Cereals	Forest
Horizontal wind velocity: 1 m s^{-1}	0.52	0.76	1.43	4.0
Horizontal wind velocity: 4 m s^{-1}	2.35	3.33	5.0	10.0

After Fowler et al., 1991.[115]

Whereas data from inferential modeling cover whole years and/or regions, micrometeorological measurements are frequently restricted to shorter periods of hours or days. Therefore, generalization of the results obtained with those measurements is limited. Most data recorded in North America are based on inferential models applied to dry deposition networks. They reflect the spatial variability of the dry deposition of $HONO_2$.[71,73,119,120] However, one should bear in mind that the application of filter packs for concentration monitoring (instead of denuders) as well as a potential influence of atmospheric chemistry on the vertical concentration profiles create a bias. Disturbances of the vertical concentration profiles of $HONO_2$ caused by locally emitted NH_3 as well as by the thermolysis of particulate NH_4NO_3 were observed.[28,51,104,121] Therefore, the application both of inferential methods and of gradient methods requires additional modeling to consider the influence of atmospheric chemistry.[51,122-125] For Europe this is of special importance as the comparatively high emission density of NH_3 results in elevated concentration levels of both NH_3 and NH_4NO_3.

The overall results can be summarized as follows: Depending on the ammonia pollution level and the surface properties of a region, the dry deposition velocity of $HONO_2$ is in the range between $1.0 \text{ cm s}^{-1} < v_D < 3.5 \text{ cm s}^{-1}$ and leads to cumulative annual fluxes in the range between $0.5 \text{ kg ha}^{-1} \text{ a}^{-1} < F_c < 3.5 \text{ kg ha}^{-1} \text{ a}^{-1}$ N. For some regions in North America high deposition rates have to be discussed (e.g., California). Besides the measurements in the dry deposition networks of the U.S. and Canada there are hardly measurements of $HONO_2$ fluxes covering longer periods. With the exception of measurements in the Netherlands, concentrations and flux data sets for Europe mostly reflect sporadic work. However, all measurements and models indicate the relative importance of $HONO_2$ in Europe and North America for acidification and eutrophication of natural and near-natural ecosystems.

14.6 CONCLUSIONS

Both HONO and $HONO_2$ are key species within the atmospheric chemistry of nitrogen and of OH formation and depletion. Their role in the complex atmospheric reaction patterns is not fully understood, and in some cases our knowledge is insufficient for application in comprehensive models. It is still to be questioned whether and to what extent HONO contributes to overall NO_y depositions. In no case can the dry deposition of $HONO_2$ be neglected in nitrogen budgets of natural and near-natural ecosystems. Although there has been progress in information over the past two decades, the statement by Finlayson-Pitts and Pitts[3] that "this is clearly an area for further work" is still valid for both species.

ACKNOWLEDGMENT

For discussion of this chapter we thank Drs. Detlev Möller, Cottbus, and Ralph Dlugi, München.

Table 14.8 Vertical Fluxes of $HONO_2$

Location	Time and Method of Measurement	Cumulative Flux Density (in kg ha^{-1} a^{-1} N)	Mean Deposition Velocity (in cm s^{-1}, without reference height)	Ref
Puszta, Hungary, re, grassland	1982, c, assumed deposition velocity, FI	5.2	2.0	154
Champaigne, IL, USA, ru, grassland	VI/82, s, DMM, FI		d: 2.5 ±0.9	103
Walker Branch Watershed, TN, USA, fo	VII/81 to VII/83, s, assumed deposition velocity, FI	Growing season: 3.0 Dormant season: 0.6	Growing season: 2.0 Dormant season: 0.5	148
Oak Ridge, TN, USA, ru	V/86 to VIII/85, IM, c, FI		0.8 (0.6–1.3)	155
State College, PA, USA, ru	V/86 to VIII/85, IM, c, FI		1.8 (1.6–2.1)	155
Argonne, IL	V/86 to VIII/85, IM, c, FI		1.9 (1.5–2.1)‘155	155
Oak Ridge, TN, USA, fo	IX/82, s, IM, FI		3.3±0.5 (2.2–6.0	156
Oak Ridge, TN, USA, fo Argonne, IL, USA, ru	IX/82, s, IM, FI 1986, IM c, FI		3.5±0.2 (2.3–4.1) 2.06	156 119
Bondville, IL, USA, ru	1986, IM c, FI		1.84	
Panola State Park, GA, USA, fo	1986, IM c, FI		1.25119	
Pennsylvania State University, PA, USA, fo, ru	1986, IM c, FI		1.97	119
West Point, NY, USA, fo	1986, IM c, FI		1.87	119
Whiteface Mountain, NY, USA, fo	1986, IM c, FI		1.88	119
Essex, Great Britain, ru, arable land	10/85 to 1/88, s, DMM, FI		0.4–7.7	109
Warren, MI, USA	1988, c, chamber experiments		0.2–1.2	157
Oak Ridge, TN, USA, fo	IV to X/87, IM, c, FI	2.4		77
Solling, Germany, fo	IV to X/87, IM c, FI	4.9		77
Oak Ridge, TN, USA, fo	VII/84 to VII/87, weekly sampling, c, IM, ac, FI	85 Total: 1.6 Spring: 0.7 Summer: 0.2 Autumn: 0.2 Winter: 0.5 86 Total: 2.1 Spring: 0.9 Summer: 0.6 Autumn: 0.8 Winter: 0.9		71

Table 14.8 continued

Location	Time and Method of Measurement	Cumulative Flux Density (in kg ha^{-1} a^{-1} N)	Mean Deposition Velocity (in cm s^{-1}, without reference height)	Ref
Argonne, IL, USA, ru, grassland	VII/84 to VII/87, weekly sampling, c, IM, ac, FI	85 Total: 3.2 Spring: 0.9 Summer: 0.6 Autumn: 0.8 Winter: 0.8 86 Total: 3.6 Spring: 1.2 Summer: 0.8 Autumn: 0.8 Winter: 0.8		71
Whiteface Mountain, NY, USA, re, fo	VII/84 to VII/87, weekly sampling, c, IM, ac, FI	85 Total: 1.4 Spring: 0.5 Summer: 0.2 Autumn: 0.4 Winter: 0.3 86 Total: 1.8 Spring: 0.5 Summer: 0.7 Autumn: 0.4 Winter: 0.2		71
State College, PA, USA, ru,	VII/84 to VII/87, weekly sampling, c, IM, ac, FI	85 Total: 3.2 Spring: 1.0 Summer: 0.8 Autumn: 0.9 Winter: 0.5 86 Total: 2.8 Spring: 0.8 Summer: 0.7 Autumn: 0.6 Winter: 0.7		71
Bondville, IL, USA, ru	VI/85 to VIII/87, weekly sampling, c, IM, ac, FI	86 Total: 2.8 Spring: 0.8 Summer: 0.7 Autumn: 0.7 Winter: 0.7		71
Panola, GA, USA, ru	VIII/85 to VIII/87, weekly sampling, c, IM, ac, FI	86 Total: 2.2 Spring: 0.7 Summer: 0.6 Autumn: 0.4 Winter: 0.5		71
West Point, NY, USA, ru	VIII/84 to VIII/87, weekly sampling, c, IM, ac, FI	86 Total: 2.6 Spring: 1.0 Summer: 0.9 Autumn: 0.5 Winter: 0.2		71
Speulderbos, the Netherlands	1988–1989, IM, c, AD	88: 2.2 89: 2.5		101
Manndorf, Germany, ru, wheat	VII/90, s, DMM, D		2.2±1.2	114

Table 14.8 continued

Location	Time and Method of Measurement	Cumulative Flux Density (in kg ha^{-1} a^{-1} N)	Mean Deposition Velocity (in cm s^{-1}, without reference height)	Ref
Halvergate, Great Britain, ru, grassland	IV/89, s, DMM, AD		0.6–5.0	114
Mt. Mitchell State Park, NC, USA, fo	V/88 to IX/89, c, IM, AD	V/88 to IX/89: 2.6±1.2 V/89 to IX/89: 3.6±2.0		78
Ulborg, Denmark, fo	10 days in V/91, s, DMM, D		4.5	151
Wülfersreuth, Bavaria, Germany, ru, fo	18 days in 1992, s, IM, WD	17	11	158
Ann Arbor, MI, USA, fo ru	1990 to 1991, c, IM, FI		1990: 1.80 1991: 1.83	120
Unionville, MI, USA, fo ru	1990 to 1991, c, IM, FI		1990: 1.63 1991: 1.63	120
Ashland, ME, USA, fo ru	1990 to 1991, c, IM, FI		1991: 2.62	120
Whiteface, NY, USA, fo re	1990 to 1991, c, IM, FI		1990: 1.60 1991: 1.57	120
Connecticut Hill, NY, USA, fo ru	1990 to 1991, c, IM, FI		1990: 1.91 1991: 1.82	120
Perkinstown, WS, USA, fo ru	1990 to 1991, c, IM, FI		1991: 1.68	120
Birch Lake, MN, USA, fo	1991 to 1993, c, weekly sampling, IM, FI	1991: 0.34 1992: 0.35 1993: 0.34	1991: 1.09 1992: 1.19 1993: 1.46	73
Cedar Creek, MN, USA, fo	1991 to 1993, c, weekly sampling, IM, FI	1991: 0.83 1992: 0.80 1993: 0.77	1991: 1.06 1992: 1.06 1993: 1.46	73
Ely, MN, USA, fo	1991 to 1993, c, weekly sampling, IM, FI	1991: 0.32 1992: 0.39 1993: 0.37	1991: 1.02 1992: 1.09 1993: 1.14	73
Finland, MN, USA, fo	1991 to 1993, c, weekly sampling, IM, FI	1991: 0.48 1992: 0.53 1993: 0.48	1991: 1.09 1992: 1.09 1993: 1.14	73
Koch, MN, USA, fo	1991 to 1993, c, weekly sampling, IM, FI	1991: 0.85 1992: 0.88 1993: 0.59	1991: 1.14 1992: 1.16 1993: 0.83	73
Marcell, MN, USA, fo	1991 to 1993, c, weekly sampling, IM, FI	1991: 0.35 1992: 0.44 1993: 0.36	1991: 1.05 1992: 1.06 1993: 1.10	73
Sandstone, MN, USA, fo	1991 to 1993, c, weekly sampling, IM, FI	1991: 0.49 1992: 0.463 1993: 0.38	1991: 0.95 1992: 1.01 1993: 0.87	73
Braunschweig, Germany, su, ru	IV to VIII/93, DMM, c, AD	1.0±0.1	1.2–2.1	97

Abbreviations: Remote: re, rural: ru, suburban: su, urban: u, forest: fo, determination of fluxes: direct micrometeorological measurements: DMM, inferential model: IM, determination of concentrations: filter: FI, annular denuder: AD, denuder: D, wet denuder: WD, measurements: continuously: c, sporadically: s, annual cycle: ac, range: ( ), maximum: max.

REFERENCES

1. Nielsen, T., Egeløv, A. H., Granby, H., and Skov, H., Observations on particulate organic nitrates and unidentified components of NO_y, *Atmos. Environ.*, 29, 1757, 1995.
2. Harrison, R. M., Peak, J. D., and Collins, G. M., Tropospheric cycle of nitrous acid, *J. Geophys. Res.*, 101(D9), 14429, 1996.
3. Finlayson-Pitts, B. J. and Pitts, J. M., *Atmospheric Chemistry: Fundamentals and Experimental Techniques*, John Wiley & Sons, New York, 1986.
4. Warneck, P., *Chemistry of the Natural Atmosphere*; International Geophysics Series, 41, Academic Press, London, 1988.
5. Pitts, J. N., Jr., Biermann, H.W., Winer, A. M., and Tuazon, E. C., Spectroscopic identification and measurements of gaseous nitrous acid in dilute auto exhaust, *Atmos. Environ.*, 18, 847, 1984.
6. Sjödin, Å., Studies in the diurnal variation of nitrous acid in urban air, *Environ. Sci. Technol.*, 22, 1086, 1988.
7. Kirchstetter, T. W,. Harley, R. A., and Littlejohn, D., Measurement of nitrous acid in motor vehicle exhausts, *Environ. Sci. Technol.*, 30, 2842, 1996.
8. Lammel, G. and Wiesen, R., Stickstoffverbindungen in der Troposphäre, *Nach. Chem. Tech. Lab.*, 44, 477, 1996.
9. Winer, A. M. and Biermann, H. W., Long pathlengths differential absorption spectroscopy (DOAS) measurements of gaseous HONO, NO_2 and HCHO in the California South Coast Air Basin, *Res. Chem. Intermediates*, 20, 423, 1994.
10. Li, S. M., Anlauf, K. G., Wiebe, H. A., Bottenheim, J. W., Shepson, P. B., and Biesenthal, T., Emission ratios and photochemical production efficiencies of nitrogen oxides, ketones, and aldehydes in the Lower Fraser Valley during the Summer Pacific 1993 Oxidant Study, *Atmos. Environ.*, 31, 2037, 1997.
11. Andrés-Hernández, M. D., Notholt, J., Hjorth, J., and Schrems, O., A DOAS study on the origin of nitrous acid at urban and non-urban sites, *Atmos. Environ.*, 30, 175, 1996.
12. Hoek, G., Mennen, M. G., Allen, G. A., Hofschreuder, P., and van der Meulen, T., Concentrations of acidic air pollutants in the Netherlands, *Atmos. Environ.*, 30, 3141, 1996.
13. Rondon, A. and Sanhueza, E., High HONO atmospheric concentrations during vegetation burning in the tropical savannah, *Tellus*, 41B, 474, 1989.
14. Raschig, F. and Prahl, W., Beitrag zur Chemie der Stickstoffoxide, *Z. Angew. Chem.*, 42, 253, 1929.
15. Chan, W. H., Nordstrom, R. J., Calver, J. G., and Shaw, J. H., Kinetic study of HONO formation and decay reactions in gaseous mixtures of HONO, NO, NO_2, H_2O and N_2, *Environ. Sci. Technol.*, 10, 674, 1978.
16. Kaiser, E. W. and Wu, C. H., A kinetic study of the gas phase formation and decomposition reactions of nitrous acid, *J. Phys. Chem.*, 81, 1701, 1977.
17. Sakamaki, F., Hatakeyama, S., and Akimoto, H., Formation of nitrous acid and nitric oxide in the heterogeneous reaction of nitrogen dioxide and water vapor in a smog chamber, *Int. J. chem. Kinet.*, 15, 1013, 1983.
18. Sjödin, Å. and Ferm, M., Measurement of nitrous acid in an urban area, *Atmos. Environ.*, 19, 985, 1985.
19. Jenkin, M. D., Cox., R. A., and Williams, D. J., Laboratory studies of the kinetics of formation of nitrous acid from the thermal reaction of nitrogen dioxide and water vapour, *Atmos. Environ.*22, 487, 1988.
20. Lammel, G. and Perner, D., The atmospheric aerosol as a source of nitrous acid in the polluted atmosphere, *J. Aerosol Sci.*, 19, 1199, 1988.
21. Lammel, G., Perner. D., and Warneck, P., Nitrous acid at Mainz: observation and implication for its formation mechanism, in *Proc. 5th European Symposium on Physico-Chemical Behaviour of Atmospheric Pollutants*, Restelli, G. and Angeletti, G., Eds., Kluwer, Dordrecht, 1990, 469.
22. Notholt, J., Hjorth, J., and Raes, F., Formation of HNO_2 on aerosol surfaces during foggy periods in the presence of NO and NO_2, *Atmos. Environ.*, 26A, 211, 1992.
23. Calvert, J. B., Yarwood, G., and Dunker, A. M., An evaluation of the mechanism of nitrous acid formation in the urban atmosphere, *Res. Chem. Intermediates*, 20, 463, 1994.
24. Becker, K. H. and Barner, I., Atmospheric chemistry relevant to urban pollution, in *Air Pollution: Monitoring and Control Strategies*, Allegrini, I. and De Santis, F., Eds., Springer, Berlin, 1996, 21.
25. Svensson, R. Ljungström, E., and Lindqvist, O., Kinetics of the reaction between nitrogen dioxide and water vapour, *Atmos. Environ.*, 21, 1529, 1987.

26. Febo, A. and Perrino, C., Prediction and experimental evidence for high air concentration of nitrous acid in indoor environments, *Atmos. Environ.*, 25A, 1055, 1991.
27. Kitto, A. M. N. and Harrison, R. M., Nitrous and nitric acid measurements at sites in South-East England, *Atmos. Environ.*, 26A, 235, 1992.
28. Neftel, A., Blatter, A., Hesterberg, R., and Staffelbach, T., Measurement of concentration gradients of HNO_2 and HNO_3 over semi-natural ecosystems, *Atmos. Environ.*, 30., 3017, 1996.
29. Lammel, G., comment on "A DOAS Study on the Origin of Nitrous Acid at Urban and Non-Urban Sites," *Atmos. Environ.*, 30, 4101, 1996.
30. Junkermann, W. and Ibusuki, T., FTIR spectroscopic measurements of surface bond products of nitrogen oxides on aerosol surfaces — implications for heterogeneous HNO_2 production, *Atmos. Environ.*, 26A, 3099, 1992.
31. Cape, J. N., Hargreaves, K. J., Storeton-West, R., Fowler, D., Colvile, R. N., Choularton, T. W., and Gallagher, M. W., Nitrite in orographic cloud as an indicator of nitrous acid in rural air, *Atmos. Environ.*, 26A, 2301, 1992.
32. Simon, P. D. and Dasgupta, P. K., Continuous automated measurement of gaseous nitrous and nitric acids and particulate nitrate and nitrate, *Environ. Sci. Technol.*, 29, 1534, 1995.
33. Graedel, T. E., Hawkins, D. T., and Claxton, D., *Atmospheric Chemical Compounds, Sources, Occurrence, and Bioassay,* Academic Press, New York, 1986.
34. Harris, G. W., Maccay, G. I., Iguchi, T., Schiff, H. I., and Schuetzle, D., Measurement of NO_2 and HNO_3 in diesel exhaust gas by tunable diode laser adsorption spectrometry, *Environ. Sci. Technol.*, 21, 299, 1987.
35. Papenbrock, T., Stuhl, F., Müller, K. P., and Rudolph, J., Measurement of gaseous nitric acid over the Atlantic Ocean, *J. Atmos. Chem.*, 15, 369, 1992.
36. Wayne, R. P., Barnes, I., Biggs, P., Burrows, J. P., Canose-Mas, C. E., Hjorth, J., Le Bras, G., Moortgat, G. K., Perner, D., Poulet, G., Restelli, G., and Sidebottom, H., The nitrate radical: physics, chemistry, and the atmosphere, *Atmos. Environ.*, 25A, 1, 1991.
37. Dimitroulopoulou, C. and Marsh, A. R. W., Modelling studies of NO_3 nighttime chemistry and its effects on subsequent ozone formation, *Atmos. Environ.*, 31, 3041, 1997.
38. Li, S.-M, Anlauf, K. G., and Wiebe, H.A., Heterogeneous night-time production and deposition of particle nitrate at a rural site in North America during summer 1988, *J. Geophys. Res.* 98(D3), 5139, 1993.
39. Schwartz, S. E. and White, W. H., Solubility equilibria of the nitrogen oxides and oxyacids in dilute aqueous solution, *Adv. Environ. Sci. Technol.*, 4, 1, 1981.
40. Wagman, D. D., Evans, W. H., Parker, V. B., Schumm, R. H., Halow, I., Bailey, S. M., Churney, K. L., and Nurral, R. L., The NBS tables of chemical thermodynamic properties: selected values for inorganic and C_1 and C_2 organic substances in SI units, *J. Phys. Chem. Ref. Data*, 11, suppl. 2, 1982.
41. Fuzzi, S., Facchini, M. C., Schell, D., Wobrock, W., Winkler, P., Arends, B. G., Kessel, M., Möls, J. J., Pahl, S., Schneider, T., Berner, A., Solly, I., Kruisz, C., Kalina, M., Fierlinger, H., Hallberg, A., Vitali, P., Santoli, L., and Tigli, G., Multiphase chemistry and acidity of clouds at Kleiner Feldberg, *J. Atmos. Chem.*, 19, 87, 1994.
42. Choularton, T. W., Wicks, A. J., Downer, R. M., Gallagher, M. W., Penkett, S. A., Bandy, B. J., Dollard, G. J., Jones, B. M. R., Davies, T.W, Gay, M. J., Tyler, B. J., Fowler, D., Cape, J. N., and Hargreaves, K. J., A field study of the generation of nitrate in a hill cap cloud, *Environ. Pollut.*, 75, 69, 1992.
43. Scholz-Seidel, C. and Dämmgen, U., Institute of Agroecology, Braunschweig, unpublished data.
44. Martin, L. R., Damschen, D. E., and Judeikis, H. S., The reactions of nitrogen oxides with SO_2 in aqueous aerosols, *Atmos. Environ.*, 15, 191, 1981.
45. Cox, R. A. and Derwent, R. G., The ultraviolet absorption spectrum of gaseous nitrous acid, *J. Photochem.*, 6, 23, 1976/77.
46. Febo, A. and Perrino, C., Measurement of high concentrations of nitrous inside automobiles, *Atmos. Environ.*, 29, 345, 1995.
47. Zimmerling, R., Dämmgen, U., Küsters, A., Grünhage, L., and Jäger, H.-J., Response of a grassland ecosystem to air pollutants. IV. The chemical climate. Concentrations of relevant non-criteria pollutants (trace gases and aerosols), *Environ. Pollut.*, 91, 139, 1996.
48. Zimmerling, R., Dämmgen, U., Küsters, A., and Wolff, D., Konzentrationen von Luftinhaltsstoffen. II. Non-criteria pollutants (N, S- und Cl-haltige Spezies), in *Untersuchungen zum Chemischen Klima in Südostniedersachsen*, Dämmgen, U., Ed., Landbauforschung Völkenrode, special issue 170, 1996, 222.

49. Cape, J. N., Hargreaves, J. J., Storeton-West, R., Jones, B., Davies, T., Colvile, R. N., Gallagher, M. W., Choularton, T. W., Pahl, S., Berner, A., Kruisz, C., Bizjoak, M., Laj, P., Facchini, M. C., Fuzzi, S., Arneds, B. G., Acker, K., Wieprecht, W., Harrison, R. M., and Peak, J. D., The budget of oxidized nitrogen species in orographic clouds, *Atmos. Environ.*, 31, 2624, 1997.
50. Allen, A. G., Harrison, R. M., and Erisman, J. W., Field measurements on the dissociation of ammonium nitrate and ammonium chloride aerosols, *Atmos. Environ.*, 23, 1591, 1989.
51. Zhang, Y., ten Brink, H., Slanina, S., and Wyers, P., The influence of ammonium nitrate equilibrium on the measurement of exchange fluxes of ammonia and nitric acid, in *Acid Rain Research: Do We Have Enough Answers?* Heij, G. J. and Erisman, J. W, Eds., Elsevier, Dordrecht, 1995, 103.
52. Slanina, J. Wild, P. J. de, and Wyers, G. P., The application of denuder systems to the analysis of atmospheric compounds, in *Gaseous Pollutants: Characterization and Cycling*, Nriagu, J. O., Ed., Advances in Environmental Science and Technology, 24, John Wiley & Sons, New York, 1992, 129.
53. Sickles, J. D,. Sampling and analysis for ambient oxides of nitrogen and related species, in *Gaseous Pollutants: Characterization and Cycling*, Nriagu, J. O., Ed., Advances in Environmental Science and Technology, 24, John Wiley & Sons, New York, 1992, 128.
54. Appel, B. R., Winer, E. M., Tokiwa, Y., and Biermann, H. W, Comparison of atmospheric nitrous acid measurements by annular denuders and differential optical absorption systems, *Atmos. Environ.*, 24A, 611, 1990.
56. Anlauf, K. G., Wiebe, H. A., Tuazon, E. C., Winer, A. M., Mackey, G. I., Schiff, H. I., Ellestad, T. G., and Knapp, K. T, Intercomparison of atmospheric nitric acid measurements at elevated ambient concentrations, *Atmos. Environ.*, 25A, 393, 1991.
57. Febo, A., Perrion, C., and Allegrini, I., Field intercomparison exercise on nitric acid and nitrate measurement (Rome, 1988): a critical approach to the evaluation of the results, *Sci. Total Environ.*, 133, 39, 1993.
58. Perrion, C., De Santis, F., and Febo, A., Criteria for the choice of a denuder sampling technique devoted to the measurement of atmospheric nitrous and nitric acids, *Atmos. Environ.*, 24A, 617, 1990.
59. Febo, A., Perron, C., and Cortiello, M., A denuder technique for the measurement of nitrous acid in urban atmospheres, *Atmos. Environ.*, 27A, 1721, 1993.
60. Keuken, M., Mallan, R. K. A. M., Woittiez, J. R. W., and Slanina, J., Verzuringsonderzoek bij ECN, Report ECN-PB-89-7, ECN, Petten, The Netherlands, 1988.
61. Zimmerling, R., Dämmgen, U., Behrens, U., and Scholz-Seidel, C., The chemical climate in East Brandenburg, Germany. II. Concentrations of air-borne N and S species in East Brandenburg, Germany, *Environ. Pollut.*, in preparation, 2000.
62. Wyers, G. P., Veltkamp, A. C., Vemeulen, A. T., Geusenbroak, M., Wayers, A., and Möls, J. J., Deposition of Aerosols to Coniferous Forest, ECN Report, ECN-C-94-105, ECN, Petten, the Netherlands, 1994.
63. Nash, T., Nitrous acid in the atmosphere and laboratory experiments on it photolysis, *Tellus*, 26, 175, 1974.
64. Brauer, M., Koutrakis, P., and Spengler, J. D., Personal exposures to acidic aerosols and gases, *Environ. Sci. Technol.*, 23, 1408, 1989.
65. Brauer, M., Ryan, P. B., Suh, J. J., Koutrakis, P., and Spengler, J. D., Measurements of nitrous acid inside two research houses, *Environ. Sci. Technol.*, 24, 1521, 1990.
66. Solomon, P. A., Salmon, L. G., Fall, T., and Cass, G. R., Spatial and temporal distribution of atmospheric nitric acid and particulate nitrate concentrations in the Los Angeles area, *Environ. Sci, Technol.*, 26, 1594, 1992.
67. Piringer, M., Ober, E., Puxbaum, H., and Kromp-Kolb, H., Occurrence of nitric acid and related compounds in the northern Vienna basin during summertime anticyclonic conditions, *Atmos. Environ.*, 31, 1049, 1997.
68. Sickles, J. E., Sampling and analysis of ambient air near Los Angeles using annular denuder systems, *Atmos. Environ.*, 22, 1619, 1988.
69. Solomon, P. A., Larson, S. M., Fall, T., and Cass, G. R., Basinwide nitric acid and related species concentrations observed during the Claremont nitrogen species comparison study, *Atmos. Environ.*, 22, 1587, 1988.
70. Grosjean, D. and Bytnerowicz, A., Nitrogenous air pollutants at a Southern California mountain forest smog receptor site, *Atmos. Environ.*, 27A, 483, 1993.

71. Meyers, T. P., Hicks, B. B., Hosker, P. R, Womack, J. D., and Satterfield, L. C, Dry deposition interferential measurement technique — II. Seasonal and annual deposition rates of sulfur and nitrate, *Atmos. Environ.*, 25A, 2361, 1991.
72. Edgerton, E. S., Lavery, T. F., and Boksleitner, R. P., Preliminary data from the USEPA dry deposition network: 1989, *Environ. Pollut.*, 75, 145, 1992.
73. Pratt, G. C., Orr, E. J., Bock, D. C., Strassman, R. L., Fundine, D. W., Twaroski, C. J., Thornton, J. D., and Meyers, T. P., Estimation of dry deposition of inorganics using filter pack data and inferred deposition velocities, *Environ. Sci. Technol.*, 30, 2168, 1996.
74. Legge, A. H., The Present and Potential Effects of Acidic and Acidifying Air Pollutants on Alberta's Environment, Critical Point I, Final Report, Prepared for the Acid Deposition Research Program by the Biophysics Research Group, Kananaskis Centre for Environmental Research, The University of Calgary, Alberta, ADRP-B-17-88, 1988.
75. Ferm, M., Samuelsson, U., Sjödin, A., and Grennfelt, O., Long-range transport of gaseous and particulate oxidized nitrogen compounds, *Atmos. Environ.*, 18, 1731, 1984.
76. Harrison, R. M. and Allen, A. G., Measurement of atmospheric HNO_3, HCI and associated species on a small network in eastern England, *Atmos. Environ.*, 24, 369, 1990.
77. Lindberg, S. E., Bredemeier, M., Schaefer, D. A., and Qi, L., Atmospheric concentrations and deposition of nitrogen and major ions in conifer forests in the United States and Federal Republic of Germany, *Atmos. Environ.*, 24A, 2207, 1990.
78. Aneja, V. P. and Murthy, A. B., Monitoring deposition of nitrogen-containing compounds in a high-elevation forest canopy, *J. Air Waste Manage.*, 44, 1109, 1994.
79. Papenbrock, T. and Stuhl, F., Measurement of gaseous nitric acid by a laser-photolysis fragment-fluorescence (LPFF) method in the Black Forest and at the North Sea coast, *Atmos. Environ.*, 25A, 2223, 1991.
80. Meixner, F. X, Müller, K. P., Aheimer, G., and Höfken, K. D., Measurements of gaseous nitric acid and particulate nitrate, in *Proceedings of the COST Action 611 Meeting "Pollutant Cycles and Transport: Modeling and Field Experiments,"* van Leeuw, F. A. A. M. and van Egmond, N D., Eds., RIVM, Bilthoven, 1985, 103.
81. Cadle, S. H., Seasonal variation in nitric acid, nitrate, strong aerosol acidity, and ammonia in an urban area, *Atmos. Environ.*, 19, 181, 1985.
82. Sirois, A. and Fricke, W., Regionally representative daily air concentrations of acid-related substances in Canada; 1983–1987, *Atmos. Environ.*, 26A, 593, 1992.
83. Puxbaum, J., Haumer, G., Mose, K., and Ellinger, R., Seasonal variation of HNO_3, HCl, SO_2, NH_3 and particulate matter at a rural site in northeastern Austria (Wolkersdorf, 240 m a.s.l.), *Atmos. Environ.*, 27A, 2445, 1993.
84. Slanina, J. and Wyers, G. P., Monitoring of atmospheric compounds by automatic denuders, *Fresenius J. Anal. Chem.*, 350, 467, 1994.
85. Dasch, J. M. and Cadle, S. H., The removal of nitric acid to atmospheric particles during a wintertime field study, *Atmos. Environ.*, 24A, 2557, 1990.
86. Foltescu, V. L., Lindgren, E. S., Isakson, J., Öblad, M., Pacyna, J. M., and Benson, S., Gas to particle conversion of sulphur and nitrogen compounds as studies at marine stations in northern Europe, *Atmos. Environ.*, 30, 3129, 1996.
87. Daum, P. H., Kelly, T. J., Tanner, R. L., Tang, X., Anlauf, K., Bottenheim, J., Brice, K. S., and Wiebe, H. A., Winter measurements of trace gases and aerosol composition at a rural site in southern Ontario, *Atmos. Environ.*, 23, 161, 1989.
88. Dämmgen, U., Zimmerling, R., Grünhage, L., Haenel, H.-D., and Jäger, H.-J., Measuring vertical flux densities of reactive S- and N-species using (slow) denuder filter sampler and a (fast) micrometeorological method (ratiometric method), in *Atmospheric Ammonia: Emission, Deposition and Environmental Impact—Poster Proceedings*, Sutton, M. A., Lee, D. S., Dollard, G. J., and Fowler, D., Eds., ITE, Edinburgh, 1996, 81.
89. Dämmgen, U. and Grünhage, L., Response of a grassland ecosystem to air pollutants. V. A toxicological model for the assessment of dose–response relationships for air pollutants and ecosystems. *Environ. Pollut.*, 101, 375, 1988.
90. Tanner, C. B., Energy balance approach to evapo-transpiration from crops, *Soil Sci. Soc. Am. J.*, 24, 1, 1960.

91. Dyer, A. J. and Hicks, B. B., Flux-gradient relationships in the constant flux layer, *Q. J. R. Met. Soc.*, 96, 715, 1970.
92. Droppo, J. G. and Doran, J. C., Experimental constraints in micrometeorological gaseous pollutant fluxes, in *Precipitation Scavenging, Dry Deposition, and Resuspension*, Vol. 2, Pruppacher, H. R., Semonin, R. G., and Slinn, W. G. N., Eds., Elsevier, New York, 1983, 807.
93. Harrison, R. M. and Kitto, A.-M. N., Evidence for a surface source of atmospheric nitrous acid, *Atmos. Environ.*, 28, 1089, 1994.
94. Hesterberg, R., Blatter, A., Fahrni, M., Rosset, M., Neftel, A., Eugster, W., and Wanner, H., Deposition of nitrogen-containing compounds to an extensively managed grassland in Central Switzerland, *Environ. Pollut.*, 91, 21, 1996.
95. Vecera, Z. and Dasgupta, P. K., Measurement of atmospheric nitric and nitrous acids with a wet effluent diffusion denuder and low-pressure ion chromatography-post column reaction detection, *Anal. Chem.*, 63, 2210, 1991.
96. Harrison, R. M. and Peak, J. D., Measurements of concentration gradients of HNO_2 and HNO_3 over a semi-natural ecosystem: discussion, *Atmos. Environ.*, 31, 2891, 1997.
97. Zimmerling, R., Dämmgen, U., Grünhage, L., Haenel, H. -D., Küsters, A., Max, W., and Jäger, H. -J., The classifying ratiometric method for the continuous determination of atmospheric flux densities of reactive N- and S-species with denuder filter systems, *Angew. Bot.*, 71, 38, 1997.
98. Wesely, M. L., Parameterization of surface resistances to gaseous dry deposition in regional-scale numerical models, *Atmos. Environ.*, 23, 1293, 1989.
99. Erisman, J. W., van Pul, A., and Wyers, P., Parametrization of surface resistances for the quantification of atmospheric deposition of acidifying pollutants and ozone, *Atmos. Environ.*, 28, 2595, 1994.
100. Erisman, J. W., Acid Deposition in the Netherlands, Report 723001002, RIVM, Bilhoven, 1991.
101. Duyzer, J. H., Verhage, H. L. M., and Weststrate, H., Monitoring Deposition of Nitrogen Dioxide on the Elspeetsche Veld Heathland, MT-TNO report, 1991.
102. Erisman, J. W., Draaijers, G. P. J., Duyzer, J. H., Hofschreuder, P., van Leeuwen, N., Römer, F. G., Ruijgrok, W., and Wyers, G. P., Particle deposition to forests: summary of results and application, *Atmos. Environ.* 31, 333, 1997.
103. Heubert, B. J., Measurement of the dry-deposition flux of nitric acid vapour to grasslands and forests, in *Precipitation Scavenging, Dry Deposition and Resuspension*, Pruppacher, H. R., Semonin, R. G., and Slinn, W. G. N., Eds., Elsevier, New York, 1983, 785.
104. Huebert, B. J. and Robert, C. H., The dry deposition of nitric acid to grass, *J. Geophys. Res.*, 90(D1), 2085, 1985.
105. Höfken, K. -D., Meixner, F., Müller, K. P., and Ehhalt, D. H., Untersuchungen zur trockenen Deposition und Emission von atmosphärischem NO, NO_2 und HNO_3 and natürlichen Oberflächen, Berichte der KFA Jülich, No. 2054, 1986.
106. Dlugi, R., Deposition of gaseous and particulate compounds from a profile method, in *Air Pollution Modeling and Its Application*; Vol. VI, Van Dop, H., Ed., Plenum, New York, 1988, 49.
107. Hanson, P. J. and Lindberg, S. E., Dry deposition of reactive nitrogen compounds: a review of leaf, canopy and non-foliar measurements, *Atmos. Environ.*, 25A, 1615, 1991.
108. Erisman, J. W., Vermetten, A. W. M., Asman, W.A. H., Waijers-Ypelaan, A., and Slanina, J., Vertical distribution of gases and aerosols: the behaviour of ammonia and related components in the lower atmosphere, *Atmos. Environ.*, 22, 1153, 1988.
109. Harrison, R. M., Rapsomanikis, S., and Turnbull, A., Land surface exchange in a chemically reactive system: surface fluxes of HNO_2, HCl and NH_3, *Atmos. Environ.*, 23, 1795, 1989.
110. Meixner, F. X., Franken, H. H., Duijzer, J. H., and Aalst, R. M. van, Dry deposition of gaseous HNO_3 to a pine forest, in *Air Pollution Modeling and Its Application*, Vol. VI, Van Dop, H., Ed., Plenum, New York, 1988, 23.
111. Dollard, G. J., Jones, B. M. R., and Davies, T. J., Dry deposition of HNO_3 and PAN, in *Field Measurements and Interpretation of Species Derived from NO_x, NH_3 and VOC in Europe*, Beilke, S., Millan, M., and Angeletti, G., Eds., Air Pollution Report 25, Commission of the European Communities, Directorate-General for Science, Research and Development, Brussels, 1990, 245.
112. Dlugi, R., Brunnemann, G., Kins, L., Köhler, E., Müller, D., Roider, G., Reußwig, K., Schween, J., Siedl, W., Seiler, Th., and Zelger, M., Charakterisierung der Chemie und Physik des anthropogenen Aerosols in der belasteten Atmosphäre, SANA-Bericht, Meteorologisches Institut der Universität München; Arbeitsgruppe für Atmosphärische Strahlung und Satellitenmeteorologie, 1993.

113. Janson, R. and Granat, L., Dry deposition of HNO_3 to the coniferous forest, in *Proceedings of the EUROTRAC Symposium 96*, Borrell, P. M., Borrell, P., Cvitãs, T., Kelly, K., and Seiler, W., Eds., Computational Mechanics Publications, Southampton, 1996, 351.
114. Müller, H., Framm, G., Meixner, F., Dollard, G. J., Fowler, D., and Possanzini, M., Determination of HNO_3 deposition by modified Bowen ratio and aerodynamic profile techniques, *Tellus*, 45B, 346, 1993.
115. Fowler, D., Duyzer, J. H., and Baldocchi, D. D., Inputs of trace gases, particles and cloud droplets to terrestrial surfaces, *Proc. R. Soc. Edinburgh*, 97B, 35, 1991.
116. Zimmerling, R., Dämmgen, U., Küsters, A., and Max, W., Flußdichten, von nichtsedimentierenden Luftinhaltsstoffen. II. S- und N-Spezies mit Denuder-Filter-Anordnungen, in *Untersuchungen zum chemischen Klima in Südostniedersachsen*, Dämmgen, U., Ed., Ladbauforschung Völkenrode, special issue 170, 170, 1996.
117. Meixner, F. X., Surface exchange of odd nitrogen oxides, *Nova Acta Leopold.*, 70(288), 299, 1994.
118. Kramm, G., Müller, H., Fowler, D., Höfken, K. D., Meixner, F. X., and Schaller, E., A modified profile method for determining the vertical fluxes of NO, NO_2, ozone, and HNO_3 in the atmospheric surface layer, *J. Atmos. Chem., 13, 265, 1991.*
119. Wesely, M. L. and Lesht, M. L., Comparison of RADM dry deposition algorithms with a site-specific method for inferring dry deposition, *Water Air Soil Pollut.,* 44, 273, 1989.
120. Brook, J. R., Sirois, A., and Clarke, J. F., Comparison of dry deposition velocities for SO_2, HNO_3 and SO_4^{2-} estimated with two inferential models, *Water Air Soil Pollut.*, 87, 205, 1996.
121. Huebert, B. J., Luke, W. T., Delany, A. C., and Brost, R. A., Measurements of concentrations and dry surface fluxes of atmospheric nitrates in the presence of ammonia, *J. Geophys. Res.*, 93D, 7127, 1988.
122. Brost, R. A., Delany, A. C., and Hubert, B. J., Numerical modelling of concentrations and fluxes of HNO_3, NH_3 and NH_4NO_3 near the surface, *J. Geophys. Res.*, 93(D3), 7137, 1988.
123. Pandis, S. N. and Seinfeld, J. J., On the interaction between equilibration processes and wet or dry deposition, *Atmos. Environ.*, 24A, 2313, 1990.
124. Kramm, G. and Dlugi, R., Modelling of the vertical fluxes of nitric acid, ammonia, and ammonium nitrate, *J. Atmos. Chem.*, 18, 319, 1994.
125. Nemitz, E., Sutton, M., Fowler, D., and Choularton, T., Application of a NH_3 gas-to-particle conversion model to measurement data, in *Atmospheric Ammonia: Emission Deposition and Environmental Impacts—Poster Proceedings*, Sutton, M. A., Lee, D. S., Dollard, G. J., and Fowler, D., Eds., ITE, Edinburgh, 1996, 98.
126. Benner, C. L., Eatough, D. J., Eatough, N. L., and Bhradwaja, P., Comparison of annular denuder and filter pack collection of HNO_3 (g), HNO_2 (g), SO_2 (g), and particulate-phase nitrate, nitrite and sulfate in the South-West Desert, *Atmos. Environ.*, 25A, 1537, 1991.
127. Bytnerowicz, A. and Riechers, G., Nitrogeneous air pollutants in a mixed conifer stand of the western Sierra Nevada, California, *Atmos. Environ.*, 29, 1369, 1995.
128. Li, S.-M., Equilibrium of particle nitrate with gas phase HONO: tropospheric measurements in the high Arctic during polar sunrise, *J. Geophys. Res.*, 99(Dl), 25469, 1994.
129. Krovetz, D. O., Sigmon, J. R., Reiter, M. A., and Lessard, L. H., An automated system for air sampling with annular denuder systems at a remote site, *Environ. Pollut.*, 58, 97, 1989.
130. Keuken, M., Mallant, R. K. A.M., Woittiez, J. R. W., and Slanina, J., Verzuringsonderzoek bij ECN, Report ECN-PB-89-7, ECN, Petten, the Netherlands, 1989.
131. Lee, H. S., Wadden, R. A., and Scheff, P. A., Measurement and evaluation of acid air pollutants in Chicago using an annular denuder system, *Atmos. Environ.*, 27A, 543, 1993.
132. Dasch, J. M., Cadle, S. H., Kennedy, K. G., and Mulawa, P. A., Comparison of annular denuders and filter packs for atmospheric sampling, *Atmos. Environ.*, 23, 2775, 1989.
133. De Santis, F., Febo, A., Perrino, C., Possanzini, M., and Liberti, A., simultaneous measurement of nitric acid, nitrous acid, hydrogen chloride and sulfur dioxide in air by means of high-efficiency annular denuders, in *Advancement in Air Pollution Monitoring Equipment and Procedures*, proceedings of the meeting held in Freiburg/Breisgau, Germany, June 2 to 6, 1985.
134. Allegrini, I., de Santis, F., di Palo, V., Febo, A., Perrino, C., Possanzini, M., and Liberti, A., Annular denuder method for sampling reactive gases and aerosols in the atmosphere, *Sci. Total Environ.*, 67, 1, 1987.
135. Buttini, P., de Palo, V., and Possanzini, M., Coupling of denuder and ion chromatographic techniques for NO_2 trace level determination in air, *Sci. Total Environ.*, 61, 59, 1987.

136. Biermann, J. W., Tauzon, E. C., Winer, A. M., Wallington, T. J., and Pitts, J. N., Jr., Simultaneous absolute measurements of gaseous nitrogen species in urban ambient air by long pathlength infrared and ultraviolet-visible spectroscopy, *Atmos. Environ.*, 22, 1545, 1988.
137. Oms, M. T., Jongjean, P. A. C., Veltkamp, A. C., Wyers, G. P., and Slanina, J., Continuous monitoring of atmospheric HCl, HNO_2, HNO_3 and SO_2 by wet-annular denuder air sampling with on-line chromatographic analysis, *Int. J. Environ. Anal. Chem.*, 62, 207, 1996.
139. Vossler, T. L., Stevens, R. K., Paur, R. J., Baumgardner, R. D., and Bell, J. P., Evaluation of improved inlets and annular denuders to measure inorganic air pollutants, *Atmos. Environ.,* 22, 1729, 1988.
140. Allen, A. G., Dick, A. L., and Davison, B. M., Sources of atmospheric methanesulphonate, non-sea-salt sulphate, nitrate and relate species over the temperate South Pacific, *Atmos. Environ.*, 31, 191, 1997.
141. Ridley, B. A., Recent measurements of oxidized nitrogen compounds in the troposphere, *Atmos. Environ.*, 25A 1905, 1991.
142. Pio, C. A., Cerqueira, M. A., Castro, L. M., and Salgueiro, M. L., Sulphur and nitrogen compounds in variable marine/continental air masses at the southwest European coast, *Atmos. Environ.*, 30, 3115, 1996.
143. Bottenheim, J. W., Barrie, L. A., and Atlas, R., The partitioning of nitrogen oxides in the lower Arctic troposphere during spring 1988, *J. Atmos. Chem.*, 17, 15, 1993.
144. Klemm, O., Talbot, R. W., Fitzgerald, D. R., Klemm, K. I., and Lefer, B. L., Low to middle tropospheric profiles and biosphere/troposphere fluxes of acidic gases in the summertime Canadian taiga, *J. Geophys. Res.*, 99(D1), 1687, 1994.
145. Brook, J. R., Di-Giovanni, F., Cakmak, F., and Meyers, T. P., Estimation of dry deposition velocity using inferential models and site-specific meteorology — uncertainty due to siting of meteorological towers, *Atmos. Environ.*, 31, 3911, 1997.
146. Geigert, M. E., Nicolaides, N. P., Miller, D.R., and Heitert, J., Deposition rates for sulfur and nitrogen to a hardwood forest in northern Connecticut, USA, *Atmos. Environ.*, 28, 1689, 1994.
147. Cadle, S. H. and Mulawa, P. A., Atmospheric summertime concentrations and estimated dry deposition rates of nitrogen and sulfur species at a smoky mountain site in North Carolina, *Proc. 81st Annual Meeting of APCA*, Dallas, TX, June 19–24, 88-119.3, 1988.
148. Lovett, G. M. and Lindberg, S. E., Dry deposition of nitrate to a deciduous forest, *Biogeochemistry*, 2, 137, 1986.
149. Brunnemann, G., Kins, L., and Dlugi, R., Physical and chemical characterization of the atmospheric aerosol: an overview of the measurements of the SANA 2 campaign at Melpitz, *Meteorol. Z.*, NF, 5, 245, 1996.
150. Sievering, H., Ender, G., Kins, L., Kramm, G., Ruoss, K., Roider, K., Roider, G., Zelger, M., Anderson, L., and Dlugi, R., Nitric acid, particulate nitrate and ammonium profiles at the Bayrische Wald: evidence of large deposition rates of total nitrate, *Atmos. Environ.*, 28, 311, 1994.
151. Anderson, H. V. and Hovmand, M. F., Ammonia and nitric acid dry deposition and throughfall, *Water Air Soil Pollut.*, 85, 2211, 1995.
152. Peake, E., MacLean, M. A, and Sandhu, H. S., Total inorganic nitrate (particulate nitrate and nitric acid) in the atmosphere of Edmonton, Alberta, Canada, *Atmos. Environ.*, 22, 2891, 1988.
153. Harrison, R. M. and Allen, A. G., Scavenging ratios and deposition of sulphur, nitrogen and chlorine species in eastern England, *Atmos. Environ.*, 25A, 1719, 1991.
154. Meszaros, E. and Horvath, L., Concentration and dry deposition of atmospheric sulfur and nitrogen compounds in Hungary, *Atmos. Environ.* 18, 1725, 1984.
155. Meyers, T. P. and Hicks, B. B., Dry deposition of O_3, SO_2 and HNO_3 to different vegetation in the same exposure environment, *Environ. Pollut.*, 53, 13, 1988.
156. Meyers, T. P., Huebert, B. J., and Hicks, B. B., HNO_3 deposition to a deciduous forest, *Boundary-Layer Meteorol.,* 49, 395, 1989.
157. Dasch, J. M., Dry deposition of sulfur dioxide or nitric acid to oak, elm and pine leaves, *Environ. Pollut.*, 59, 1, 1989.
158. Peters, K. and Bruckner-Schatt, G., The dry deposition of gaseous and particulate nitrogen compounds to a spruce stand, *Water Air Soil Pollut.*, 85, 2217, 1995.

CHAPTER 15

Effects of Sulfur Dioxide and Acid Deposition on Chinese Crops

Cao Hongfa, Jianmin Shu, Yingwa Shen, Yingxin Gao, Jixi Gao, and Linbo Zhang

CONTENTS

15.1 INTRODUCTION

Since the 1980s, China's economy has been developing rapidly. With increasing coal consumption, the emission of acidic gases has greatly increased. The area affected by acid rain has continually extended in southern China, until about 40% of the area in the country is now affected by acid rain. The whole region of southern China other than the Yangzi River is the heavily affected by acid rain, including Sichuan, Guizhou, Guangdong, Guangxi, Hunan, and Zhejiang Provinces. There are two areas with a high-concentration of sulfur dioxide (SO_2) in northern and southern China. In the urban areas of Guiyang, Taiyuan, Beijing, and Urumchi, SO_2 concentration in air is as high as 0.1 ppm in winter.

1-56670-341-7/00/$0.00+$.50

The damage by acid deposition to plants is one of the serious environmental problems confronting China.[1,2] This chapter reports the results of various studies conducted to investigate the effects of acid deposition/acid gas SO_2 on growth and yield of vegetables, cash crops, and cereal crops.

The vegetables, cash crops, and cereal crops, that are commonly cultivated in the area severely impacted by acid deposition were chosen as plant materials. Quantity and frequency of acid rain treatments were based on the trend of natural deposition in Zhejiang Province, which has suffered from serious acid rain problems. There were five acidity treatments of simulated acid rain (SAR), having pH values of 5.6, 4.6, 4.2 or 4.0, 3.6 or 3.4, and 2.8. Each treatment in which the ratio of SO_4^{2-} to NO_3^- was 8.32:1.68 contained the main ions of natural rain, such as Ca^{2+}, Mg^{2+}, K^+, Na^+, Mn^{2+}, and Cl^-. A stock solution for SAR was prepared and then diluted into solutions of different pH by adding double-distilled water just before the treatment. A continuous stirred tank reactor (CSTR) test system with open-top chambers of 3-m diameter and 2.4-m height was used for SO_2 fumigation. Pure gas of SO_2 was diluted through a dynamic gas-distributing facility (Model DP-3, manufactured by Institute of Environmental Chemistry of Academia Sinica), then pumped into the CSTR. The concentration of SO_2 varied less than 10% in the CSTR.[3-5] The experiments for the interactive effect of acid deposition and SO_2 were based on the method of orthogonal test in CSTR.[5]

15.2 VISIBLE DAMAGE OF SO_2 AND ACID DEPOSITION SINGLY AND IN COMBINATION ON AGRICULTURAL CROPS

The crops species studied included wheat (*Triticum aestivum*), maize (*Zea mays*), barley (*Hordeum vulgare*), rice (*Oryza sativa*), soybean (*Glycine max*), cotton (*Gossypium hirsutum*), bean (*Phaseolus vulgaris*), rape (*Brassica pekinensis* var. *oleifera*), cucumber (*Cucumis sativus*), tomato (*Lycopersion esculentum*), Chinese cabbage (*Brassica pekinensis*), and spinach (*Spinacia oleracea*). Crops were sprayed with SAR treatment of 15 mm for 10 min by artificial shower every day. Each treatment was repeated 10 times. The seedlings were covered by plastic sheets to protect them from the interference of natural rain.

SAR treatment did not cause any visible injury symptoms initially. But injury symptoms gradually appeared and were aggravated with time. For sensitive crops, injury lesions became visible within a few hours after spraying with high-acidity SAR, whereas for most other crops the injury symptoms appeared only after 24 h. The visible symptoms became steady within 48 to 72 h after spraying. The acute injury symptoms appeared as small white sear spots, mainly concentrated near veins. These were different from "patchlike" symptoms caused by SO_2 acute injury. At very low pH, injury symptoms were yellowish or white strips, and most of the mesophyll tissue was damaged. Soybean and Chinese cabbage also showed curled leaf tip and withered margin along with visible lesions.

The threshold of visible injury on leaves was determined by 5% injured leaf area. Injury thresholds for SAR varied among the crops. It was 3.0 pH for wheat, maize, barley, and spinach; 2.5 pH for Chinese cabbage, soybean, bean, rape, cucumber, tomato and cotton; and 2.0 for rice.

The sensitivity of crops to SO_2 was also influenced by various factors, such as development stage, leaf age, season, and photoperiod. A vast range of sensitivity to SO_2 was found between different crop species and their cultivars.[5] Based on CSTR experiments, the relative sensitivity of seedlings exposed to SO_2 was categorized into three types: sensitive, midsensitive or intermediate, and tolerant. Barley, cotton, soybean, cucumber, spinach, carrot, and bush red pepper were sensitive species; wheat, bean ,and rape were mid-sensitive; rice, maize, and potato were tolerant. The dose–response of five crops to SO_2 were studied and results were shown in Table 15.1. Dose–response refers to the ratio of responding individuals with symptoms to the total population tested under a certain exposure dose. Exposure dose is the product of SO_2 concentration and time exposed.

Table 15.1 Dose–Response of Selected Crop Species to SO_2

Crop Species	SO_2 (ppm)	Visible Leaf Injury (%)				
		1 h	2 h	4 h	6 h	8 h
Wheat	0.25	—	—	<5	<5	<5
	0.35	—	<5	<5	<5	5
	0.59	<5	<5	5	8	10
Barley	0.51	—	<5	<5	<5	5
	0.98	<5	<5	5	10	10
Soybean	0.40	—	—	—	<5	<5
	0.61	<5	<5	<5	<5	8
Bean	0.21	—	—	<5	<5	<5
	0.41	<5	<5	<5	5	8
Cotton	0.25	—	—	<5	<5	<5
	0.50	—	—	<5	5	6

The percentage visible injury on leaves increased with increasing SO_2 dose. However, the contribution of SO_2 concentration or exposure time to leaf injury was not the same. The results suggest that the concentration of SO_2 affected the visible injury percentage more seriously than exposure time ($p < 0.01$). It was most significant for cotton among the crops investigated. The relative sensitivity of the crops to SO_2 was classified according to their acute injury threshold values, based on the dose of SO_2 where 5% leaf injury appeared (Table 15.2). The relative sensitivity of crop species to combined effects of SO_2 and SAR treatment showed bean to be the most sensitive; tomato, soybean, and rape to be midsensitive; and rice, wheat, and cotton to be tolerant.

Table 15.2 Acute Injury Threshold Values of Crops to SO_2 under Sensitive Conditions

Exposure Period (h)	SO_2 Concentrations at 5% Visible Injury (ppb)		
	Sensitive	Intermediate	Tolerant
1.0	2.88–3.43	4.95–5.78	≥11.0
2.0	1.86–3.12	3.89–4.69	≥10.0
4.0	1.67–2.29	1.69–2.86	≥ 8.5
6.0	1.17–2.00	1.48–1.86	≥ 7.0
8.0	0.88–1.43	0.85–1.45	≥ 5.7

15.3 PHYSIOLOGICAL RESPONSES OF AGRICULTURAL CROPS TO SO_2 AND ACID DEPOSITION, SINGLY AND IN COMBINATION

15.3.1 Effects on Chlorophyll Content

The results are shown in Table 15.3. For most crops, the chlorophyll content declined with SAR treatment. With the decrease in pH of acid rain, the decline in chlorophyll content increased. Among the crops studied, tomato was found to be the most sensitive. For the vegetable and cash crops, the ratio of chlorophyll *a*/*b* was higher than the control, but for other crops the ratio did not change significantly. The effects of acid rain on the composition of chlorophyll depended on the type of crops.

The chlorophyll content in monocotyledonous plants increased when treated with SO_2 (Table 15.4). But for the dicotyledonous plants and cash crops, the chlorophyll content decreased when treated with SO_2. The effects of SO_2 on chlorophyll *b* was more deleterious than on chlorophyll *a*.

Table 15.3 Effects of SAR on Chlorophyll Content (mg g^{-1}) of Different Crop Plants

Crops	Parameters	Control	SAR pH			
		5.6	4.6	4.1	3.6	2.8
Wheat	Chlorophyll content	1.49	1.48	1.32	1.30	1.14
	Ratio of chl. *a*/*b*	0.71	0.72	0.71	0.71	0.71
Rice	Chlorophyll content	1.93	—	2.03	2.04	2.00
	Ratio of chl. *a*/*b*	0.79	—	0.78	0.79	0.78
Rape	Chlorophyll content	1.28	1.10	1.18	1.31	1.32
	Ratio of chl. *a*/*b*	3.12	3.35	3.24	3.12	3.21
Tomato	Chlorophyll content	3.05	2.66	2.17	2.25	1.80
Soybean	Chlorophyll content	4.01	3.95	3.79	3.76	3.62
	Ratio of chl. *a*/*b*	2.11	2.21	2.35	2.36	2.29
Cotton	Chlorophyll content	1.97	1.94	1.90	1.73	1.63

Table 15.4 Effects of SO_2 Treatment on Chlorophyll Content (mg g^{-1}) of Selected Crop Plants

Crops	Items	Control	SO_2 Concentration (ppm)			
			0.05	0.1	0.15	0.25
Wheat	Chlorophyll content	0.92	1.99	1.02	1.48	1.38
Rape	Chlorophyll content	1.28	0.96	1.17	1.16	1.02
	Ratio of chl. *a*/*b*	3.12	3.23	3.34	3.33	3.36
Tomato	Chlorophyll content	3.05	2.06	2.52	2.03	1.87
Soybean	Chlorophyll content	4.01	3.94	3.53	3.45	3.16
	Ratio of chl. *a*/*b*	2.11	2.23	2.80	2.59	2.43
Cotton	Chlorophyll content	1.97	1.56	1.59	1.55	1.05

Total chlorophyll content declined with increase of SO_2 concentration and acidity of SAR. The effect of acid rain and SO_2 in combination on chlorophyll was a synergistic response; i.e., the effect of acid rain and SO_2 in combination (a) was greater than the effect of acid rain (b) added to the effect of SO_2 (c). Further analysis showed that SO_2 contributed more than acid rain in causing adverse effects on chlorophyll (Table 15.5).

Table 15.5 Effects of SAR and SO_2 Treatment in Combination on Chlorophyll Content of Cotton Leaves (mg g^{-1})

SO_2 (ppm)	SAR pH				
	5.6	4.6	4.1	3.6	2.8
0	1.97	1.94	1.90	1.73	1.62
0.5	1.56	1.57	1.34	1.30	0.87
0.1	1.59	1.41	1.29	1.25	0.82
0.15	1.55	1.56	1.31	0.92	0.64
0.25	1.05	1.04	1.01	0.93	0.48

15.3.2 Effects on Cell Membrane Permeability

The cell membrane permeability of plant leaves increased when treated with SAR (Table 15.6). The increases of K^+ amount (Y) has a power function relation to the pH value (X):

Tomato: $Y = 10.2477 - 0.7043X$ $(r = -0.973)$
Carrot: $Y = 5.8827 - 0.8132X$ $(r = -0.998)$
Cotton: $Y = 16.9122 - 1.5993X$ $(r = -0.901)$

Table 15.6 Effects of SAR on Cell Membrane Permeability, Amino Acid Content, and Leaf Extract pH

Characteristics/Plants	SAR pH				
	5.6	4.6	4.1	3.6	2.8
Cell Membrane Permeability (%)					
Tomato	6.48	6.87	7.13	7.78	8.40
Carrot	1.36	2.14	2.53	2.88	3.67
Cotton	8.70	8.20	13.80	11.60	12.60
Amino Acid (mg g^{-1})					
Wheat	4.17	4.99	4.32	10.21	5.54
Leaf Extract pH					
Pinto bean	6.38	6.23	6.23	6.08	6.00

Cell membrane permeability increased when concentration of SO_2 was less than 0.1 ppm (Table 15.7). When concentration of SO_2 was higher than 0.1 ppm, the leaves of tomato were injured seriously, but the flux ratio of K^+ was decreased. The increase of K^+ amount (Y) has a power function relation to the SO_2 (X):

Tomato: $Y = 1.2842 + 0.0064X \quad (r = 0.984)$
Cotton: $Y = 7.6838 + 0.0378X \quad (r = 0.898)$

Table 15.7 Effects of SO_2 on Cell Membrane Permeability, Amino Acid Content, and Leaf Extract pH

Characteristics/Plants	SO_2 (ppm)				
	0	0.5	0.1	0.15	0.25
Cell Membrane Permeability (%)					
Tomato	6.48	6.60	8.84	6.46	6.29
Carrot	1.36	1.42	2.00	2.31	2.86
Cotton	8.70	10.60	9.90	11.10	18.90
Amino Acid (mg g^{-1})					
Wheat	4.04	8.19	4.21	5.94	5.17
Leaf Extract pH					
Pinto bean	6.38	6.28	6.23	6.21	6.11

Similar to the effects of acid rain and SO_2 alone, the effects of these in combination on cell membrane permeability were found to depend on exposure intensity. With increase in exposure intensity, the membrane permeability increased. The combined effects were synergistic, and SO_2 had more stronger effects compared with acid rain.

15.3.3 Effects on Amino Acid

Amino acid content was measured in wheat leaves (see Tables 15.6 and 15.7). The amount of amino acid in leaves increased to pH 3.6, but declined at pH 2.8. Amino acid content in wheat leaves decreased initially due to SO_2 exposure (Table 15.7). But after prolonged exposure it increased as compared with the control.

15.3.4 Effects on Peroxidase Activity

The effects of acid rain on peroxidase activity of two crop plants were studied. Peroxidase activity increased sharply in response to SAR in plants with high basic peroxidase activity, while in the crops with lower basic peroxidase activity first declined and then an increment was observed after prolonged exposure. Effects of SO_2 on peroxidase activity varied among plants. For plants with high basal activity, the response of peroxidase activity was stronger. The response was not obvious in plants with low basic peroxidase activity. This response was similar to that of the effects of acid rain on peroxidase activity. Acid rain and SO_2 in combination decreased the peroxidase activity, which suggests an antagonistic response.

15.3.5 Effects on Sugar Content

The sugar content in rape decreased when treated with SAR, but, at pH 2.8, the sugar content was higher than with other treatments. The sugar content increased with increasing SO_2 concentration up to 0.15 ppm, but, when SO_2 concentration increased to 0.25 ppm, the sugar content declined in rape plants. In the case of soybean, SO_2 did not significantly affect sugar content.

15.3.6 Effects on Leaf Extract pH

The pH value of soybean leaf extract decreased as the acidity of acid rain (see Table 15.6) and SO_2 concentration increased (see Table 15.7) when treated individually. The leaf extract pH declined when plants were exposed to acid rain and SO_2 in combination. At pH > 3.4 and SO_2 < 0.1 ppm, the interactive effect was antagonistic or additive, but at pH ≤ 3.4 and SO_2 ≥ 0.1 ppm, the combination effect was synergistic. Further analysis showed that acid rain is the main factor affecting the pH of the leaf.

15.4 EFFECTS OF SO_2 AND ACID DEPOSITION, SINGLY AND IN COMBINATION, ON GROWTH OF CROP SPECIES

There are two ways, direct and indirect, by which acid rain exerts its influence on crops. Acid rain may directly affect the assimilation capacity and productivity of crops through contact with assimilate organs and reproductive organs of crops, or the growth and productivity of crops may be indirectly impacted by reduction of soil fertility.[6-8] Because the growth stage of crops, which usually lasts months, is comparatively short, the indirect effect is neglected when the effect of acid rain on crop growth is studied.

The experiments with crop species showed diverse effects of SAR on growth of crops (Table 15.8). The biomass of 10 crop seedlings, out of 12, including rape, cabbage, carrot, pinto bean, tomato, soybean, cotton, wheat, and maize, reduced with increase in acidity of SAR. SAR had a more significant impact on the biomass of vegetables than that of cereal crops and cash crops. SAR at pH 4.1 and 3.5 led to reduction in cabbage biomass by 17 and 19%, respectively, while at pH 3.6 the biomass of vegetables, such as rape, pinto bean, carrot, and tomato, decreased between 11% to 19%. At the same treatment, however, the biomass of cash crops decreased by less than 10%, and the biomass of cereal crops decreased between 6 and 21%.

The difference in magnitude of reduction in biomass of various crops indicates different sensitivities of crops to SAR. In terms of biomass reductions of crops at pH 4.1, cabbage is the most sensitive crop; tomato, carrot, maize, and cotton belong to mid-sensitive category; and rice and wheat as tolerant crops.

It is necessary to determine the injury threshold value of SAR effects on growth and yield for assessment of standards for environmental management. According to the commonly accepted

Table 15.8 Effects of SAR on Biomass (g plant^{-1}) of Different Crop Species

Crops		pH of SAR Treatment				
		5.6	4.6	4.1	3.6	2.8
Vegetables	Rape	1.36	0.64	0.60	0.63	0.55
	Cabbage	2.95	2.96	2.45	2.41	2.16
	Pinto bean	44.07	39.61	35.13	31.91	26.57
	Carrot	33.2	35.6	31.6	28.8	28.6
	Tomato	12.2	13.6	11.4	10.8	9.6
Cash crops	Peanut	38.5	37.9	41.2	38.4	38.6
	Soybean	0.82	0.83	0.82	0.79	0.72
	Cotton	5.07	4.90	4.78	4.55	4.50
Cereal crops	Barley	2.30	2.25	2.01	1.79	1.89
	Wheat	3.1	3.2	3.3	2.9	2.8
	Rice	4.95	4.85	5.12	5.05	4.89
	Maize	176.4	169.2	167.4	152.6	144.5

visible threshold value of atmospheric pollutant, acidity level may be taken as the growth threshold value of SAR at which the growth and yield reduced by 5%. The biomass damage threshold value for SAR treatment was calculated as pH 5.24 for pinto bean, 5.04 for carrot, 5.01 for rape and cabbage, 4.49 for soybean, 4.44 for barley, 4.36 for tomato and cotton, 4.10 for maize, and less than 2.8 for rice and peanut.

Nine species of crops, including rape, soybean, tomato, carrot, pinto bean, cotton, wheat, barley, and rice, were exposed to SO_2 at five levels, i.e., 0, 0.05, 0.1, 0.15, and 0.25 ppm, for their whole growth duration. Among all crops, vegetable crops were most sensitive to SO_2, followed by cash crops and then cereal crops (Table 15.9). When treated with SO_2 at 0.05 or 0.25 ppm, biomass of pinto bean reduced by 22 and 54%, that of tomato by 20 and 45%, respectively. Sulfur dioxide at low concentration benefited the growth of crops for a short time initially. After 2-week treatment of SO_2, the growth of barley seedling accelerated, and relative growth rate (RGR) increased. As sulfur is one of the essential elements for plant growth, a small amount of it in air may be a supplement resource of sulfur to promote plant growth. But the plants were injured at higher concentration when sulfur accumulation in plants exceeded the demand.

Table 15.9 Effects of Different Concentration of SO_2 on Biomass (g plant^{-1}) of Selected Crop Species

Crops	SO_2 Concentration (ppm)				
	Control	0.05	0.1	0.15	0.25
Pinto bean	12.27	9.52	6.86	6.34	5.65
Rape	1.36	0.44	0.59	0.56	0.44
Tomato	12.2	9.8	10.5	8.9	6.7
Carrot	8.4	8.1	7.5	6.4	5.5
Soybean	28.9	24.5	23.7	19.6	18.2
Cotton	50.7	40.3	44.6	28.7	31.0
Barley	2.30	2.42	1.27	1.99	1.96
Wheat	35.11	34.21	32.05	30.47	29.14

The growth threshold value of SO_2 is defined as the concentration at which the biomass of crop treated with SO_2 decreased by 5%.[9,10] To establish crop growth damage threshold values for SO_2, equations of linear regression between biomass of crops and concentration of SO_2 were calculated. From the calculated figures, it is clear that the growth threshold value of SO_2 varied among different crops, as it was lower for vegetables than for cash crops and cereal crops. Biomass threshold values of SO_2 were 0.072, 0.093, 0.098, 0.133, 0.140, 0.174, 0.186, and 0.199 mg m^{-3} for pinto bean, carrot, rape, tomato, barley, soybean, wheat, and cotton, respectively.

Among all levels of treatment, combined treatments of SAR and SO_2 led to increase in height, leaf area, and aboveground biomass of cereal crops, but the underground biomass reduced. In combined treatment, in general, detrimental effects exerted by SO_2 on biomass was ameliorated (Table 15.10). When concentration of SO_2 was lower than 0.1 ppm, biomass of tomato treated by SAR and SO_2 in combination was larger than corresponding controls. When the concentration of SO_2 was between 0.1 and 0.25 ppm, the biomass of tomato treated at higher acidity of SAR was lower than corresponding control, but higher than that of individual treatment of SO_2.

Table 15.10 Interactive Effects of SAR and SO_2 on Biomass of Different Crop Species (g $plant^{-1}$)

Crop Species	SAR pH + SO_2 Concentration (ppm)				
	5.6 + 0	4.6 + 0.05	4.1 + 0.1	3.6 + 0.15	2.8 + 0.25
Barley	33.56	43.12	33.64	37.22	30.50
Soybean	44.40	40.80	39.84	35.64	32.42
Cotton	50.72	33.00	25.50	16.50	12.30
Tomato	12.2	12.5	10.9	9.6	7.6
Pinto bean	5.09	7.54	5.84	3.58	2.59
Rape	2.72	1.89	1.76	2.13	1.80

Further analysis of soybean leaf area indicated that interactive effect (a) is lower than the sum (b + c) of single effects when concentration of SO_2 was less than 0.15 ppm. When concentration of SO_2 was equal to or larger than 0.15 ppm, there was a relation of a > (b + c) which means additive interactive effect of SAR and SO_2. Biomass reduction of soybean and cotton treated by SAR and SO_2 was less than the sum of the corresponding single treatments which suggests that interactive effect on cash crops was antagonistic. The effect of SAR and SO_2 was maximum on cash crops, followed by vegetables and then cereal crops.

15.5 EFFECTS OF SO_2 AND ACID DEPOSITION, SINGLY AND IN COMBINATION, ON YIELD AND QUALITY OF AGRICULTURAL CROPS

The experiments indicated that SAR also affected the crop quality to some degree. It significantly decreased the quantity of vitamin C in tomato, while increased the quantities of deoxidized sugar, total sugar and acid numbers. The protein content of soybean seeds decreased from 2.6 to 11.8% at increasing acidity. A significant positive correlation (r = 0.87) was observed between increase in acidity and decrease in protein content.

The analysis of yield parameters showed that SAR mainly affected the hundred grain weight (HGW) of soybean and thousand grain weight (TGW) of wheat. Among various pH treatments, there was no significant difference in productive pod numbers and grain numbers per plant. But the HGW decreased as the pH value of the acid rain decreased. Compared with the productive spike number and the grain number per spike, grain weight in wheat was found more sensitive to SAR. The TGW decreased linearly as the hydrogen ion concentration in the acid rain increased ($r = 0.86$, $P < 0.05$). There was a linear correlation between the percentage decrease in wheat yield and the percentage decrease in TGW ($r = 0.90$). At pH 2.5, not only was the grain weight affected negatively, but also the grain number per spike was affected significantly.

Crop yield is the end point indicator of the integrated effects of SAR on crops. It is the most important indicator of the effect of acid rain on agricultural yield and economic loss. Vegetables were the most sensitive to acid rain, followed by cash crops and then cereal crops (Table 15.11). At a treatment of pH 4.6, yield reductions in vegetables ranged from 5 to 12%. The yield of soybean and cotton did not decrease until a treatment of pH 4.1, while wheat and barley did not show reductions in yield until pH 3.6. Yield deduction threshold (YDT) of 5% decrease in crop yield subjected to SAR showed linear regression equations for different test crops (Table 15.12).

Table 15.11 Yield Response (g plant^{-1}) of Various Crop Species to SAR

Crops		SAR pH				
		5.6	4.6	4.1	3.6	2.8
Vegetables	Rape	31.89	29.33	28.48	26.09	24.50
	Pinto bean	44.07	39.61	35.13	31.91	36.57
	Carrot	112.1	98.6	90.4	88.1	85.8
	Tomato	94.7	90.4	88.1	84.5	85.0
Cash crops	Soybean	41.00	39.88	39.20	38.54	37.06
	Cotton	30.94	30.14	29.66	28.12	27.66
Cereal crops	Barley	35.69	35.20	33.12	33.04	31.53
	Wheat	23.04	23.30	22.37	21.86	19.74
	Rice	26.20	27.24	27.01	27.12	26.90

Table 15.12 Yield Deduction Threshold of Crops to SAR and SO_2

Crop	SAR pH	SO_2 Concentration (ppb)
Tomato	4.36	46.9
Cotton	4.37	69.7
Barley	4.44	49.0
Soybean	4.49	60.9
Wheat	4.59	65.1
Rape	5.01	34.3
Carrot	5.04	32.6
Pinto bean	5.25	25.2

The yield of crops also declined due to SO_2 exposure and the decline increased as SO_2 concentration increased (Table 15.13). Among the nine test crops, at as low as 0.05 ppm of SO_2 the yield of rape, pinto bean, tomato, carrot, soybean, cotton, and barley decreased by 11, 15, 14, 15, 4, 4, and 11%, respectively, while the yield of wheat and rice did not decrease. In terms of yield reductions, vegetables were most sensitive to SO_2, cash crops being the second, and cereal crops seemed relatively insensitive. Pinto bean was the most sensitive to SO_2 among vegetable crops. At 0.05 ppm SO_2 treatment, its yield decreased by 15%. Barley among cereal crops was also sensitive and its yield decreased by 11% at 0.05 ppm SO_2. The YDTs obtained at 5% reduction in yield of tested crops due to SO_2 exposure are listed in Table 15.12.

The intensity of the effects of SO_2 on yield components was higher for pod number per plant, length per pod, and weight per pod. These were significantly correlated to SO_2 concentration (r = 0.99, 0.98, and 0.85, respectively, $P < 0.01$). Soybean was sensitive to SO_2 with respect to pod number per plant and HGW. The correlation coefficients were -0.978 ($P < 0.01$) and -0.795 ($P < 0.05$), respectively. Among the yield component parameters, yield was most sensitive to grain number per spike ($r = 0.95$, $P < 0.05$). It was implied that main effect of SO_2 on yield of leguminous plants was to reduce the legume pod number. It was considered that SO_2 might affect the procreation procedure of plants. There were similar results in some other papers.[12-14]

Table 15.13 Yield Responses (g plant^{-1}) of Various Crop Species to SO_2

Crops		SO_2 Concentration (ppm)				
		Control	0.05	0.1	0.15	0.25
Vegetables	Rape	31.89	28.52	26.22	23.17	20.44
	Pinto bean	44.07	33.15	30.33	25.25	22.38
	Tomato	94.7	87.8	82.0	73.9	70.6
	Carrot	112.1	95.9	82.0	78.8	69.5
Cash crops	Soybean	41.00	39.28	37.64	34.80	32.92
	Cotton	30.94	29.42	28.16	26.94	25.42
Cereal crops	Barley	35.69	31.92	29.68	27.25	26.83
	Wheat	23.04	23.38	31.54	20.72	18.96
	Rice	26.2	26.1	25.7	25.1	23.5

The sensitivity order of crops to SAR and SO_2 combinations was vegetable > cash crops > cereal crops, while among the vegetable crops, leguminous vegetable > rhizomatic vegetable > leafy vegetables > fruit vegetable (Table 15.14). SO_2 was the main factor causing yield reductions in combined treatment with SAR.

Table 15.14 Effect of SAR and SO_2 Combinations on Percentage Reduction in Crop Yield

Crop Species	SAR pH+SO_2 (ppm)			
	4.6 + 0.05	4.1 + 0.1	3.6 + 0.15	2.8 + 0.25
Wheat	—	—	2.66	23.57
Rice	—	—	8.26	11.30
Barley	—	8.36	12.59	12.76
Soybean	0.95	26.15	36.10	44.85
Cotton	21.58	36.47	50.76	66.87
Rape	32.24	37.75	40.45	44.15
Pinto bean	26.58	28.94	33.32	44.84
Tomato	9.07	15.08	24.37	37.55
Carrot	22.03	25.87	39.79	45.34

Effects of combined treatment on yield reduction of pinto bean, tomato, and carrot were less than the sum of those of two treatments individually, suggesting an antagonistic action. For cotton at a middle treatment level (pH: 4.6 to 3.6, SO_2: 0.05 to 0.15 ppm), the combined impact was synergistic. However, at low treatment levels, or at high treatment levels (pH < 2.8 or SO_2 > 0.25 ppm), the interactive impact was antagonistic.

15.6 CONCLUSIONS

The studies conducted to evaluate the effects of different levels of SAR and SO_2, singly and in combination, on variety of plants of China clearly showed that plants are adversely affected by the above treatments even at levels that are commonly present in some areas. Both SAR and SO_2 treatments led to development of injury symptoms above a particular threshold. Physiological functions, levels of metabolites, biomass accumulation, and yield components showed unfavorable alterations in response to individual and combined treatments of SAR and SO_2. The level of response varied with plant species, levels of SAR or SO_2, individual and combined effects of treatments, and measured response variable. For yield response to SAR and SO_2, vegetable crops were found to be most sensitive, followed by cash crops and then cereal crops.

REFERENCES

1. Hongfa, C., Air pollution and its effects on plants in China, *Acta Ecol. Sin.*, 10, 7, 1990.
2. Yanyun, L., Jianmin, S., and Yingxin, G., Effects of simulated acid rain and SO_2 on growth and yield of vegetables, *Acta Sci.Circumstantiae*, 11, 327, 1991.
3. Hongfa, C., Yingxin, G., and Jianmin, S., Study on simulated acid precipitation effects on growth and yield of agricultural crops, *Acta Phytoecol. Geobot. Sin.*, 13, 56, 1989.
4. Hongfa, C. and Taylor, O. C., Dose-injury relationship to SO_2 in plants, *Acta Ecol. Sin*, 6, 114, 1986.
5. Yanyun, L., Hongfa, C., Jianmin, S., and Yingxin, G., SO_2 dose-response of five species of crops and their acute injury threshold, *China Environ. Sci.*, 9, 183, 1989.
6. Junhua, W., Effects of acid rain on growth of several vegetable species, in *Acid Rain Analects*, China Environ. Sci. Press, 1989, 451.
7. Yugu, C. and Zongying, H., Effects of acid rain on rape, in *Acid Rain Analects*, China Environ. Sci. Press, Beijing, 1989, 445.
8. Xuping, G., Hongfa, C., and Jianmin, S., Responses of 105 species of plants to simulated sulfuric acid rain, *China Environ. Sci.*, 7, 16, 1987.
9. Jianmin, S. Hongfa, C., Yingxin, G., and Yanyun, L., Effect of long-term exposure to low concentration of SO_2 on growth and yield of *Phaseolus vulgaris*, *China Environ. Sci.*, 7, 9, 1989.
10. Hongfa, C., Jianmin, S., and Yingxin, G., Study on dose-response between SO_2 and plants, *Res. Environ. Sci.*, 1, 52, 1988.
11. Zhang, Q., Effect of simulated acid rain on growth of greengrocery and lettuce, *China Environ. Sci.*, 5, 31, 1985.
12. Hongfa, C., Jianmin, S., and Yanyun, L., Effects of simulated acid rain and SO_2 in combination on several vegetables, *Res. Environ. Sci.*, 5, 29, 1992.

Chapter 16

The Use of Calibrated Passive Monitors to Assess Crop Loss due to Ozone in Rural Locations

Victor C. Runeckles and Patricia A. Bowen

CONTENTS

16.1 INTRODUCTION

Networks of passive monitoring devices have been and continue to be used in many studies throughout the world to provide information about ambient air quality. Their simplicity makes them admirably suited both to their use in regions where the lack of readily available electrical power restricts the use of monitoring instruments[1] and to establishing a greater density of monitoring sites than is usually feasible with monitoring instruments. Although passive monitors can provide information on the spatial variations in the levels of a pollutant in the ambient air, their very nature precludes their ability to determine short-term temporal (hourly) variations.

In the case of ozone (O_3), phytotoxicity to a range of crops and other plant species has been described in numerous studies following the early demonstrations of plant injury in California in 1950 by Middleton et al.[2] Since then, O_3 has been shown to be a major pollutant throughout the world, usually based on evidence from networks of instrumented O_3-monitoring stations. However, most instrumented networks concentrate on air quality in urban rather than rural or wilderness areas.

To permit the assessment of the impact of O_3 on crops and other vegetation, it is necessary to use O_3 response functions of the species of interest. Major studies such as the U.S. National Crop Loss Assessment Program (NCLAN)[3] and similar studies in Europe and elsewhere have led to the

1-56670-341-7/00/$0.00+$.50

publication of O_3 loss functions for many crops and other species.[4] Concurrently with these studies, a large number of indexes were evaluated for suitability as measures of the exposure to which the plants were subjected and hence could be used as the independent variable in response functions.[5-7] However, the indexes and functions were developed from studies with continuous instrumental monitoring of different experimentally introduced O_3 levels to which plants were exposed in field, greenhouse, or growth chamber situations.

The present chapter describes the application of crop loss functions based on weekly passive monitor data and their use in assessing the impacts of O_3 on crops in locations adjacent to passive monitor sites in rural areas.

16.2 PASSIVE OZONE MONITORS

The earliest report of the use of a passive monitor for O_3 appears to have been by the Swiss chemist Schönbein in the mid-19th century, using test papers impregnated with potassium iodide.[8] Since then, there have been numerous reports of passive monitors developed using KI,[9] and other reactants including alizarin,[10] curcumin,[11] dimethylbiacrylidene,[12] dipyridylethylene,[13] indigo,[10,14] indigo carmine,[11,15,16] methylbenzothiazolinone,[17] nitrite,[18] and phenoxazine.[11,19] All of these are relatively specific for O_3, but show some reactivity toward other atmospheric oxidants. Nitrite-based samplers are considered to be specific for O_3 and have been proposed for wide-scale use.[20] Indigo-based samplers are somewhat reactive towards oxidants such as formaldehyde, nitrogen dioxide (NO_2), and peroxyacetylnitrite. However, of these, only NO_2 is thought likely to introduce a significant positive bias of up to 15% at the concentrations likely to occur in photochemically polluted air.[11]

16.3 PASSIVE OZONE MONITOR NETWORKS

Recent years have seen an increasing number of reports of the use of networks of passive monitors to obtain information about ambient O_3 levels. For example, Grosjean et al.[21] surveyed 46 mountainous forest and desert locations in Southern California using indigo carmine samplers.

For 2-week exposures, typical values were 40 to 80 ppb and exhibited elevational variations and temporal variations throughout the year. For exposures up to about 15 days, a high correlation (r^2 = 0.954) between indigo carmine color loss and concurrent instrumental determinations of cumulative exposure was found, suggesting little interference by other ambient oxidants. Indigo carmine monitors have also been used to reveal the extent of O_3 pollution in parts of Poland[16] and the Ukraine.[22]

A study in two wilderness areas in New Hampshire and Vermont using nitrite-based samplers observed weekly cumulative exposures ranging from 22 to 44 ppb.[1] Concurrent instrument monitoring at the sites led to a regression between passive and instrument data with an adjusted r^2 value of 0.76 ($P > 0.001$). However, the slope of the regression indicated over-estimation by the passive monitors of about 4%.

Such studies have demonstrated the usefulness of passive monitors for measuring average ambient O_3 concentrations. While they may help in understanding longer-term variations, they cannot reveal the short-term dynamics of exposure. Such short-term dynamics are considered to be more important determinants of plant response than longer-term averages.[23]

16.4 THE FRASER VALLEY NETWORK

The lower Fraser Valley of British Columbia covers an area of approximately 200 km^2. A network of over 20 instrumental O_3 monitors is maintained by the Greater Vancouver Regional District and the British Columbia Ministry of Environment (Figure 16.1). However, only six (T12, T15, T16, T17, T25, and T29; Figure 16.1) are located away from centers of population, and none of these is in a truly agricultural location. Agricultural production is an important local economic resource and includes the production of berry crops such as strawberry, raspberry, blueberry, and cranberry, a wide range of vegetables, and corn, hay, and pasture to support the local dairy industry. Several of the crops grown in the valley are ranked as being "sensitive" to O_3 and hence the impacts of ambient O_3 downwind of Vancouver and its suburbs are of concern to producers.

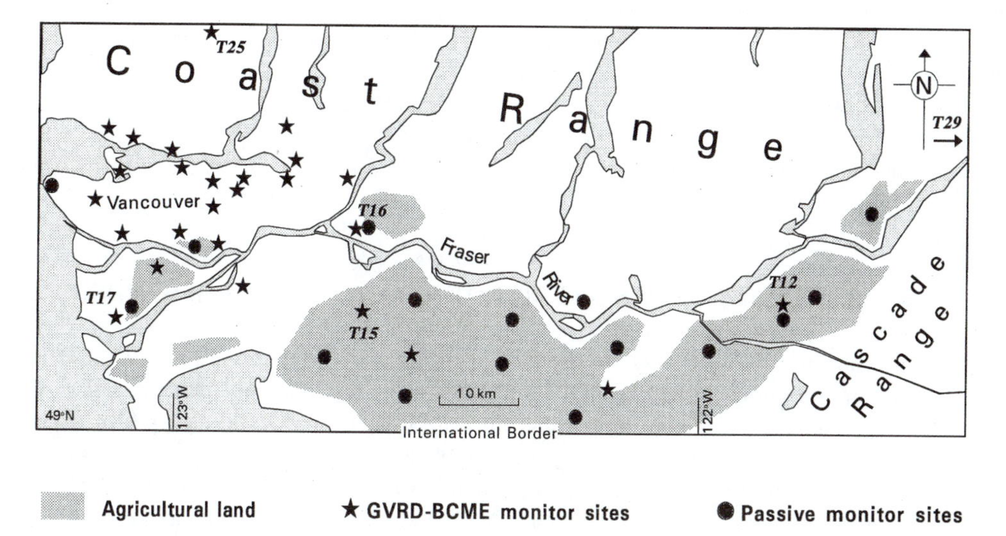

Figure 16.1 Map of the Fraser Valley showing instrument monitor and passive monitor locations in 1995 and 1996.

In 1995 and 1996, a network of passive O_3 monitors was set up and maintained throughout most of the growing seasons (see Figure 16.1). Each monitor consisted of an indigo-impregnated filter paper disk supported above a Teflon diffusion membrane within a polystyrene dish. The monitors faced downward, and were mounted beneath PVC shelters.[24] Two monitors were located at each site in 1995 and four in 1996. The monitors were changed at approximately 168-h (7-day) intervals. Preparation of the impregnated papers employed immersion for 2 min in a 2 g l^{-1} aqueous suspension of indigo in an ultrasonic bath at 30°C, followed by drying in an oxidant-free atmosphere at 50°C under partial vacuum. The process was repeated twice. The dried papers were stored between waxed paper in an airtight container containing activated charcoal. The papers were kept in darkness until installed at the monitoring sites.

For analysis, the papers were removed from the holders and sealed in flasks containing 5 ml ethanol. After shaking for 5 min at 60°C, the elutes were filtered. Absorbances were determined at 408 nm, the λ_{max} for isatin, the indigo reaction product.[25] Absorbances for unexposed control papers prepared at the same time were subtracted from exposed paper values.

16.5 EVALUATION OF THE PASSIVE MONITORS

The methodology used had been previously shown to yield cumulative O_3 exposures in good agreement with instrumental measurements.[24] Confirmation of their performance was obtained by including the passive monitors in field experiments conducted on the campus of the University of British Columbia (the most westerly passive sampling site shown in Figure 16.1). These studies used an open-air zonal air pollution system (ZAPS). The facility was a development of an earlier ZAPS,[26] and provided exposures to nine levels of O_3 ranging from ambient to approximately twice ambient. Each O_3 treatment was distributed over a 10 × 10 m plot. Four passive monitors were located within each plot, the air over which was sampled continuously from four adjacent points for O_3 measurement using a time-share system feeding into Dasibi Model 1008AH O_3 monitors.

Good agreement was obtained between the passive monitor data and the cumulative O_3 exposures. Regressions of the mean weekly data are shown in Figure 16.2, for each year. Although the number of passive samplers was doubled in 1996, the data showed more variability than was observed in 1995. For 1995, the adjusted r^2 value was 0.884 ($P < 0.0001$) while in 1996 it was 0.596 ($P < 0.0001$). The regressions revealed no curvilinearity, and the closeness of their origins to zero suggested little interference from other ambient oxidants. The adjusted r^2 value of 0.758 for the pooled data for the 2 years is similar to that observed by Manning et al.[1] using a smaller number of nitrite-based monitors.

16.6 CALIBRATION OF PASSIVE MONITOR DATA TO CROP RESPONSES

Reference has already been made to the inability of passive samplers to provide information about short-term variations in ambient O_3 levels. However, this does not prevent their use for assessing crop responses; it merely means that they are unable to provide the types of exposure data that can be immediately utilized by many of the published crop response functions. A condensed summary of O_3 exposure indexes is presented in Table 16.1.[5] Cumulative indexes such as SUMxx and AOTyy, where xx = 06 (i.e., a 0.06 ppm threshold) and yy = 40 (i.e., a 40 ppb threshold), have been favored by American and European workers, respectively, as providing the best overall indexes for use in crop response functions. However, total cumulative exposure (TOTDOSE, Table 16.1) has been reported to be an appropriate index for certain crops.[6] Since passive sampler response is closely related to TOTDOSE (Figure 16.2), exposures in the ZAPS facility provided the means for calibrating passive monitor data in relation to crop response.

Response functions for four crops (broccoli, lettuce, orchardgrass, and strawberry) were developed. The crops were grown using normal local practices in 2.5 × 5 subplots within the 10 × 10 m plots subjected to each O_3 treatment. No statistically significant relationships between passive O_3 exposures and growth and yield parameters of broccoli or lettuce were observed in either year. However, significant relationships were observed for some cuttings of orchardgrass and on strawberry yields in both years. For example, in 1996, the fourth cut of orchardgrass showed significant decreases in dry weight with increased exposure, whether assessed by passive monitoring (Figure 16.3a) or by determination of total exposure (Figure 16.3b). In both cases, although the linear regressions had significant coefficients ($P < 0.05$), extrapolation to zero is inappropriate as a means of estimating threshold values from which losses can be computed. Weibull functions were therefore calculated to obtain estimates of the threshold yield values. For the passive sampler situation, a percent yield loss Weibull function was then calculated as shown in Figure 16.4a. The function is

$$\% \text{ loss} = 100[1 - \exp\{-(\text{ozone}/3.77)^{13.4}\}], \qquad 16.1$$

where O_3 is corrected mean passive monitor absorbance ($r^2_{cor} = 0.86$).

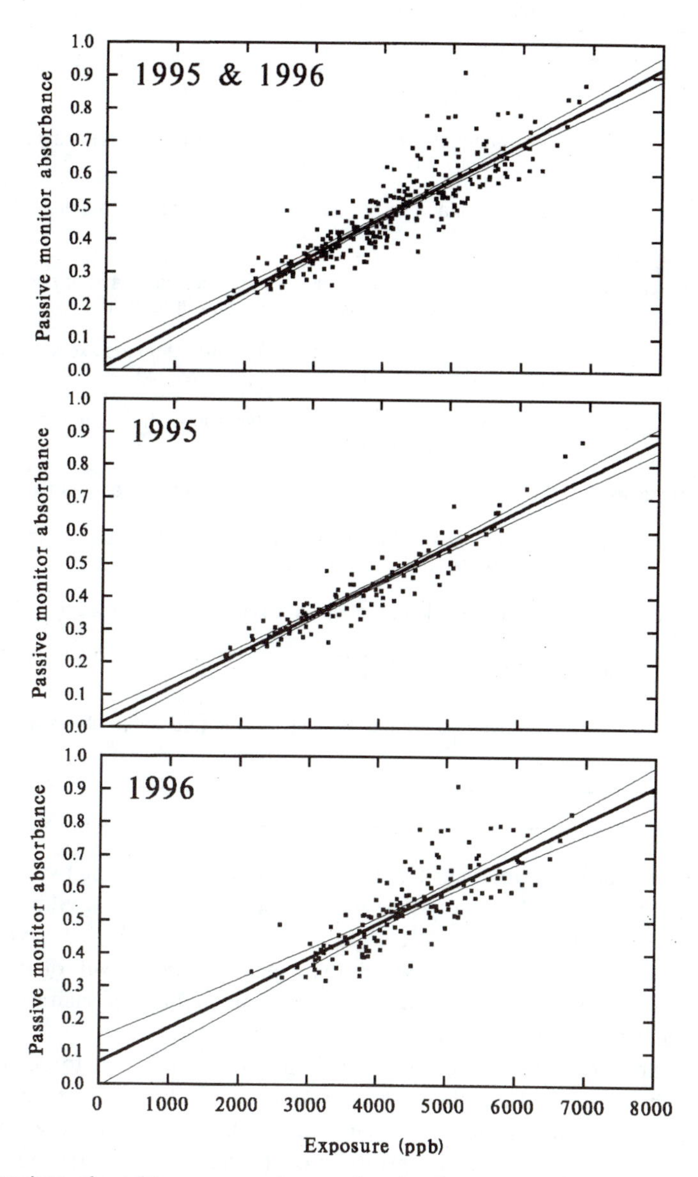

Figure 16.2 Regressions of weekly mean passive monitor absorbances and corresponding cumulative exposures observed in the ZAPS facility in 1995 and 1996 (0.95 confidence limits).

Loss of strawberry fruit was also found to be significant (data not shown). For the passive data case, the linear regression probability was 0.03, while for total exposure it was 0.01. The Weibull regression led to the loss curve shown in Figure 16.4b based on the function:

$$\% \text{ loss} = 100[1 - \exp\{-(\text{ozone}/8.59)^{3.04}\}];\ (r^2_{\text{cor}} = 0.57). \qquad 16.2$$

16.7 ESTIMATION OF CROP LOSSES IN THE FRASER VALLEY

In both graphs presented in Figure 16.4, losses of the crops estimated from Equations 16.1 and 16.2 for passive data obtained from sites in the Fraser Valley are shown. Since passive absorbances

Table 16.1 A Summary of Types of O_3 Exposure Indexes

Type	Examples
One event	The maximum hourly mean for a year or season, such as the current U.S. secondary air quality standard
Mean	The seasonal mean of 7-h daily means (M7); the seasonal mean of the daily 1-h maxima (M1)
Cumulative	The seasonal sum of hourly means, i.e., total exposure (TOTDOSE)
Cumulative weighted:	
Concentration based	The seasonal sum of hourly mean concentrations at or above a threshold (SUMxx), e.g., SUM06 for a 0.06 ppm threshold
	The seasonal sum of hourly mean concentrations exceeding a threshold (AOTyy), e.g., AOT40 for a 40 ppb threshold
	The seasonal sigmoidally weighted sum of hourly mean concentrations (SIGMOID).
Number of episodes	the sum of hours with means at or above a threshold (HRSxx)
	The sum of episodes (days) with means at or above a threshold (NUMEPxx or DAYSxx)
	The average length of episodes at or above a threshold (AVGEPxx)
Multicomponent	Indexes incorporating other environmental variables or phenological weighting for specific crops
Respite time	The average number of days between episodes above a threshold (DAYBETxx)

Adapted from Lee et al., 1989[5]

were accumulated over different intervals for the two crops, the data for the two sets of valley sites do not coincide. In the case of orchardgrass, these clearly show that losses would have been minimal at any site. However, for strawberry, estimated losses reach about 8%. Such assessments are, of course, specific to the time intervals over which the exposures and growth of the crops occurred on the campus plots. Nevertheless, they demonstrate the potential use of such calibration methods in providing estimates of losses in the field.

Although the exposures achieved on the ZAPS plots were sufficient to demonstrate adverse yield effects on orchardgrass and strawberry, they were by design, however, not able to demonstrate the same at the low end of the range of exposures to which plants are exposed in many parts of the world. Indeed, improvement in the air quality in the Fraser Valley has resulted in few observations of hourly mean O_3 concentrations exceeding 100 ppb in recent years. The relatively modest levels of O_3 observed in 1995 and 1996 are nevertheless reflected in the passive data obtained from valley sites.

Ozone pollution in the valley is transported in the westerly daytime flow of air from Greater Vancouver, and, with the narrowing of the valley to the east imposed by topography, greater O_3 levels have been reported for the eastern half. The passive network data provide some confirmation of this as shown in Figure 16.5. This shows a plot of the seasonal accumulated passive monitor responses against longitude of the monitor sites. Although there is considerable vertical scatter partly due to the variations in latitude of the sites, the regression lines suggest an increase from west to east. These regressions exclude the campus site, which has a history of slightly elevated O_3 levels, in spite of being upwind of Vancouver.

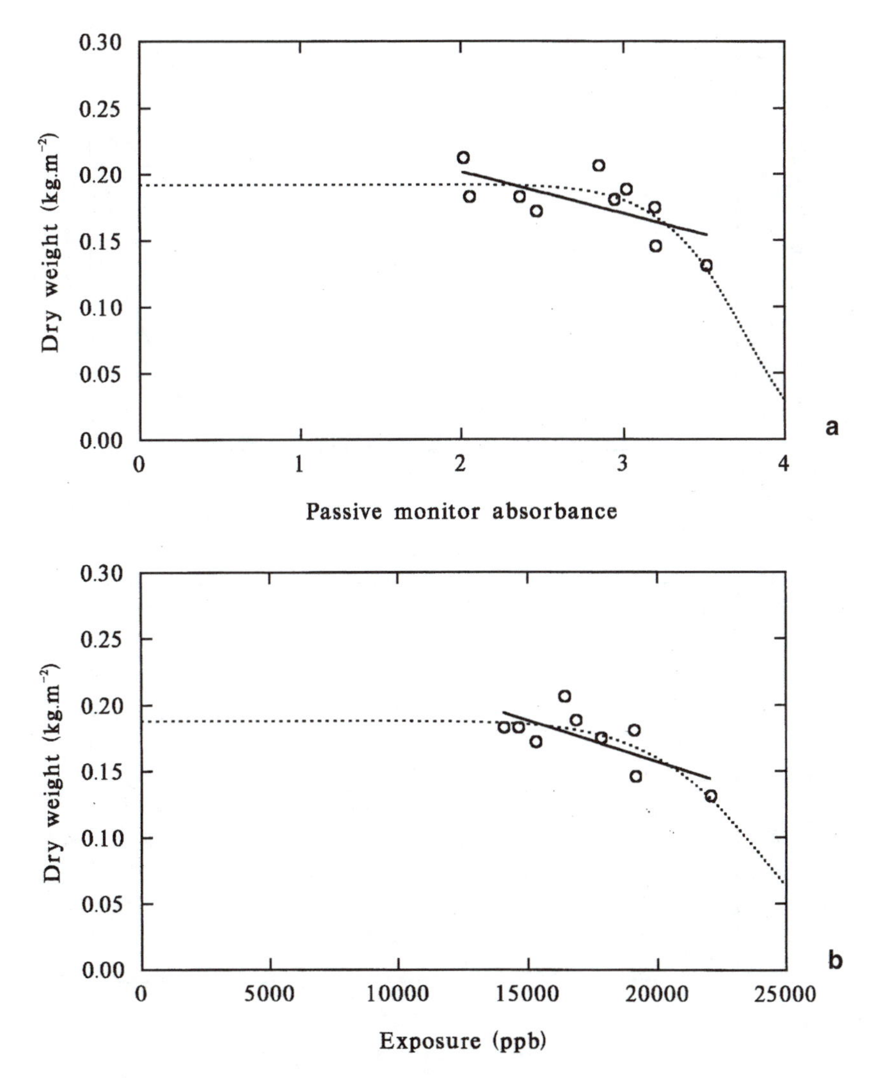

Figure 16.3 Response to O_3 of dry weight yield of orchardgrass (1996, fourth cut):(a) based on passive monitor absorbance; (b) based on exposure (TOTDOSE).

16.8 SUMMARY

The use of crop-calibrated passive monitors in the Fraser Valley has demonstrated that such methods can be used to estimate losses to vegetation in regions in which ambient O_3 levels can only be estimated by atmospheric modeling. Where topographic and other features limit the precision of such models, and in regions where little information about ambient O_3 exists, the use of calibrated passive monitors can provide a simple means for assessing impact on a range of species *in situ*.

ACKNOWLEDGMENTS

The authors thank Tim Williams, whose experience in preparing the monitors was invaluable, and Heidi Rempel, who managed the network and performed the analyses.

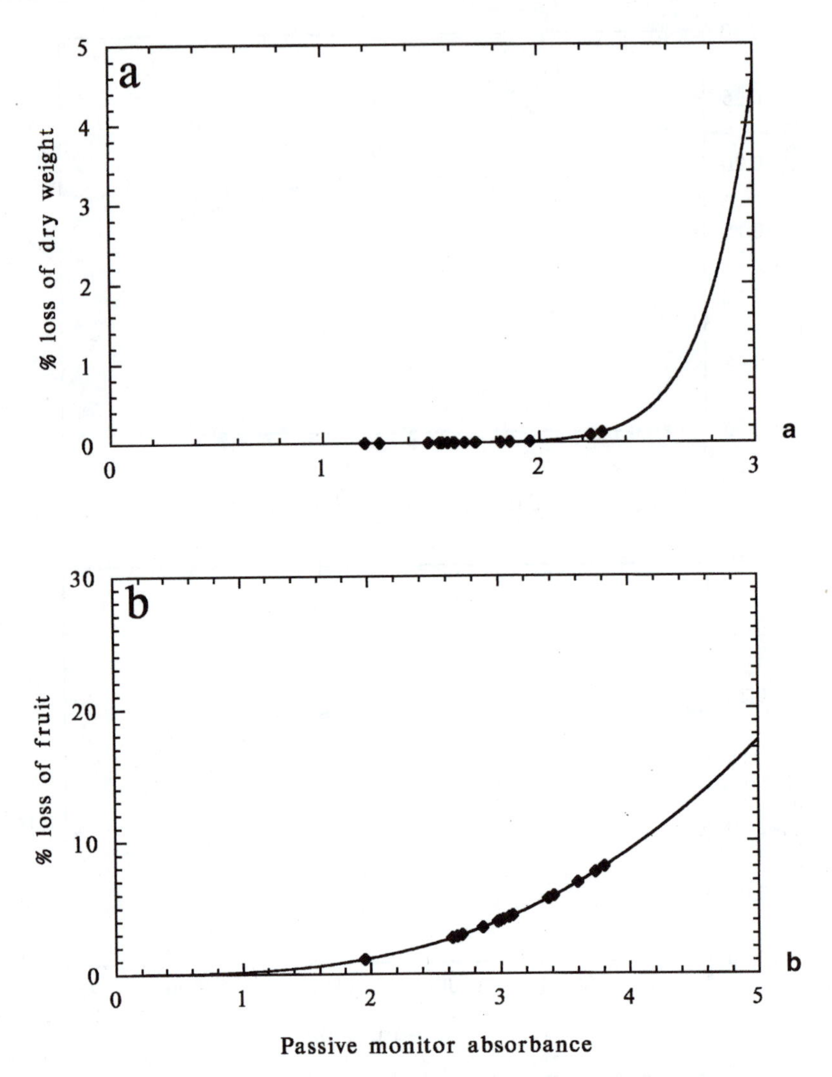

Figure 16.4 Weibull percent loss curves for 1996 for (a) orchardgrass dry weight (fourth cut) and (b) number of strawberry fruits.

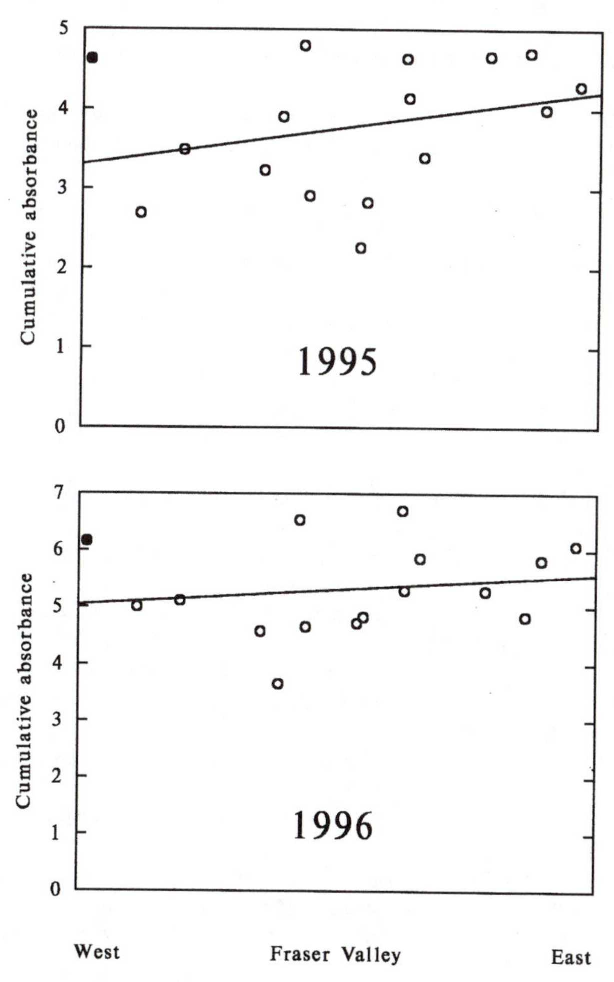

Figure 16.5 Total seasonal cumulative mean passive monitor responses observed in the Fraser Valley in 1995 and 1996 in relation to longitude.

REFERENCES

1. Manning, W. J., Krupa, S. V., Bergweiler, C. J., and Nelson, K. I., Ambient ozone (O_3) in three Class I wilderness areas in the northeastern USA: measurements with Ogawa passive samplers, *Environ. Pollut.*, 91, 399, 1996.
2. Middleton, J. T., Kendrick, J. B., Jr., and Schwalm, H. W., Injury to herbaceous plants by smog or air pollution, *Plant Dis. Rep.*, 34, 245, 1950.
3. Heck, W. W., Cure, W. W., Rawlings, J. O., Zaragoza, L. J., Heagle, A. S., Heggestad, H. E., Kohut, R. J., Kress, L. W., and Temple, P. J., Assessing impacts of ozone on agricultural crops: I. Overview, *J. Air Pollut. Control. Assoc.*, 34, 729, 1984.
4. Heagle, A. S., Kress, L. W., Temple, P. J., Kohut, R. J., Miller, J. E., and Heggestad, H. E., Factors influencing ozone dose-yield response relationships in open-top field chamber studies, in *Assessment of Crop Loss from Air Pollutants*, Heck, W. W., Taylor, O. C., and Tingey, D. T., Eds., Elsevier Applied Science, London, 1988, 141.

5. Lee, E. H., Tingey, D. T., and Hogsett, W. E., Interrelation of Experimental Exposure and Ambient Air Quality Data for Comparison of Ozone Exposure Indexes and Estimating Agricultural Losses, Report for U.S. Environmental Protection Agency, Office of Air Quality Planning and Standards, Research Triangle Park, NC, 1989, 98.
6. Musselman, R. C., McCool, P. C., and Younglove, T, Selecting ozone exposure statistics for determining crop yield loss from air pollutants, *Environ. Pollut.*, 53, 63, 1988.
7. Rawlings, J. O., Lesser, V. M., and Dassel, K. A., Statistical approaches to assessing crop losses, in *Assessment of Crop Loss from Air Pollutants*, Heck, W. W., Taylor, O. C., and Tingey, D. T., Eds., Elsevier Applied Science, London, 1988, 389.
8. London, J., The observed distribution of atmospheric ozone and its variations, in *Ozone in the Free Atmosphere*, Whitten, R. C. and Prasad, S. S., Eds., Van Nostrand Reinhold, New York, 1985, 11.
9. Kanno, S. and Yanagisawa, Y., Passive ozone/oxidant sampler with coulometric determination using I_2/nylon charge-transfer complex, *Environ. Sci. Technol.*, 26, 744, 1992.
10. Grosjean, D., Grosjean, E., and Williams, E. L., Fading of artists' colorants by a mixture of photochemical oxidants, *Atmos. Environ.*, 27, 765, 1993.
11. Grosjean, D. and Hisham, M. W. M., A passive sampler for atmospheric ozone, *J. Air Waste Manage. Assoc.*, 42, 169, 1992.
12. Surgi, M. R. and Hodgeson, J. A., 10,10'-Dimethyl-9,9'-biacrylidene impregnated film badge dosimeters for passive ozone sampling, *Anal. Chem.*, 57, 1737, 1985.
13. Monn, C. and Hangartner, M., Passive sampling for ozone, J. Air Waste Manage. Assoc., 40, 357, 1990.
14. Werner, H., Die ozonkerze—ein passiver Integrator zur flächendeckenden Abschaätzung der Ozonimmission, *Proc. 14th Int. Meeting for Specialists in Air Pollution Effects on Forest Ecosystems*, IUFRO, Interlaken, Switzerland, 1989, 547.
15. Bergshoeff, G., Lanting, R. W., van Ham, J., Prop, J. M. G., and Reijnders, H. F. R., Spectrophotometric determination of ozone in air with indigo disulphonate, *Analyst*, 109, 165, 1984.
16. Bytnerowicz, A., Manning, W. J., Grosjean, D., Chmielewski, W., Dmuchowski, W., Grodzinska, K., and Godzik, B., Detecting ozone and demonstrating its phytotoxicity in forested areas of Poland: a pilot study, *Environ. Pollut.*, 80, 301, 1993.
17. Lambert, J. L, Paukstelis, J. V., and Chiang, Y. C., 3-Methyl-2-benzothiazolinone acetone azine with 2-phenylpropanol as a solid passive monitor for ozone, *Environ. Sci. Technol.*, 23, 241, 1989.
18. Koutrakis, P., Wolfson, J. M., Bunyaviroch, A., Froelich, S. E., Hirano, K., and Mulik, J. D., Measurement of ambient ozone using a nitrite-coated filter, *Anal. Chem.*, 65, 209, 1993.
19. Lambert, J. L., Liaw, Y.-L., and Paukstelis, J. V., Phenoxazine as a solid monitoring reagent for ozone, *Environ. Sci. Technol.*, 21, 503, 1987.
20. Mulik, J. D., Varns, J. L., Koutrakis, P., Wolfson, M., Williams, D., Ellenson, W., and Kronmiller, K., The passive sampling device as a simple tool for assessing ecological change, in *Measurement of Toxic and Related Air Pollutants, Proc. USEPA/AWMA Int. Symp.*, Vol. 1, Air and Waste Management Association, Pittsburgh, PA, 1991, 285.
21. Grosjean, D., Williams II, E. L., and Grosjean, E., Monitoring ambient ozone with a network of passive samplers: a feasibility study, *Environ. Pollut.*, 88, 267, 1995.
22. Blum, O., Bytnerowicz, A., Manning, W., and Popovicheva, L., Ambient tropospheric ozone in the Ukrainian Carpathian mountains and Kiev region: detection with passive samplers and bioindicator plants, *Environ. Pollut.*, 98, 299, 1997.
23. Kickert, R. N. and Krupa, S. V., Modeling plant response to tropospheric ozone: a critical review, *Environ. Pollut.*, 70, 271, 1991.
24. Williams, T. P. W., Use of Passive Monitors to Assess Plant Bioindicators for Ground-Level Ozone, M.Sc. thesis, University of New Brunswick, Fredericton, 1994, 131.
25. Mangini, A. and Passerini, R., Absorption spectra in the isatin series, *Chem. Abstr.*, 46, 350, 1952.
26. Runeckles, V. C., Wright, E. F., and White, D., A chamberless field exposure system for determining the effects of gaseous air pollutants on crop growth and yield, *Environ. Pollut.*, 63, 61, 1990.

CHAPTER 17

Wild Plant and Crop Plant Species for *In Situ* Microspore Analysis of a Polluted Environment

G. Murín and K. Miéieta

CONTENTS

1-56670-341-7/00/$0.00+$.50

17.1 INTRODUCTION

For evaluation of environmental chemicals we need competent monitoring and indication.[1] The best tools are bioindicators because characterization of ecotoxicology factors through chemical analysis is limited.[2] Higher plants have already shown their ability to fulfill all requisite conditions for the bioindication of genotoxicity of environmental chemicals.[3,4] Nevertheless nearly all methods were *in vivo* tests.[5-9] There is no doubt that in this condition they could hardly compete with routine methods of testing mutagens and/or carcinogens using bacterial and mammalian materials, although plant tests are cheaper and approximately 60 to 70% of hazardous chemicals have the same effects on plant tests as on bacterial and mammalian tests.[3,10]

Unfortunately, the basic advantage, as we consider it — *in situ* bioindication of polluted environment via wild plant and crop plant specie — was utilized only partly in the past. The theoretical base for it was established in reports of De Serres,[4] Mulcahy,[11] and Grant and Zura.[10] Several authors also reported *in situ* studies with formerly *in vivo* standardized plant species as *Tradescantia*,[12-14] *Vicia faba*,[15] or both.[16] For field tests tobacco Bell W_3 was also used.[17] Interesting results were obtained from radioactive contamination in Chernobyl on *Arabidopsis*[18] and phytotoxic and mutagenic effects of polluted waters on *Vicia sativa*.[5] Only a few reports were about *in situ* cytogenetic monitoring of vegetation affected by pollutants such as herbicides,[19] polymers,[20] or photochemical smog.[21] After nearly 10 years and more than 15 reports from several Slovak, one Austro-Slovakian, and one Kuwaitian project, we would like to summarize here our results from a long-term study of the ability of wild and crop plant species to serve as suitable bioindicators of polluted environment. With the background of our previous experiences with *in vivo* tests of phytotoxic and mutagenic effects of polluted waters[5,22] and the WHO/IPCS collaborative study for the standardization of certificated plant tests on the genotoxic effects of xenobiotics,[23] we focused our interest on *in situ* biondication. An initial report by Murín[24] stated the basic criteria for selection of plant bioindicators. It led us consequently to the need of proving this method in practical cases. Finally, we are able to present the results from biomonitoring of different sources of pollution from different regions of Slovakia and Kuwait. Furthermore, the new possibility to test our environment in retrospective biomonitoring by means of herbal samples has also shown promising results.

17.2 MATERIALS AND METHODS

The basic criteria for using pollen grains as a marker of polluted environment were described by Murín.[24,25] They are as follows:

1. The selected species must be diploid (with pollen grains haploid), to avoid high pollen abortion which is common in polyploids.
2. The plants should produce well-developed and viable pollen grains under common climatic conditions (pollen abortion less than 5%).
3. The plants should have a common distribution and grow also on urban and industrial habitats.
4. Blooming seasons of various plant species should be in some sequence to cover all year seasons: spring, summer, autumn, and winter, respectively.
5. The determination of selected plant species in question should be easy and indubitable for common practice.
6. It is advisable to select both the terrestrial and aquatic plant species.
7. It is an advantage to choose also (if possible) plant species with persistent pollen tetrads to distinguish physiological and genetic damage of the pollen grains easily.

For crop plant species, particularly, all information about the selected cultivar, its agrochemical treatment during growth period, and weather conditions during whole vegetation have to be considered for possible influence to the final results.[26]

The broad spectra of plant species allow us to utilize all levels of sensible parts of plants for detection of environmental pollution. They are pollen grains/microspores, micronuclei in tetrads, and chromosomal aberrations/mitotic abnormalities in meioses and mitoses.

17.2.1 Pollen Grains/ Microspores

Young and closed flowers and flower buds from a sufficient number of randomly selected individual plants (minimum 10 individuals) are collected and fixed in a mixture of ethanol:acetic acid (3:1). After 24 h, the fixing solution is replaced by 75% ethanol and the samples then stored until further processed. The use of pollen as an indicator of mutagenicity and phytotoxicity must be precisely defined as the pollen has high sensitivity, which very readily responds to altered conditions — such as extreme temperature, aridity, nutrient deficiency, etc.[11] Therefore, in evaluating the induced toxicity within polluted areas it is necessary to recognize the natural abortivity of the pollen. The frequency of pollen abortivity is influenced by the concentration of the pollutant emissions during the period of meiosis, which is different in each species. Therefore, every selected species has to be screened for three seasons (shortly before its blooming period), to avoid possible influences of extreme weather conditions such as drought or fluctuation of temperature.

17.2.1.1 Preparation of Slides for the Evaluation of Pollen Abortion

Young and closed flowers are removed from the fixing solution, washed, and the anthers excised. The pollen grains are pushed out from each anther with a needle, using a slight pressure. The debris is removed from the slide. The pollen grains were stained with 0.5% aceto-carmine or 0.05% aniline blue in lactophenol.[27]

In some plant species, the mature pollen grains occurred in tetrads and this characteristic offers a simple way of distinguishing between genetic and physiological toxicity of the pollutants in question. The abortion of one, two, or three pollen grains in the pollen tetrad usually indicates genetic damage. Physiological damage usually affects whole flowers, anthers, and all four pollen grains in tetrads. The genetic damage of all four pollen grains is so rare that if, in the course of evaluation, we include them among the physiological events it will have no influence on the statistical significance.[11] Pollen grains were evaluated for size, form, and staining ability, with deviations considered as evidence for lack of viability (abortion) or the possible occurrence of chromosomal aberrations (anomalies).

17.2.1.2 Evaluation of Abortion of the Pollen Grains in Pinuses

Male strobili of *Pinus sylvestris* and *P. nigra* from different areas are collected. On each area 10 tips of the young branches with microstrobules from five individuals each of both species are fixed in an ethanol:acetic acid (3:1) mixture. The fixation solution is replaced after 24 h by 75% ethanol. Slides are prepared following the standard process.[24] From microstrobules, the pollen grains are pressed out on a slide and stained by 0.05% of aniline blue.[27] After stain penetration, the grains are covered by cover glass and scored under a microscope. From each area, 3000 pollen grains are evaluated per species.

17.2.2 Micronuclei in Tetrads

Flower buds containing early-stage tetrads of meiotic microspore mother cells are used for the evaluation of micronuclei. The four encased cells of the early tetrad usually remain together during slide preparation, whereas the later tetrad stage is inadequate for micronucleus scoring.

The young flower buds are macerated for 5 to 10 min in a mixture of ethanol:hydrochloric acid (1:1). The required materials have to be collected soon after finishing tetragenese, when an exina of tetrads is not yet developed. Only under this condition, sufficient microscope evaluation is available. After maceration, the flower buds are washed in distilled water. On the slide, the pollen tetrads are dissected out from the anthers. All the debris should be removed from the slide. The tetrads are stained with acetic-orcein. Staining could be supported by repeated heating over a flame, or on a hot plate, and then pressing gently under several layers of filter paper. About 3000 tetrads have to be evaluated from 10 to 20 flowers of different plants, i.e., 150 to 300 tetrads per slide.

17.2.3 Mitoses

For evaluation of chromosomal aberrations and mitotic abnormalities root tips cells or branches with young needles (in pines) from five individuals per species are fixed in ethanol:acetic acid (3:1) mixture. The fixation solution is replaced after 24 h by 7 % ethanol. Samples are hydrolyzed for 10 min in the concentrated hydrochloric acid:ethanol (1:1) mixture at room temperature. Squash preparations are made using meristematic tissues. The chromosomes are stained in 1% aceto-orcein.[27] Chromosomal aberrations (CA) in ana- and telophases (fragments, bridges, fragments and bridges) are distinguished from mitotic abnormalities (MA = lagging chromosomes, multipolar spindles, and c-mitosis) because of their different origin (sources of CA are affected chromosomes, while MA are due to injured mitotic apparatus, particularly mitotic spindle). For each tested area 300 ana- and telophases of all species have to be scored.

17.2.4 Meioses

Preparation of slides for evaluation: For the analysis of meiosis, immature anthers were used. Slide preparation and evaluation are the same as above; a propion-orcein stain is used. From ana- and telophases, the frequency of bridges, fragments, and mitoclastic events (lagging chromosomes, multipolar spindles, and c-mitosis) have to be determinated.

17.2.5 Statistical Analysis

Statistical analysis between samples from a check area and polluted areas is recommended to be conducted following the procedure of Amphlett and Delow,[28] which is based on the Poisson distribution. For data, an analysis of variance (ANOVA) and Wilcoxin sign-rank (one-sided) test are also useful. All evaluations have to be done under blind conditions.

17.3 RESULTS

17.3.1 Theoretical Background

An initial report by Murín,[24] followed by a later one,[25] which stated the basic criteria for selection of plant biondicators from the regional flora for monitoring of an environmental pollution, was the essential impulse for our series of studies. At the first stage of our study we were focused at the theoretical background of the new method and its comparison with existing methods. From that time there exists some reports from our laboratory dealing more deeply with the methodology of the plant test and its ability to serve in bioindication and genotoxic risk assessments especially on emission territories of industrial complexes.[29-34]

17.3.2 Sources of Selected Bioindicators

Although most of our research was done with wild plant species, there were, in particular, two group of species which also invited our interest.

17.3.2.1 Crop Plant Species

A few successful studies were done using crop plant species as suitable bioindicators of polluted environment[35-37] based on the initial study of Mièieta.[15] Field experiments showed the ability of 21 species (from one to three cultivares for each) to fulfill specific criteria of bioindicators. For example, *Vicia faba* L. was used for comparison of the IPCS/WHO-recommended *in vivo* test[23] and the *in situ* tests of commercial herbicides Topogard, Dual, Ladob, Avadex, and Basagran. While tests of separate herbicides *in vivo* showed no increase of chromosomal aberrations, certain combinations of these herbicides in routine agriculture field treatments of *V. faba* L. seeds/seedlings led to the significant increase of the frequency of chromosomal aberrations. The set of evaluation criteria used showed the lowest effect in the basic test of vitality, with gradual significant results from abortivity of pollen grains, to chromosomal aberrations and sister chromatid exchanges, to the highest differences in micronucleus test.

17.3.2.2 Arborescent species

Significant part of our work was dedicated to implementation of the arborescent species to the list of suitable bioindicators of Slovak flora.[38-40] Studies with some of these species have not only shown very interesting data, but have also cast new light on the method applied in this study.

***Pinus sylvestris* L. and *P. nigra* Arn.** — For heavy polluted areas the introduction of such arborescent species as *P. sylvestris* L. and *Pinus nigra* Arn. is very useful for their ability to survive under such conditions. In the samples collected around an aluminum plant at Ziar nad Hronom in addition to the abortive pollen grains of *P. sylvestris* L. and *P. nigra* Arn. a higher amount of altered pollen grains, with diameters approximately one third of the pollen grains in the check samples, was also observed. Around a smelting plant at Banská Štiavnica, Central Slovakia, altered pollen grains also had a larger number (three to four) of air bags (sacculus aereus), or appearance of continual bags at the equatorial level.[41] We regarded these pollen grains as unreduced, diploid pollen grains. The tendency of their frequency was analogous to those observed for abortive pollen grains.[33] The pollen grains of nearly the same shape were already observed in *P. pinea* L. collected from Pisa town center affected mainly by pollution of Pb, S, and O_3 (photochemical smog).[21] Cytogenetic analysis, where chromosomal aberrations and especially mitoclastic events in ana- and telophasic cells of the meristem of pine-needle bases occurred, confirmed this tendency.[41] The interesting occurrence of the highest amount of mitoclastic events, especially multipolar spindles, can explain the higher amount of unreduced pollen grains in trees affected by emissions from the smelting plant.[41]

Pinus mugo — Among the selected arborescent species *P. mugo* was found as an interesting bioindicator especially for polluted areas in cities, as this species is usually used for "green stripes" in mobile containers situated near crossroads with heavy traffic.[40]

Other arborescent species are also under investigation and have shown promising data for implementation into a list of suitable bioindicators. They are, *Robinia pesudacacia*, *Taxus baccata*, *Salic caprea*, *Cornus mas*, *Corylus avellana*, *Populus alba*, and *Ligustrum vulgare* (Table17.1). *Thuja occidentalis*, *Juniperus communis*, *Juniperus virginiana*, *Negundo aceroides*, *Larix decidua*, and *Picea excelsa,* require further investigations.

17.3.3 Sources of Evaluated Environemntal Pollution

Parallel to previous studies our effort was focused on the utilization of this method for different types of pollution.

Table 17.1 List of Species of Slovak Native Flora Selected as Suitable Bioindicators in Order of Their Blooming Periods

Species	S	H	B
Corylus avellana L.	1.4 ± 0.3	7.1 ± 0.8 R	III–IV
Taxus baccata L.	0.2 ± 0.2	3.3 ± 0.5 R	III–IV
Cornus mas L.	2.7 ± 0.3	10.7 ± 1.5 R	III–IV
Salix casprea L.	3.9 ± 1.1	20.4 ± 4.4 R	III–IV
Populus alba L.	3.1 ± 1.2	7.7 ± 1.5 R	III–IV
Bellis perennis L.	2.0 ± 0.4	—	III–X
Lamium purpureum L.	3.0 ± 1.0	3.4 ± 0.3 A	III–X
Juniperus communis L.	1.7 ± 0.3	4.2 ± 1.2 R	IV–V
Juniperus virginiana L.	1.6 ± 0.2	5.6 ± 1.2 R	IV–V
Biota orientalis (L.) Endl.	1.8 ± 0.3	6.9 ± 1.4 R	IV–V
Negundo aceroides Moench	2.1 ± 0.3	8.2 ± 0.9 R	IV–V
Anthriscus silvestris (L.) Hoffm.	3.3 ± 0.9	—	IV–VI
Lamium maculatum L.	2.9 ± 0.2	5.8 ± 0.6 A	IV–IX
Pinus sylvestris L.	3.1 ± 0.3	5.9 ± 0.1 Sm	IV–VI
Pinus nigra L.	3.0 ± 0.3	7.6 ±0 .3 S	V–VI
Robinia pseudoacacia L.	4.0 ± 0.3	9.0 ± 1.2 A	V–VI
Iris pseudoacorus L.	1.0 ± 0.3	—	V–VII
Cerinthe minor L. [a]	0.3 ± 0.3	5.9 ± 0.6 A	V–VII
Chelidonium majus L. [a]	2.0 ± 0.3	5.0 ± 0.4 A	V–VIII
Chamomilla recutita (L.) Rauchert	1.8 ± 0.3	9.9 ± 0.8 A	V–VIII
Nuphar lutea (L.) Sm.	1.2 ± 0.4	—	V–VIII
Melilotus alba Medic. [a]	1.0 ± 0.3	7.1 ± 0.7 A	V–VIII
Melilotus officianalis (L.) Pallas	1.4 ± 0.4	4.6 ± 0.4 A	V–IX
Ballota nigra L.	2.0 ± 0.6	5.7 ± 1.6 N	VI–VII
Ligustrum vulgare L.	0.5 ± 0.2	0.9 ± 0.2 A	VI–VII
Consolida regalis S.F. Gray	1.8 ± 0.4	2.1 ± 0.6 N	VI-VIII
Chaerophyllum bulbosum L.	2.2 ± 0.3	—	VI–VIII
Typha angustifolia L.	5.0 ± 0.8	—	VI–VIII
Alisma plantago-aquatica L.	1.6 ± 0.4	—	VI–IX
Calystegia sepium R. Br.	1.6 ± 0.2	4.4 ± 0.6 A	VI–IX
Carduus acanthoides L.	1.2 ± 0.2	3.9 ± 0.8 N	VI–IX
Conium maculatum L.	1.0 ± 0.4	—	VI–IX
Chenopodium hybridum L.	0.2 ± 0.1	0.4 ± 0.1 N	VI–IX

Table 17.1 continued

Melandrium pratense (Rafn) Roeh.	2.0 ± 0.4	—	VI–IX
Raphanus raphanistrum L. [a]	5.0 ± 0.8	16.5 ± 1.5 A	VI–IX
Saggitaria sagittifolia L.	2.0 ± 0.7	—	VI–IX
Sinapsis arvensis L.	4.4 ± 0.8	7.9 ± 0.7 A	VI–IX
Stachys annua L.	3.9 ± 0.4	4.4 ± 0.5 N	VI–IX
Calluna vulgaris (L.) Hull	4.9 ± 0.2	27.4 ± 6.1 A	VI–X
Conyza canadensis (L.) Cronq.	0.1 ± 0.1	0.2 ± 0.1 N	VI–X
Mercurialis annua L.	1.6 ± 0.4	3.9 ± 0.9 N	VI–X
Berteroa incana (L.) DC.	1.2 ± 0.3	2.5 ± 0.4 N	VI–XI
Daucus carota L. [a]	4.4 ± 1.2	4.9 ± 0.3 A	VIvXI
Linaria vulgaris Mill. [a]	4.6 ± 1.1	7.9 ± 0.4 N	VI–XI
Trifolium pratense L. [a]	3.8 ± 0.6	9.9 ± 0.9 A	VI–XI
Artemisia vulgaris L.	2.4 ± 0.7	2.6 ± 0.5 A	VII–VIII
Typha latifolia L.	1.0 ± 0.3	—	VII–VIII
Pastinaca sativa L. [a]	3.4 ± 0.4	3.5 ± 0.3 A	VII–VIII
Kochia scoparia (L.) Schrad.	0.2 ± 0.2	0.4 ± 0.1 N	VII–IX
Cichorium intibus L.	1.0 ± 0.2	1.2 ± 0.2 N	VII–X
Picris hieracioides L.	0.4 ± 0.2	1.6 ± 0.4 N	VII–X

Note: S = spontaneous frequency (%) of abortiveness of pollen grains; H = highest observed frequency (%) of abortiveness of pollen grains in polluted environment from emissions of heavy industries: A = aluminum plant, N = nickel plant, R = oil refinery, Sm = smelting plant; B = blooming period (months).

[a] These plant species were tested near the nuclear power plant Jaslovske Bohunice on sites with increased radiation upto 322 kBq/kg of 137_{Cs}, but without significant impact on abortiveness of pollen grains.

17.3.3.1 Heavy metals

The basic effort was focused on studies of the different sources of pollution in Slovakia — nickel plant dumps at Sereï,[42] a smelting plant in Banská Štiavnica,[34] and an aluminum factory at Ziar nad Hronom.[43] As a check, we used samples from Botanical Garden of Comenius University in Bratislava.

17.3.3.1.1 Nickel plant dumps at Sereï — In some cases, frequency analysis of abortivity of pollen grains can show significant results, which cannot be confirmed by cytogenetic analysis. For example, biomonitoring of a highly polluted area around the nickel plant dumps (polluted with solid flue dusts of waste smelting, which contain heavy metals; mainly extractable parts of Ni, Cr, Co) at Sereï, South Slovakia, demonstrated the sensitivity of the testing method.[42] From the 20 species of native flora found in this area and chosen according to the basic criteria, five species (*Ballota nigra*, *Trifolium pratense*, *Mercurialis annua*, *Carduus acanthoides*, and *Picris hieracioides*) showed sensitivity to nickel and demonstrated significant increase of the abortivity of pollen grains.[42] Contrary to the general presumptions, genotoxic effects were not confirmed through tetrad analysis in *Typha latifolia*, nor through cytogenetic analysis in *in vivo Vicia faba* and *V. sativa* tests.[42]

17.3.3.1.2 Smelting plant in Banská Štiavnica — The possible adaptation and/or tolerance of the exposed population cannot be neglected. In some cases indicators, which statistically significantly detect the presence of genotoxic factors *in situ*, after growth in a contaminated environment through several vegetation periods, are less sensitive. Moreover, the frequency of their mutations as compared with controls could become insignificant.[18] In our study with arborescent species (*Pinus nigra* and *P. sylvestris*) we also had to deal with a possible adaptive response, especially at the time of chronic exposure of selected species in the polluted area near the smelting plant.[40] In the locality of Banská Štiavnica during germination pollen grains of both species tolerated higher concentrations of Pb in the cultivation media in comparison with control.[44] Nearly the same results were found in *Picea abies* (L.) Karst. in 20 populations with different degrees of industrial pollution at Slovenia.[45]

17.3.3.1.3 Aluminum factory at Ziar nad Hronom — Of the suitable species (see Table 17.1), *Artemisia vulgaris*, *Calystegia sepium*, *Chelidonium majus*, *Daucus carota*, *Lamium purpureum*, *Matricaria chamomilla*, *Melilotus albus*, *Melilotus officinalis*, *Pastinaca sativa*, *Raphanus raphanistrum*, and *Trifolium pratense*, all grown in Ziar nad Hronom area, were useful for sampling and scoring. Among the 11 analyzed species, no significant differences between the samples from polluted and control areas were observed with *A. vulgaris*, *D. carota*, *L. purpureum*, and *P. sativa*. This verifies the already stated necessity that, if we would, for different reasons, choose only these indication species, we could, by not adhering to methodical principle, deduce distorted and mistaken conclusions. It does not mean that these species could not be sensitive to other types of pollution. In the vicinity of the aluminum plant in the Ziar nad Hronom area, we found evidence of an increase in abortion in other indicators from our group, especially by *Calystegia sepium*, *Chelidonium majus*, *Matricaria chamomilla*, *Melilotus albus*, *Melilotus officinalis*, *Raphanus raphanistrum*, and *Trifolium pratense*.

17.3.3.2 Polycyclic Aromatic Hydrocarbons

The aluminum factory at Ziar nad Hronom, Central Slovakia produced emissions containing extremely high concentrations of fluorides, SO_2, As, PCB, and polycyclic aromatic hydrocarbons (PAH). The composition and amounts of these emissions are so extensive that a presumption of not only toxic but also mutagenic effects on living organisms is reasonable. The location of the factory is in a basin, with a frequent occurrence of atmospheric inversions, which increase the concentrations of emissions. Contents of metals as Al, As, and Cr in soil are 127 to 156 $mg{\cdot}kg^{-1}$, 40 to 160 $mg{\cdot}kg^{-1}$, and 60 to 160 $mg{\cdot}kg^{-1}$, respectively. This includes PAH that evaporate from the electrode paste in Söderberg process, consisting of coal tar and pitch.[46] Benzo(a)pyrene, an important carcinogen is a typical component of PAH. Territory within the radius of 10 to 15 km from the plant is threatened. The location of the factory in a basin and a frequent occurrence of atmospheric inversions implies increased concentrations of such emissions in the given territory as fenantren = 2 $mg{\cdot}kg^{-1}$, fluoranten = 3.3 $mg{\cdot}kg^{-1}$, pyren = 1.7 $mg{\cdot}kg^{-1}$, chryzen = 1.4 $mg{\cdot}kg^{-1}$, and perylen = 3.1 $mg{\cdot}kg^{-1}$.[43]

In the biological material (*Calluna vulgaris*) used for the study 16 priority PAHs were analyzed and quantified. In species with persisting mature pollen in tetrads (*C. vulgaris*) evaluation allows differentiation of pollen abortivity caused genetically from that caused physiologically. The principle of differentiation is based on the information that the physiological lesion usually affects all the inflorescences, flowers, anthers, and within the anthers the entire tetrads. If there is in the tetrad one, two, or three abortive grains, it is a case of genetic lesion. Genetic lesion of all four pollen grains is rare; hence, if it is classified as a physiological type, the overall resulting value remains unaffected.[11] In our opinion, mainly the components of PAH[46,47] are responsible for the increase of the frequency of abortive pollen and genotoxic effect, as confirmed by the results of tetrad analysis in *C. vulgaris*.[43]

17.3.3.3 Radionuclides

The study of bioindication of radioactive spots in the vicinity of the nuclear power plant Jaslovské Bohunice was accomplished in cooperation with the Institute of Nuclear Power Plants Research, Tranava (Slovakia) and the Institute of Tumorbiology, University of Wien, Austria.[48,49] The study showed a limit in sensitivity of wild plant species. Although there were selected soil samples with radioactivity at a large scale (0.067, 0.15, 2.38, 9.5, 45.5, and 322 kBq/kg 137_{Cs}) and in the pollen grains of *Raphanus raphanistrum* L., *Linaria vulgaris* Mill., *Trifolium pratense* L., *Cerinthe minor* L., *Lamium purpureum* L., *Chelidonium majus* L., *Melilotus alba* Medic., *Pastinaca sativa* L., and *Daucus carota* L. grown at these radioactive spots no higher abortiveness was found than at control areas. This result was confirmed also by a cytogenetic test where ana- and telophases (500 for each species) from root tips of *Roegneria canina*, *Cerinthe minor*, *Dactylis glomerata*, *Festuca pratensis*, *Arrhenatherum elatior*, *Poa pratensis*, *Lysimachia nummularia*, *Symphytum officinale*, and *Solidago canadensis* were evaluated without finding significant differences between radioactive spots and control areas.[49]

17.3.3.4 Crude Oil

For a general demonstration of the usefulness of this method a collaborative study with the Department of Botany and Microbiology of Kuwait University was accomplished.[50,51] The necessity of the proper time and method for collection and consequent successful testing in designated areas was particularly shown in this study. In all, 58 species of native Kuwait flora (e.g., *Gastroctyle hispida*, *Atractylis carduus*, *Malva parviflora*, *Plantago boissieri*, *Reseda arabica*, *Datura innoxia*) from 18 families (e.g., Asclepiadaceae, Cistaceae, Convolvulaceae, Geraniaceae, Liliaceae, Orobanchaceae, Rutaceae, and Zygopyllaceae) were evaluated with respect to the abortiveness of their pollen grains.[51] These plants were collected in April and May 1992 (and again in February and April 1993) from the most-polluted area (Burgan), i.e., the area most affected by the environmental impact of the Gulf War. In parallell, the same species were collected from more or less clean areas (Shuweich, Qurtuba, Khaldiya a Andalus). Environmental pollution resulting from burning the oil wells in the Kuwaiti desert during the Gulf War has been under genotoxic investigation. The cytogenetic assays performed to detect mutagenicity of oil pollution included pollen abortion analysis, pollen mother cell (PMC), and mitotic analysis for chromosomal aberrations.

Sterility of anthers, the presence of meiotic irregularities, and inhibition of growth were the main indicators of disturbance. An unusual kind of cytokinesis was observed which led to the formation of four pollen grains without the usual cell plate formation.

Out of 58 native plant species, 15 were selected as bioindicators according to certain criteria. Some were sensitive and could be used for short-term pollution testing, while others were resistant and could only be used for long-term pollution testing.[50]

17.3.3.5 Retrospective Biomonitoring

For a long time we were interested in comparing current results with a historical ecological background of a particular area. Finally, we were able to develop a method of retrospective biomonitoring of polluted environment by means of herbarium samples. Some results of this method have already been presented.[52,53] The new utility of herbarium samples for retrospective biomonitoring of polluted environment was tested by means of tetrad analysis of *Calluna vulgaris* during the last 100 years in the selected regions of the Slovakia. We were able to stain the old pollen grains from herbarium samples which showed (in comparison with the equally old pollen grains of herbarium samples from control areas) significant differences of pollen abortions in relation to the place and years of collections. Interesting is the dynamics of impact of polluted environment on pollen tetrads, for example, the heavy polluted area of Ziar nad Hronom with a peak in the year 1965 showing 56% of abortiveness. Surprising results with relatively high abortiveness were observed in High Tatras which is used as a recreation area. Possible explanation is that this area is affected by long-range pollution from heavy industrial areas around Czech and Polish coal mines (Ostrava, Katowice). This test gives us an opportunity to compare the impact of polluted environment 100 years retrospectively.

17.4 CONCLUSIONS

To date, the plant species used for practical tests of polluted environment are mostly cultivated ones, which are only partially useful in field conditions. Therefore, our laboratory specialized in the preparation of test using wild and crop plant species as bioindicators. The resulting method is efficient and easily applicable. Based on the data collected during these studies we are able to present a list of suitable bioindicators from species of Slovak flora (and Kuwaiti flora, too) as an operative, effective, and orientative front-line tool in ecology and ecotoxicology studies. This test could be used at the first step to recognize the current situation in regions of interest. It allows one to decide afterwards (with results from this test) if it is worth checking the genotoxic impact by other tests. This may be considered one of the main values of the present method, which could thus be applied for broad practical use.

This method could be useful, under specific conditions, for other regions as well. The local flora of these regions through selection of species according to the basic criteria laid down and their utilization in line with the recommendations made in this study could serve as a reliable source of plant biondicators.

ACKNOWLEDGMENT

We want to express our gratitude to Prof. Robert Murray Davis, University of Oklahoma, for his valuable comments on the manuscript.

REFERENCES

1. Bromberg, S. M., Identifying ecological indicators: an environmental monitoring and assessment program, *J. Air Pollut. Control. Assoc.*, 40, 976, 1990.
2. Holsen, T. M., Chaberski, C. M., Khalii, N. R., Scheff, R. A., and Keil, C. B., The Composition of Landfill Gas and Its Impact on Local Ambient Quality, Report of Project OSWR-02-006, Office of Solid Waste Research of University of Illinois, Center of Solid Waste Management and Research, 1991, 32.
3. Constantin, M. J. and Owens, E. T., Introduction and perspectives of plant genetic and cytogenetic assays, *Mutat. Res.*, 99, 1, 1982.
4. De Serres, F. J., Utilization of higher plant systems as monitors of environmental mutagens, *Environ. Health Perspect.*, 27, 3, 1978.
5. Murín, A., Simultaneous test of phytotoxic and mutagenic effects of polluted waters and herbicidal chemicals, *Biologia* (Britislava), 39, 15, 1984.
6. Plewa, M. J., Specific-locus mutation assays in *Zea mays*, *Mutat. Res.*, 99, 317,1982
7. Posthumus, A. C., Higher plants as indicators and accumulators of gaseous air pollution, *Environ. Monit. Assess.*, 3, 263, 1983.

8. Van°t Hof, J. and Schairer, L. A., *Tradescantia* assay system for gaseous mutagens, *Mutat. Res.*, 99, 303, 1982.
9. Vig, B. K., Soybean [*Glycine max* L. (Merill)] as a short-term assay for study of environmental mutagens, *Mutat. Res.*, 99, 339, 1982
10. Grant, W. F. and Zura, K. D., Plants as sensitive *in situ* detectors of atmospheric mutagens, in *Mutagenicity-Horizon in Genetic Toxicology*, Heddle, J. A., Ed., Academic Press, New York, 1982, 407.
11. Mulcahy, D. L., Pollen tetrads as indicators of environmental mutagenesis, *Environ. Health Perspect.*, 37, 91, 1981.
12. Cebulska-Wasilewska, A., *Tradescantia* stamen-hair mutation bioassay on the mutagenicity of radioisotope-contaminated air following the Chernobyl nuclear accident and 1 year later, *Mutat. Res.*, 270, 23, 1992.
13. Knasmüller, S., Kim, T.-W., and Ma, T.-H., Synergistic effect between tannic acid and X-rays detected by the *Tradescantia*-micronucleus assay, *Mutat. Res.*, 270, 31, 1992.
14. Sandhu, S. S., Gill, B. S., Casto, B. C., and Rice, J. W., Application of *Tradescantia* micronucleus assay for *in situ* evaluation of potential genetic hazards from exposure to chemicals at a wood-preserving site, *Hazard.Waste Hazard. Mater.*, 8, 257, 1991.
15. Mièieta, K., Zvýšená aberantnost' chromozómov u *Vicia faba* L. vplyvom kombinácie niektorých herbicídov po ošetrení v pol'ných kultúrach, *Agriculture* (Nitra), 12, 1094, 1987.
16. Grant, W. F., Lee, H. G., Logan, D. M., and Salamone, M. F., The use of *Tradescantia* and *Vicia faba* bioassays for the *in situ* detection of mutagens in an aquatic environment, *Mutat.Res.*, 270, 53, 1992.
17. Posthumus, A. C. and Tonnejick, A. E. G., Monitoring of effect of photo-oxidants on plants, in *Monitoring of Air Pollutants by Plants. Methods and Problems*, Steubing, L. and Jäger, H.-J., Eds., Dr. W. Junk Publishers, The Hague, 1982, 115.
18. Abramov, V. J. and Schevchenko, V. V., Genetic consequences of radioactive contamination for populations of *Arabidopsis*, *Sci. Total Environ.*, 112, 19, 1992.
19. Tomkins, D. J. and Grant, W. F., Monitoring natural vegetation for herbicide-induced chromosomal aberrations, *Mutat. Res.*, 36, 73, 1976.
20. Sharma, C. B. S. R. and Raju, D. S. S., *In situ* cytogenetic monitoring of vegetation around Hindustan polymers factory, Vishakhapatnam, India, *Environ. Mutat.*, 5, 409, 1983.
21. Cela Renzoni, G., Viegi, L., Stefani, A., and Onnis, A., Different *in vitro* germination responses in *Pinus pinea* pollen from two localities with different levels of pollution, *Ann. Bot. Fennici*, 27, 85, 1990.
22. Murín, A., Koleková, A., Váchová, M., and Regula, Š., Jednoduchá orientaèná metóda na urèenie fytotoxicity chemicky zneèistených odpadových vôd pomocou modelových organizmov vyších: Simultaneous test of phytotoxic and mutagenic effects of polluted waters and herbicidal chemicals, *Biológia*, 35, 937, 1980
23. Kanaya, N., Gill, B. S., Grover, I. S., Murín, A., Osiecka, R., Sandhu, S. S., and Andersson H. C., *Vicia faba* chromosomal aberration assay, *Mutat. Res.*, 310, 231, 1994.
24. Murín, A., Flowers as indicators of mutagenicity and phytotoxicity of polluted environment, *Biológia* (Bratislava), 42, 447, 1987.
25. Murín, A., Basic criteria for selection of plant bioindicators from the regional flora for monitoring of an environmental pollution, *Biológia* (Bratislava), 50, 37, 1995.
26. Sheoran, I. S. and Saini, M. S., Drought-induced male sterility in rice: changes in carbohydrate levels and enzyme activities associated with the inhibition of starch accumulation in pollen, *Sex Plant Reprod.*, 9, 161, 1996.
27. Darlington, C. D and La Cour, L. F., *The Handling of Chromosomes*, George Allen and Unwin, London, 1976, 182.
28. Amphlett, G. E. and Delow, G., Statistical analysis of the micronucleus test, *Mutat. Res.*, 128, 161, 1984.
29. Mièieta, K., Bioindication of mutagenous effects of polluted environment by means of higher plants, *Zivot. Prostr.*, 24, 267, 1990.
30. Mièieta, K., Aspekty adaptácie bioindikátorov vyšších rastlín pri detekcii genotoxicity zneèisteného zivotného prostredia, *Prùm. Toxikol.*, 19, 32, 1992.
31. Mièieta, K. and Murín, G., Higher plants in bioindication and monitoring of genotoxicity of the polluted environment [in Slovak], in *Monitoring of Environment*, Marko, J., Ed., MZP SR, Bratislava, 1993, 81.

32. Mièieta, K. and Murín, G., Bioindication of genotoxicity by means of species of regional flora in monitoring of polluted environment, in *Monitoring of Biota*, Eliáš, P., Ed., SAV, Bratislava, 1993, 41.
33. Mièieta, K. and Murín, G., The response of some bioindicators, species of Slovakian flora, from genotoxic impact of polluted environment, *Popul. Biol.*, 4, 85, 1996.
34. Mièieta, K. and Murín, G., Wild plant species in practice use for bioindication of polluted environment, *Ekológia* (Bratislava), 16, 193, 1997.
35. Mièieta, K., Zvýšená frekvencia sesterských chromatidových výmien u *Vicia faba* L. vplyvom kombinácie herbicídov Tobogard, Dual, Ladob, *Agriculture* (Nitra), 5, 438, 1988.
36. Mièieta, K., Cultivated plants as indicators of phytotoxicity and mutagenity of contaminated environment, *Agriculture* (Nitra), 35, 122, 1989.
37. Mièieta, K. and Murín, G., Example of using cultivated plants as a suitable bioindicators of polluted environment, in *Proc. Progress in Crop Sciences from Plant Breeding to Growth Regulation*, Ördög, V., Szigeti, J., and Pulz, O., Eds., Pannon University of Agricultural Sciences, Masonmagyarovar, Hungary, 1997, 25.
38. Mièieta, K. and Murín, G., Dreviny v bioindikácii a monitoringu genotoxicity v zneèistenom zivotnom prostredí, *Prum. Toxikol. Ekotoxikol.*, 22, 42, 1995.
39. Mièieta, K. and Murín, G., Microspore analysis for genotoxicity of a polluted environment, *Environ. Exp. Bot.*, 36, 21, 1996.
40. Mièieta, K. and Murín, G., Three species of genus *Pinus* suitable as bioindicators of polluted environment, *Water Air Soil Pollut.*, 104, 413, 1998.
41. Mièieta, K. and Murín, G., The use of *Pinus sylvestris* L. and *Pinus nigra* Arn. as a bioindicator species for environmental pollution, *Proc. of the First IUFRO Cytogenetics Working Party*, University of Zagreb, 1997, 253.
42. Uhríková, A. and Mièieta, K., *In situ* bioindication of genotoxic effect using the species of native flora in the vicinity of the nickel plant, *Biológia* (Bratislava), 50, 65, 1995.
43. Mièieta, K., Ostrovský, I., and Murín, G., Native flora in the bioindication of genotoxicity of PAH , in *Organics Xenobiotics and Plants (Impact, Metabolism and Toxiclogy)*, Proc. of 4th International Workshop, Bach, T., Collins, C., Keith, G., Rether, B., Schroder, P., and Weiss, P., Eds., Fed. Environ. Agency, Vienna, 1997 (unpublished data).
44. Holub, Z. and Ostrolucká, M. G., Tolerance of pollen from trees to chronic effect of Pb and acidity, *Biológia* (Bratislava), 48, 331, 1993.
45. Bavcon, J., Druškeviè, B., and Papeš, D., Germination of seeds and cytogenetic analysis of the spruce in differently polluted areas of Slovenia, *Phyton* (Austria), 33, 267, 1992.
46. Maòkovská, B., Puera, R., and Huttunen, S., Deposition of air-borne pollutants on the surface of spruce needles around the aluminium smelting, *Ekológia*, 7, 291, 1988.
47. Thrane, K. E., Ambient air concentrations of polycyclic aromatic hydrocarbons, fluoride, suspended particles and particulate carbon in areas near aluminium production plants, *Atmos. Environ.*, 21, 617, 1987.
48. Murín, G. and Mièieta, K., Radiation impact on the selected wild plant species in the vicinity of nuclear power-plant Jaslovské Bohunice, *Folia Biol.*, 42, 122, 1996.
49. Murín, G., Mièieta K., Knasmüller, S., and Steinkellner, J., Bioindication of radioactive-contaminated sites near nuclear power plant Jaslovské Bohunice by means of wild plant species, Zivotné prostredie (Bratislava), 3, 140, 1996.
50. Malallah, G., Afzal, M., Murín, G., Murín, A. and Abraham, D., Genotoxicity of oil pollution on some species of Kuwaiti flora, *Biológia* (Bratislava), 51(1), 18, 1997.
51. Murín, G. and Malallah, G., Biomonitoring of polluted environment in Kuwait after 100-days war, *Zivot. Prostr.* (Bratislava), 1, 30, 1994.
52. Mièieta, K. and Murín, G., Odraz genotoxickej zát'aze prostredia v tetrádovej analýze *Calluna vulgaris* za posledných sto rokov vo vybraných regiónoch Slovenska, in *Populaèná biológia rastlín III*, Eliáý, P., Ed., Sekos, 1994, 68.
53. Mièieta, K. and Murín, G., New utility of herbal samples for retrospective biomonitoring of polluted environment, *Zivotné prostredie* (Bratislava), 5, 262, 1996.

CHAPTER 18

Phytomonitoring in Industrial Areas

Sharad B. Chaphekar

CONTENTS

18.1 INTRODUCTION

Degradation of the environment has almost seemed the order of the day, ever since industrialization began. In developing countries, this problem is much more serious, where pollution control and environmental management as an integrated practice have taken root only in recent times. Landscaping, harvesting of local resources for construction of industrial structures, transportation

1-56670-341-7/00/$0.00+$.50

of material and people, development of residential habitats, air and water pollution, and solid waste disposal keep taking place "in tandem" in the process of industrial development. Each of the steps from clearing of land, through construction, and to operation of industry has its own contribution to environmental degradation. Although this degradation is perceptible in the form of general barrenness and pollution of land, water, and air, this chapter focuses on the last one, i.e., air pollution. Assessment of air pollution with the help of plants is the main focus of this effort. It is envisaged that an understanding of the acuteness and distribution of pollutants in industrial areas is the key to planning a strategy for air pollution abatement in particular and environmental management of industrial areas in general. Early warning of industrial air pollution using indicator plants — phytomonitoring — combined with other efforts for environmental management, e.g., "Zoning Atlas for Siting of Industries,"[1] "Guidelines for Development of Green Belts in Industrial Areas,"[2] etc., would help in achieving environmental management effectively.

18.2 POLLUTION INDICATORS AND PHYTOMONITORING

18.2.1 Advantages of Phytomonitors

A plant indicator indicates a stress situation in its environment. As per the *Oxford Dictionary*, *indicator* is a device that "indicates the condition of...". A *monitor*, on the other hand, is "a device for checking, or warning" about a situation. *To monitor* means "to maintain regular surveillance over..." Regular evaluation of environmental quality or health is called monitoring and, when plant indicators are employed as tools, the term *phytomonitoring* is used.

Some serious efforts for phytomonitoring of the regional environment have been made in different places. Notable one is by Posthumus,[3] who monitored levels of air pollutants, using physical and chemical methods over a large number of sites, forming a network across the Netherlands. Transplants of indicator plants like alfalfa, gladiolus, etc. were placed at those sites and related plant growth or injury due to air pollution levels at respective sites was assessed. Excellent maps of air quality in the Netherlands have been prepared as an output of the exercise. On the basis of the experience gathered during the work, he has stated the advantages of phytomonitoring:

1. Phytomonitoring provides integrated effects of all environmental factors including air pollution and weather conditions.
2. It gives direct effects of pollution on living organisms.
3. When phytomonitoring is combined with physicochemical monitoring, the relationship between pollutant concentration and its effects on plants can be traced.
4. The work shows the possibility of determining trends in the occurrence and intensity of effects of several air pollutants on plants.
5. Accumulation of pollutants by plants can be determined.
6. The phytomonitoring exercise can serve as an early warning system to enable timely measures to control pollution and reduce its disastrous effects.

18.2.2 Interference in Indication

Visual symptoms of injury are preceded by effects on growth and yields of plants. Concentration, duration and frequency of pollutants to which plants are exposed, combined with environmental factors like temperature, humidity, soil moisture, light, nutrient conditions, age of tissues, etc. are involved with phytomonitoring.[4] As such, the exercise leads to giving not only the state of air pollution in the area, but also the general environmental health there. Ting and Heath[5] have also stated that plants differ in their response to air pollution under different nutritional conditions and age. Sensitivity of plants to SO_2 was affected by nutritional conditions and application of herbicides.[6]

Intraspecific variations in response to air pollution were also reported.[6] Jain [7] has also reported that response of plant species to air pollutants differed during their sapling and mature stages of life. Treshow[8] has cautioned against indiscriminate use of phytomonitoring for assessment of environmental health on the grounds that many unrelated stresses could cause symptoms resembling air pollution–induced injury. Chemical analyses of plant tissues for sulfur and fluorides have been recommended by him to supplement phytomonitoring results, possibly along with histopathological studies of the phytomonitors.

18.2.3 Industrialization and Air Pollution

Air pollution due to industries takes place during two phases — during the construction phase and during operational phase of the industry. During the first phase, it is mostly dust and cement that contribute, sometimes very heavily, to air pollution in the immediate vicinity where the industry is being set up and construction of roads and building structures is under way. Emissions from trucks and lorries and sometimes diesel-generating sets used for drilling, pile foundations, etc. also contribute substantially to air pollution during this phase,[9] which may continue for some months to some years. The problem is serious in localized areas of quarries and stone crushers, transportation routes, loading and unloading of construction material, and the construction site itself.

The nature of the air pollution during the operational phase of a project depends on the nature of the processes used in the factory. The pollutants are mainly SO_2, NO_x and HF as primary emissions and oxidants like O_3, PAN, smog, etc. as secondary ones, formed in the industrial atmosphere due to atmospheric reactions. Ammonia and chlorine are also sometimes emitted, because of lack of maintenance, leakage, and accidents. Ethylene is an important emission from refineries and other organic chemical processes. Dust gets emitted from several types of factories — fertilizers, cement, thermal power plants, and metal smelters, the particulate from the last one being metallic in nature (Table 18.1). Any of these emissions coming in contact with plants leaves its typical impressions on the foliar surfaces, in the form of visible injury. Symptoms of foliar injury are specific to the pollutants. Flowers and fruits also sometimes exhibit injury symptoms. Sensitive plants showing typical injury symptoms (Table 18.1) are often recommended for use as indicator species. Halbwach[4] and Taylor[10] have listed plants that are sensitive to primary and secondary air pollutants. Halbwach[4] has cited examples of very sensitive plants which get injured at as low pollutant concentration as 2 $\mu g\ m^{-3}$ HF or 50 $mg\ m^{-3}$ SO_2, for 24 h.

18.2.4 Mechanism of Injury

For plants to be useful as good indicators of air pollution, it is necessary that they exhibit injury symptoms clearly and specifically. Pollutant gases enter leaf tissues, mostly through stomatal apertures[11] or through cuticle.[12] SO_2 was found more soluble in cuticle than in water and SO_2 permeated *Citrus* cuticle via the lipophilic phase.[12]

Numerous studies have been carried out to understand biochemical, enzymatic, and intracellular reactions taking place within cells, leading to inhibition of photosynthesis and impairment of other metabolic processes, as a result of absorption of pollutants by plants.[13-15] All the studies indicate a relationship between pollutants and the upsetting of metabolic balance of plants, e.g., breakage of the chlorophyll molecule and degradation of chloroplasts due to SO_2[16]; severe plasmolysis, loss of water from degenerating cell wall due to O_3[17]; interaction of fluoride with enzymes and nutrients, leading to effects on oxygen uptake, cell wall formation, starch synthesis[18]; etc.

18.2.5 Magnitude of Injury and Pollution Dose

Qualitative relationship between pollution dose and magnitude of injury cannot, however, be traced. No sound mathematical relationship between contents of (pollutant) elements in leaves and

Table 18.1 Industrial Sources of Major Air Pollutants and Plant Injury Symptoms Caused by Them

Serial Number	Pollutant Gas	Sources	Plant Injury Symptoms on Leaves
1	Sulfur dioxide (SO_2)	Coal, fuel oil, and petroleum combustion	Interveinal chlorosis, necrosis, bronzing; red brown dieback and banding in pines.
2	Nitrogen oxides (NO_x)	High-temperature combustion of fuels in power plants; internal combustion engine	Interveinal bronzing as by SO_2; red-brown distal necrosis in pines
3	Hydrogen fluoride (HF)	Phosphate rock processing; smelting of iron, aluminum; brick and ceramic works; glass manufacturing	Tip and margin necrosis
4	Ethylene (C_2H_4)	Incomplete combustion of fuels; auto exhausts	Chlorosis, necrosis, abscission, premature defoliation, curling
5	Particulates (SPM)	Incomplete combustion; fly ash, cement factories; metal smelters; auto exhausts	Surface deposition, serious effects according to chemical nature of dust
6	Chlorine (Cl)	Leakage in storage tanks, transfer pipes, transport	Chlorosis, upper-surface fleck, distal necrosis in pines.
7	Ammonia (NH_3)	Leakage in fertilizer factories, transfer pipes, transport	Interveinal necrosis as by SO_2, distal necrosis in pines.
8	Ozone (O_3)	Atmospheric reactions in urban and industrial areas	Upper-surface flecks, needle point chlorotic dots, chlorosis, necrosis in pines
9	Peroxy-acetyl nitrate (PAN)	Atmospheric reactions in industrial areas	Lower surface bronzing, chlorosis, early senescence
10	Smog	Atmospheric reactions in industrial areas	Chlorotic banding on grass blades, chlorosis, necrosis as with O_3

injury can be obtained.[19] Effects of very low concentration of pollution over long periods on plants may possibly be studied with the help of electron microscopy and biochemical investigations of lipid metabolism and of enzymatic changes characteristic of stress metabolism.[20] It is difficult to trace more quantitative aspects of pollution levels and their relationship to leaf injury. Two reasons probably make it impossible to quantify pollution levels on the basis of the magnitude of injury to plants.

First, atmosphere in industrial area is almost never polluted by a single pollutant. A thermal power plant using coal as fuel emits particulate, SO_2, and NO_x, and a fertilizer manufacturing plant releases into its atmosphere SO_2, NO_x, HF, particulate and, sometimes, ammonia. More commonly, we have industrial centers, where several industries have been established together for the convenience of the development of infrastructure facilities. Several pollutants have been recorded in such areas.[21] Similar examples are reported in at least eight major industrial centers in India alone. The formation of secondary pollutants as a result of atmospheric reactions contributes oxidants to the same air. Mixtures of pollutants may have synergistic or additive effects on plants. Antagonistic effects, although theoretically possible, are rarely reported, at least with reference to air pollution. Reduced yields in several plants are reported, due to synergism in pollutants like $SO_2 + NO_2$, $SO_2 + O_3$ and SO_2 + HF.[21-23] Microscopic annular rings of bleached leaf surface were recorded in Chembur in humid Bombay (Mumbai), due to deposition on tree crowns of hot dust particles that were emitted from the stack of a thermal power plant, that cooled, and that gathered condensed water vapor and acidic gases during their descent.[24] It is common to observe leaves of roadside trees coated with dust, soot and oily deposits, and lead.[25]

Second, the environmental complex, being variable in space and time, often dictates the sensitivity of plants toward pollutants. All the plants of an area were found to be severely damaged due to air pollutants emitted by a factory on a stormy night.[26] Detailed investigations revealed that plants downwind from the factory were affected more severly than those elsewhere in the region. During that night, stormy winds were blowing from southwest to northeast. The field observation was later confirmed with the help of controlled fumigation under simulated windy conditions.[27] The role of external environmental and internal physiological, nutritional, and developmental conditions of plants in controlling their pollution sensitivity to some extent was repeatedly stressed by different workers.[4,28] These findings have an important bearing on the detailed design of phytomonitoring programs.

18.3 CHOICE OF INDICATORS FOR PHYTOMONITORING

18.3.1 Considerations

Under the complex situation of environmental variables and variables in the internal characters of plants, it is desirable that phytomonitor plant species are selected judiciously to arrive at a correct reflection of the pollution status of an area. Meticulous observations of the chosen species need to be planned for reliable data collection. The following considerations may prove useful for this purpose.

18.3.1.1 Easy Availability

Phytomonitors should be easily available in the area to be monitored. Some simulation experiments to confirm responses of plants to individual pollutants and to combinations of pollutants are desirable to ensure that the chosen plant species are consistent in their response. Plants found widely suitable as indicators are alfalfa for SO_2, gladiolus for HF, tobacco (Bel W_3) for O_3, etc. A large number of plant species are already identified in different places, by different researchers (Table 18.2).

18.3.1.2 Strains and Age

Phytomonitors should belong to the same strain/cultivar and age. Quite often, population differentiation occurs within a species distributed in different parts of a region. Some of these populations respond differently to local stresses. An SO_2-tolerant population of *Lolium perenne* was identified by Bell et al.[29] A similar observation was made in wheat by Verma and Agrawal,[6] who detected differences in sensitivity of different cultivars toward pollution. Development of tolerance to stress, including pollution stress, is stated as a step in adaptive evolution.[30] Such an evolutionary process may be fairly rapid.[31] While choosing a plant for monitoring, care should be taken to ensure that only the most sensitive strain is selected and it has to be used for a complete set of phytomonitoring effort. Populations multiplying for generations in a stressed area should preferably be avoided. All monitors in a batch will have to be of the same age, since plants of different ages were found to be different in their responses to SO_2, during experiments on tropical tree species,[7] as well as herbaceous cluster beans.[32]

Table 18.2 Plant Indicators Useful as Phytomonitors of Industrial Air Pollution

Herbs, Seasonals	Shrubs, Trees
Arachis hypogea	*Mangifera indica*
Trigonella foenum-graceum	*Nerium odorum*
Spinacea oleracea	*Syzygium cuminii*
Raphanus sativus	*Polyalthia longifolia*
Helianthus annuus	*Dalbergia sissoo*
Petunia sp.	*Gossypium* spp.
Medicago sativa	*Terminalia catappa*
Glycine max	*Bougainvillaea spectabilis*
Cicer arietinum	*Pinus* spp.
Pisum sativum	*Abies* spp.
Dolichos lablab	*Ginkgo biloba*
Tithonia diversifolia	*Vitis vinifera* (climber)
Vicia faba	*Betula* spp.
Commelina benghalensis	*Tilia* spp.
Lycopersicum esculentum	*Quercus* spp.
Nicotiana tabacum	*Ailanthus* spp.
Phaseolus aureus	*Ficus religiosa*
Gladiolus sp.	*Ficus* spp.
Triticum aestivum	*Platanus* spp.

18.3.1.3 Simplicity for Observation

Phytomonitors should be simple to observe and interpret. During the opening address to the 7th International Bioindicator Symposium at Kuopio, Finland in 1992, Sisula[33] emphasized that the "Administration" needed bioindicators that were specific to a pollutant (or a group of pollutants) ... and of an early-warning type. Bioindicators must also be relatively inexpensive and easy to use in extensive surveys. It was further stated that sophisticated indicators requiring highly qualified expertise could hardly be helpful in routine monitoring and surveys.

The description of injury symptoms is specific for pollutants like SO_2, NO_2, HF, O_3, PAN, etc., but for mixtures of pollutants it is hardly so. Acute pollution levels also lead to large-scale scorching and defoliation. Meticulous observations of individual sample leaves then prove to be less significant, even less relevant, than methodical observation of many leaves on different twigs exposed differentially to incoming pollutants. Ratios of damaged to total leaf area, injured leaves to total leaves, defoliation on windward to leeward sides, etc. are necessary to quantify for proper evaluation of pollution status of the area, as in case of a region overshadowed by a coal-based thermal power plant in the coastal area north of Mumbai.[34]

18.3.1.4 Specificity

Phytomonitors should be specific to the stresses to be evaluated. Ideally, each type of stress should be indicated by a specific injury symptom by the monitor. Under conditions of pollutant mixtures, acute pollution, episodic concentrations of pollutants, and interfering factors of nutrition and/or environment, specificity of injury symptoms is hardly maintained. A slight refinement to monitoring could be possible by adding the parameter of chemical contents of tissues, at least in case of pollution by SO_2 and HF. The same is not applicable in the case of injury due to NO_x. Injury symptoms due to NO_x are similar to those due to SO_2, but unlike sulfur, nitrogen is not stored in injured tissues. Some air pollutants like O_3, PAN, ethylene, etc. degenerate inside plants.[8,17]

18.3.1.5 Multiple Species

More than one phytomonitor species should be selected for more reliable assessment. It is logical to assume that observation of more monitor species would enhance the reliability of the phytomonitoring effort. This has been recommended right from the time of development of the concept of plant indicators.[35] A high level of experience of fieldwork is a prerequisite for this purpose.

Naturally growing plant communities are also good indicators of their environment. Reduction in species diversity under pollution stress due to loss of sensitive species over a period of time is reported. Singh et al.[36] found that tree cover and species richness improved with distance from pollution source, while the number of even herbaceous species reduced near the pollution source, where tree cover was almost nonexistent.

18.4 PASSIVE VS. ACTIVE PHYTOMONITORING

18.4.1 Passive Monitoring

Phytomonitoring is sometimes distinguished into two convenient types — passive, where plants already growing in the field are used for observations, and active, plants that are cultivated for the purpose of phytomonitoring itself. Numerous examples of both types of monitoring are available. Thomas[37] discussed injury to plants around industries, as evidence of the polluted environment prevailing there. Leaf area and biochemical contents of leaves of trees were used as parameters to substantiate pollution status of the region around Shaktinagar Thermal Power Plant, India.[38] Variations in morphological characters like stomatal indexes, trichome length and density, etc. were also used for phytomonitoring in several plant species growing in industrial areas.[39] Reports on needle injury in pines and foliar damage in various tree species have been reported by several workers from temperate and cold countries, as well as from other parts of the world.[8,11,40-42] A brief review of air pollution–induced injury in plants, from the viewpoint of pollution indication, mainly from studies in India was presented by Chaphekar.[43]

18.4.2 Active Monitoring

There are several limitations in passive phytomonitoring, specifically, variations in soils, nutrition, water drainage, exposure, etc., where the selected phytomonitors are growing. Species of trees and their varieties may also differ, making comparison of data from different sites unsuitable. Results of this phytomonitoring are essentially qualitative and site specific. To overcome the problems mentioned above, active phytomonitoring is suggested. In this, plants are mostly herbaceous, potted annuals or seasonal. These are planted in identical pots, with identical soil mixture, watered identically, and even exposed in comparable fashion as far as possible. The seeds used are of uniform quality or plants are propagated from common stock to ensure genetic uniformity.[44] The candidate plants are studied intensively, under controlled conditions, preferably using fumigation chambers for their responses to pollutant gases of known dosages. Data thus collected are useful for comparison at a later stage, when laboratory-grown plants are exposed at monitoring sites for a predetermined period and studied for response. Differences in growth due to exposure to even sublethal doses of pollutants are recorded and computed for pollution levels prevailing at the sites during the period of exposure. With this technique, estimated pollution levels can be expressed on a three or four point, semiquantitative scale, specifically, very high to "nil" pollution at two extremes with one or two grades of pollution between the extremes.

The system is advantageous as it provides better and relatively more accurate information on pollution levels at monitoring sites. Disadvantages of this monitoring exercise are that it needs exhaustive planning and research support at all the stages of work, i.e., selection of candidate species, fabrication and standardization of fumigation chamber, uniform culturing of standardized phytomonitor, its distribution for exposure to ambient air and timely collection for analyses, and computation of data. With reference to sustained primary productivity, especially of important food or commercial crops, cultivated at the interface of industries and agriculture, utility of this method of monitoring has no substitute. Also, in all probability, this method is also an asset in town and industrial estate planning with an adequately designed approach. Preparation of air quality maps of Bombay city in the 1980s was one such effort involving intensive planning.[45] Several studies of this type can be cited, where useful data generation about environmental quality in industrial and other areas has been carried out.[3,45-48]

18.4.3 Lower Plants as Phytomonitors

A series of interesting studies in phytomonitoring of air quality in industrial areas, with the help of lower plants like mosses and lichens, have been carried out. These plants are sensitive to air pollutants due to the absence of protective cuticle or wax layer on their thalli. As a result, on exposure to polluted air over a period of time, the more sensitive species either get eliminated or exist with feeble growth rates. Intensive phytosociological surveys of lichens carried out at a point of time show absence or low growth of sensitive species. Based on these observations made on epiphytic lichens to maintain uniformity of habitat conditions, DeSloover and LeBlanc[49] have prepared maps of the Sudbury area, Ontario, in Canada. In these maps, five zones of atmospheric purity indexes have been marked, with suitable recommendations for further zone-wise development of the town. Other attempts in a similar effort for phytomonitoring with epiphytic lichens are those by Seward[50,51] and many others. Gilbert[52] suggested use of mosses for similar purpose.

18.4.4 Dendroecological Approach

Based on the premise that plant growth suffers under pollution stress, in both the vertical and radial dimensions, studies on the width of annual rings of trees were carried out by Keller[53] under conditions of experimental fumigations, with promising results. Miller[54] reported studies on forest trees in San Bernardino mountain region, in the U.S. for vertical and radial growth of shoots as related to pollution by oxidants. Suppression of growth in this case was found to be serious. Dendroecology as a phytomonitoring tool is expected to emerge powerfully in the near future, especially since it involves a nondestructive, inexpensive method of using only tubular probes and microscopes, to get important information about tree growth in urban-industrial, as well as forest areas, in time and space.

18.4.5 Mango Tree — A Phytomonitor of Promise

A number of trees have been used for monitoring air quality, such as pines, ash, elm, etc., in temperate countries. Corresponding to the high diversity of species in a country like India, the number of trees studied for air pollution effects is also high. However, when several studies are considered together, one realizes that the tree species that commands more attention has been the popular fruit tree, mango (*Mangifera indica*). From studies in Bombay on the west coast,[55] through Ujjain in central India,[56] to the Gangetic region in north India,[36] mango tree emerges as the most consistently observed tree for air pollution effects as well as for capturing dust from the polluted environment. Progressive reduction in leaf area,[36] reduction in leaf injury with increasing distance from pollution source along a 2-km-long transect,[55] biochemical contents of leaves like chlorophylls, ascorbic acid, amino acids, GDH (glutamate dehydrogenase), and GS (glutamine synthetase)[56] showed the reliability of mango as a consistent indicator of air pollution. The species is widely distributed in the country, in the wild in the forests of peninsular India and under cultivation everywhere from the southern peninsula to the Himalayan foothills, and from coast to coast in the peninsula. It is possible to control the varietal choice in an area, since it is traditionally cultivated for a particular type of fruit from a particular area, and also since most cultivation is done by grafting, budding, air layering, root cutting, or marcotting. Its high sensitivity to air pollution makes it a matter of concern whenever industries are planned in predominantly agricultural or forest areas.[57] It was on this background that the species forms a major component of green belts being cultivated in new industrial establishments. Here, mango is expected to provide fruit, a sink for pollutants, and a reliable phytomonitor of the environmental quality.[58]

18.5 SOME EXAMPLES OF PHYTOMONITORING

As already stated, one of the best examples of phytomonitoring of a vast region is presented by Posthumus,[46] where he distributed potted plants of uniform nature, grown in identical medium, and exposed in identical fashion, all over the Netherlands. Results obtained after a period of exposure at those sites were plotted for differences in growth. Instrumental monitoring of air quality was also done simultaneously and the two sets of results were corroborated. Maps of the country based on the results show clean to polluted areas. As part of a multistate air pollution monitoring program in the eastern U.S., Jacobson and Feder[59] studied atmospheric oxidant concentration and foliar injury to tobacco plants which were used as indicators. Tobacco Bel W3 and watermelon plants were used to assess potential impact of oxidants in Greece and Spain. Davison et al.,[60] however, preferred the use of *Pinus halepensis* as a biomonitor for assessing the actual impact of oxidants on vegetation and crops in about 100 km around Athens and between Valencia and Terragona in Spain.

Proline contents in leaves/needles of *Betula pendula*, *Fraxinus excelsior*, *Tilia cordata*, *Pinus sylvestris*, and *Taxus baccata* were found to be good markers for biomonitoring of polluted urban Slovakia.[61] A large-scale approach much simpler to work on was undertaken by Galstova and Vasina[62] for a forest monitoring survey in the northwest part of Russia, as part of the project on long-range transmission of air pollutants in Europe. The simple parameters employed were defoliation of pine needles and needle longevity. In addition, epiphytic lichen flora and chemical (including heavy metals) burden of mosses were studied to prepare pollution maps of the region, which could be related to pollution sources there. Diversity of species of higher plants was used as a parameter by Bohac et al.[63] for ecological planning in agricultural settlements in Havron territory of the Czech Republic.

Several urban areas have also been evaluated for their air pollution status, using phytomonitors. In one of the earliest studies in India, this author[55] reported a gradient of air pollution downwind of a fertilizer factory in suburban Bombay, based on decreasing leaf injury in several plant species, with increasing distance from pollution source. Trees like *Syzygium cumini*, mango, coconut, *Terminalia catappa*, and herbaceous ruderals like *Hygrophila auriculata* and *Xanthium strumarium* were used as indicators in this study.

Agrawal and Agrawal[64] reviewed the biomonitoring studies conducted at Varanasi and Mirzapur districts of Uttar Pradesh, India, where several plants were used as phytomonitors. Leaves of plants were analyzed for injury, chlorophyll pigments, sulfur contents, leaf extract pH, relative water content (RWC), peroxidase activity, and biomass (accumulation). Air pollution tolerance index (APTI) values were calculated on the basis of total chlorophylls, ascorbic acid content, leaf extract pH, and RWC.[65] Plants with high APTI values were more tolerant, whereas those with low APTI values were sensitive. Efforts were made to demarcate with precision APTI values for sensitivity, tolerance, and intermediate responses of plants toward pollution stress. A wider testing of the concept would help its confirmation or improvisation for possible universal applicability.

Herbaceous plant species like *Helianthus annuus* var. japanese miniature, *Commelina benghalensis*, *Crotalaria juncea*, and *Cyamopsis tetragonolobus* were studied in controlled fumigation experiments, and the first two were selected for phytomonitoring of air quality in Bombay city, because of the consistency of the response.[45] A 4-week exposure in different parts of the city of these phytomonitors showed foliar injury of different levels, from death, through necrosis, to growth reduction, in different parts of the city. Not only did old industrial areas and roadside areas with heavy vehicles showed worse injury, even death, of phytomonitors within a week, but a few places revealed very high pollution levels along seafront traffic arteries which had never before attracted the attention of the State Pollution Control Board.

Zn, Pb, and Cd were determined from vegetation samples around a smelting complex in Avonmouth area of Britain, to realize that the metals were distributed from the smelter in air currents, to distances as far as 10 to 15 km.[66] Klumpp et al.[67] identified *Hemerocallis* as a suitable bioindicator of fluoride pollution in the industrial belt of Cubatao, Brazil. A drop in photosynthetic activity measured by fluorescence spectrum of canopy, biochemical contents, and metal accumulation were considered suitable parameters for phytomonitoring of air pollution in Turin, Italy.[68] Numerous surveys of foliar injury to plants around new industrial areas in western India showed widespread problems of air pollution in the predominantly agricultural and forested regions.[69] Severe foliar injury in morning glory in Kantoh district of Japan was found to correspond directly with O_3 concentrations in the respective regions.[70]

Ruderal vegetation and traffic-island ornamentals in Bombay were studied for Pb pollution caused by vehicular traffic. Analyses of leaf washings showed Pb values as high as 164 mg g^{-1} of leaf tissue.[25] Plant species like *Ipomoea carnea*, *Hyptis svaveolens* (both ruderals), and *Pithecellobium dulce*, *Lantana camara*, *Ervatamia divaricata* (all ornamentals) were found efficient in capturing dust loaded with automobile exhaust Pb.

Lichens and mosses have been successfully used for monitoring air pollution and heavy metal fallout in industrial areas, especially in Europe and Canada, e.g., Brumelis and Nikodamus in Latvia,[71] Ruhling in northern Europe,[72] and others. The high sensitivity of lichens to air pollution is considered to be due to the unprotected (naked of protective cuticle, wax, etc.) thalli losing water on exposure to pollutant gases.[28] Lichens were also employed as monitors of air quality in the metropolitan zone of Palermo[73] and Varese[74] in Italy. A major limitation in employing this technique elsewhere is mainly climatic, especially in dry and semidry tropics which prevents year-round growth of these lower plants.

18.6 FUTURE PROSPECTS FOR PHYTOMONITORING

Phytomonitoring is admittedly a simple and relatively inexpensive tool for rapid and continual surveys of environmental quality. For widespread use of this technique of monitoring air quality, especially with reference to sustainability of crops and forests, it is necessary that authentic reference material be produced in abundance. It should be easily available, readable, and utilizable by large fleets of field-workers. Pictorial atlases of visible injury to locally available plants need to be prepared along the lines of those from the eastern U.S.[40] and the Mojave Desert.[75] Need is felt more urgently in the developing world where the spread of industrialization is rapid and environmental controls are relatively liberal or remain to be enforced. Even in countries where forests are major contributors to the national economy, continuous, organized phytomonitoring of air quality is an urgent necessity.

There is an adequate knowledge base in the subject of phytomonitoring of air pollution. Problems of water and soil pollution as well as of land degradation are also serious enough to warrant immediate attention for standardization of phytomonitoring, which should be employed on the scale demanded by the widespread prevalence of the problem. A large number of studies in the field should prove to be a step in the right direction for balanced developmental activity coupled with conservation of nature.

REFERENCES

1. Biswas, D. K., Zoning atlas for siting of industries, *Parivesh Newsl.*, Central Pollution Control Board, New Delhi, 3(1), 2, 1996.
2. Kapoor, R. K., Chaphekar, S. B., and Gupta, V. K., Guidelines for Development of Green Belts, Final Report to Central Pollution Control Board, New Delhi, 1996.
3. Posthumus, A., General philosophy for the use of plants as indicators and accumulators of air pollutants and as biomonitors of their effects, *VI World Congr. Air Qual.*, Paris, 2, 555, 1983.
4. Halbwach, G., Organismal responses of higher plants to atmospheric pollutants, SO_2 and HF, in *Air Pollution and Plant Life*, Treshow, M., Ed., John Wiley & Sons , New York, 1984, 175.
5. Ting, I. P. and Heath, R. L., Responses of plants to air pollutant oxidants, *Adv. Agron.*, 27, 89, 1975.
6. Verma, M. and Agrawal, M., Interactive effects of SO_2, herbicide and fertility regime on two cultivars of wheat, *Plant Physiol. Biochem.*, 23, 53, 1996.
7. Jain, N., Assessment of Few Tropical Tree Species Against Air Pollutants, Ph.D. thesis, Vikram University, India, 1993.
8. Treshow, M., Diagnosis of air pollution effects and mimicking symptoms, in *Air Pollution and Plant Life*, Treshow, M., Ed., John Wiley & Sons, New York, 1984, 97.
9. Anonymous, A Handbook of Environmental Impact Assessment Guidelines, Department of Environment, Ministry of Science, Technology, and Environment, Kuala Lumpur, 1995.
10. Taylor, O. C., Organismal responses of higher plants to atmospheric pollution:photochemical and others, in *Air Pollution and Plant Life*, Treshow, M., Ed., John Wiley & Sons, New York, 1984, 215.
11. Treshow, M. , *Environment and Plant Response*, McGraw-Hill, New York, 1970.
12. Lendzian, K. J., Permeability of plant cuticles to gaseous air pollutants, in *Gaseous Air Pollutants and Plant Metabolism*, Koziol, M. J. and Whatley, F. R., Eds., Butterworths, New York, 1984, 49.
13. Darrall, N. M., The effect of air pollutants on physiological processes in plant, *Plant Cell Environ.*, 12, 1, 1989.
14. Ormrod, D. P., Black, V. J., and Unsworth, M. H., Depression of net photosynthesis in *Vicia faba* L. exposed to SO_2 and O_3, *Nature*, 291, 585, 1981.
15. Hallgren, J. E., Photosynthetic gas exchange in leaves affected by air pollutants, in *Gaseous Air Pollutants and Plant Metabolism*, Koziol, M. J. and Whatley, F. R., Eds., Butterworths, New York, 1984, 147.
16. Rao, D. N. and LeBlanc, F., Effects of SO_2 on lichen algae, with special reference to chlorophyll, *Bryologist*, 69, 69, 1965.
17. Smith, W. H., *Air Pollution and Forests*, Springer-Verlag , New York, 1981.
18. Weinstein, L. H., Fluoride and plant life, *J. Occup. Med.*, 19, 49, 1977.
19. Garsed, S. G., Uptake and distribution of pollutants in the plant and residence time of active species, in *Gaseous Air Pollutants and Plant Metabolism*, Koziol, M. J. and Whatley, F. R., Eds., Butterworths, New York, 1984, 83.
20. Huttunen, S. and Soikkeli, S., Effects of various gaseous pollutants on plant cell ultrastructure, in *Gaseous Air Pollutants and Plant Metabolism*, Koziol, M. J. and Whatley, F. R., Eds., Butterworths, New York , 1984, 117.

21. Runeckles, V., Impact of air pollutant combinations in plants, in *Air Pollution and Plant Life*, Treshow, M., Ed., John Wiley & Sons, New York, 1984, 239.
22. Agrawal, M., Nandi, P. K., and Rao, D. N., Ozone and sulphur dioxide effects on *Panicum miliaceum* plant, *Bull. Torrey Bot. Club*, 110, 435, 1983.
23. Yu, S.-W., Air pollution problems and the research conducted on the effects of gaseous pollutants on plants in China, in *Gaseous Air Pollutants and Plant Metabolism,* Koziol, M. J. and Whatley, F. R., Eds., Butterworths, New York, 1984, 49.
24. Chaphekar, S. B., Plant indicators: the concept and new additions, *Int. J. Ecol. Environ. Sci.*, 4, 45, 1978.
25. Joshi, N. C., Experiments in Phytomonitoring of Urban Atmosphere, Ph.D. thesis, University of Bombay, India, 1991.
26. Chaphekar, S. B., An evidence of high pollution level in Chembur, *Scavenger*, 9, 22, 1979.
27. Giridhar, B. A. and Chaphekar, S. B., Effect of wind and air pollution on plants, in *Proc. 80th Annu. Meeting of APCA*, New York, 87, 366, 1987.
28. Anderson, F. K. and Treshow, M., Responses of lichens to atmospheric pollution, in *Air Pollution and Plant Life*, Treshow, M., Ed., John Wiley & Sons, New York, 1984, 259.
29. Bell, J. N. B., Ayazloo, M. and Wilson, G. B., Selection of sulphur-dioxide tolerance in grass populations in polluted areas, in *Urban Ecology, Proc. 2nd European Ecol. Symp.*, Berlin, Lee, J. A. and Seward, M. R. D., Eds., Blackwell Scientific, London, 1982, 171.
30. Bradshaw, A. D., Pollution and evolution, in *Effects of Air Pollutants on Plants*, Mansfield, T. A., Ed., Cambridge University Press, London, 1976, 134.
31. Ormrod, D. P., Impact of trace element pollution on plants, in *Air Pollution and Plant Life*, Treshow, M., Ed., John Wiley & Sons, New York, 1984, 291.
32. Boralkar, D. B., A Contribution to the Study of the Effects of Industrial Air Pollutants on Plants, Ph.D. thesis, University of Bombay, India, 1979.
33. Sisula, H., Bioindicators in international environmental co-operation: a users view, in *Bioindicators of Environmental Health, Proc. 7th Int. Bioindicators Symposium*, Munawar, M., Hanninen, O., Roy, S., Munawar, N., Karenlampi, L., and Brown, D., Eds., SPB Academic Publishers, the Netherlands, 1995, 9.
34. Madav, R., Study of Responses of Plants to Different Types of Environmental Stresses, Ph.D. thesis, University of Pune, India, 1997.
35. Clements, F. E., *Plant Indicators*, Carnegie Inst. Wash., Pub. 290, 1920.
36. Singh, J., Agrawal, M., and Narayan, D., Effect of power plant emissions on plant community structure, *Ecotoxicology*, 3, 110, 1994.
37. Thomas, M. D., Gas damage to plants, *Annu. Rev. Plant Physiol.*, 2, 293, 1951.
38. Agrawal, M., Singh, S. K., Singh, J., and Rao, D. N., Biomonitoring of air pollution around urban and industrial sites, *J. Environ. Biol.*, Sp. Issue, Bioindicators, Chaphekar, S. B., Ed., Academy of Environmental Biology, Muzzafarnagar, India, 1991, 211.
39. Ahmad, K. J., Yunus, M., Singh, S. N., Srivastava, K., Singh, N., Pandey, V., and Misra, J., Air pollution and plants, *CSIR News*, (New Delhi), 41, 15, 1991.
40. Jacobson, J. S. and Hill, A. C., *Recognition of Air Pollution Injury to Vegetation : A Pictorial Atlas*, APCA, Pittsburgh, PA, 1970.
41. Taylor, O. C., Oxidants Air Pollution Effects on a Western Coniferous Forest Ecosystem, Task C Report EP-R3-73-043B, SAPRC, Riverside, CA, 1973, 189.
42. USDA Forest Service, Air Pollution Damages Trees, State & Private Forestry, Upper Darby, PA, 1973, 32
43. Chaphekar, S. B., Air pollution and plants, *Indian Rev. Life Sci.*, 2, 41, 1982.
44. Chaphekar, S. B., Protocols for phytomonitoring in industrial areas in a tropical monsoonal region., in *Bioindicators of Environmental Health, Proc. 7th Int. Bioindicators Symposium*, Munawar, M., Hanninen, O., Roy, S., Munawar, N., Karenlampi, L., and Brown, D. , Eds., Kuopio, Finland, SPB Acad. Publ., the Netherlands, 1995, 195.
45. Chaphekar, S. B., Boralkar, D. B., and Shetye, R. P., Plants for air monitoring in industrial areas, in Trop. *Ecology and Development, Proc. V Int. Symp.*, Furtado, J. I., Ed., International Society for Tropical Ecology, Kuala Lumpur, 1, 1980, 669.

46. Posthumus, A. C., Monitoring levels and effects of air pollutants, in *Air Pollution and Plant Life*, Treshow, M., Ed., John Wiley & Sons, New York, 1984, 73.
47. Chaphekar, S. B., Ratnakumar, M., and Bhavanishankar, V., Biomonitoring of industrial air pollution with plants, *Proc. Symp. Biomonitoring State of Environment*, INSA, New Delhi, 1985, 258.
48. Pandey, J. and Agrawal, M., Evaluation of air pollution phytotoxicity in a seasonally dry tropical urban environment using three woody perennials, *New Phytol.*, 126, 53, 1994.
49. DeSloover, J. and LeBlanc, F., Mapping of atmospheric pollution on the basis of lichen sensitivity, in *Proc. Symp. Rec. Adv. Trop. Ecol.*, Misra, R. and Gopal, B., Eds., International Society for Tropical Ecology, Varanasi, India, 1968, 42.
50. Seward, M. R. D., Lower plants and urban landscape, *Urban Ecol.*, 4, 217, 1979.
51. Seward, M. R. D., Lichen ecology of changing urban environment, in *Urban Ecology, Proc. 2nd European Ecology Symposium*, Lee, J. A. and Seward, M. R. D., Eds., Blackwell Scientific, London, 1982, 181.
52. Gilbert, O. L., Bryophytes as indicators of air pollution in Tyne Valley, *New Phytol.*, 67, 15, 1968.
53. Keller, T., The effect of a continuous springtime fumigation with SO_2 and CO_2 uptake and structure of the annual ring in spruce, *Can. J. For. Res.*, 1980 (Also in Smith, W. H., Air Pollution and Forests, Springer-Verlag, New York, 1981, 298).
54. Miller, P. R., Photochemical Oxidant Air Pollutant Effects on a Mixed Conifer Forest Ecosystem, EPA-600/3-77-104, U.S. EPA, Corvallis, OR, 338, 1977.
55. Chaphekar, S. B., Effect of atmospheric pollution on plants in Bombay, *J. Biol. Sci.*, 15, 1, 1972.
56. Pawar, K. and Dubey, P. S., Air pollution research review of contributions of Vikram University, in *Perspective in Environmental Botany*, Vol. I, Rao, D. N., Ahmad, K. J., Yunus, M., and Singh, S. N., Eds., Print House, Lucknow, India, 1985, 101.
57. Rao, D. N., *Mangifera indica*, a bioindicator of pollution in the tropics, in *Proc. 22nd Int. Geog. Cong.*, Montreal, 1972, 272.
58. Chaphekar, S. B., Environmental Management at Thermal Power Plant, Dahanu, Report for 1989–1997 to BSES, Mumbai, India, 1997.
59. Jacobson, J. S. and Feder, W. A., A regional network for environmental monitoring, atmospheric oxidant concentrations and foliar injury to tobacco indicator plants in the eastern U.S., Bulletin No.604, 1974.
60. Davison, A. W., Velissarion, D., Barnes, J. D., Gimeno, B., and Inclan, R., The use of Aleppo pine, *Pinus halepensis* Mill as a bioindicator of ozone stress in Greece and Spain, in *Bioindicators of Environmental Health*, Munawar, M., Hanninen, O., Roy, S., Munawar, N., Karenlampi, L., and Brown, D., Eds., SPB Acad. Publ., the Netherlands, 1995, 63.
61. Supuka, J., Biomonitoring in urban Slovakia by the accumulation of proline and inorganic elements in tree leaves, in *Bioindicators of Environmental Health*, Munawar, M., Hanninen, O., Roy, S., Munawar, N., Karenlampi, L., and Brown, D., Eds., SPB Acad. Publ., the Netherlands, 1995,133.
62. Galstova, N. I. and Vasina, T. V., The use of bioindication for estimation of pollution in forest ecosystems of the Leningrad region, in *Bioindicators of Environmental Health*, Munawar, M., Hanninen, O., Roy, S., Munawar, N., Karenlampi, L., and Brown, D., Eds., SPB Acad. Publ., the Netherlands, 1995, 141.
63. Bohac, J., Kubes, J., Fuch, R., and Curnova, A., The use of biomonitoring of ecological planning and ecological policy in agricultural settlements, in *Bioindicators of Environmental Health*, Munawar, M., Hanninen, O., Roy, S., Munawar, N., Karenlampi, L., and Brown, D., Eds., SPB Acad. Publ., the Netherlands, 1995, 155.
64. Agrawal, M. and Agrawal, S. B., Use of plants in air pollution monitoring, in *Environmental Issues and Programmes*, Mohan., I., Ed., Ashish Publ. House, New Delhi, India, 1989, 165.
65. Singh, S. K. and Rao, D. N., Evaluation of plants for their tolerance to air pollution, in *Proc. Symp. Air Pollut. Control*, New Delhi, 1983, 218.
66. Little, P. and Martin, M. H., A survey of Zn, Pb and Cd in soil and natural vegetation around a smelting complex, *Environ. Pollut.*, 3, 241, 1972.
67. Klumpp, G., Klumpp, A., and Domingos, M., *Hemexocallis* as bioindicator of fluoride pollution in tropical countries, *Environ. Monit. Assess.*, 35, 27, 1995.
68. Matta, A. and Nicolotti, G., Plants and air pollution in a city, *Inquinamento*, 38, 58, 1996.

69. Patil, A. P., Productivity Studies in Plants with Reference to Air Pollution, Ph.D. thesis, University of Bombay, India, 1990.
70. Furukawa, A., Defining pollution problems in the Far East — a case study of Japanese air pollution problems, in *Gaseous Air Pollutants and Plant Metabolism*, Koziol, M. J. and Whatley, F. R., Eds., Butterworths, New York, 1984, 49.
71. Brumelis, G. and Nikodemus, O., Biological monitoring in Latvia using moss and soil: problems in the partitioning of anthropogenic and natural effects, in *Bioindicators of Environmental Health*, Munawar, M., Hanninen, O., Roy, S., Munawar, N., Karenlampi, L., and Brown, D., Eds. , SPB Acad. Publ., the Netherlands, 1995, 123.
72. Ruhling, A., Atmospheric heavy metal deposition in Europe estimated by moss analysis, in *Bioindicators of Environmental Health* , Munawar, M., Hanninen, O., Roy, S., Munawar, N., Karenlampi, L. and Brown, D., Eds., SPB Acad. Publ., the Netherlands, 1995, 115.
73. Giovenco, A., Ottonello, D., Dia, G., and Orecchio, S., Lichens and air pollution: air quality in the metropolitan zone of Palermo, *Inquinamento*, 38, 48, 1996.
74. Zocchi, A., Roella, V., and Calanari, D., Air quality assessment in the Varese area using epiphytic lichens, *Ing. Ambientale*, 25, 78, 1996.
75. Thompson, C. R., Olszyk, D. M., Kats, G., Bytnerowicz, A., Dawson, P. J., Wolf, J., and Fox, C. A., *Air Pollutant Injury on Plants of The Mojave Desert,* Southern California Edison Co., Rosemead, 1984, 5.

CHAPTER 19

Statistical Baseline Values for Chemical Elements in the Lichen *Hypogymnia physodes*

James P. Bennett

CONTENTS

19.1 INTRODUCTION

Although it is common knowledge that many lichens are very sensitive to air pollutants, it is equally well known that some are able to absorb unusually large amounts of chemical elements from atmospheric or other sources with little obvious harm. These large amounts appear to result from several factors: the absence of a cuticle barrier, the ability to immobilize chemicals in the cell wall, internal particulate entrapment, and the failure of most field studies to measure intercellular chemical concentrations.[1-4] In other words, although lichens are unable to prevent chemicals from entering their thalli, they are able to prevent the chemicals from getting inside cells, so routine chemical analyses often report high thallus concentrations that are not actually the cellular concentrations which are most important. Performing such tissue analyses is difficult and not commonly done. Most chemical analysis studies of lichens report whole-thallus concentrations, not internal cellular concentrations. Nevertheless, this has not stopped scientists from attempting to determine threshold concentration levels for the harmful effects of some chemicals, even though the concentrations being reported appear unusually high for those species studied. It is possible that the actual tissue concentrations that are harmful to lichens are actually much lower than the reported thallus concentrations, but the thallus concentrations are used to determine harmful effects.

When a lichen dies, it usually disintegrates fairly quickly (1 to 2 weeks).[5] This makes it difficult to locate moribund specimens in the field and to sample them to determine toxicity levels of

1-56670-341-7/00/$0.00+$.50

chemicals. Some attempts have been made to do this by sampling along transects downwind of known pollutant sources, these attempts are reviewed by Tyler.[4] The focus of such work is on determining a toxicity threshold, i.e., what concentrations in the thallus are the levels beyond which the lichen can no longer survive? D. N. Rao[3] presented some toxicity thresholds for several species in 1977, and Folkeson and Andersson-Bringmark[6] extended this knowledge with work in Sweden in 1988. Toxicity, however, may be the result of high increase factors in addition to exceedances of toxicity thresholds.[7,8] It might be that a suite of chemicals that increase twofold or more, regardless of the toxicity levels, may overwhelm the lichen physiologically and bring about a fatal decline.

To explore this in more detail, it was decided to assemble as much data as possible on the chemical element concentrations in *Hypogymnia physodes* reported in the literature and test the hypothesis that significant differences between background and enriched chemical concentrations were the result of a doubling of the concentrations. In addition, because of concern that has been raised about all the factors that affect the use of chemical element concentrations in lichens for biomonitoring,[9] a statistical approach to determine threshold levels was attempted if sufficient data could be aggregated for the particular element. The hypothesis that enriched values would be those that exceed the 95% confidence interval of the mean or median of all the values reported in the literature would then be tested.

19.2 SELECTION OF SPECIES

Hypogymnia physodes was selected because it is the most-studied lichen for air pollution effects,[10] and because it is considered intermediate in overall pollution tolerance (although see Pirintsos et al.[11] for a discussion of an instance of this species being sensitive), thus allowing sampling to be performed in many places throughout its range. The species has a circumpolar arctic and temperate region distribution.[12]

Many lichen-biomonitoring scientists measure the chemical element concentrations in samples collected for various reasons in the field and then compare them to reports of concentrations from other studies. Often the comparisons end up being somewhat subjective because the ranges of concentrations reported in the literature are so large. Resource management specialists for public lands want to know if the concentrations found in lichens in their areas are unusually high or not. This study will help these people to establish for *H. physodes* the normal ranges for many elements from sites throughout its range. This is possible because of the large number of studies on this species.

19.3 METHODOLOGICAL APPROACH

Since the mid-1960s almost 75 field studies have been reported on chemical element concentrations in *H. physodes* and more are being published every year.[13-85] Although some studies have been reported prior to the 1960s, the laboratory methods to analyze the chemicals are generally not comparable to modern methods, and the reporting of the results is often inadequate and difficult to use. Such studies were not considered for this analysis.

The studies were all reviewed to determine the chemical elements measured, the methods of measurement, whether or not the lichens were cleaned of debris, whether or not the lichens were washed prior to analysis, and the geography of the field sampling in order to determine the type of field exposure. Some studies were large-scale studies of entire countries, some were of natural areas, some were in and around cities, and some were in proximity to point sources of pollutants. Whatever the type of study, sample sites were classified into two categories: background and enriched. A background site was considered any site that was not in a city, not near a point source, remote geographically, and not under any kind of local pollutant influence. Enriched sites were the opposite of these criteria. It was immediately evident that the definitions of background and enriched were somewhat subjective, because they varied from study to study. But because there were so many studies, the sheer number of data points that were available for both categories was encouraging because it allowed enough data to describe the ranges of values for them.

Descriptive statistics were calculated for each element and a determination was made if the distributions were normal or skewed. Confidence intervals around the means and medians were then calculated accordingly. The hypothesis that values beyond the upper 95% confidence interval were enriched was tested by comparing these values with the enriched means calculated above.

It also became evident that very few studies reported chemical concentrations of known anthropogenic elements (e.g., Pb) that were near zero, the value they should be without any air pollutants present. This meant that the background averages would be biased by ambient levels of pollutants that appear to be a problem worldwide. In an attempt to determine if near zero values for these elements could be obtained, four samples of *H. physodes* collected in the North American Arctic region were obtained from the University of Wisconsin Herbarium and analyzed for some of these elements using methods described earlier.[7] These values were incorporated into the database and analyzed with the literature values. Details on the four samples are as follows:

Year Collected	Collector	Location
1958	A. J. Sharp	Valley of Mancha Creek, Northern Alaska
1962	J. D. Lambert	Reindeer Station, Mackenzie District, Northwest Territories
1964	J. W. Thomson and J. A. Larsen	Reindeer Station, Mackenzie District, Northwest Territories
1964	J. W. Thomson and J. A. Larsen	Canoe Lake, Richardson Mountains, Northwest Territories

For each study, the lowest reported concentration for any element was tabulated as the background value, and the highest concentration as the enriched value. Averages were only used if individual data points were not reported. In a few studies, the background and enriched values had to be determined by back-calculations from means, standard deviations, and/or coefficients of variation. Concentrations that were reported as below the detection limits of measurement instruments were included in analyses by using values that were one half the detection limit values.

19.4 VARIATIONS IN ELEMENT CONCENTRATIONS

Thallus concentrations of 58 elements have been reported for *H. physodes* in 75 separate studies at the time of this writing. Of these elements, 21 had 10 or more data points, and were selected for further analysis (Table 19.1). The remaining 37 elements, most of which had only one observation, include Ba, Br, C, Ce, Cl, Co, Cs, Dy, Er, Eu, F, Ga, Gd, H, Hf, Ho, La, Li, Lu, Mo, Nb, Nd, Pr, Rb, Sb, Sc, Se, Sm, Sn, Sr, Tb, Th, Tm, U, Y, Yb, and Zr. Means, standard deviations, and minima of 19 elements from the four Arctic samples collected from the 1960s are shown in Table 19.2.

More than half of the studies reported that the lichens had been cleaned of debris, none reported that they had not been cleaned, and it could not be determined if cleaning had been performed in the remaining studies (Table 19.3). Most of the studies reported that the lichens were not washed in water to remove deposited soil particles. Two studies reported that they compared washed and unwashed samples, which are not included in Table 19.2. A small fraction of the studies did not report whether or not the lichens were washed. These data indicated that neither the inclusion of contaminating debris nor the dilution of chemicals by washing are likely to be significant factors in this analysis.

The number of data points for the 21 elements selected for analysis ranged from 10 for Ti to 84 for Fe and Zn. Five of the elements, K, Ca, Mg, P, and As, had normally distributed data, while

Table 19.1 Descriptive Statistics (ppm) of 21 Chemical Elements in *H. physodes* (in descending order by sample size)

	Fe	Zn	S	Pb	Cd	Cu	Cr
N	84	84	81	76	71	71	64
Minimum	114.00	7.00	42.00	0.05	0.00	0.80	0.30
Median	760.25	88.00	980.00	22.80	0.80	6.02	1.35
Mean	1,821.82	246.10	1,210.54	71.75	1.52	16.81	22.15
SD	3,605.58	722.93	754.43	172.11	2.97	36.67	125.23
Maximum	21,410.00	5,621.00	3,404.00	1,034.00	21.70	199.60	1,000.00
95% CI Mean	1,039–2,604	89–403	1,044–1,377	32–111	0.82–2.22	8.1–25.5	-9.1–53.4
95% CI Median	595–953	80–95	832–1116	18–31	0.6–1	5.2–6.7	1.1–1.8

	Ni	K	Mn	Al	Ca	Mg	P
N	64	63	63	62	61	59	48
Minimum	0.00	240.00	13.00	119.00	388.00	210.00	311.10
Median	2.10	3,000.00	120.20	540.00	12,298.00	711.00	718.50
Mean	6.84	3,051.63	184.98	727.09	13,713.33	777.40	897.19
SD	20.07	1,152.57	167.97	756.11	10,726.45	420.59	555.71
Maximum	150.00	6,348.00	693.00	5,500.00	50,827.00	3,000.00	3,296.00
95% CI Mean	1.8–11.9	2,761–3,342	143–227	535–919	10,966–16,461	668–887	736–1059
95% CI Median	1.7–2.4	2,658–3,437	91–158	418–646	7,631–16,374	634–776	592–923

	Na	B	Hg	N	V	As	Ti
N	46	42	29	15	14	12	10
Minimum	9.30	0.20	0.07	4,000.00	1.50	0.15	18.00
Median	32.00	2.60	0.31	11,300.00	20.00	2.08	42.20
Mean	93.76	4.06	9.30	15,413.33	98.57	2.50	126.28
SD	190.67	5.14	35.19	10,372.28	168.57	2.05	159.70
Maximum	928.00	31.49	188.20	34,000.00	578.00	7.10	490.00
95% CI Mean	37–150	2.5–5.7	-4.1–22.7	9,669–21,157	1.2–195.9	1.2–3.8	12–240
95% CI Median	26–39	1.7–3.4	0.2–0.6	7,500–24,701	5.6–85.3	1–3.9	22.7–217

Note: Data are from 75 elemental studies reported in the literature. SD = standard deviation, and CI = confidence interval.

Table 19.2 Means, Standard Deviations, and Minima of Element Concentrations in *H. physodes* from Four 1960s Arctic Samples

	Concentration(ppm)		
Element	Mean	Standard Deviation	Minimum
Al	666	599	191
As	(BDL)		
B	52.3	26.4	31.5
Ca	32,619	24,404	2,030
Cd	0.85	0.57	0.35
Co	1.17	(N = 1; 3 BDLs)	1.17
Cr	3.05	1.56	1.65
Cu	2.87	2.40	1.16
Fe	557	221	248
K	2,010	746	961
Mg	922	453	456
Mn	117	82	18
Mo	(BDL)		
Na	173	204	47
Ni	1.58	1.90	2.37
P	585	189	**311**
Pb	**(BDL)**		
S	450	125	313
Zn	56.9	17.6	37.8

Note: Detection limits for As, Co, Mo, and Pb were 0.3, 0.02, 0.02, and 0.10 ppm, respectively. The N for Ni = 2 because two samples were below detection limit (BDL) of the instrument. P and Pb values (shown in bold print) were the lowest of any values in the literature.

Table 19.3 Frequencies of Lichen Cleaning and Washing in 75 Lichen Studies

	Yes	No	Unknown
Cleaned of debris	40	0	35
Washed in water	6	52	15

the remaining 17 elements were highly skewed to the right, as shown by the means being greater than the medians (Table 19.1). When data are not normally distributed, the mean is not a good measure of the central tendency of the data because it is biased toward the skewed values. In these cases the median is a better measure of the central tendency of the data. If the median is a better measure of centrality, then this value could be considered the overall "background" value for these elements, because the extreme high values, which are presumably the result of pollution, are not included in the calculation. Following this logic, the mean K, Ca, Mg, P, and As values would represent the overall background for those elements, while the median would for the remaining 17 elements.

Therefore, it would appear reasonable to conclude that values lying beyond the 95% confidence interval of the mean and the median, respectively, are values that are beyond the background central tendency, and therefore represent extreme values. For example, with Fe, any values beyond 953 ppm (the upper limit of the 95% confidence interval of the median) would be significantly greater than the median of 760, and perhaps greater than normal. Likewise, the comparable upper limits for the other elements are (all in ppm) 95 for Zn, 1116 for S, 31 for Pb, 1 for Cd, 6.7 for Cu, 1.8 for Cr, 2.4 for Ni, 158 for Mn, 646 for Al, 39 for Na, 3.4 for B, 0.6 for Hg, 24,701 for N, 85 for V, 217 for Ti, 3342 for K, 16,461 for Ca, 887 for Mg, 1059 for P and 3.8 for As.

19.5 STATISTICAL ANALYSIS OF DATA

The descriptive statistics in Table 19.1 are based on the entire set of data for each element. In Table 19.4, the data have been grouped by the background and enriched categories and the means calculated. This is done to determine how close observed values in polluted areas are to the predicted values from the confidence intervals above. In addition, the means are presented in descending order by the percent increase between the background and the enriched, and a "t" test was performed to the test the difference between the two categories. Enriched means that are between 57 and 613% of the background means are statistically significantly different, except for Na. The enriched mean for B is only 13% more than the background mean and is not significantly different. The enriched means for V, Hg, Cr, and Ti are much, much greater than the background means and, due to the high variability, are not significantly different statistically.

It could be argued at this point, after having said that 17 of the 21 elements were not normally distributed, that t tests of differences between means would not be appropriate statistical tests. This problem was explored by calculating tests of significant differences between background and enriched medians (Mood's median test). Of the 21 pairs of medians, 20 (As being the exception) were significantly different at the 0.05 probability level. However, all differences between background and enriched medians were much, much smaller than the differences between the means. This would lead the analysis toward the conclusion that smaller differences between background and enriched "treatments" were meaningful, which is illogical. Given that this lichen appears to absorb large amounts of elements, it would not help to establish enriched threshold levels at concentrations that are not much greater than the background concentrations. The use of the t test was adopted for the purposes of this study, which was to find biological differences that were meaningful and statistically different.

Of the 15 elements with significantly different background and enriched means, all but N, As, Ca, P, K, and possibly S have the means very close to the upper and lower values of the confidence

Table 19.4 Background and Enriched Means of 21 Chemical Elements in *H. physodes*, Percent Changes and T Test Probabilities

	V	Hg	Cr	Ti	Ni	Pb	Zn
Background average	16.9	0.253	2.11	26.4	**1.72**	**19.5**	**73**
(*n*)	(8)	(15)	(33)	(5)	(33)	(39)	(43)
Enriched average	207.5	19.00	43.50	226.2	**12.28**	**126.9**	**427**
(*n*)	(6)	(14)	(31)	(5)	(31)	(37)	(41)
Percent change	11,270	7,413	1,960	757	613	552	482
T probability	0.088	0.18	0.21	0.068	0.045	0.0088	0.03

	Fe	Cu	Cd	N	Mn	As	S
Background average	621	5.96	0.562	7162	85.2	1.33	738
(*n*)	(43)	(37)	(37)	(8)	(32)	(7)	(41)
Enriched average	3,081	**28.63**	**2.561**	24,843	**288**	**4.15**	1,695
(*n*)	(41)	(34)	(34)	(7)	(31)	(5)	(40)
Percent change	396	381	356	247	238	213	130
T probability	0.0025	0.014	0.0072	0.0004	0	0.05	0

	Al	Ca	Na	P	Mg	K	B
Background average	**448**	**8,411**	62	590	534	**2,380**	3.82
(*n*)	(32)	(31)	(24)	(24)	(30)	(32)	(21)
Enriched average	1,025	19,192	128.4	**1,204**	**1,029**	**3,745**	4.30
(*n*)	(30)	(30)	(22)	(24)	(29)	(31)	(21)
Percent change	129	128	107	104	93	57	13
T probability	0.0034	0	0.25	0.0001	0	0	0.77

Note: See text for definitions of background and enriched. Elements are presented in descending order of percent change. Significantly different means at the 0.05 probability level are shown in bold print.

limits of the mean of the entire data set. The others are similar to the confidence limits of the medians. This would suggest that although values beyond the upper limit of the confidence interval of the median are significantly different, those values that are beyond the upper limit of the confidence interval of the mean are actually closer to values observed in field studies in polluted areas. Therefore, although the confidence interval of the median makes more sense statistically, to be on the safe side, it would be better to classify values beyond the confidence interval of the mean as extreme, or polluted values. This would only apply to the 15 significant elements in Table 19.4. The elements V, Hg, Cr, Ti, Na, and B need more study before such limiting values can be determined. The limiting values (with some rounding) for the 15 elements are shown in Figure 19.1. Concentrations observed beyond these values are statistically greater than 95% of all reported values of each element, and they match the mean of the enriched values in the literature. There is a very good chance these values are abnormally enriched.

It is also evident from Table 19.4 that not all enriched values on average are double the background values. Only two come close: P and Mg. K is enriched when it increases about half, while for the remainder it takes a factor of 1.3 up to 6 to be associated with enrichment. These factors are in general agreement with Tyler's[4] factors of 5 to 10 and Seaward's[8] of 3 to 5. It would appear that it takes more than double the thallus concentrations to classify the lichen as coming from an enriched site, but whether or not this means toxicity would occur is uncertain.

Element concentrations in the four Arctic samples from the 1960s had the lowest values overall for P and Pb, the latter being essentially zero. There is no reason to expect Pb background levels to be above zero anyway, so the historical values should be regarded as the true background. Unfortunately, Pb is now distributed worldwide in the atmosphere and biosphere, and it may well

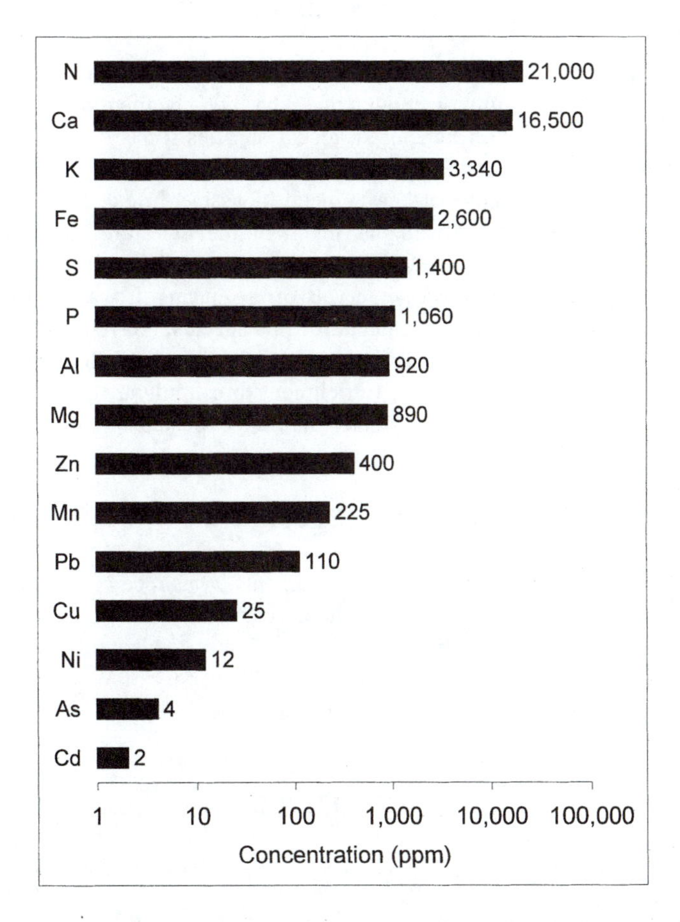

Figure 19.1 Enrichment threshold concentrations (ppm) for 15 chemical elements for *H. physodes.* These values are the rounded upper 95% confidence interval around the mean of all the values from the literature studies. Values beyond these would exceed the mean of values in the literature for that element with 95% confidence that they do not equal the mean.

be impossible to find samples with no Pb in them. The enriched threshold value for Pb in Figure 19.1, therefore, is biologically incorrect, but realistically correct for our time period. The value of 110 ppm for Pb is about twice what Seaward[8] found over 20 years ago in 1974, but is significantly less than the range of 718 to 918 ppm that Rao et al.[3] suggested in 1977.

The threshold of Cd of 2 ppm is half the value of 5 ppm of Rao et al.,[3] but the Zn level of 400 ppm is much more than their value of 168 to 233 ppm, and four times the value of 101 ppm of Seaward.[8] This may be because of the very high values that Folkeson and Andersson-Bringmark[6] reported: 1900 ppm was the survival threshold in their study. The Fe threshold of 2600 ppm is comparable to Seaward's value of 3150 ppm.[8] The threshold level of 25 ppm for Cu, however, is significantly less than the values in the hundreds reported by Rao et al.[3] in 1977 and Folkeson and Andersson-Bringmark[6] in 1988, even though these value are for the same species. Of the 71 Cu values in the literature, all but three of them are below 100 ppm, so these other threshold values are highly unusual.

19.6 CONCLUSIONS

Hypogymnia physodes is a cosmopolitan lichen species that is found in many industrialized countries around the world. It is common enough that it has become the most widely studied species in the monitoring of air quality with lichens. In recent decades it has been the subject of at least 75 published studies reporting chemical concentrations of 58 elements. In an attempt to address the concerns about the use and abuse of heavy metal bioassays in environmental monitoring,[9] a synoptic approach was taken by analyzing as much of the literature data statistically. Of the elements, 21 have been reported 10 or more times and were analyzed in this study to determine baseline values. By using 95% confidence intervals of the means, it was statistically possible to determine enrichment thresholds for 15 elements for this species: Al, As, Ca, Cd, Cu, Fe, K, Mg, Mn, N, Ni, P, Pb, S, and Zn. High variability of data for the other elements did not permit a statistical determination of enriched values. Enrichment thresholds are generally associated with increase factors of two to six, which were greater than the expected factor of doubling.

ACKNOWLEDGMENTS

This chapter has benefited from the excellent comments it has received from M. R. D. Seaward (University of Bradford, U.K.) and C. M. Wetmore (University of Minnesota, U.S.). I wish to express my gratitude to them for taking the time to review it. I also acknowledge the generous cooperation I have received from J. W. Thomson of the University of Wisconsin Herbarium, who willingly allowed me to analyze samples from the collection.

REFERENCES

1. Brown, D. H. and Brown, R. M., Mineral cycling and lichens: the physiological basis, *Lichenologist*, 23, 293, 1991.
2. Nash III, T. H., Nutrients, elemental accumulation and mineral cycling, in *Lichen Biology*, Nash III, T. H, Ed., Cambridge University Press, Cambridege, U. K., 1996, chap. 8.
3. Rao, D. N., Robitaille, G., and LeBlanc, F., Influence of heavy metal pollution on lichens and bryophytes, *J. Hattori Bot. Lab.*, 42, 213, 1977.
4. Tyler, G.,Uptake, retention and toxicity of heavy metals in lichens, *Water Air Soil Pollut.*, 47, 321, 1989.
5. Wetmore, C. M., Lichen decomposition in a black spruce bog, *Lichenologist*, 14, 267, 1982.
6. Folkeson, L. and Andersson-Bringmark, E., Impoverishment of vegetation in a coniferous forest polluted by copper and zinc, *Can. J. Bot.*, 66, 417, 1988.
7. Bennett, J. P., Dibben, M. J. and Lyman, K. J., Element concentrations in the lichen *Hypogymnia physodes* (L.) Nyl. after 3 years of transplanting along Lake Michigan, *Environ. Exp. Bot.*, 36, 255, 1996.
8. Seaward, M. R. D., Some observations on heavy metal toxicity and tolerance in lichens, *Lichenologist*, 6, 158, 1974.
9. Seaward, M. R. D., Use and abuse of heavy metal bioassays in environmental monitoring, *Sci. Total Environ.*, 176, 129, 1995.
10. Bennett, J. P. and Buchen, M. J., BIOLEFF: three databases on air pollution effects on vegetation, *Environ. Pollut.*, 88, 261, 1995.
11. Pirintsos, S. A., Vokou, D., Diamantopoulos, J., and Galloway, D. J., An assessment of the sampling procedure for estimating air pollution using epiphytic lichens as indicators, *Lichenologist*, 25, 165, 1993.
12. Thomson, J. W., *Lichens of the Alaskan Arctic Slope*, University of Toronto Press, Toronto, 1979.
13. Addison, P. A. and Puckett, K. J., Deposition of atmospheric pollutants as measured by lichen element content in the Athabasca oil sands area, *Can. J. Bot.*, 58, 2323, 1980.

14. Bennett, J. P., Abnormal chemical element concentrations in lichens of Isle Royale National Park, *Environ. Exp. Bot.*, 35, 259, 1995.
15. Bennett, J. P. and Wetmore, C. M., Chemical element concentrations in four lichens on a transect entering Voyageurs National Park, *Environ. Exp. Bot.*, 37, 173, 1997.
16. Bruteig, I. E., The epiphytic lichen *Hypogymnia physodes* as a biomonitor of atmospheric nitrogen and sulphur deposition in Norway, *Environ. Monit. Assess.*, 26, 27, 1993.
17. Bylinska, E. A., Marczonek, A., and Seaward, M. R. D., Mercury accumulation in various components of a forest ecosystem influenced by factory emissions, *Proc. Int. Symp. Urban Ecology*, Didim, Turkey, 1991, 75.
18. Evans, C. A. and Hutchinson, T. C., Mercury accumulation in transplanted moss and lichens at high elevation sites in Quebec, *Water Air Soil Pollut.*, 90, 475, 1996.
19. Farkas, E., Lokos, L., and Verseghy, K., Lichens as indicators of air pollution in the Budapest agglomeration. I. Air pollution map based on floristic data and heavy metal concentration measurements, *Acta Bot. Hung.*, 31, 45, 1985.
20. Folkeson, L., Interspecies calibration of heavy-metal concentrations in nine mosses and lichens: applicability to deposition measurements, *Water Air Soil Pollut.*, 11, 253, 1979.
21. Gailey, F. A. Y., Smith, G. H., Rintoul, L. J., and Lloyd, O. L. Metal deposition patterns in central Scotland, as determined by lichen transplants, *Environ. Monit. Ass.*, 5, 291, 1985.
22. Garty, J. and Ammann, K., The amounts of Ni, Cr, Zn, Pb, Cu, Fe and Mn in some lichens growing in Switzerland, *Environ. Exp. Bot.*, 27, 127, 1987.
23. Garty, J., Kauppi, M., and Kauppi, A., Accumulation of airborne elements from vehicles in transplanted lichens from urban sites, *J. Environ. Qual.*, 25, 265, 1996.
24. Halonen, P., Hyvarinen, M., and Kauppi, M., Emission related and repeated monitoring of element concentrations in the epiphytic lichen *Hypogymnia physodes* in a coastal area, W. Finland, *Ann. Bot. Fenn.*, 30, 251, 1993.
25. Herzig, R., Liebendorfer, L., Urech, M., and Ammann, K., Passive biomonitoring with lichens as a part of an integrated biological measuring system for monitoring air pollution in Switzerland, *Int. J. Environ. Anal. Chem.,* 35, 43, 1989.
26. Holopainen, J. K., Mustaniemi, A., Kainulainen, P., Satka, H., and Oksanen, J., Conifer aphids in an air-polluted environment. I., *Environ. Pollut.*, 80, 185, 1993.
27. Jeran, Z., Byrne, A. R., and Batic, F., Transplanted epiphytic lichens as biomonitors of air-contamination by natural radionuclides around the Zirovski VRH uranium mine, Slovenia, *Lichenologist*, 27, 375, 1995.
28. Kanerva, T., Sarin, O., and Nuorteva, P., Aluminium, iron, zinc, cadmium and mercury in some indicator plants growing in south Finnish forest areas with different degrees of damage, *Ann. Bot. Fenn.*, 25, 275, 1988.
29. Kansanen, P. H. and Venetvaara, J., Comparison of biological collectors of airborne heavy metals near ferrochrome and steel works, *Water Air Soil Pollut.*, 60, 337, 1991.
30. Kauppi, M. and Mikkonen, A., Floristic versus single species analysis in the use of epiphytic lichens as indicators of air pollution in a boreal forest region, northern Finland, *Flora*, 169, 255, 1980.
31. Kauppi, M. and Halonen, P., Lichens as indicators of air pollution in Oulu, northern Finland, *Ann. Bot. Fenn.*, 29, 1, 1992.
32. Kubin, E., A survey of element concentrations in the epiphytic lichen *Hypogymnia physodes* in Finland in 1985–86, in *Acidification in Finland*, Kauppi, P., Anttilla, P., and Kenttamies, K., Eds., Springer-Verlag, Berlin, 1990, 421.
33. Kytomaa, A., Nieminen, S., Thunbeberg, P., Haapala, H., and Nuorteva, P., Accumulation of aluminum in *Hypogymnia physodes* in the surroundings of a Finnish sulphite-cellulose factory, *Water Air Soil Pollut.*, 81, 401, 1995.
34. Laaksovirta, K. and Olkkonen, H., Epiphytic lichen vegetation and element contents of *Hypogymnia physodes* and pine needles examined as indicators of air pollution at Kakkola, W. Finland, *Ann. Bot. Fenn.*, 14, 112, 1977.
35. Laaksovirta, K. and Olkkonen. H., Effects of air pollution on epiphytic lichen vegetation and element contents of a lichen and pine neeedles at Valkeakoski, S. Finland, *Ann. Bot. Fenn.*,16, 285, 1979.
36. Laaksovirta, K., Olkkonen, H., and Alakuijala, P., Observations on the lead content of lichen and bark adjacent to a highway in southern Finland, *Environ. Pollut.*, 11, 247, 1976.

37. Lackovicova, A., Martiny, E., Pisut, I., and Stresko, V., Element content of the lichen *Hypogymnia physodes* and spruce needles in the industrial area of Rudnany and Krompachy, *Ekologia*, 13, 415, 1994.
38. Lippo, H., Poikolainen, J., and Kubin, E., The use of moss, lichen and pine bark in the nationwide monitoring of atmospheric heavy metal deposition in Finland, *Water Air Soil Pollut.*, 85, 2241, 1995.
39. Lodenius, M. and Laaksovirta, K., Mercury content of *Hypogymnia physodes* and pine needles affected by a chlor-alkali works at Kuusankoski, SE Finland, *Ann. Bot. Fenn.*, 16, 7, 1979.
40. Lodenius, M., Regional distribution of mercury in *Hypogymnia physodes* in Finland, *Ambio*, 10, 183, 1981.
41. Lodenius, M. and Tulisalo E., Environmental mercury contamination around a chlor-alkali plant, *Bull. Environ. Contam. Toxicol.*, 32, 439, 1984.
42. Lodenius, M. and Tulisalo E., Open digestion of some plant and fungus materials for mercury analysis using different temperatures and sample sizes, *Sci. Total Environ.*, 176, 81, 1995.
43. Lounamaa, K. J., Studies on the content of iron, manganese and zinc in macrolichens, *Ann. Bot. Fenn.*, 2, 127, 1965.
44. Lupsina, V., Horvat, M., Jeran, Z., and Stegnar, P., Investigation of mercury speciation in lichens, *Analyst*, 117, 673, 1992.
45. Manninen, S., Huttunen, S., and Torvela, H., Needle and lichen sulphur analyses on two industrial gradients, *Water Air Soil Pollut.*, 59, 153, 1991.
46. Markert, B. and Wtorova, W., Inorganic chemical investigations in the forest biosphere reserve near Kalinin, USSR, *Vegetatio*, 98, 43, 1992.
47. Nygard, S. and Harju, L., A study of the short range pollution around a power plant using heavy fuel oil by analysing Vanadium in lichens, *Lichenologist*, 15, 89, 1983.
48. O'Hare, G. P., Lichens and bark acidification as indicators of air pollution in west central Scotland, *J. Biogeog.*, 1, 135, 1974.
49. Pakarinen, P. and Hasanen, E., Mercury concentrations of bog mosses and lichens, *Suo*, 34, 17, 1983.
50. Pakarinen, P., Kaistila, M., and Hasanen, E., Regional concentration levels of vanadium, aluminum and bromine in mosses and lichens, *Chemosphere*, 12, 1477, 1983.
51. Palomaki, V., Tynnyrinen, S., and Holopainen, T., Lichen transplantation in monitoring fluoride and sulphur deposition in the surroundings of a fertilizer plant and a strip mine at Siilinjarvi, *Ann. Bot. Fenn.*, 29, 25, 1992.
52. Pfeiffer, H. N. and Barclay-Estrup, P., The use of a single lichen species, *Hypogymnia physodes*, as an indicator of air quality in northwestern Ontario, *Bryologist*, 95, 38, 1992.
53. Pilegaard, K., Heavy metals in bulk precipitation and transplanted *Hypogymnia physodes* and *Dicranoweisia cirrata* in the vicinity of a Danish steelworks, *Water Air Soil Pollut.*, 11, 77, 1979.
54. Pilegaard, K., Rasmussen, L., and Gydesen, H., Atmospheric background deposition of heavy metals in Denmark monitored by epiphytic cryptogams, *J. Appl. Ecol.*, 16, 843, 1979.
55. Pintaric, M., Turk, R., and Peer, T., Vergleichende Untersuchungen uber den Ca, Mg und K Gehalt von Flechten und ihrem Substrat von Kalk und Silikatstandorten. Biblio, *Lichenologist*, 57, 363, 1995.
56. Sarkela, M. and Nuorteva, P., Levels of aluminum, iron, zinc, cadmium and mercury in some indicator plants growing in unpolluted Finnish Lapland, *Ann. Bot. Fenn.*, 24, 301, 1987.
57. Showman, R. and Long, R. P., Lichen studies along a wet sulfate deposition gradient in Pennsylvania, *Bryologist*, 95, 166, 1992.
58. Sochting, U., Lichens as monitors of nitrogen deposition, *Crypt. Bot.*, 5, 264, 1995.
59. Solberg, Y. J., Studies on the chemistry of lichens. IV. The chemical composition of some Norwegian lichen species, *Ann. Bot. Fenn.*, 4, 29, 1967.
60. Steinnes, E. and Krog, H., Mercury, arsenic and selenium fall-out from an industrial complex studied by means of lichen transplants, *Oikos*, 28, 160, 1977.
61. Takala, K., Kauranen, P., and Olkkonen, H., Fluorine content of two lichen species in the vicinity of a fertilizer factory, *Ann. Bot. Fenn.*, 15, 158, 1978.
62. Takala, K. and Olkkonen, H., Lead content of an epiphytic lichen in the urban area of Kuopio, east central Finland, *Ann. Bot. Fenn.*, 18, 85, 1981.
63. Takala, K. and Olkkonen, H., Titanium content of lichens in Finland, *Ann. Bot. Fenn.*, 22, 299, 1985.
64. Takala, K., Olkkonen, H., Ikonen, J., Jaaskelainen, J., and Puumalainen, P., Total sulphur contents of epiphytic and terricolous lichens in Finland, *Ann. Bot. Fenn.*, 22, 91, 1985.

65. Takala, K., Olkkonen, H., and Salminen, R., Iron content and its relations to the sulphur and titanium contents of epiphytic and terricolous lichens and pine bark in Finland, *Environ. Pollut.*, 84, 131, 1994.
66. Taylor, R. J. and Bell, M. A., Effects of SO_2 on the lichen flora in an industrial area northwest Whatcom County, Washington, *Northwest Sci.*, 57, 157, 1983.
67. Tynnyrinen, S., Palomaki, V., Holopainen, T., and Karenlampi, L., Comparison of several bioindicator methods in monitoring the effects on forest of a fertilizer plant and a strip mine, *Ann. Bot. Fenn.*, 29, 11, 1992.
68. Vestergaard, N. K., Stephansen, U., Rasmussen, L., and Pilegaard, K., Airborne heavy metal pollution in the environment of a Danish steel plant, *Water Air Soil Pollut.*, 27, 363, 1986.
69. Wetmore, C. M., Lichens and Air Quality in Acadia National Park, Plant Biology Department, University of Minnesota, St. Paul, 1984, 26.
70. Wetmore, C. M., Lichens and Air Quality in Voyageurs National Park, Chemical Analysis Supplement, Plant Biology Department, University of Minnesota, St. Paul, 1984, 6.
71. Wetmore, C. M., Lichens and Air Quality in Isle Royale National Park, Plant Biology Department, University of Minnesota, St. Paul, 1985, 41.
72. Wetmore, C. M., Lichens and Air Quality in Delaware Water Gap Recreation Area, Plant Biology Department, University of Minnesota, St. Paul, 1987, 31.
73. Wetmore, C. M., Lichens and Air Quality in Boundary Waters Canoe Area of the Superior National Forest, Plant Biology Department, University of Minnesota, St. Paul, 1987, 28.
74. Wetmore, C. M., Lichens and Air Quality in Pictured Rocks National Lakeshore, Plant Biology Department, University of Minnesota, St. Paul, 1988, 25.
75. Wetmore, C. M., Lichens and Air Quality in Sleeping Bear Dunes National Lakeshore, Plant Biology Department, University of Minnesota, St. Paul, 1988, 25.
76. Wetmore, C. M., Lichens and Air Quality in Apostle Islands National Lakeshore, Plant Biology Department, University of Minnesota, St. Paul, 1988, 28.
77. Wetmore, C. M., Lichens and Air Quality in White Mountain National Forest Wilderness Areas, Plant Biology Department, University of Minnesota, St. Paul, 1989, 39.
78. Wetmore, C. M., Lichens and Air Quality in Grand Portage National Monument, Plant Biology Department, University of Minnesota, St. Paul, 1992, 22.
79. Wetmore, C. M., Lichens and Air Quality in Chequamegon National Forest, Rainbow Lake Wilderness Area, Plant Biology Department, University of Minnesota, St. Paul, 1993, 25.
80. Wetmore, C. M., 1992 Elemental Analysis of Boundary Waters Canoe Area Lichens of the Superior National Forest, Plant Biology Department, University of Minnesota, St. Paul, 1993, 16.
81. Wetmore, C. M., 1992 Elemental Analysis of Lichens from Grand Portage National Monument, Plant Biology Department, University of Minnesota, St. Paul, 1993, 5.
82. Wetmore, C. M., 1993 Elemental Analysis of Lichens from Grand Portage National Monument, Plant Biology Department, University of Minnesota, St. Paul, 1994, 5.
83. Wetmore, C. M., Lichens and Air Quality in Lye Brook Wilderness of the Green Mountain National Forest, Plant Biology Department, University of Minnesota, St. Paul, 1995, 36.
84. Wetmore, C. M., 1993 Elemental Analysis of Lichens of the White Mountain National Forest Wilderness Areas, Plant Biology Department, University of Minnesota, St. Paul, 1995, 24.
85. Wetmore, C. M. and Bennett, J. P., 1995 Lichen Studies in Apostle Islands National Lakeshore, Plant Biology Department, University of Minnesota, St. Paul, 1996, 26.

CHAPTER 20

Monitoring Air Pollutant Deposition in the Arctic with a Lichen by Means of Microscopy and Energy-Dispersive X-ray Microanalysis: A Case Study

Richard F. E. Crang

CONTENTS

20.1 INTRODUCTION

While many regions of the Arctic may be considered relatively free from the impacts of air pollution, a growing number of sites are subject to the assault of either chronic or episodic pollutions emanating from point sources. Cold air masses, particularly during the prolonged winters, aid in the containment of pollutants over regions important for the sustenance of native wildlife and regional human populations, as well as for many migratory species. It has been recognized that Arctic ice sheets contain evidence of times in which there have been elevated pollutants, dating back to over six millennia ago.[1,2] Episodal events, such as the eruption of Mount St. Helens in 1980 showed a wide-ranging impact on the morphological state of lichens in the state of Washington.[3] Perhaps the most notable present-day point source involving air pollution in the Arctic emanates from a cluster of nickel- and copper-smelting plants encircling the northern Siberian city of Noril'sk. In 1992 it was estimated that the total annual airborne sulfur output from the Noril'sk

1-56670-341-7/00/$0.00+$.50

region was 2.3 million metric tons, equivalent to the entire output from all sources in Canada. During the months of the extended winter, effluent from the Noril'sk site mostly follows a gradient to the northeast along the Taimyr Peninsula,[4] a region of permafrost. It is generally believed that atmospheric deposition along this gradient is sufficiently dispersed upon reaching the area around Lake Taimyr that there should be no contamination of consequence in that region.

This case study is aimed specifically at a microscopic and surface elemental analysis of *Cetraria cucullata*. This lichen has not only long been considered a bioindicator of environmental pollution,[5,6] but is an important part of the Arctic food pyramid, serving as a major nutritional source for animals of the region such as the caribou. Comparative data for these ecological studies are important for the economic impact of air pollution on these fragile ecosystems and serve as a unique opportunity to study organisms from some of the few remaining sites on Earth where anthropogenic activities are minimal. Light microscopic information affords an opportunity to conduct morphometric studies on the relative composition of the algal (*Chlorococcus* sp.) and fungal symbionts in relation to the available air space within the cortex of the lichen thallus, and to changes in the medulla (outer layers) of this relatively simple organism. Scanning electron microscopy (SEM) and energy-dispersive X-ray (EDX) microanalysis have previously been successfully employed in the study of other contaminated lichen thalli[7] and in the detection of metals in mycorrhizae.[8] Martin et al.[9] also showed that EDX microanalysis revealed the progressive accumulation of various elements in lichens from soils of mud volcano activity. The results of this case study reflect on the suitability of *C. cucullata* as a bioindicator species in monitoring air pollution deposition.

20.2. STUDY SITES

Lichen samples of *C. cucullata* were obtained from two field sites: within a 15 km radius of the city of Noril'sk (69.5° N latitude by 88° E longitude) representing a region of high atmospheric pollution, and from a remote northeastern region of the Taimyr Peninsula adjacent to Lake Taimyr (75° N latitude by 108° E longitude) representing an environmentally "clean" site. The city of Noril'sk and the surrounding communities have a resident population over 250,000 and are geographically located 330 km above the Arctic Circle. The Lake Taimyr site is located approximately 800 km to the northeast of Noril'sk (Figure 20.1).

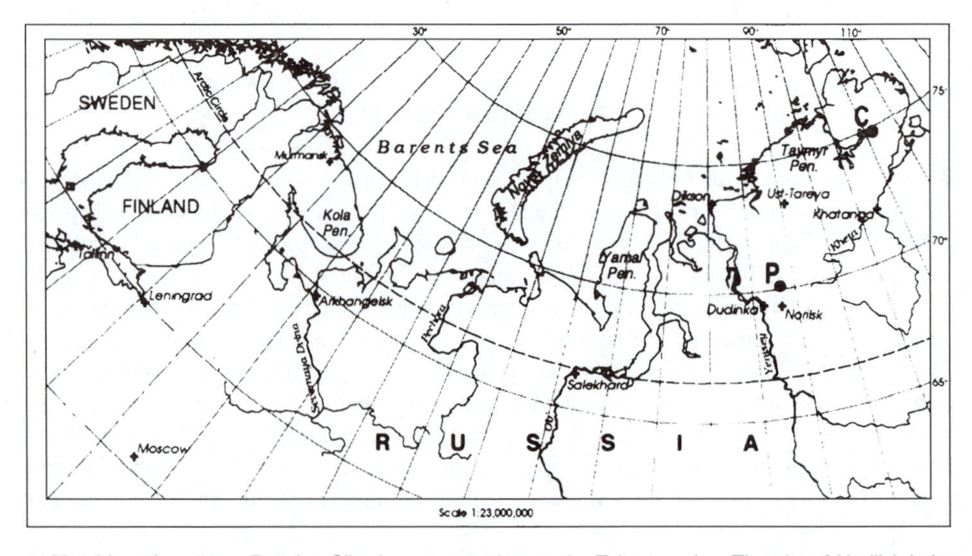

Figure 20.1 Map of northern Russian Siberia concentrating on the Taimyr region. The city of Noril'sk is located approximately 330 km north of the Arctic Circle (P). Note the "clean" collection site in the remote region of the Taimyr Peninsula near Lake Taimyr (C).

20.3 RESEARCH APPROACH

Intact lichens were removed from their substrates at the collection sites with Teflon-coated forceps, air-dried, and stored in glass vials or individually sealed Ziploc™ bags for transporting. Upon arrival in the laboratory, specimens to be stored long term were introduced into a desiccator containing silica gel as the drying agent. Samples to be prepared for light and/or transmission electron microscopy were rehydrated over the course of 3 to 4 days in a controlled clean humidity chamber (80% RH), but those for EDX microanalysis were maintained in the desiccated state. In most cases the specimens were cut or broken into three segments representing the distal (top) portion of the lichen thallus, the midregion, and the basal (bottom) portion to the site of attachment. The lichen thallus is fruticose (Figure 20.2), and develops a semi-cylindrical body with inner and outer surfaces. These surfaces, as well as broken edge surfaces, were individually examined by means of EDX studies from segments representing distal, mid, and basal portions of the thalli.

Figure 20.2 SEM of *C. cucullata* portion of thallus showing representative surfaces analyzed with EDX. 35x.

Specimens for microscopy were prepared by immersion in 2.5% glutaraldehyde and 1.0% paraformaldehyde in a sodium cacodylate buffer (0.2 *M*) at pH 6.8, postfixed in 2% buffered osmium tetroxide, dehydrated, and embedded. Samples for light microscopy (1-μm thick sections) were stained using a modified toluidine blue and basic fuchsin protocol.[10] Both uranium/lead-stained and unstained ultrathin sections were observed with a transmission electron microscope operating at 80 kV.

Transverse sections of the thallus viewed with light microscopy and showing both cortical (outer) and medullary (inner) regions were analyzed for basic morphometry using standardized methods employing an overlay grid for point counting.[11] Relative percent volumetric data was obtained from the analyses using the technique of Crang and McQuattie[12] as applied to foliar structures.

For SEM and EDX, samples of distal, mid, or basal segments were oriented on carbon-coated stubs for observation at magnifications up to 3000 (at an accelerating voltage of 15 kV). For EDX specifically, attempts were made to use surfaces oriented normal to the electron beam with standardized geometry between the stage of the SEM and the EDX detector (148 eV resolution). Multiple spectral acquisitions were each collected for a period of 100 live time seconds. X-ray data were analyzed by taking the peak-to-background (P/B) ratios obtained for all elements and subtracting 1.0 since that positive value would represent no peak present.

20.4 GROWTH HABIT

Cetraria cucullata specimens from the Siberian collection sites averaged 10 to 12 cm in height and were estimated to be in excess of 50 years old. The growth habit of *C. cucullata* was noticeably affected by the proximity to Noril'sk, presumably from air pollution originating at the nickel- and copper-smelting sites. The lichen appearance was indicated by a shorter, shrubbier thallus morphology. The distal portion of the thalli were more branched in the samples subjected to high levels of pollution, and their pigmentation sometimes appeared slightly darker (often gray, as opposed to creamy light green), even in desiccated specimens. While they sometimes revealed macroscopic differences in appearance due to exposure to atmospheric pollutants, such differences were not always apparent.

20.5 ANATOMY AND MORPHOMETRY

Transverse sections of the lichen thalli viewed with light microscopy, or in fractured view as observed with SEM, typically appeared to show a more-compacted mycelium in the inner medullary region from samples subjected to pollution (Figure 20.3a) as opposed to those from "clean" sites on the Taimyr Peninsula (Figure 20.3b). In order to obtain a nonbiased assessment, point-counting morphometry which was employed using light micrographs, revealed a 16% relative increase in medullary cell volume occupied in the polluted samples, while there was a decrease in the amount of medullary air space and in the amount of cell volume occupied by the cortical tissue (Figure 20.4). The average transverse thickness of the thalli as measured with light microscopy did not reveal a significant difference between the "clean" and polluted specimens. Microscopic examinations of the thalli from polluted specimens always showed obvious damage. In this regard, and certainly with respect to the evident propensity of *C. cucullata* to bind elemental agents from the environment, the lichen may well be considered a suitable bioindicator species.

The ability of *C. cucullata* to bind soil elements poses an interesting adaptation to environmental conditions. Bradley et al.[13] showed that plants with ericoid mycorrhizae could successfully colonize soils containing high levels of heavy metals, such as Cu and Zn, but that nonmycorrhizal plants could not survive. It was believed that the mycorrhizal fungus possessed an intracellular ability to bind the metals. In lichens, metal ions taken up from solution are reversibly bound through an ion exchange process.[14] The occurrence of functional chemical groups suitable for metal binding and having a pKa between 2 and 6 has also been demonstrated by these workers. Gadd and Griffiths[15] have also shown that metal binding to organic materials, precipitation, complexation, and ionic interactions can all be heavily influenced by environmental conditions.

a.

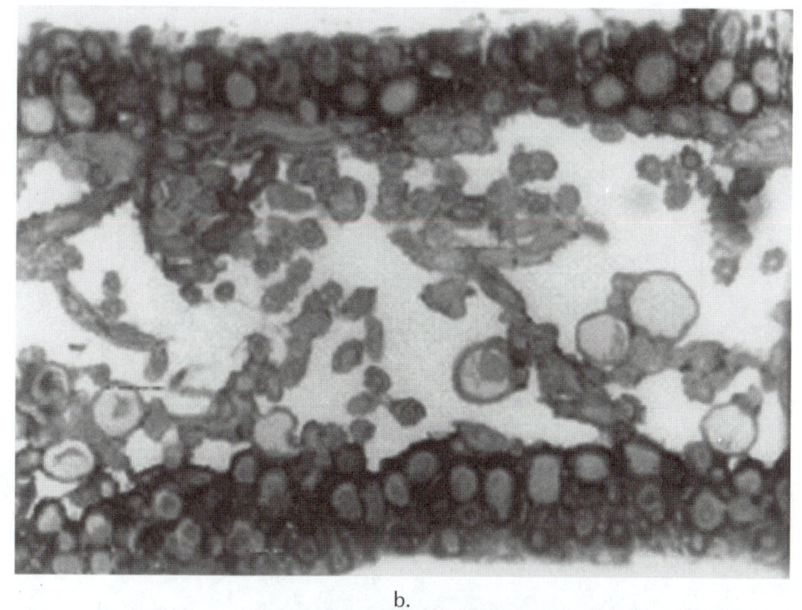

b.

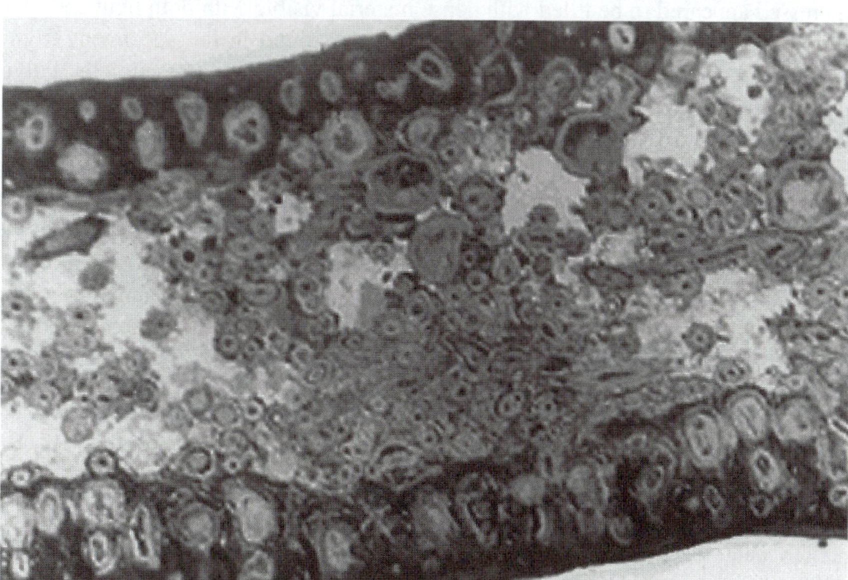

Figure 20.3 Light microscope cross sections of *C. cucullata* thallus from the distal regions representing typical structure of (a) "clean" site specimens and (b) polluted site specimens. Note the greater density of cells within the medulla and reduced air space in the polluted material (See Figure 20.4) 230x.

20.6 ULTRASTRUCTURE OF LICHEN SPECIMENS FROM POLLUTED SITES

Views from transmission electron microscopy showed that both algal and fungal cells from the medullary region of polluted specimens had broken membranes, loss of structural integrity of organelles, and dense accumulations within their protoplasm (Figures 20.5 and 20.6). Structurally, the photobiont cells appeared to be most affected, with early damage to the chloroplasts of *Chlorococcus* sp. In the extracellular spaces of the medulla, large regions where fungal cells were

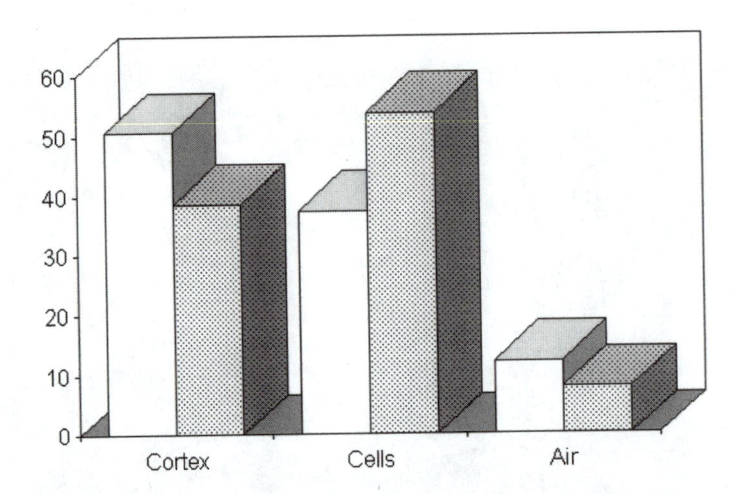

Figure 20.4 Average relative percent volumes of cortex and cells and air space from the medulla of specimens collected at "clean" site (clear bars) and polluted site (shaded bars). (Morphometric data were determined by point-counting from micrographs. Only the distal portions of *C. cucullata* thalli were utilized in these analyses.)

closely appressed appeared to be filled with dense material visible with both light and transmission electron microscopy. These deposits were almost always found in the specimens from polluted sites, and could represent accumulations of polyphenolics from the stressed cells. Polyphenolics have also been documented to be accumulated and secreted in a wide variety of plant and fungal specimens affected by various environmental stress mechanisms.[16]

It is possible that at least some of the intracellular and intercellular dense deposits found in the medulla fungal cells of *C. cucullata* may have been formed under the influence of the enhanced metal environment, and even served as a site of containment of the metals. Thus, as in the case of mycorrhizal fungi, the fungal component of *C. cucullata* may protect its algal symbiont from metal toxicity. Even at that, this study indicated that the algal cells were highly labile in the presence of the polluting factors as indicated by the evident chloroplast thylakoid disruption and loss of plastid starch (Figure 20.5).

20.7 ELEMENTAL COMPOSITION OF LICHEN SPECIMENS BASED ON X-RAY MICROANALYSIS

From EDX data, the average P/B ratios of inorganic elemental material incorporated into the lichen thalli revealed mostly the same elements present in the distal, mid, and basal portions of the thalli collected from "clean" sites. The most prevalent (i.e., P/B - 1 > 0.25) elements were Al, Ca, Cl, Fe, K, Ni, P, S, and Si (Figure 20.6).

While Ca, Cl, K, P, and S may (at least in part) be considered native elemental components of the tissues, the elevated levels of Al, Ca (to some extent), Fe, Ni, and Si are most likely due to soil contamination over the surfaces of the thalli. The soils of the Taimyr region are rich in these elements as well as Cu which was also found in smaller quantities in all of the samples examined from the "clean" sites. Aluminum and silicon are likely components of an aluminum silicate from the soil, and these elements (particularly Si) were most abundant. Other elements present, but at a reduced extent in the "clean" site specimens, were magnesium, titanium, and vanadium. Elevated levels of silicon are most likely not a matter of serious concern as that element is typically not toxic even in high concentrations, and it has been found to be a prevalent component of plant cell walls, as well as in pollen.[17]

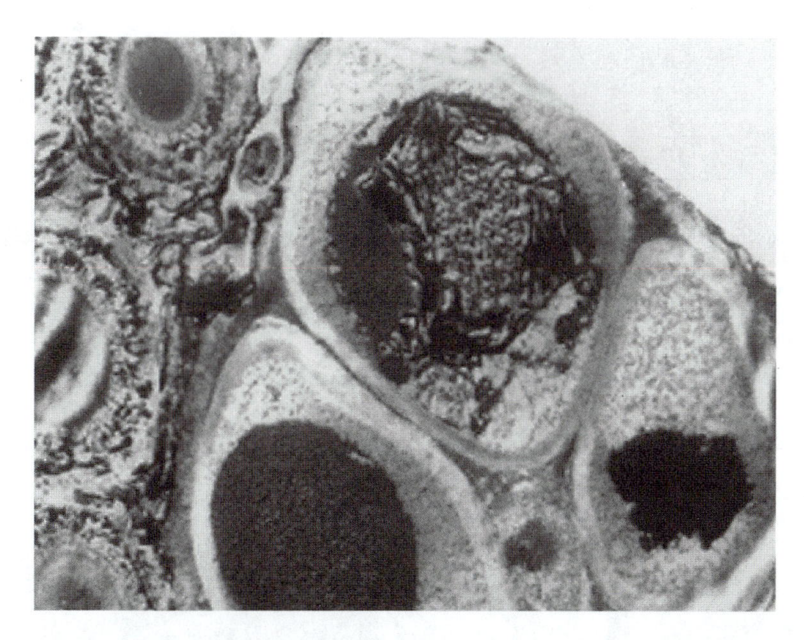

Figure 20.5 Specimen from polluted site. Transmission electron micrograph of *C. cucullata* medullary region showing both algal and fungal cells. Note the deteriorated status of the algal chloroplast thylakoids. Furthermore, fungal cells possess electron-dense deposits in and adjacent to the cell walls. 2200x.

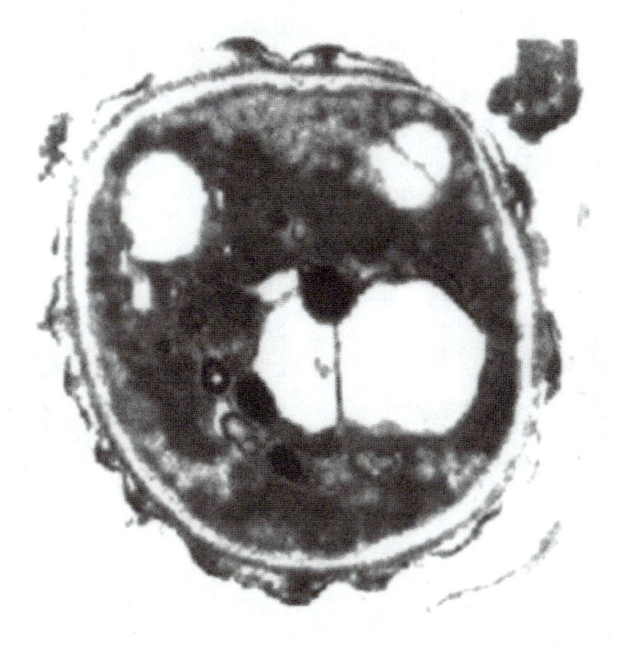

Figure 20.6 Transmission electron micrograph showing medulla fungal cell of *C. cucullata* collected from polluted site. The accumulation of dense bodies in the cytoplasm, the degradation of membrane structures, and the general loss of cytoplasmic detail is a characterstic response. 21,000x.

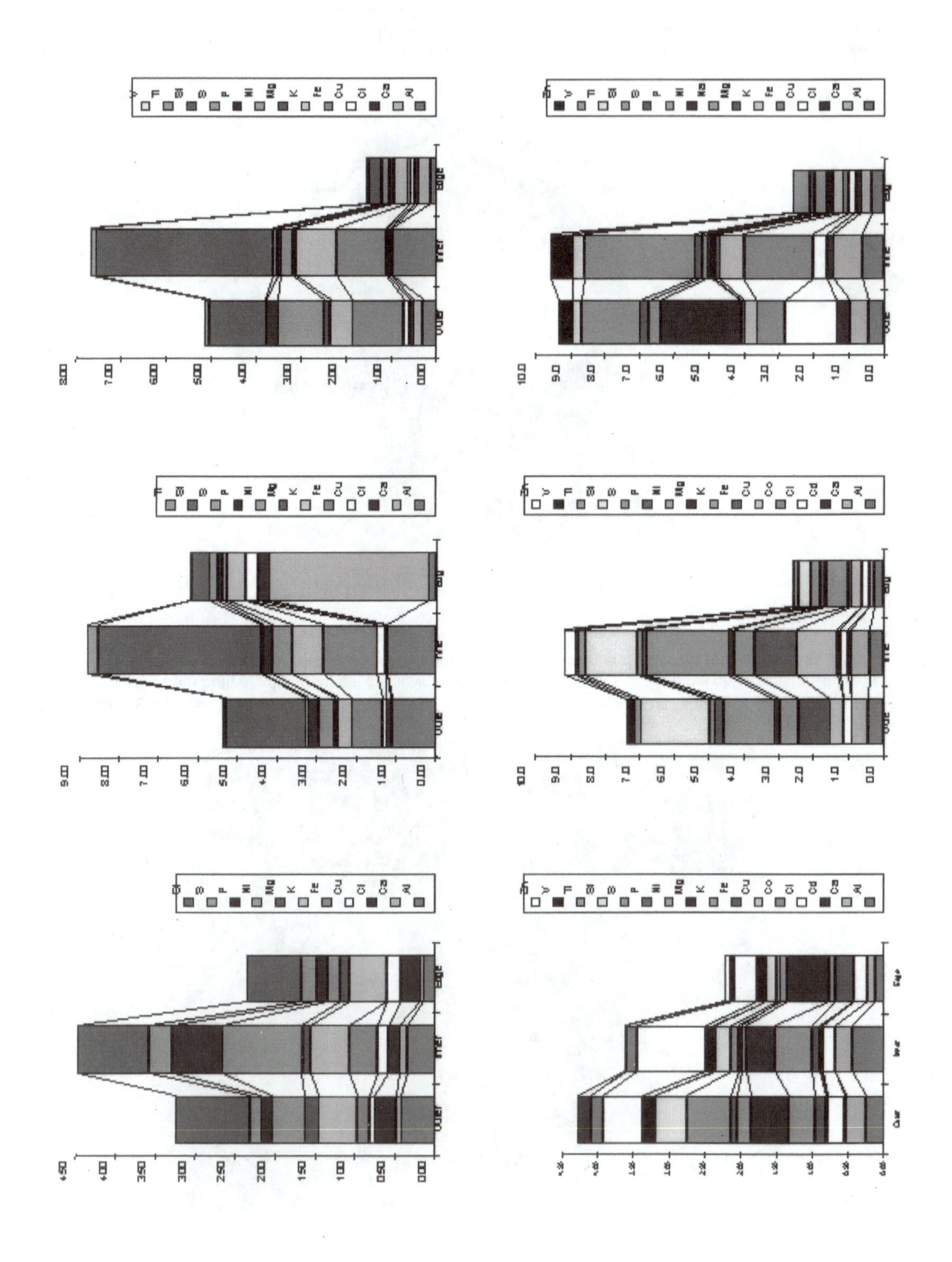

The concentration of these elements as reflected by their P/B ratios, consistently showed the highest levels from the inner surfaces at distal, mid, and basal segments of the thalli, with Si being the most prevalent element. Detection of elements from broken edges of the thalli showed varying results (Figure 20.7). Relatively little qualitative variation of major elements occurred throughout the three levels of sampling within the thalli, although quantitatively, the distal portion showed the lowest incorporation of the elements. The small amounts of Ti and V were only found in the mid and basal portions of the thalli.

In the samples taken from the polluted sites near Noril'sk, there were substantially greater numbers of elements represented at all levels throughout the height of the thalli, but the most prevalent elements (same standard used as above) were the same as in the "clean" site specimens. It was the incorporation of elements equal to or less than a P/B - 1 of 0.25 which included Cd, Co, Cr, Na, Pb, Ti, V, and Zn. Except for Ti and V, none of these elements were found from the "clean" site specimens. As with the "clean" site specimens, these elements increased quantitatively in the basal direction, and with the inner surfaces increasing the most (although at a lower rate than in the "clean" samples). All of the contaminating elements appear to be in line with known emissions from nickel- and copper-smelting plants. Although, in general, a greater quantitative assessment of most elements was determined from the basal portions of thalli, the greatest qualitative diversity of contaminating elements was found from the inner and outer surfaces of the distal portion, followed by the midregion, and with the least diversity from the basal portions (Figure 20.7).

Although some previous studies have been conducted to show the phytotoxic effect of metals such as Al, Cd, Co, Cr, Cu, Mn, Ni, and Zn on plants hydroponically grown,[18,19] there have been no studies on the toxic effect which particulates may have on plant tissues, or their means of absorption. Conner and Meredith[20] have addressed a variety of ways in which plants have developed mechanisms of resistance to Al toxicity including its chelation with citrate or EDTA, or when cells are provided with elevated levels of phosphates. Using nuclear magnetic resonance spectroscopy, Pfeffer et al.[21] also showed that plant root systems may sequester toxic metals such as Al and Mn through compartmentalization with inorganic phosphates. It is not so much the amount of phosphates that were present in the root cells that would bind the metals, but their physiochemical state which would allow for entrapment of relatively high concentrations of metal ions. It is, in fact, noted that the polluted samples also showed a higher level of P (Figure 20.7). Although this study was directed at a general assessment of the surface-bound and native elemental composition of *C. cucullata*, the intracellular dynamics that entrap metals should be studied more intensively.

In SEM studies on the surfaces of *C. cucullata* from the "clean" sites, a few specimens were observed which possessed particles of distinctive structural features as that of fly ash (Figure 20.8).[22] EDX analysis showed such particles to be primarily composed of Al, Si, and Ti. Fly ash particles from samples at the "clean" site indicated that the air pollution deposition gradient from the Noril'sk site geographically extended much farther than previously anticipated. This may also explain the traces of contaminating elements in the EDX data such as Ti and V at the "clean" collection sites. It is for this reason that these sites are designated in this chapter with quotation marks.

Figure 20.7 Average elemental composition of surfaces from *C. cucullata* collected at "clean" site on distant Taimyr Peninsula, and from site within 15 km of Noril'sk that was heavily polluted. Units on the vertical axis represent average P/B -1 values for the elements which are shown in alphabetical rank (from bottom to top). Note that vertical axes may represent different ranges of P/B -1 values. Data from each set of elements represent a minimum of nine EDX spectral readings (three sites per sample, three samples). Edge represents broken surfaces which exposed the inner medullary region of the thallus in each of the segments above.

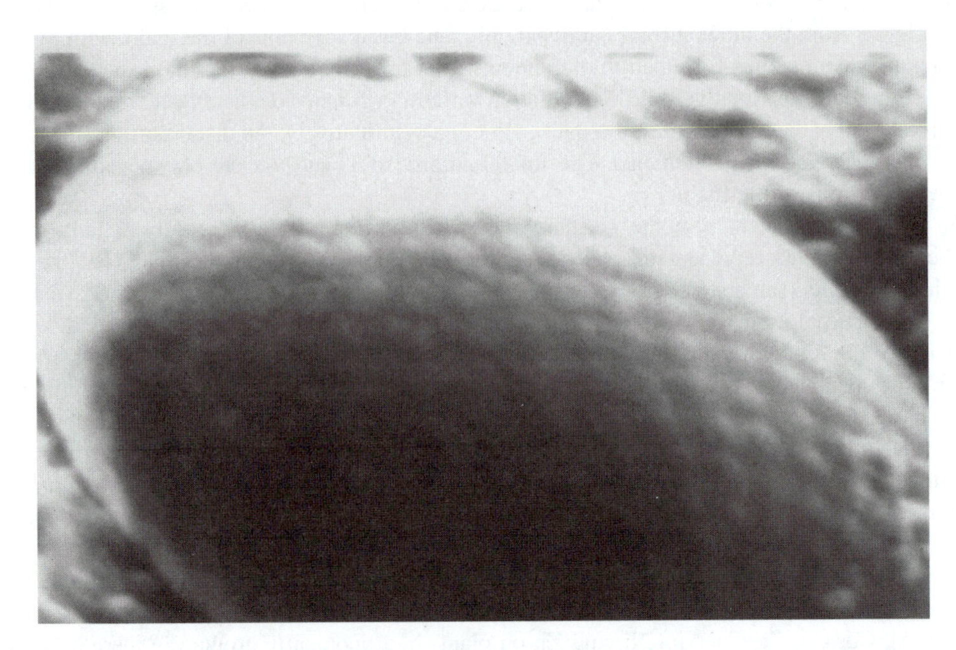

Figure 20.8 SEM view of particle representing presumed fly ash found on outer surface of *C. cucullata* basal region of thallus from "clean" site on Taimyr peninsula. 2200x.

20.8 SUMMARY AND CONCLUSIONS

Airborne effluent from nickel/copper-smelting plants in the Noril'sk district of northern Siberia impacts on a region extending across the length of the Taimyr Peninsula in the Arctic. Deposition from this source has been shown by means of SEM and EDX microanalysis to accumulate on the surfaces of the lichen *C. cucullata*, primarily over the basal portions, and on both inner and outer surfaces of the thallus. The overall amount of inorganic accumulation within the tissues did not appear changed under conditions of pollution, but that composition included pollutant elements. Pollutants increased on the outer surfaces in the mid-portions and to the basal portions of the lichen, but not at the distal portions. Morphometric data from light microscopy have shown more-compacted tissues and reduced air space within the lichen thallus medulla after exposure to pollutants, but the cortex appeared relatively unaffected. In addition, light and electron microscopy have revealed substantial structural cell damage to both algal and fungal components of polluted lichen specimens. Specimens collected from polluted sites possessed intercellular materials of enhanced density and, to a lesser extent, dense cytoplasmic materials could also be observed. In appearance, the dense accumulations appear similar to reports of polyphenolic accumulations. It is not known to what extent intra- and extracellular sites within the lichen tissues may also serve as microscopic regions of metal containment.

Clearly, an understanding of the ability of lichens such as *C. cucullata* to not only serve as bioindicators, but to accumulate metals from the environment will require further study of the interaction of soil and other habitat factors with the lichen. Such studies will also be of considerable significance in aiding our understanding to monitor the environmental impact of anthropogenically generated air pollutants. The fact that *C. cucullata* serves as a "sink" for so many elements of environmental pollution appears to make it a useful agent for chemical analysis as well as a morphological tool of bioindication and bioaccumulation.

ACKNOWLEDGMENTS

This case study review was supported by the International Plant and Pollution Laboratory, Bowling Green, Ohio, and the U.S. EPA Environmental Research Laboratory, Corvallis, Oregon. The author also wishes to acknowledge the assistance of Drs. Dixon Landers, Juri Martin, and Reginald Noble in the preliminary review of this manuscript. Technical laboratory assistance by Zheng Wu is also gratefully acknowledged, as is the collection of specimens by Dr. Jesse Ford.

REFERENCES

1. Hong, S., Candelone, J. -P., Patterson, C. C., and Boutron, C. F., Greenland ice evidence of hemispheric lead pollution two millennia ago by Greek and Roman civilizations, *Science*, 265, 1841, 1994.
2. Sturges, W. T., Ed., *Pollution of the Arctic Atmosphere*, Elsevier, London, 1991.
3. Moser, T. J., Swafford, J. R., and Nash III, T. H., Impact of Mount St. Helens' emissions on two lichen species of south-central Washington, *Environ. Exp. Bot.*, 23, 321, 1983.
4. Vlasova, T. M., *Int'l Conf. on the Role of the Polar Regions in Global change: Proceedings of a Conf., June 11–15*, 1990, Ureller, G., Wilson, C. L., and Sevevin, B. A. B., Eds., 1991, 423.
5. Sernander, P., *Stockholms Natur.*, Uppsala, 1926.
6. Nash, T. H., III, and Wirth, V., Eds., *Lichens, Bryophytes and Air Quality*, J. Cramer, Berlin, 1988.
7. Martin, J. L., Noble, R. D., and Schwab, D. W., Lichen and moss surface analysis using scanning electron microscopy and energy dispersive X-ray spectroscopy, *Proc. Estonian Acad. Sci. Ecol.*, 2, 81, 1992.
8. Wasserman, J. L., Mineo, L., and Majumdar, S. K., Detection of heavy metals in oak mycorrhizae of northeastern Pennsylvania forests, using X-ray microanalysis, *Can. J. Bot.*, 65, 2622, 1987.
9. Martin, J., Martin, L., Tamm, K., Kazachevsky, A., Atnashev, V., Alexeyev, V., and Alexeyeva, N., Element accumulation in lichens, mosses and soils connected with mud volcano activity, in *Air Pollution Effects on Vegetation Including Forest Ecosystems, Proc. 2nd U.S.–USSR Symposium*, Noble, R. D., Martin, J. L., and Jensen, K. F., Eds., USDA/EPA,1989, 205.
10. Hoffmann, E. O., Flores, T. R., Coover, J., and Garrett II, H. B., Polychrome stains for high resolution light microscopy, *Lab. Med.*, 14, 779, 1983.
11. Toth, R., An introduction to morphometric cytology and its application to botanical research, *Am. J. Bot.*, 69, 1694, 1982.
12. Crang, R. E. and McQuattie, C. J., A quantitative light microscopic technique to assess the impact of air pollutants on foliar structure, *Trans. Am. Microsc. Soc.*, 106, 164, 1987.
13. Bradley, R., Burt, A. J., and Read, D. J., The biology of mycorrhiza in the Ericaceae: VIII. The role of mycorrhizal infection in heavy metal resistance, *New Phytol.*, 91, 197, 1982.
14. Nieboer, E., Puckett, K. J., and Grace, B., The uptake of nickel by *Umbilicaria muhlenbergii*: a physicochemical process, *Can. J. Bot.*, 54, 724, 1976.
15. Gadd, G. M. and Griffiths, A. J., Microorganisms and heavy metal toxicity, *Microbial Ecol.*, 4, 303, 1978.
16. Zobel, A. M., Brown, S. A., Kuras, M., Crang, R., Wierzbecka, M., and Nighswander, J. E., Some secondary metabolites altered by heavy metals, in *Heavy Metals in the Environment*, Toronto, 1, 214, 1993.
17. Crang, R. E. and May, G., Evidence for silicon as a prevalent elemental component in pollen wall structure, *Can. J. Bot.*, 52, 2171, 1974.
18. Harrison, S. J., Lepp, N. W., and Phipps, D. A., Uptake of copper by excised roots. IV. Copper uptake from complexed sources, *Z. Pflanzenphysiol.*, 113, 445, 1984.
19. Taylor, G. J. and Foy, C. D., Differential uptake and toxicity of ionic and chelated copper in *Triticum aestivum*, *Can. J. Bot.*, 63, 1271, 1985.
20. Conner, A. J. and Meredith, C. P., Simulating the mineral environment of aluminum toxic soils in plant cell culture, *J. Exp. Bot.*, 36, 870, 1985.
21. Pfeffer, P. E., Tu, S.-I., Gerasimowicz, W. V., and Cavanaugh, J. R., *In vivo* ^{31}P NMR studies of corn root tissue and its uptake of toxic metals, *Plant Physiol.*, 80, 77, 1986.
22. Hayes, T., Biophysical aspects of scanning electron microscopy, *Scanning Electron Microsc.*, 1, 1, 1980.

Chapter 21

Phytochelatins and Metal Tolerance

Rajesh K. Mehra and Rudra D. Tripathi

CONTENTS

21.1 INTRODUCTION

Living organisms employ a range of mechanisms to resist/tolerate heavy metal stress. A tight control over the homeostasis of even essential metal ions such as copper and zinc is necessary for proper physiological functioning of cells. The regulation of intracellular concentrations of metal ions occurs by energy-dependent efflux of metal ions, changes in their redox status, binding to cell wall, extracellular or intracellular sequestration of metal ions by peptides/proteins or other biomolecules, and finally their storage within vesicular compartments.[1-7] Eukaryotes regulate the concentrations of reactive free metal ions by intracellular sequestration. The chelated metal ions may also be stored within intracellular compartments such as vacuoles and tertiary lysosomes. Glutathione (GSH),[8] SH-related phytochelatins (PCs), and cysteine-rich metallothioneins (MTs)[6,7] are the main metal-sequestering molecules. Organic acids may also participate in metal ion sequestration in plants. GSH constitutes the primary line of defense against metal toxicity in most organisms. Inhibition of GSH synthesis by specific chemical agents sensitizes the cells to Cd and possibly other metal ions.[8,9] Similarly, mutations in GSH synthesis also sensitize cells to metals.

1-56670-341-7/00/$0.00+$.50

An additional line of defense against toxic metals is provided by MTs in animals, some yeast, and possibly plants.[6,7] Animals, yeast, and possibly plants lacking functional MT genes have significantly reduced the ability to tolerate metals, particularly Cd.[10] It is important to note that yeasts do not appear to use MTs for sequestering Cd. The detoxification role of MTs in yeasts appears to be limited to Cu(I) sequestration. No evidence has yet been obtained regarding PC biosynthesis in animals. MT-like genes and their transcripts have been reported from a number of plant species.[11-13] Some of the plant MT genes are also induced in response to metal.[12-14] Heterologous expression of the plant MTs in *Saccharomyces cerevisiae* increased the metal tolerance of a metal-sensitive strain. These results were interpreted to indicate that plant MTs can bind copper and are likely involved in Cu tolerance in *Arabidopsis thaliana*.[12-13] In fact, five MT genes have been characterized from *A. thaliana*. The position of each gene is conserved and they encode for two groups of proteins, MT1 and MT2. Four of the MT genes MT1a, MT1c, MT2a, and MT2b are expressed while the fifth gene, MT1b, is inactive in response to Cu.[13] Additional evidence has been presented to indicate that reduced expression of MT genes sensitizes plants to Cu.

There appears to be considerable evidence to support the involvement of PCs in the management of heavy metal stress in plants. However, metal tolerance in plants may not always correlate with PC levels as factors other than these peptides can contribute to metal tolerance. PCs were originally discovered in *Schizosaccharomyes pombe* by Hayashi and Mutoh[15] who named them cadystins. Zenk and co-workers independently found these peptides in plants (see Zenk[16]) and algae and named them phytochelatins (PCs). The term phytochelatins is now the most commonly employed nomenclature for these peptides. Analogs of PCs differing in the C-terminal residue have been detected in plants.[17,18] These PC variants probably also chelate metals and are likely synthesized from the corresponding GSH variants.[16] However, mechanisms of their synthesis might be different from those of PCs as PC synthase did not seem to catalyze the synthesis of these isoforms of PCs.[16]

Biochemical and genetic analysis of metal-resistance pathways in the yeast *Candida glabrata* uncovered dual synthesis of PCs and MTs in this opportunistic pathogen.[19] However, the synthesis of these two classes of metal-sequestering molecules occurred in a metal-specific manner. Cu(II) induced MTs, whereas PCs were produced in response to Cd(II). *Neurospora crassa* and *S. cerevisiae* were also later shown to exhibit Cd-specific production of PCs, Cu-induced MT synthesis is well known in *N. crassa* and *S. cerevisiae*.[6,7,21] As discussed above, evidence now exists regarding induction of MT-like genes in plants in response to Cu and possibly other metal ions.[11,13] The purification and characterization of metalloforms of MT-like proteins have not been accomplished satisfactorily yet. Characterization of purified plant MTs can resolve issues concerning the roles of these proteins in metal detoxification in plants. It seems likely that further studies will show similarity in yeast and plants regarding metal specificity in the synthesis/induction of PCs and MTs.

Although PCs have been studied extensively in metal-exposed plants, not all metal-tolerant plants exhibit increased production of PCs. These peptides do not constitute exclusive mechanisms for metal tolerance in plants. It must be realized that the acquisition of metal tolerance may depend on several different factors such as reduced uptake, sequestration by organic acids, and vesicular storage. However, it seems logical to assume that increased expression of PCs should result in enhanced metal sequestration and thus increased metal tolerance. As detailed later, the formation of sulfide complexes of PCs may be of greater significance in the acquisition of metal resistance.[22] Besides, an interesting finding was the fact that labile sulfite is also a component of PC complexes at levels several-fold higher than those of sulfide ion.[22] This chapter primarily discusses the currently known aspects of PC biosynthesis, the metal-binding characteristics of these peptides, and their role in the formation of nanocrystalline (NC) semiconductors within yeast or plant cells. More information on this subject can also be found in recent reviews.[16,23-25]

21.2 BIOSYNTHESIS AND DEGRADATION OF PCs

The biosynthesis of PCs has generally been studied by treating plants/yeasts with selected metal ions followed by analyses of metal-binding components by a variety of chromatographic techniques. Quantitative PC analyses in crude extracts is generally carried out by HPLC assays coupled with pre- or post-column derivatization of separated components. Some studies have also been carried out using radioactive precursors or partially purified enzyme PC synthase.

21.2.1 Biosynthesis of PCs

It is now generally agreed that PCs are synthesized from GSH. Inhibition of GSH synthesis by buthionine sulfoximine (BSO) or other agents sensitizes yeast and plant cells to Cd toxicity presumably due to lack of PC production.[22,33] The treatment of animal cells with BSO also results in increased Cd sensitivity.[8] Animal cells apparently do not synthesize PCs. Consequently, BSO-induced Cd sensitivity does not necessarily mean that it is due to inhibition of PC biosynthesis. It is important to distinguish between the tolerance provided by GSH alone and a combination of GSH and PCs. Quite clearly, GSH mutants did not produce PCs in *S. pombe*[26] or in *C. glabrata*.[27] The most important mutants described in Cd tolerance pathway came from *A. thaliana*.[28,29] The isolation of both GSH and PC synthase mutants was sensitive to Cd. It was demonstrated that the PC-deficient mutants contained negligible contents of PC synthase. However, it was also shown that GSH did not make a big contribution to Cd tolerance. These observations on *A. thaliana* are in contrast to most other studies on plants, animals, and yeast which demonstrate clearly the significance of GSH in Cd tolerance.[9] *S. cerevisiae* mutants defective in the transport of Cd–GSH complexes into the vacuole were sensitive to Cd. During 24-h exposure of mercury, both leaves and roots of submerged plants *Vallisneria spiralis* accumulated higher levels of GSH probably showing a role of GSH in detoxification of Hg besides synthesizing lower levels of PCs. However, during 96-h exposure, only PCs played a role in Hg detoxification.[30] During exposure of Hg, a wetland plant *Bacopa monnieri* accumulated higher levels of GSH, which was responsible for the Hg detoxification (Table 21.1). During exposure of lead, both *Hydrilla verticillata* and *V. spiralis* synthesized PCs as a Pb(II)-detoxifying system, showing no direct role of GSH in Pb detoxification in these plants.[31] In submerged plant *H. verticillata*[25,32] and water lettuce *Pistia stratiotes*[33] Cd tolerance was conferred by synthesis of PCs. In response to BSO, PC synthesis declined due to inhibition of GSH biosynthesis in these plants.

Variants of GSH with different C-terminal residues were found in some plants.[16] These plants synthesize PCs that have the same C-terminus as the corresponding form of GSH. These studies further supported the notion that GSH or its variants are indeed PC precursors. Plants synthesizing homoglutathione (Glu-Cys-Ala) in place of GSH produced PCs that had Ala at their C-terminus.[34] Radioactivity in GSH was found to appear in newly synthesized PCs *in vivo*.[35] PC deficiency in GSH-defective *S. pombe*,[26] *S. cerevisiae*,[36] *C. glabrata*,[27] or *A. thaliana*[28,29] mutants strongly supports the thesis that GSH or its precursor Glu-Cys is involved in the synthesis of PCs. Four sets of PC peptides have been discovered; all are homologous to GSH only differing in C-terminal residue. Most prevalent and of universal distribution appears to be $(\gamma\text{-glutamyl cysteinyl})_n$ glycine ($n = 2$ to 11).[15,16,36-39] The other homologous compounds were one which contained β-alanine instead of glycine and were derived from homo-GSH and named homophytochelatins $(\gamma\text{-glu-cys})_n$-β-ala ($n =$ 2 to 7), (h-PC)[34] and were prevalent in order Fabales (tribe Phaseoleae).[16] Interestingly, species of tribe Omonideae, Trifolieae, Viciaceae, Astragalaceae, and Loteae showed the presence of both PC and h-PC.[16] Members of family Poaceae contained hydroxymethyl-GSH $(\gamma\text{-glu-cys})_n$-ser peptides that were named hydroxymethyl PCs.[17] Another set of peptides such as $(\gamma\text{-gly-cys})_n$ Glu was induced by Cd in maize.[18] However, clear biochemical or genetic data regarding PC biosynthesis are still not available.

Table 21.1 Metal Tolerance Mediated through MTs, PCs, and GSH

Plant	Metal tolerance	Chemical Species	Ref.
Algae			
Thalassiosira weissflogi (marine diatom)	Cd tolerance	PC_2 and PC_3 were involved in laboratory and field experiments	73
T.weissflogi	Cd^{2+},Pb^{2+} and Ni^{2+} tolerance	PCs PCs at very low 1 n*M* and 1 nM>	73
Chlamydomonas	Hg (ii) tolerance	Both PC and GSH	75
Euglena gracilis	Cd tolerance	PC_2,PC_3,PC_4,PC_5	74, 76
Fragilaria crotonesis	Cd tolerance	PC_2,PC_3,PC_4,PC_5	74, 76
Chlamydomonas reinhardtti	Cd tolerance	PC_2,PC_3,PC_4,PC_5	74, 76
Sargassum muticum	Cd tolerance	PC_2,PC_3,PC_4,PC_5	74, 76
Porphyridium cruentum	Cd tolerance	PC_2,PC_3,PC_4,PC_5	74, 76
Fungi/Yeast			
Schizosaccharomyces pombe	Cd(II) detoxification	GSH	9
	Cd (II) detoxification	GSH	63
	Cd tolerance	Purine biosynthetic genes	77
	Cd tolerance	Homologous PCs	37
Candida glabrata	Cd (II) tolerance	Cd-sulfide crystallites	60
	Cd (II) tolerance	GSH coated	
S. pombe	Cd(II) tolerance	Cd-S crystallites	58
	Cu(I) tolerance	Cadystins γ-glutamyl peptides	
C. glabrata	Cu(I) tolerance	MTs	52
	Cd (II) tolerance	PCS	19
S. pombe	Cd (II) tolerance	type protein (the vacuolar membrane transporter for PCs)	40, 41
Saccharomyces cerevisiae	Cd tolerance	Vacuolar transporter yeast cadmium factor (YCF1) protein	42
S. cerevisiae, strain DTY 167 and isogenic wild type strain DTY165	Cd tolerance	Transmembrane YCF1 mediates Mg ATP energized vacuolar accumulation of $(GS)_2$ Cd complex; DTY167, harboring a deletion of YCF1 gene is Cd^{2+} hypersensitive; YCF1 does not mediate transport of $(\gamma\text{-glu-cys})_n$ Gly	43
S. cerevisiae	Cd tolerance	*Arabidopsis* coding for an ABC type transmembrane multidrug-resistant protein (At MRP3) exhibts high homologies to YCF1; like MRP1 and YCF1, AtMRP3 confers Cd tolerance to Cd sensitive strain DTY168; complemented with this gene, DTY168 (At-MRP3) showed high tolerance to Cd	64
	Cd tolerance	GSH-deficient mutants are unable to synthesize PCs and are hypersensitive to Cd	36

Table 21.1 continued

Plant	Metal tolerance	Chemical Species	Ref.
Bryophytes/Pteridophytes (Aquatic Mosses)			
Rhyncostegium riparoides	Zn, Cd ,and Pb	PC–metal complexes were isolated in response to Zn and Cd	71
Bryophytes and pteridophytes	Cd tolerance	PCs	69
Gymnosperms			
Picea rubens *Abies balsamea*	Heavy metal detoxification	PC as of heavy metal exposure, correlated with forest decline	72
Acer sp.	Zn detoxification	PCs involved in metal detoxification in mining area; the angiosperm *Silene cucubalus* was more Zn tolerant but had low level of PCs suggesting involvement of an additional tolerance mechanism	34
Higher plants/(Angiosperms)			
Tomato cell lines of *Lycopersicum esculentum*	Cd^{2+} tolerance	Phytochelatins	65, 66
Cell suspension culture of *Rauwolfia serpentina*	Cd, Cu, tolerance	PCs	38
Tomato cells	Cd resistance	PCs	67
Root cultures of *Rubia tinctorum*	Cd tolerance	PCs and desglycyl peptides	68
Various plants species	Cd tolerance	PCs	69
Pisum sativum (Sweet pea)	Cd tolerance	PCs and homophytochelatins	70
Zea mays seedings	Cd detoxification	Glutamyl cysteine glutamic acid	24
Z. mays	Cd detoxification	Three families of thiol peptides	18
Datura inoxia	Cd detoxification	PCs binding to Cd	78
Nicotiana rustica var. *pavonii*	Cd detoxification	PC_3 and PC_4 were bound to CdBPs localized in vacuole	49
Bacopa monneri	Hg-tolerance	GSH seems to be involved	79
Various plant species	Cd tolerance	PCs	69
Silene vulgaris	Cd tolerance	Poly-γ-glutamyl cysteine glycine	4
L. esculentum	Cd-tolerance	Cd-$(\gamma\text{-EC})_n$G peptide complex	81
P. sativum (sweet pea)	Cd tolerance	PCs and homophytochelatins	70
R. serpentina	Cd tolerance	Two high-molecular-weight Cd–PC complexs, one low Cd–PC complex, all the three were different in chain length.	82

Table 21.1 continued

Plant	Metal tolerance	Chemical Species	Ref.
Silene cucubalus	Cd tolerance	High- and low-molecular-weight PC–metal complexes could not be separated	82
	Cd tolerance	Cd–PC complex, sulfide is incorporated in low molecular weight making high-molecular -weight complexes	4
S. vulgaris	Cd tolerance	PCs are synthesized and PC synthase demonstrated	80
L. esculentum	Cd detoxification	Cd–PC complex	67
	Cd tolerance	PCs, purification of PC synthase from various plant parts.	46
Brassica juncea	Cd detoxification	Cd-PC complex.	88
Arabidopsis ecotypes	Copper tolerance	Both nonprotein thiol (NPT) and MTs were expressed by Cu; corelation between tolerance and NPT was not significant; both MT1 and MT2 were expressed; however, MT2 was Cu inducible; Cu tolerance was associated with MT2	83
A. thaliana	Cu tolerance, mercury tolerance less	MT2 gene mutated in Cup 1 mutant	84
	Cu tolerance.	Five MT genes with four active MT1a, MT1c, MT2a, and MT2b, MT1b being inactive; the locations of these empty genes were different from that of CAD1 gene involved in Cd tolerance in this plant.	13
	Cd tolerance	CAD1 gene (PC synthase system) CAD1 mutant (much less PC synthase) CAD1 mutant are PC deficient due to defect in PC synthase system and are hypersensitive to Cd	28
	Cd tolerance	Due to defective GSH biosynthesis CAD2 mutants are PC deficient and sensitive to Cd	29
a.) Monocots: Maize, oat, barley,rice b.) Dicots: Azukibean,cucumber, lettuce, pea, raddish, sesame, and tomato	More Cd tolerance in cereals (monocots than dicots)	Cereals root synthesized Cd–PC complex in cytoplasm; this complex was not found in this fraction in dicots	85
Pistia stratiotes	Cd tolerance	PC_2, PC_3, and unidentified thiols	33
Hydrilla verticillata	Cd tolerance	PC_2, PC_3, and unidentified thiols	25
	Pb tolerance	PCs, PC–Pb complex was found	31

Table 21.1 continued

Plant	Metal tolerance	Chemical Species	Ref.
Vallisneria spiralis	Hg tolerance	PCs, PC–Hg complexes were found	30
H. verticillata	Hg tolerance	PCs, PC–Hg complexes were involved	30
V. spiralis	Hg tolerance	PCs, GSH also seem to be involved	30
Many plant species	Pb tolerance	PCs and other tolerance modes	87

Various strategies adopted by various categories of plants, including yeasts, conferring metal tolerance have been depicted in Table 21.1. The heavy metal tolerance based on a number of ostensibly disparate observations appears to be ubiquitous in various groups of plants, specifically, algae, fungi, bryophytes, pteridophytes, gymnosperms, and angiosperms. The prevalent detoxification mechanism seems to be induction of PCs. Heavy metal tolerance in fungi has been largely attributed to MTs.[21] Studies with *S. pombe* (the fission yeast), *C. glabrata*, *S. cerevisiae*, and *N. crassa* also implicated PCs to impart tolerance. While Cu stimulates MTs, Cd induces PCs in *C. glabrata*.[19] *S. pombe* appears to be good model in which PCs have been demonstrated[40] with further characterization of HMT1 gene, a six-transmembrane span-single nucleotide binding fold (single half) ATP-binding cassette mediating transport of both PCs and Cd–PCs.[40,41] Further, another ABC type of transporter, yeast Cd factor protein (YCF1), responsible for Cd tolerance in *S. cerevisiae* has been demonstrated.[42,43] Interestingly, the transmembrane transporter YCF1 mediates uptake of (GS)2-Cd but not $(\gamma\text{-glu-cys})_n$-Gly.[43]

21.2.1.1 PC Synthase and the Mechanism of PC Biosynthesis

As suggested by Hayashi et al.,[44] PC biosynthesis may proceed by a variety of reactions

$$ECG + ECG \rightsquigarrow (EC)_2G + G \tag{21.1}$$

$$ECG + (EC)_2G \rightsquigarrow (EC)_3G + G \tag{21.2}$$

$$(EC)_2G + (EC)_2G \rightsquigarrow (EC)_3G + (EC)_4G \tag{21.3}$$

$$nEC \rightsquigarrow (EC)_n \tag{21.4}$$

$$(EC)_n + G \rightsquigarrow (EC)_n\,G \tag{21.5}$$

$$ECG + EC \rightsquigarrow (EC)_2G \tag{21.6}$$

Grill et al.[45] isolated an enzyme designated PC synthase that catalyzed the reactions shown in Reactions 21.1 through 21.3. PC synthase was activated by metals in the following order: Cd(II)>Ag(I)>Bi(III)>Pb(II)>Zn(II)> Cu(II)>Hg(II)> Au(I). The *Silene cucubalus* PC synthase had an Mr of 95,000 and appeared to be composed of four subunits. A 50 KDa dimer was also active in the synthase activity. Howden et al.[28] isolated CAD1 mutant of *A. thaliana* which was deficient in forming Cd–PC complexes, while GSH biosynthesis was comparable to that of the wild type. It was inferred that CAD1 gene was the structural gene for PC synthase. This indeed is the single most important proof in favor of a role for PC synthase in PC production. It was also shown that CAD1 mutant did not exhibit any PC synthase activity *in vitro*. PC synthase has recently been isolated from tomato (*Lycospersicum esculentum*) cell cultures and plants. Cadmium-resistant tomato cells growing in the medium containing 6 m*M* Cd^{2+} have 65% more PC synthase activity as compared with unselected cells. PC synthase is also present in the roots and stem of tomato plants, but not in leaves and fruits.[46] Tomato enzyme (PC synthase) uses GSH or PCs as substrate. However, it does not use GSSG or g-glu-cys as substrate.

The nonpurified enzyme preparation from *Pisum sativum* synthesizes $(\text{g-Glu-Cys-})_2$ b-Ala from h-GSH and also $(\text{g-Glu-Cys-})_2$ Ser from g-Glu-Cys-Ser, as well as producing $(\text{g-Glu-Cys})_n$Gly from GSH during Cd exposure,[16] showing that Pea enzyme has a broad substrate choice. It was observed that PC synthase from *Silene* culture cells could not synthesize $(\text{g-Glu-Cys-})_n$ b-Ala from h-GSH.[16] The distribution of enzyme in various organs of tomato plants and cells indicates that PC synthase may have additional functions in plant metabolism.[46] Clearly, more research has to address possibly various forms and functions of PC synthase from various sources. A model of PC biosynthesis, its transport, storage, and degradation is depicted in Figure 21.1.

21.2.2 Degradation of PCs

Much less is known about the degradation mechanisms for PCs. However, it was noted early that some forms of PCs isolated from both plants and yeast lacked the terminal glycine. These so-called desGly variants of PCs have been demonstrated at many instances.[16,21-23] *C. glabrata* is not known to synthesize PCs longer than $(EC)_2G$. Furthermore, $(EC)_2$ occurs at very high concentrations in this yeast following exposure to Cd for extended periods of time. The desGly variants increased in concentration with time following Cd treatment indicating that these were products of catabolic pathway.[47] This idea is also supported by following observations: (1) *C. glabrata* cells exposed to a single dose of Cd for 36 h contained much more $(EC)_2$ than $(EC)_2G$, whereas the cells exposed to $CdSO_4$ for 36 h but in two doses (the metal salt being added at time 0 and 18 h and the cells analyzed 18 h after the second addition) showed predominantly $(EC)_2G$ and a much smaller amount of $(EC)_2$ (RKM, unpublished results), and (2) significant reduction in the proportion of desGly PCs occurred at higher Cd concentrations in both wild-type and Cd-resistant *C. glabrata*.[48] These results suggest that increased Cd concentration enhanced PC biosynthesis resulting in the predominance of $(EC)_2G$ over the desGly component. Several experiments suggest that Cd–PCs are transported to vacuoles.[49] Cd may dissociate under acidic conditions in vacuole releasing apoPCs which may then be degraded by vacuolar peptidases. Further research is needed to understand both the biosynthesis and degradation of PCs.

21.3 METAL-BINDING CHARACTERISTICS OF PCs

PCs, because of the presence of –SH groups of cysteine are expected to bind a variety of metal ions. An understanding of the roles played by PCs in metal sequestration depends upon answers to the following important questions:

1. How are metal ions presented to PCs in the cells?
2. Does the metal-binding capacity of PCs significantly increase with their chain length?
3. Can shorter-chain PCs transfer metal ions to longer-chain PCs during biosynthesis?

Spectroscopic and chromatographic methods have been developed to address these issues.

The purified Cu–PCs exhibited strong luminescence at room temperature indicating the presence of solvent-shielded Cu(I)–thiolate clusters in these peptides.[50] The Cu(I)-binding stoichiometries have been estimated to be 1.25, 2.0, and 2.5 for PC_2, PC_3, and PC_4, respectively.[51] Cu(I)-binding stoichiometry of PCs decreased slightly with the increasing pH of the reconstitution assay. Another

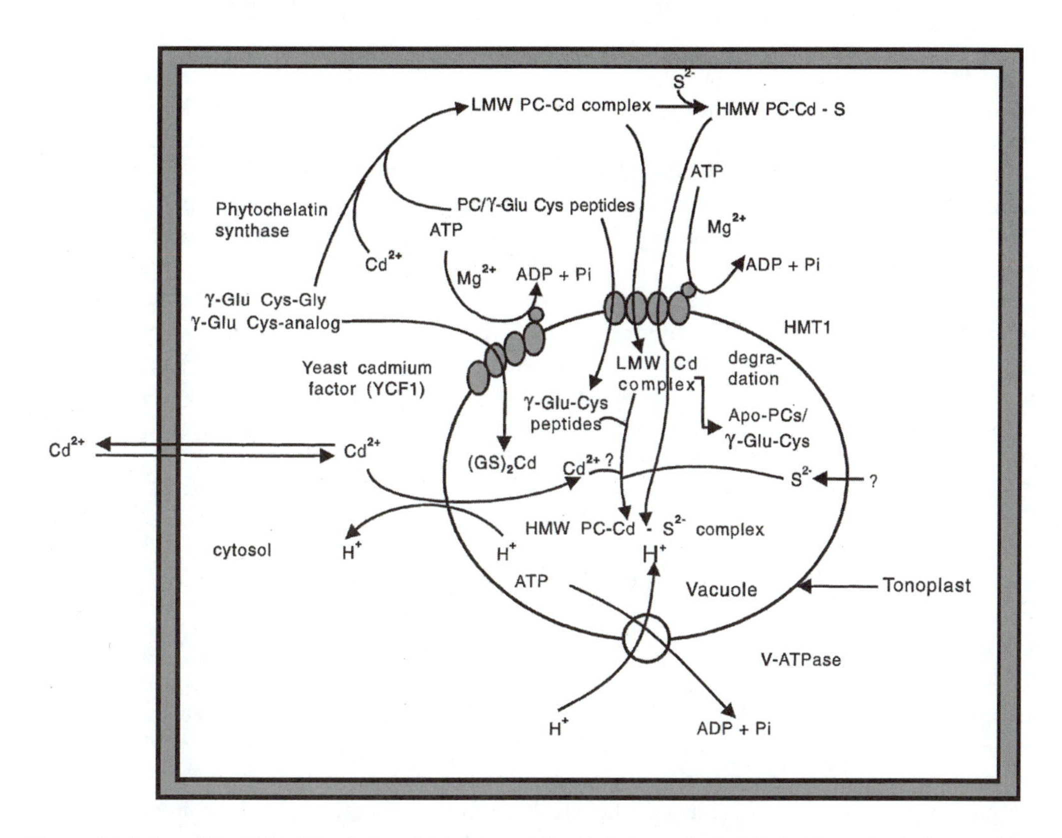

Figure 21.1 A model of Metal (cadmium) tolerance mediated through phytochelatin biosynthesis in yeast/plant cells. Associated transport processes are also depicted. Cd^{2+} taken up by the cell activates the constitutive PC-synthase which uses γECG or related molecules to form (γEC)nG peptides or PCs which chelate cytoplasmic metal ions (Cd^{2+} or other heavy metal cations) to form the LMW PC-Cd complex. LMW PC-Cd from the cytoplasm is actively transported across the tonoplast through the transporter HMT1 into vacuole. Within the cytosol S^{2-} is added to form stable HMW PC-Cd-S (Storage form) which may be actively transported through the transmembrane HMT1 across tonoplast in the vacuole. In addition , within the vacuole S^{2-} may be added to LMW PC-Cd to form HMW PC-Cd-S complex? Alternatively, LMW PC-Cd can be finally degraded to release Cd^{2+} and Apo-PCs can be degraded using peptidase. Because of higher ratio of Cd^{2+} to PCs in HMW complex *vis-a-vis* LMW complex., additional Cd^{2+} may be augmented by Cd^{2+}/H^+ antiporter? Apo-PCs may be also transported across tonoplast through HMT1. The vacuolar ATPase generating proton gradient and transport of S^{2-} across vacuolar membrane are obscure. An alternative pathway of Cd detoxification involving glutathione $(GS)_2$ Cd, through vacuolar transporter yeast cadmium factor protein (YCF1) is also depicted. (After Ortiz et al.,[41] Rauser,[24] Mehra et al.,[48] Li et al.,[42,43] Singh et al.[87]).

interesting feature of Cu(I)–PCs is that these might exist as oligomeric structures.[52] Ag(I) binds –SH groups more strongly than Cu(I). At neutral pH, PC_2, PC_3, and PC_4 were found to bind 1.0, 1.5, and 4.0 equivalents of Ag(I), respectively. Under acidic conditions (pH 5 and below) PCs formed complexes that exhibited Ag(I)/–SH ratios close to 1.0.[53]

Among divalent cations, the binding of Cd(II), Pb(II), and Hg(II) to PCs has been studied in detail.[54-57] Most of the studies on Cd(II) binding have concentrated on the sulfide complexes. Pb(II) coordination in PCs depends on chain length, a feature distinct from that observed for MTs.[6,55] Multiple complexes of Hg(II) were also formed with PCs.[56] This finding is similar to that observed for MTs.[6]

21.4 GSH-MEDIATED TRANSFER OF METAL IONS TO PCs

Several lines of evidence suggest that GSH is the initial metal-sequestering molecule within the cell.[8] It has been shown that the exposure of *C. glabrata* to Cd leads to the formation of Cd–GSH prior to the formation of Cd–PCs.[47] Cd–GSH complexes appear to predominate in the rich medium.[58] We have observed that the conversion of the Cd–GSH into PCs occurs in both the synthetic defined medium and rich complex medium, although the rate of conversion is slower in the rich medium (RKM, unpublished data). It, therefore, follows that GSH must be able to donate metal ions to PCs as these are synthesized. There are significant differences between GSH and PCs in their optical spectra and/or HPLC retention times. These differences could be used to demonstrate metal transfer reactions between GSH and PCs[51] by using fluorescence assays to demonstrate facile transfer of Cu(I) from GSH to PCs. Difference spectroscopy was used to show that GSH also mediates transfer of Cd(II)[54] and Pb(II)[55] into PCs. Similarly, reverse-phase HPLC assays were able to show transfer of Hg(II) from GSH into PCs.[56] The studies discussed in the previous sections suggest a model in which GSH-bound metals are sequentially transferred to PCs as these are synthesized; e.g., Hg(II) is shown to transfer from smaller to larger PCs.

21.5 SULFIDE COMPLEXES OF PCs

Characterization of cadmium sulfide nanocrystallites (CdS NCs) isolated from *S. pombe* or *C. glabrata* showed the particles to consist of Cd, PCs, and sulfide.[58-60] The PC-capped CdS NCs isolated from *S. pombe* and *C. glabrata* were characterized by size exclusion chromatography (gel permeation chromatography), transmission electron microscopy (TEM), X-ray detection (XRD), ultraviolet/visible (UV/VIS) absorption spectroscopy, and composition analysis.[58-60] *C. glabrata* CdS NCs were ~20 ± 3 Å in diameter, whereas those from *S. pombe* were ~18 Å in diameter.[58-60] This size difference in the two complexes was explained by the variations in sulfide incorporation. Ratios of sulfide to Cd were 0.7 and 0.6 for CdS from *C. glabrata* and *S. pombe*, respectively. Both CdS NCs were determined to be near monodisperse, spherical, and homogeneous particles by XRD analysis.[58] The bond length of CdS in both complexes was shorter on average by about 3 to 6% than that of bulk CdS. This was attributed to surface-bonding changes in small particles. The lattice structures of CdS from *C. glabrata* and *S. pombe* were suggested to be a rock salt and an intermediate structure between rock salt and zinc blende, respectively.[58] *C. glabrata* CdS NCs of 20 Å were suggested to form a complex consisting of core 85 CdS units coated with approximately 30 peptides of predominantly $(Glu\text{-}Cys)_2Gly$ and $(Glu\text{-}Cys)_2$ at the surface.[58] *C. glabrata* CdS particles were more monodisperse than typical CdS NCs synthesized from capping with small organic groups or hexametaphosphate (HMP). Both yeasts also produced larger extracellular CdS particles. The particles were shown to have zinc blende structure and were 295 Å in diameter.[58] However, detailed characterizations of capping materials were not reported for extracellular CdS particles.

The synthesis of NCs is difficult mainly because of their coalescing tendency into larger particles or precipitation as bulk materials.[54,58-62] However, native CdS NCs isolated from *S. pombe* did not easily coalesce and CdS NCs capped with $(Glu\text{-}Cys)_3Gly$ or $(Glu\text{-}Cys)_4Gly$ were more resistant to accretion than those capped with $(Glu\text{-}Cys)_2Gly$.[59] *C. glabrata* CdS NCs were less stable than those of *S. pombe* at extreme pH. Acidification of *S. pombe* CdS NCs to pH 4 sharpened the UV/VIS absorption peak presumably due to the removal of low-sulfide-containing particles.[59] Further acidification of this complex resulted in minimal redshifting in absorption spectra. Exposure of *C. glabrata* CdS NCs to 65°C led to a time-dependent redshifting in absorption spectra.[59] The same temperature caused a sharpening of the absorption peak in *S. pombe* CdS NCs. This might have occurred due to the removal of low-sulfide-containing particles leaving high-sulfide-containing particles in the complex. *C. glabrata* CdS NCs were also more sensitive to chelators such as EDTA than the *S. pombe* CdS NCs.[58]

PCs were very effective in controlling the size of CdS NCs or preventing accretion.[3-5] Furthermore, the PC-capped CdS NCs protected NCs from oxygen-radical-mediated dissolution, whereas some of the polymers showed the disintegration of NCs by radicals.[54,61,62] The effect of the type of peptide bond on the formation of PC-capped CdS NCs has been studied *in vitro* with $(Glu\text{-}Cys)_4Gly$ and a mouse MT hexapeptide (Lys-Cys-Thr-Cys-Cys-Ala), both having normal peptide bonds between adjacent residues.[59] Addition of sulfide to the MT hexapeptide led to a redshifting in absorption peak up to 295 nm. Excess sulfide did not further shift the red peak. The maximal ratio of sulfide to Cd(II) in these complexes was 0.45. Titration of sulfide into Cd(II)–$(Glu\text{-}Cys)_4Gly$ also led to a redshifting in absorption peak up to near 300 nm. In contrast, maximal redshifting was up to 320 nm in the CdS capped with $(Glu\text{-}Cys)_4Gly$.[58,59] Although NCs formed by –peptide were smaller than those formed by –bond, there was only a little difference in the amount of sulfide incorporated into the respective complexes.

Optoelectronic properties of NCs are largely dependent on both the size of the particle and the properties of capping materials.[54,61,62] Consequently, photochemical properties of PC-capped CdS NCs may be easily modulated by using different residues other than glutamic acid in peptide, changing the number of torsion angles within the dipeptide repeat, and the separation distance between the cysteinyl thiols or carboxyl groups.[54,61,62] It was suggested that the resistance of the PC-capped CdS NCs to accretion or dissolution could be useful in photosensing, photosynthetic, and photoredox applications.[58] GSH-capped CdS NCs have also been identified from *C. glabrata*.[60] When cultured in a rich nutrient medium, the yeast formed intracellular CdS particles coated with a mixture of GSH and γ-glutamylcysteine dipeptide. In contrast, cultures in synthetic minimal medium yielded CdS particles coated with a mixture of $(Glu\text{-}Cys)_2Gly$ and $(Glu\text{-}Cys)_2$.[58] Ratios of sulfide to Cd(II) were more variable in GSH-capped CdS NCs than PC-capped particles.[60] The molar ratios of sulfide to Cd(II) ranged from 0.49 to 0.13 in fractions separated from Sephadex G-50 column. Particles having a high ratio of sulfide to Cd(II) showed bigger size, higher Cd(II)-to-GSH ratio, and greater anionic character.

The optical properties of the eluent fractions of GSH-capped CdS particles showed more heterogeneity compared with those of $(Glu\text{-}Cys)_2Gly/(Glu\text{-}Cys)_2$-, $(Glu\text{-}Cys)_3Gly$-, and $(Glu\text{-}Cys)_4$-capped CdS particles.[54,58-60] Maximal redshifting of GSH-capped CdS particles (S/Cd(II) = 0.49) was nearly 370 nm, whereas PC-capped particles did not show absorption shoulders to the red of 320 nm. GSH-capped CdS particles containing low sulfide could readily be converted to high-sulfide-containing particles by *in vitro* titration of NCs with inorganic sulfide. Similar changes were not observed in GSH–Cd(II) complexes without sulfide.[60] Rather, the yellowish bulk (CdS) was developed. Only samples containing a Cd(II)-to-GSH ratio below 0.1 could be titrated with sulfide to form CdS NCs.[60] Cysteine as a free amino acid was also as effective as GSH in serving as a matrix for CdS NCs formation at a Cd(II)-to-cysteine ratio below 0.1.[60] Photoactivity of CdS NCs isolated from yeasts was determined by photoreduction of methylviologen under anaerobic and alkaline conditions. Photoexcitation of these CdS NCs in the presence of methylviologen resulted in the reduction of the dye indicated by the appearance of the characteristic absorption peaks near 375 and 600 nm.[58]

21.5.1. *In Vitro* Assembly of Sulfide Complexes

The incorporation of inorganic sulfide into Cd–PCs is readily accomplished *in vitro* by mixing Cd–PCs with aqueous solutions of sodium sulfide.[54,59] The CdS–PCs formed *in vitro* appear to be indistinguishable from those formed *in vivo*. The sulfide-to-Cd ratios of *in vivo* formed Cd–PCs can be increased by the titration of additional sulfide *in vitro*.[59] The sulfide-to-Cd ratios in the *in vitro* prepared samples depend on a variety of factors such as pH, temperature, and the chain length of PCs. A net effect of increasing the sulfide-to-Cd ratios is an increase in Cd-to-PC ratios indicating that the metal-binding capacity of PCs is typically increased upon sulfide incorporation. PC_2, the smallest of the PCs, typically incorporated ~0.8 sulfide ions per Cd(II).[48] Studies carried out at near neutral pH showed that sulfide-to-Cd ratios may reach close to unity in PC_2 samples.[54] PC_3 and PC_4 appear to limit the amount of sulfide incorporated.[54,59] It has been suggested that the amount of sulfide incorporated may depend on the affinity of Cd(II) for SH groups of the PC involved. Thus, PC_2 with lower affinity for Cd(II) incorporates significantly more sulfide than PC_4 which presumably has higher affinity for Cd(II). MTs with much greater affinity for Cd(II) are unable to incorporate any sulfide. Our most recent studies show that Zn–GSH and Zn–PCs are also capable of forming NCs with photochemical properties.[62]

21.5.2 Semiconductor Properties of GSH– or PC–Metal Sulfides

The optical spectral characteristics of CdS–PCs indicated that these complexes had the characteristics of nanometer-sized CdS crystallites.[58,60] The NC nature of CdS–PC complexes was confirmed by the reduction of methylviologen.[59] Additionally, electron microscopic and XRD studies showed that the sizes of these were typically in the 2 nm range.[60] GSH also forms CdS crystallites.[58]

GSH-capped CdS crystallites are presumably formed prior to the synthesis of PCs. Studies have shown that PCs are able to displace GSH from GSH-capped CdS crystallites.[54] It is curious to note that PCs by themselves are not able to form crystallites larger than 2 nm but replace GSH from larger particles without changing the size of the particle. In a series of experiments, Bae and Mehra[54] have studied the formation of GSH and PC-capped CdS crystallites. These studies show that there is a considerable size distribution in GSH-capped CdS crystallites as determined by size-fractionation studies. However, PCs appear to form generally uniformly sized particles. The size of the particles formed is dictated by the amount of sulfide titrated. The largest particles are formed at the highest amounts of titratable sulfide. Thus, the size heterogeneity observed in the GSH-capped CdS crystallites isolated from yeast cultures might occur at any intracellular sulfide concentration. In contrast, the size distribution observed in the PC-capped crystallites is probably due to the formation of these complexes at different time points when the intracellular levels of sulfide are different. Clearly, the stationary-phase cultures of *S. pombe* showed both the highest levels of sulfide and the preponderance of largest-sized CdS particles. A Cd-resistant strain of *C. glabrata* exhibited Cd-stimulated production of sulfide and consequently the sulfide-to-Cd ratios of 0.8 in Cd–PC complexes.

21.6 CONCLUSIONS AND FUTURE RESEARCH

There is reasonable evidence to support the suggested roles of PCs in Cd detoxification in yeast and plants. The identification of MTs raises the questions regarding their roles in metal tolerance mechanisms in plants. Although plant MTs can confer metal tolerance in heterologous systems, it is unclear if these proteins also participate in metal detoxification in plants. Efforts to clone the putative PC synthase have not been fruitful so far. The Cad1 mutant of *A. thaliana* is a good reagent in pursuit of the PC synthase gene. Much of the research on PC synthase enzyme has been concentrated on *S. cucubalus* partly due to the difficulty in assaying this enzyme in other systems. Microcharacterization techniques might prove useful for identification of the protein and subsequent cloning of the genes. There has been rather very limited research on the degradation of PCs. It is important to understand the fate of Cd and of PCs once these have been synthesized. Does Cd remain within the cell or is it pumped out? Initial studies on the metal-binding properties of PCs have shown interesting results. PCs appear to be unique models for studying the metal-binding characteristics of peptides with continuous repeats of Cys-X-Cys. Very little has so far been done on the nuclear magnetic resonance (NMR) and other optical studies on PCs. Although GSH and

PCs play crucial roles in metal detoxification reactions, the process of formation of metal–sulfide–PC complexes appears to be more important in metal accumulation. The identification of gene(s) important for metal-stimulated sulfide production might pave the way for the construction of plants for phytoremediation purposes.

ACKNOWLEDGMENTS

RKM thanks members of his laboratory for sharing the data with him. This work was supported in part by funds from the Air Force Office of Scientific Research (AFOSR), University of California Toxic Substances Program and University of California Riverside Academic Senate. RDT is thankful to the Director of the National Botanical Research Institute, Lucknow, for providing necessary laboratory facilities.

REFERENCES

1. Baker, A. J. M. and Walker, P. L., Ecophysiology of metal uptake by tolerant plants, in *Heavy Metal Tolerance in Plants: Evolutionary Aspects*, Shaw, A. J., Ed., CRC Press, Boca Raton, FL, 1990, 155.
2. Brown, M. T. and Hall, I. R., Metal tolerance in fungi, in *Heavy Metal Tolerance in Plants: Evolutionary Aspects*, Shaw, A. J., Ed., CRC Press, Boca Raton, FL, 1990, 95.
3. Shaw, A. J., Ed., *Heavy Metal Tolerance in Plants: Evolutionary Aspects*, CRC Press, Boca Raton, FL, 1990, 95.
4. Verkleij, J. A. C. and Schat, H., Mechanisms of metal tolerance in higher plants, in *Heavy Metal Tolerance in Plants: Evolutionary Aspects*, Shaw, A. J., Ed., CRC Press, Boca Raton, FL, 1990, 95.
5. Silver, S., Plasmid-determined metal resistance mechanisms: range and overview, *Plasmid*, 27, 1, 1992.
6. Stillman, M. J., Shaw III, F. C., and Suzuki, K. T., *Metallothioneins*, VCH, Weinheim, 1992.
7. Winklemann, G. and Winge, D. R., Eds., *Metal Ions in Fungi*, Marcel Dekker, New York, 1994.
8. Singhal, R. K., Anderson, M. E., and Meister, A., Glutathione, a first line of defense against cadmium toxicity, *FASEB J.*, 1, 220, 1987.
9. Coblenz, A. and Wolf, K., The role of glutathione biosynthesis in heavy metal resistance in the fission yeast *Schizosaccharomyces pombe*, *FEMS Microbiol. Rev.*, 14, 303, 1994.
10. Masters, B. A., Kelly, E. J., Quaife, C. J., Brinster, R. L. and Palmiter, R. D., Targeted disruption of metallothionein I and II genes increases sensitivity to cadmium, *Proc. Natl. Acad. Sci. U.S.A.*, 91, 584, 1994.
11. Robinson, N. J., Tommey, A. M., Kuske, C., and Jackson, P. J., Plant metallothioneins, *Biochem. J.*, 295, 1, 1993.
12. Zhou, J. and Goldsbrough, P. B., Functional homologs of fungal metallothioneins from *Arabidopsis thaliana*, *Plant Cell*, 6, 875, 1994.
13. Zhou, T. and Goldsbrough, P., Structure, organization and expression of Metallothionein gene family in *Arabidopsis thaliana*, *Mol. Gen. Genet.*, 248, 318, 1995.
14. Snowden, K. C., Richards, K. D., and Gardner, R. C., Aluminum-induced genes: induction by toxic metals, low calcium, and wounding and pattern of expression in root tips, *Plant Physiol.*, 107, 341, 1995.
15. Hayashi, Y and Mutoh, N., Cadystin (phytochelatin) in fungi, in *Metal ions in Fungi*, Winkelmann, G. and Winge D. R., Eds., Marcel Dekker, New York, 1994, 311.
16. Zenk, M. H., Heavy metal detoxification in higher plants—a review, *Gene*, 179, 21, 1996.
17. Klapheck, S., Fliegner, W., and Zimmer, I., Hydroxymethyl phytochelatins [(γ-glutamylcysteine)*n*-serine] are metal-induced peptides of Poaceae, *Plant Physiol.*, 104, 1325, 1994.
18. Meuwly, P., Thibault, P., Schwan, A. L., and Rauser, W. E., Three families of thiol peptides are induced by cadmium in maize, *Plant J.*, 7, 391, 1995.
19. Mehra, R. K., Tarbet, E. B., Gray, W. R., and Winge, D. R., Metal-specific synthesis of two metallothioneins and (γ-glutamyl peptides in *Candida glabrata*, *Proc. Natl. Acad. Sci. U.S.A.*, 85, 8815, 1988.

20. Kneer, R., Kutchan, T. M., Hochberger, A., and Zenk, M. H., Saccharomyces cerevisiae and *Neurospora crassa* contain heavy metal sequestering phytochelatin, *Arch. Microbiol.*, 157, 305, 1992.
21. Mehra, R. K. and Winge, D. R., Metal ion resistance in fungi — molecular mechanisms and their regulated expression, *J. Cell. Biochem.*, 45, 30, 1991.
22. Steffens, J. C., The heavy metal-binding peptides of plants, *Annu. Rev. Plant Physiol. Plant Mol. Biol.*, 41, 553, 1990.
23. Rauser, W. E., Phytochelatins, *Annu. Rev. Biochem.*, 59, 61, 1990.
24. Rauser, W. E., Phytochelatins and related peptides — structure, biosynthesis, and function, *Plant Physiol.*, 109, 1141, 1995.
25. Tripathi, R. D., Yunus, M., and Mehra, R. K., Phytochelatins and phytometallothioneins: the potential of these unique metal detoxifying systems in plants, *Physiol. Mol. Biol. Plants*, 2, 101, 1996.
26. Mutoh, N. and Hayashi, Y. Isolation of mutants of *Schizosaccharomyces pombe* unable to synthesize cadystin, small cadmium-binding peptides, *Biochem. Biophys. Res. Commun.*, 151, 32, 1988.
27. Hunter, T. C. and Mehra, R. K., A role for HEM2 in Cd tolerance, *J. Inorg. Biochem.*, 69, 293, 1998.
28. Howden, R., Goldsbrough, P. B., Andersen. C. R., and Cobbett, C. S., Cadmium-sensitive, cad1 mutants of *Arabidopsis thaliana* mutants are phytochelatin deficient, *Plant Physiol.*, 107, 1059, 1995.
29. Howden, R., Andersen, C. R., Goldsbrough, P. B., and Cobbett, C. S., A cadmium-sensitive glutathione-deficient mutant of *Arabidopsis thaliana*, *Plant Physiol.*, 107, 1067, 1995.
30. Gupta, M., Tripathi, R. D., Rai, U. N., and Chandra, P., Synthesis of phytochelatins and glutathione levels in *Hydrilla verticillata* (l.f.) and *Vallisneria spiralis* Royle under mercury stress, *Chemosphere*, 37, 785, 1998.
31. Gupta, M., Rai, U. N., Tripathi, R. D., and Chandra, P., Lead induced changes in glutathione and phytochelatin in *Hydrilla verticillata* (l.f.) Royle, *Chemosphere*, 30, 2011, 1995.
32. Tripathi, R. D., Rai, U. N., Gupta, M., and Chandra, P., Phytochelatin synthesis and glutathione levels under cadmium stress in aquatic weed *Hydrilla verticillata* (l.f.) Royle, *Bull. Environ. Contam. Toxicol.*, 56, 505, 1996.
33. Rai, U. N., Tripathi, R. D., Gupta, M., and Chandra, P., Phytochelatin synthesis and glutathione levels under cadmium stress in water lettuce *Pistia stratiotes*, *J. Environ. Sci. Health*, 30, 537, 1995.
34. Grill, E., Gekeler, W., Winnaceker, E.-L., and Zenk, M. H., Homophytochelatins are heavy metal binding peptides of homo GSH containing Fabales, *FEBS Lett.*, 205, 47, 1986.
35 Berger, J. M., Jackson, P. J., Robinson, N. J., Lujan, L. D., and Delhaize, E., Studies of product-precursor relationship of poly (g-glutamylcysteinyl) glycine biosynthesis in *Datura innoxia*, *Plant Cell Rep.*, 7, 632, 1989.
36 Ohtake, Y., Satoce, A., and Yabucchi, S., Isolation and charcterization of glutathione biosynthesis deficient mutant of *Saccharomyces cerevisiae*, *Agric. Biol. Chem.*, 54, 3145, 1990.
37. Grill, E., Gekeker, W., Winnecker, E. L., and Zenk, M. H., Synthesis of seven different homologous phytochelatins in metal exposed *Schizosaccharomyces pombe* cells, *FEBS Lett.*, 205, 47, 1986.
38. Grill, E., Winnacker, E. L., and Zenk, M. H., Phytochelatin a class of heavy metal binding peptides from plants are functionally analogous to metallothioneins, *Proc. Natl. Acad. Sci. U.S.A.*, 84, 439, 1987.
39. Mehra, R. K., Biosynthesis and metal binding characteristics of phytochelatins, in *Pollution Stress, Indication, Mitigation, and Conservation*, Yunus, M., Ed., Kluwer Academic, Dordrecht, the Netherlands, 1998, in press.
40. Ortiz, D. F., Kreppel, L., Speiser, D. M., Scheel, G., Mcdonald, G., and Ow, D., Heavy metal tolerance in fission yeast requires an ATP-binding cassette-type vacuolar membrane transporter, *EMBO J.*, 11, 3491, 1992.
41. Ortiz, D. F., Ruscitti, T., McCue, K. F., and Ow, D. W., Transport of metal-binding peptides by HMT1, a fission yeast ABC-type vacuolar membrane protein, *J. Biol. Chem.*, 270, 4721, 1995.
42. Li, Z. S., Szchypka, M., Lu, Y. P., and Thiele, D. J., The yeast cadmium factor protein (YCF1) is a vacuolar glutathione S-conjugate promp, *J. Biol. Chem.*, 271, 6813, 1996.
43. Li, Z. S., Lu, Y. P., Zhen, R., Sizeypka, M., Theeke, D. J., and Rea, P. A., A new pathway of vacuolar cadmium sequestration in *Saccharomyces cerevisiae*: YCF1 catalysed transport of bis (glutathionato) cadmium, *Proc. Natl. Acad. Sci. U.S.A.*, 94, 43, 1997.
44. Hayashi, Y., Nakagawa, C. W., Mutoh, N., Isobe, M., and Goto, T., Two pathways in the biosynthesis of cadystins $(\gamma\text{-EC})_nG$ in the cell free system of the fission yeast, *Biochem. Cell Biol.*, 69, 115, 1991.

45. Grill, E., Loffler, S., Winnacker, E.-L., and Zenk, M. H., Phytochelatins, the heavy metal-binding peptides of plants, are synthesized from glutathione by a specific γ-glutamylcysteine dipeptidyl transpeptidase (phytochelatin synthase), *Proc. Natl. Acad. Sci. U.S.A.*, 863, 6838, 1989.
46. Chen, J., Zhou, J., and Goldsbrough, P. B., Characterization of phytochelatin synthase from tomato, *Physiol. Plant.*, 101, 105, 1997.
47. Barbas, J., Santhanagopalan, V., Blaszczynski, M., Ellis, W. R., Jr., and Winge, D. R., Conversion in the peptides coating cadmium:sulfide crystallites in *Candida glabrata, J. Inorg. Biochem.*, 48, 95, 1992.
48. Mehra, R. K., Mulchandani, P., and Henter, T. C., Role of Cds quantum crystallites in Cd resistance in *Candida glabrata, Biochem. Biophys. Res. Commun.*, 200, 1193, 1994.
49. Vogeli-Lange, R. and Wagner, G. J., Subcellular localization of cadmium and cadmium-binding peptides in tobacco leaves — implications of a transport function for cadmium binding peptides, *Plant Physiol.*, 92, 1086, 1990.
50. Reese, R. N. and Winge, D. R., Sulfide stabilization of the cadmium-(γ-glutamyl peptide complex of *Schizosaccharomyces pombe, J. Biol. Chem.*, 263, 12832, 1988.
51. Mehra, R. K. and Mulchandani, P., Glutathione mediated transfer of Cu(I) into phytochelatins, *Biochem. J.*, 307, 697, 1995.
52. Mehra, R. K. and Winge, D. R., Cu(I) binding to the *Schizosaccharomyces pombe* (γ-glutamyl peptides varying in chain length), *Arch. Biochem. Biophys.*, 265, 381, 1988.
53. Mehra, R. K., Tran, K., Scott, G. W., Mulchandani, P., and Saini, S. S., Ag(I)-binding to phytochelatins, *J. Inorg. Biochem.*, 61, 125, 1996.
54. Bae, W. and Mehra, R. K., Metal-binding characteristics of a phytochelatin analog $(Glu\text{-}Cys)_2Gly$, *J. Inorg. Biochem.*, 68, 201, 1997.
55. Mehra, R. K., Kodati, V. R., and Abdullah, R., Chain-length dependent Pb(II)-coordination in phytochelatins, *Biochem. Biophys. Res. Comm.*, 215, 730, 1995.
56. Mehra, R. K., Miclat, J., Kodati, R. V., Abdullah, R., Hunter, T. C., and Mulchandani, P., Optical spectroscopic and reverse-phase HPLC analyses of Hg(II)-binding to phytochelatins, *Biochem. J.*, 314, 73, 1996.
57. Strasdeit, H., Duhme, A.-K., Kneer, R., Zenk, M. H., Hermes, C., and Nolting, H.-F., Evidence for discrete $Cd(SCys)_4$ units in cadmium phytochelatin complexes from EXAFS spectroscopy, J. *Chem. Soc. Chem. Commun.*, 16, 1129, 1991.
58. Dameron, C. T., Smith, B. R., and Winge, D. R., Glutathione-coated cadmium-sulfide crystallites in *Candida glabrata, J. Biol. Chem.*, 264, 17355, 1989.
59. Dameron, C. T. and Winge, D. R., Characterization of peptide-coated cadmium-sulfide crystallites, *Inorg. Chem.*, 29, 1343, 1990.
60. Dameron, C. T., Reese, R. N., Mehra, R. K., Kortan, A. R., Carrol, P. J., Steigerwald, M. L., Brus, L. E., and Winge, D. R., Biosynthesis of cadmium sulphide quantum semiconductor crystallites, *Nature*, 338, 596, 1989.
61. Bae, W. and Mehra, R. K., Properties of glutathione and phytochelatin-capped CdS bionanocrystallites, *J. Inorg. Biochem.*, 69, 33, 1998.
62. Bae, W., Abdullah, R., Henderson, D., and Mehra, R. K., Characteristics of glutathione-capped ZnS nanocrystallites, *Biochem. Bipophys. Res. Commun.*, 237, 16, 1997.
63. Glaeser, H., Coblenz, A., Kruckczeck, R., Ebert-jurg, A., and Wolf, K., Glutathione metabolism and heavy metal detoxification in *Schizosaccharomyces pombe*, isolation and characterization of glutathione deficient cadmium sensitive mutants, *Curr. Genet.*, 19, 207, 1991.
64. Tommasini, R., Vogt, E., Fromenteae, M., Hortensteiner, S., Matile P., Amrthein, N., and Martinola, E., An ABC-transporter of *Arabidopsis thaliana* has both glutathione conjugate and chlorophyll catabolite transport activity, *Plant J.*, 13, 773, 1998.
65. Scheller, H. V., Huang, B., Hatch, E., and Goldsbrough, P. B., Phytochelatin synthesis and glutathione levels in response to heavy metals in tomato cells, *Plant Physiol.*, 85, 1031, 1987.
66. Gupta, S. G. and Goldbrough, P. B., Phytochelatin accumulation and cadmium tolerance in selected tomato cell lines, *Plant Physiol.*, 97,306, 1991.
67. Steffens, J. C., Hant, D. F., and Williams, B. G., Accumulation of non protein metal binding polypeptides $(\gamma\text{-glutamyl cysteingl})_n$-glycine in selected cadmium resistant tomato cells, *J. Biol. Chem.*, 261, 13879, 1986.

68. Kubota, H., Sata, K., Yamada, T., and Maitent, T., Phytochelatins (class III metallothioneins) and their desglycyl peptides induced by cadmium in normal root cultures of *Rubia tinctorum* L., *Plant Sci.*, 106, 157, 1995.
69. Gekeler, W., Grill, E., Winnacker, E. L, and Zenk, M. H., Survey of the plant kingdom for the ability to bind heavy metals through phytochelatins, *Z. Naturforsch.*, 44, 361, 1989.
70. Klapcheck. S., Schlumz, S. and Berman, L., Synthesis of Phytochelatins and homophytochelatins in *Pisum sativum* L., *Plant Physiol.*, 107, 515, 1995.
71. Jackson, P. J., Robinson, N. J., and Whitton, B. A., Low molecular weight metal complexes in the fresh water moss *Rhynchostegium riparoides* exposed to elevated concentrations of Zn, Cd and Pb in the laboratory and field, *Environ. Exp. Bot.*, 31, 359, 1991.
72. Gawel, J. E., Ahner, B. A., Freidlan, A. J., and Morel, F. M. M., Role of heavy metals in forest decline indicated by phytochelatin measurements, *Nature*, 381, 64, 1996.
73. Ahner, B. A, Price, N. M., and Morel, F. M. M., Phytochelatin production by marine phytoplankton at low free metal ion concentrations: laboratory studies and field data from Massachusetts Bay, *Proc. Natl. Acad. Sci. U.S.A.*, 91, 8433, 1994.
74. Gekeler, W., Grill, E. Winracker, E.-L., and Zenk, M. H., Algiae Squester Heads metals via synthesis of photoelectric complexes, *Arch. Microbiol.*, 50, 197 1988.
75. Howe, G. and Merchant, S., Heavy metal activated synthesis of peptides in *Chlamydomonas reinhardtii*, *Plant Physiol.*, 98, 127, 1992.
76. Reddy, G. N. and Prasad, M. N. V., Heavy metal binding proteins/peptides occurrence, structure and function. A review, *Environ. Exp. Bot.*, 30, 251, 1993.
77. Speiser, D. M., Ortiz, D. F., Kreppel, L., Scheel, G., McDonald, G., and Ow, D. W., Purine biosynthetic genes are required for cadmium tolerance in *Schizosaccharomyces pombe*, *Mol. Cell. Biol.*, 12, 5301, 1992.
78. Delhaize, E., Jackson, P. J., Lajan, L. D., and Robinson, N. J., Poly(γ-glutamyl cysteinyl) glycine synthesis in *Datura innoxia* and binding with cadmium, *Plant Physiol.*, 89, 700, 1989.
79. Sinha, S., Gupta, M., and Chandra, P., Bioaccumulation and biochemical effects of mercury in plants, *Bacopa monnieri*, *Environ. Toxicol. Water Qual.*, 11, 105, 1996.
80. de Knecht, J. A., van Baren, N., Bookum, W. M. T., Sang, H. W. W. F., Koevoets, P. L. M., Schaat, H., and Verkleij, J. A. C., Synthesis and degradation of phytochelatins in cadmium-sensitive and cadmium-tolerant *Silene vulgaris*, *Plant Sci.*, 106, 9, 1995.
81. Reese, R. N., White, C. A., and Winge, D. R., Cadmium-sulfide crystallites in Cd-$(\gamma\text{-EC})_n$G peptide complexes from tomato, *Plant Physiol.*, 98, 225, 1992.
82. Kneer, R. and Zenk, M. H., The formation of Cd–phytochelatin complexes in plant cell cultures, *Phytochemistry*, 44, 69, 1997.
83. Murphy, A. and Taiz, L., Comparison of metallothionein gene expression and nonprotein thiols in ten *Arabidopsis* ecotypes, *Plant Physiol.*, 109, 945, 1995.
84. Vliet, C. V., Anderson, V. V., and Cobett, C. S., A copper sensitive mutant of *Arabidopsis thaliana*, *Plant Physiol.*, 109, 871, 1995
85. Inouhe, M., Minomiya, S., Tohoyama, H., Joho, M., and Murayama, T., Different characteristics of the roots in cadmium-tolerance and Cd-binding complex formation between mono and dicotyledonous plants, *J. Plant Res.*, 107, 201, 1994.
86. Grill, E., Winnacker, E.-L., and Zerk, M. H., Occurrence of heavy metals binding phytochelatins in plants growing in a mining refuse area, *Experientia*, 44, 539, 1988.
87. Singh, R. P., Tripathi, R. D., Sinha, S. K., Maheshwari, R., and Srivastava, H. S., Responses of higher plants to lead contaminated environment, *Chemosphere*, 34, 2467, 1997.
88. Spesier, D., Abrahmson, S. L., Banuelos, G., and Ow, D., *Brassica juncea* produces a phytochelatin–cadmium–sulfide complex, *Plant Physiol..*, 99, 817, 1992.

Index

A

B

D

H

I

J

K

L

Q

R

S

T

U

V

W

Y

Z

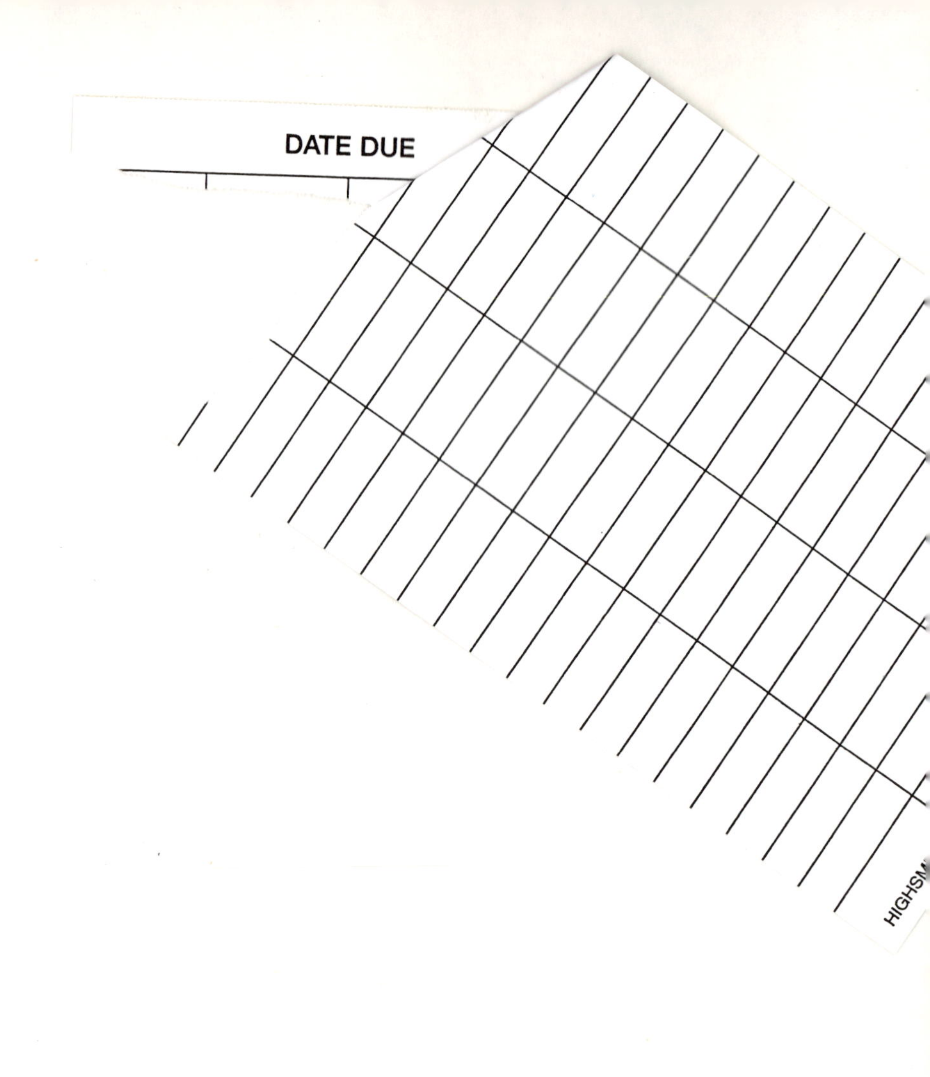
DATE DUE
HIGHSM